FUNDAMENTALS OF ANALYTICAL FLAME SPECTROSCOPY

FUNDAMENTALS OF ANALYTICAL FLAME SPECTROSCOPY

C Th J ALKEMADE and R HERRMANN

Translated from German by

R AUERBACH and PAUL T GILBERT Jr

A HALSTED PRESS BOOK

JOHN WILEY & SONS

New York

Library of Congress Cataloging in Publication Data

Alkemade, Cornelis Theodorus Joseph.
    Fundamentals of analytical flame spectroscopy.

    "A Halsted Press book."
    Includes bibliographical references and index.
    1. Flame spectroscopy.   I. Herrmann, Roland, joint author.   II. Title.
    QD96.F5A44   1979   543'.085   79-4376
    ISBN 0-470-26710-0

Published in the USA by Halsted Press,
a division of John Wiley & Sons, Inc., New York

Printed in Great Britain by J W Arrowsmith Ltd, Bristol

Contents

Preface

Analytical flame spectroscopy is a simple, rapid and versatile physical method of quantitative chemical analysis. The method is now well established and has found widespread application in the analysis of (usually) metallic elements in solution. The solution is nebulised into a flame where the solvent and solute are vaporised and decomposed. The strength of a characteristic line (or band) in the emission, absorption or fluorescence spectrum of the element in the flame is taken as a measure for its solution concentration. There are correspondingly three main branches of analytical flame spectroscopy: flame emission, atomic absorption, and atomic fluorescence spectroscopy. Instead of flames, electrothermal devices have now also come into general use for 'atomising' the sample in atomic absorption spectroscopy.

Most of the numerous books that have appeared in this field deal with only one or two of the above-mentioned branches. Many of them contain a fairly detailed description of the instrumental and methodological aspects and of the performance of practical analyses of various elements in different materials; the underlying physical and physico-chemical principles are usually presented in a more condensed form.

This book treats extensively and exclusively the *fundamentals* of flame spectroscopy as a general method of chemical analysis. It is not intended to be a textbook on isolated topics in atomic spectroscopy or combustion research; it treats the fundamentals only in relation to their significance for chemical analysis. The book deals primarily with the use of flames in combination with a pneumatic nebuliser as a means of vaporising and exciting liquid samples. However, many of the fundamental aspects and general relationships discussed also apply to non-flame atomisers, which are occasionally referred to in the text. Flames with or without metal vapours have been studied for a long time and they are among the best understood excitation sources. The strong interest in flames in other disciplines of technology and science has contributed greatly to this understanding. Besides, phenomena encountered in analytical flame spectroscopy have often triggered off special studies that are of fundamental interest too.

Flame spectroscopy is a relative method that is calibrated by means of reference samples of known composition. At first glance this method looks attractively simple to apply, as—ideally speaking—it is sufficient that the relation between optical signal and solution concentration be reproducible and unambiguous. A theoretical understanding of the sample transformations and the various processes and reactions in the flame would then not be a prerequisite for performing a correct chemical analysis. However, complications and interferences of various kinds often arise which would cause large systematic errors in the analysis if they were not attended to. To overcome these difficulties in a purely empirical way, that is, without the help of theoretical understanding, a tedious preliminary study of possible errors has to be made for every analytical task and every instrument. The resulting procedure of analysis will be unduly detailed and of very limited scope, for the analyst working out a method without knowledge of the fundamentals may easily overlook some essential point, while the causes of the errors will remain obscure to him.

But an analyst who is reasonably familiar with the fundamentals of the flame spectroscopic method will have a guideline at hand for dealing more efficiently with his analytical problems. He will readily comprehend the otherwise perplexing multiplicity of possible error sources, since a sound theoretical basis will make it easier for him to classify and check them. He is then in a better position to judge whether and to what extent a procedure worked out in another situation can be transferred to the case under consideration. He is able to anticipate which features or measures may be taken over, which additional points should be heeded, and which can be ignored. And he is less likely to report on the particular features and problems of his own instrument or procedure as if they had general validity.

Insight into the physical and chemical fundamentals will also provide guidelines in the search for optimum operating conditions concerning sensitivity, accuracy, detection limit, shape of the analytical curve, freedom from interferences, etc. The unremitting striving towards a better understanding of the processes involving metal species in flames (and associated hot-gas systems) has contributed greatly to the improvement of the method and even to the development of new analytical possibilities. The recent studies of the interaction of metal atoms in flames with an intense, tunable laser field are a striking example. The text incorporates the most relevant results of such fundamental studies in atomic spectroscopy, the physics of collisions, chemical kinetics, combustion research, etc.

The purpose of this book is to provide the practising analyst with an adequate theoretical understanding of analytical flame spectroscopy. At the same time it is intended to serve as a 'state-of-the-art' report and reference source for those who are engaged in research work in this and adjacent fields. The student, the beginning analyst and the non-specialist will find a general introduction into the field in the first two chapters; this will help them to obtain a better grasp of the analytical significance of the subjects treated in subsequent chapters. To understand these chapters a basic knowledge of physics and physical chemistry as taught to the average analytical chemist suffices. That basic knowledge which is essential for comprehension of the subject matter is recapitulated and illustrated by simple examples. A recapitulation is presented particularly in regard to the dissociation and ionisation equilibria treated in Chapter 5, and the atomic and molecular spectra treated in Chapter 6.

The main emphasis is on a clear presentation of the underlying physical principles and fundamental relationships. Extensive mathematical derivations and detailed presentations of theoretical or experimental numerical results are avoided, since these usually tend to obscure the fundamental principles. The validity of detailed calculations or concrete experimental results is often overestimated, as they usually relate to idealised conditions or particular situations that are hardly met in practice.

The three branches of analytical flame spectroscopy (emission, absorption and fluorescence) have much in common, both in a theoretical and in an instrumental respect. They are treated here 'synoptically', that is, in close relation to each other or in clear juxtaposition within the same chapter. We feel that in this way their specific merits and limitations will stand out more clearly. Besides, the unified treatment of their common aspects avoids undue repetitions.

Anyone who is acquainted with the literature on analytical flame spectroscopy and atomic spectroscopy will realise that the List of References (nearly 1000) presented in this book is by no means exhaustive. We have included only those entries that have been actually referred to in the text or that are of general importance (such as atlases and tables, bibliographies, books and conference proceedings; see references 1–153). References have been inserted in the text for the purpose of supporting particular arguments and statements or giving credit to original work, or simply for supplementary reading. The selection of references is often rather arbitrary and should not be interpreted as being the result of critical evaluation; the selection does not imply that the authors of this book necessarily agree fully with the contents. Anyway, we believe that the List of References includes most of the important papers published before 1978.

We have not hesitated to include some references to the older literature, although some may seem obsolete. During the tumultuous development of atomic absorption spectroscopy since the mid-1960s several papers have been published on work which was believed to be original but which was in fact a repetition of (unknown) older work in flame emission spectroscopy.

In order to enlarge the general usefulness of the book we have added in the Appendix an Atlas of 49 Spectrograms, a Table of 7000 wavelengths of spectral lines and bands observable in flames, and an extensive Glossary (with the equivalent German terms) explaining theoretical and practical terms used in flame spectroscopy. Last but not least, the extensive, detailed Index may help the reader to use this book effectively as a reference work too.

In our terminology we have kept strictly to the international recommendations on the nomenclature of analytical flame spectroscopy and associated non-flame procedures issued by the IUPAC in 1976. We hope that this will promote the general acceptance of a consistent, unambiguous and clear terminology among flame spectroscopists. The use of different terms for the same item or even of deceptive terms has caused confusion in the past. To help the reader to identify some of the older, now abandoned terms, we have included these in the Glossary too.

The original manuscript was in German but was translated into English after Adam Hilger, fortunately, had undertaken to publish the book and sell it to a wider market. Differences in style may still be observed between Chapters 1 and 2, on the one hand, and Chapters 3–9, on the other, which were drafted by the second and the first mentioned author, respectively. Further differences in style to be found in Chapters 1–4, 9 and the Glossary, and Chapters 5–8, are due to the fact that these parts had two different translators. We believe that these differences in personal taste will not detract from the inner coherence and intelligibility of the text. It should be noted that the authors and the editor have made some additions and changes to the translated texts which have not been checked by the translators.

The authors are grateful for the dedicated work done by the translators. They wish to express their appreciation of the numerous expert comments received from Mr Paul T Gilbert Jr in the course of his conscientious translation of Chapters 1–4, 9 and the Glossary. His comments have substantially improved the readability of the book. The enthusiasm with which Mr Gilbert has also prepared the extensive Index has added considerably to the value of the book as a whole.

Several colleagues have helped with the preparation of tables, etc. Their valuable contributions are duly acknowledged in the captions.

Authors and publishers who have given permission to reproduce or redraw figures and spectrograms from their publications are thanked for their kind cooperation. Full credit is of course given to the sources in the captions and introductions.

Without the patient, dedicated help from a fine team of accurate typists, this book could never have been completed. We thank all of them most cordially.

C Th J Alkemade
R Herrmann

1 Introduction

1.1. Spectrochemical Analysis

On gaining the required energy, every free atom in the gas phase can emit radiation of definite wavelengths or frequencies. The discrete wavelengths of this emission, which appear in the spectrum as bright lines on a darker background, characterise the emitting atom. This fact is the basis of qualitative *atomic emission spectroscopy* (AES). Every free atom can also absorb incoming radiation at discrete wavelengths. The wavelengths of absorption (often seen as dark lines on a bright background in the spectrum) similarly characterise the atom. This is the basis of qualitative *atomic absorption spectroscopy* (AAS). The absorption spectrum has fewer lines than the emission spectrum since the resonance lines of the emission spectrum (see Glossary) are practically the only ones that can also occur as absorption lines.

Atomic fluorescence, finally, can be utilised for spectrochemical analysis. In this method the atoms absorb incoming radiation at their characteristic wavelengths (the resonance lines) but the absorption itself is not observed directly. After a very brief interval some of the atoms re-emit the absorbed energy. This characteristic atomic emission induced by external irradiation is called atomic fluorescence and is the basis of qualitative *atomic fluorescence spectroscopy* (AFS).

At a given line the magnitude of the total emission, absorption, or fluorescence of many atoms of the same kind in a given volume of space is a measure of the number of free atoms in the observed volume of space and hence of their concentration there. If by suitable means, for example by use of a pneumatic nebuliser (see Glossary), the analytical sample in solution form can be brought into the observed space, vaporised and dissociated in a reproducible manner, then the magnitude of the total emission, absorption, or fluorescence is also a measure of the concentration of the element in the original sample provided that the method can be calibrated. This serves as the basis for *quantitative* atomic spectroscopy.

If the *analyte* (the element sought, that is, the element of interest in a chemical analysis) occurs in the observed space largely or entirely combined as molecules or radicals in the gas phase rather than as free atoms, then the emission or absorption of any of its bands (see Glossary) can in principle be employed for spectrochemical analysis. But the usefulness of molecular absorption spectroscopy is greatly limited by the fact that molecular absorption in the gas phase is much weaker than atomic absorption (see § 6.2.2). For the same reason molecular fluorescence spectroscopy in flames is of no analytical significance. In *flame emission spectroscopy* (FES), on the other hand, there are many analytical procedures that make use of molecular emission.

1

If atomic emission or fluorescence is to occur, an electron of the outermost shell of the atom must first be raised to a more energetic orbit by gaining energy (see § 6.2.1). This process is called *excitation* and the atom is said to enter an *excited state*. From the higher (more energetic) orbit, the electron can fall back to an orbit of lower energy with emission of radiation with quantised energy (a photon). In the simplest case the atom returns to the state of lowest energy, the *ground state*. If, on the other hand, atomic absorption is to occur, an atom in the ground state must take up energy from the incident radiation, raising the electron to a higher orbit. The energy absorbed from the radiation is subsequently given off by the electron to the surroundings, usually not as radiation but in some other form such as vibrational energy through collision with a flame molecule. The absorbed photon is thus lost as heat and the incident radiation is weakened. In AFS use is made of the fact that an atom excited by absorption of a photon may return to the ground state by re-emission of a photon in a random direction. The intensity of fluorescence depends on the probability of deactivation of the excited atom by photon emission and not by collision with a flame molecule. Such collisions are called quenching collisions because they quench the fluorescence.

In each of these three methods of atomic spectroscopy use can be made of a flame (see chapter 3) to convert the sample to atomic vapour, a process called *atomisation* (see Glossary). That is, the substance is vaporised and dissociated in the flame (see chapters 4 and 5). In the emission method the heat provided by the flame also serves for excitation (see chapter 6). The flame can be replaced by other devices such as the electric furnace, the electric arc, and the spark. Flames are convenient and popular for practical work, however, because their shape, size and temperature are easily reproduced and held constant and because they are almost unaffected by the introduced sample.

Quantitative flame spectroscopy is basically a relative method capable of calibration for determining the concentration of an element in a sample. The calibration is done by means of reference samples (see Glossary and chapter 8) of known composition. If the sample is wholly consumed, the quantity (e.g. in mg) of analyte present in the total sample is obtained by time integration of the emission or absorption signal, suitably calibrated. In practice, however, it is usually only the concentration (e.g. in mg l^{-1}) of the analyte in the sample that is required. The concentration is derived from the stationary intensity of the emission, absorption, or fluorescence while the sample solution is introduced at a constant rate (expressed for example in ml min^{-1}).

Generally speaking, flame spectroscopic analysis gives information on the concentration or quantity of an element regardless of the chemical bonding of the element in the sample. In special cases, however, it may provide information on the chemical bonding and the associated anions.

1.2. Apparatus

A diagram of the basic apparatus for flame spectroscopic analysis by AES, AAS and AFS is shown in figure 1. In the upper panel of the figure it is assumed that the

2

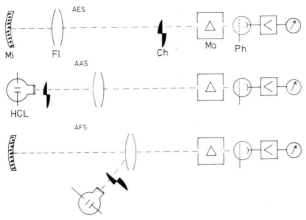

Figure 1. Apparatus for flame analysis by atomic emission (AES), atomic absorption (AAS) and atomic fluorescence spectrometry (AFS). Mi – mirror; Fl – flame; Ch – chopper; Mo – monochromator; Ph – phototube; HCL – hollow-cathode lamp.

sample solution is nebulised and enters the flame, where the solvent evaporates. The residual salt particles melt and vaporise and the vapour dissociates into free atoms, which are finally excited to emission. The emitted light (which is often subjected to modulation, i.e. periodic interruption by a chopper; see Glossary) is isolated from unwanted parts of the spectrum by the spectral apparatus and is then measured photoelectrically. If the photocurrent is weak, it is amplified by an alternating-current amplifier. The higher the atomic concentration is, the stronger the emission and the greater the deflection of the indicating meter.

In the absorption method (middle panel of figure 1) the same operations occur but a so-called background light source is also present. This is usually a hollow-cathode lamp whose cathode consists of the same element as the analyte. Let us take magnesium as an example. In the lamp several lines of magnesium are excited to emission by an electric discharge. This radiation passes through the flame to the spectral apparatus, where a resonance line from the lamp is isolated and its intensity measured photoelectrically. The meter should read 100 per cent transmittance (zero absorbance; see Glossary), the lamp intensity being assumed constant. If a sample solution containing magnesium is now supplied to the nebuliser, free magnesium atoms will appear in the flame and will to some extent weaken the resonance line of the background source (atomic absorption). The meter will no longer read 100 per cent transmittance but a lower value, depending on the concentration of magnesium. The thermal (atomic) magnesium emission simultaneously originating in the flame itself can be suppressed in various ways. One way is modulation of the lamp radiation before it enters the flame, together with the use of an alternating-current amplifier so that the unmodulated emission of the flame is not detected.

Atomic fluorescence (lower panel of figure 1) is less commonly used for analysis than atomic emission or atomic absorption. As in AAS, use is made of an auxiliary

3

light source – the so-called primary light source – that excites the atoms at their resonance frequency. They consequently emit a (secondary) fluorescence radiation, which, especially in the case of ultraviolet resonance lines, is much stronger than their thermal radiation. Fluorescence also occurs of course in the atomic absorption method, but its intensity is generally negligible compared to that of the background source radiation, because (i) the probability of re-emission is usually small in the flames commonly used in AAS, and (ii) the re-emission is not directional but isotropic. In AFS the flame gases are often selected with a view to raising the probability of re-emission (see § 6.8). In the analytical utilisation of fluorescence (lower panel of figure 1), the primary source radiation, in contrast to absorption spectroscopy, enters the flame obliquely and does not reach the photodetector. The fluorescence radiation appearing within a selected solid angle is measured in the same way as in the emission method.

In each of the three methods there is a read-out device associated with the spectrometer. This can be an analogue meter indicating current or voltage, a digital meter or printer, or a strip-chart recorder. In some cases the meter can be calibrated directly in units of concentration.

1.3. Elements Determinable by Flame Spectrometry

The periodic table in figure 2 is marked to show which elements can be determined by either of the two methods of flame spectroscopy (FES and AAS) most commonly used at present, and which elements can be determined by flame emission spectroscopy only. Atomic fluorescence spectroscopy should in principle

Periods	IA	IIA	IIIB	IVB	VB	VIB	VIIB	VIII B			IB	IIB	IIIA	IVA	VA	VIA	VIIA	VIIIA
1	H																	He
2	Li	Be											B	C	N	O	F	Ne
3	Na	Mg											Al	Si	P	S	Cl	Ar
4	K	Ca	Sc	Ti	V	Cr	Mn	Fe	Co	Ni	Cu	Zn	Ga	Ge	As	Se	Br	Kr
5	Rb	Sr	Y	Zr	Nb	Mo	Tc	Ru	Rh	Pd	Ag	Cd	In	Sn	Sb	Te	I	Xe
6	Cs	Ba	La	Hf	Ta	W	Re	Os	Ir	Pt	Au	Hg	Tl	Pb	Bi	Po	At	Rn
7	Fr	Ra	Ac															

Lanthanides	Ce	Pr	Nd	Pm	Sm	Eu	Gd	Tb	Dy	Ho	Er	Tm	Yb	Lu
Actinides	Th	Pa	U	Np	Pu	Am	Cm	Bk	Cf	Es	Fm	Mv	No	Lr

☐ E   ◻ E+A   ▨ not with normal direct flame methods

Figure 2. The periodic table, showing elements that can be determined by flame emission and atomic absorption spectroscopy. Included in the emission method are methods based on molecular emissions and chemiluminescence effects. However, indirect methods and methods employing unusual instruments and flames are excluded; if these were to be included, nearly all elements could be determined.

be applicable to essentially the same elements as those that can be determined by atomic absorption, but in practice it is still limited to fewer elements.

1.4. Comparison of Flame Spectrometry with Chemical Methods

The advantages and disadvantages of the emission and absorption methods, taken jointly, are listed below with respect to classical chemical methods. The comparison applies to atomic fluorescence as well.

The following are the chief advantages of flame spectrometry over chemical methods:

(i) Most metals can be determined in a single solution without chemical separation. The state of combination of the metal with anions, etc, plays a subordinate role. The observed lines and also many bands (see below) are specific for these elements or for their compounds that form in the flame.

(ii) A sequence of analyses by flame spectrometry of similar samples takes much less time than the corresponding chemical analyses. Flame spectrometry is therefore well suited for the routine determination of series of similar samples. The more tedious and cumbersome the corresponding chemical determination is, the greater is the saving in time. This advantage is particularly marked with the alkali metals, which are easily determined by flame spectrometry; their quantitative chemical determination (for example, of potassium as tetraphenylborate) is very troublesome.

(iii) The amount of material required is often less than that for chemical micro- or ultramicroanalysis. This advantage is utilised in special applications in biology, medicine and criminology.

(iv) The concentration limit of detection (see Glossary) is in many cases better than in any chemical method. For example, sodium can be detected by AES at a concentration of 10^{-11} g ml^{-1}, and calcium by AAS at 10^{-9} g ml^{-1}.

(v) The accuracy is generally good, sometimes in fact better than that of chemical methods (e.g. for lithium), especially at low concentrations. Accuracy is retained even at relatively low concentrations except for those close to the detection limit. Flame spectrometry is therefore exceptionally well suited for investigating impurities, trace elements, etc, in various materials.

(vi) Apart from the original development of the method, the analysis can be done by a technician.

(vii) The preparation of samples is usually easier.

(viii) There are special advantages in certain determinations; for example, ammonia formed from unstable organic substances may interfere with the chemical determination of potassium.

The disadvantages of flame spectrometry, compared with chemical methods, are the following:

(i) Since in all spectrochemical methods the measurements are relative, reference samples must be used for comparison. If the reference samples are in error, then the analysis will also be in error.

(ii) Costlier apparatus is needed.

(iii) Initial establishment of a method for a given application needs more work, for example, in the preparation of reference solutions, analytical curves, etc. It is usually justified only when many similar determinations are to be carried out or when the corresponding chemical method is very tedious, as is the case for the rare earths.

(iv) About 75 elements can be determined directly by flame spectrometry provided they are present in sufficient concentration. Unless flame methods are combined with chemical or other methods, the flame usually gives no information about the other elements and radicals and only very limited information about the kind of chemical bonding of the analyte in the original sample.

(v) Interferences of concomitants (see Glossary) may cause difficulties (see chapter 9). They depend, moreover, upon instrumental characteristics such as flame temperature and nebuliser type. In difficult cases flame spectrometry can lead to considerable error, especially when the method or apparatus is unsuitable. Nevertheless, these interferences can be alleviated more easily and quickly than those occurring in chemical methods of analysis. Each new type of sample generally demands the development of a new flame spectroscopic procedure. Methods described in the literature can only be used under the right conditions.

The methods of flame spectrometry can thus replace chemical methods of analysis only in part. But in many cases the flame methods offer great advantages and can be recommended as methods of choice.

Before undertaking an analysis by flame spectrometry it should be considered whether some other analytical method might not be better or quicker. Procurement of equipment needed for flame spectrometry will only be justified when its advantages can be put to full use, but if the equipment is already at hand and if the method is fully understood, flame spectrometry may often be used in cases where its advantages are less apparent.

1.5. Comparison of Flame Emission with Flame Atomic Absorption Spectrometry

For the comparison of flame emission with flame atomic absorption spectrometry we assume that the sample entering the flame vaporises and dissociates completely. In the final section, these two flame methods will be briefly contrasted with non-flame methods.

The following points list the advantages that the emission method has over the absorption method:

(i) The emission method, requiring simpler equipment, will be preferred when the outlay for such additional components as hollow-cathode lamps and their power supply, needed for atomic absorption, is a deterrent.

(ii) Emission is preferable when the concentration of analyte is low and the detection limit is better than in absorption. This is true of most of the alkali metals.

Even for elements of high excitation energy the emission method may offer lower detection limits when chemiluminescence can be utilised, as with phosphorus or sulphur (see § 6.7).

(iii) Even when the detection limit is poor, emission is preferable when the background source needed to determine the analyte by atomic absorption is not, or not readily, obtainable (e.g. boron, carbon, cerium, osmium, thorium), or when the lamp has poor operating characteristics.

(iv) The emission method is advantageous when many elements in the same sample are to be determined by scanning the appropriate wavelength range and recording the spectrum. In the absorption method this can only be done to a limited extent by using multi-element hollow-cathode lamps. If the same thing is attempted with a spectral-continuum background source (see Glossary) in the absorption method, a great deal of sensitivity will be lost with the usual instruments.

(v) Emission is preferred when there are considerable variations in the concentration of the analyte (or analytes, if several are to be determined together). For good precision in the absorption method, the absorbance (see Glossary) must remain within a fairly narrow range between about 0·1 and 0·6 except in special cases. AES is more flexible with respect to the concentration range of the sample since the amplifier gain can be adjusted as required.

(vi) The flame emission method may be more useful when the analyte is present in the flame as a compound that has too high a dissociation energy to dissociate sufficiently. Bands of the molecule or radical can then be used for emission analysis but there will be no good atomic absorption lines in the flame. Another advantage of using bands is the absence of self-absorption curvature in their analytical curves (see § 8.3).

(vii) A further advantage of emission is the generally wider choice of atomic lines as well as molecular bands. For an element atomic absorption is observable only at a few lines.

(viii) For AES the usual chemical flames can be replaced by other simple devices such as special electric 'flames' or gas discharges of high electronic temperature, for example the direct-current plasma jet, the inductively coupled high-frequency plasma, and the glow discharge (see the literature at the end of the chapter). We return to the non-flame methods in § 1.6.

The advantages of the absorption over the emission method are apparent under the following circumstances:

(i) When the concentration is low, absorption is advantageous if it offers better detection limits.

(ii) AAS is preferred when the sample solution contains organic constituents such as protein or alcohol in high, variable concentrations, affecting the flame temperature and spectrum, and when their chemical separation, as by ashing, is too troublesome. It is important to note that emission depends critically (exponentially) upon the flame temperature (compare § 6.4), while atomic absorption is, to a first approximation, unaffected by it when complete vaporisation and dissociation of the sample can be assumed.

(iii) AAS offers a further advantage when the flame background radiation in the emission method impairs the detection limits. This occurs when an incandescent reducing flame is needed to atomise the analyte and when the spectral apparatus has only moderate resolving power. By modulating the light of the background source in the absorption method, the absorption signal can be readily separated electronically from the unmodulated component of the thermal flame–background radiation without the help of a costly monochromator.

(iv) The absorption method is better when the matrix of the sample interferes (in emission) by affecting the flame temperature or the background emission. For the reasons given in (ii) such interference is much weaker or absent in absorption and can be rather easily avoided, for instance, by modulation. Reference solutions are then easier to prepare for absorption because not all the constituents of the sample have to be exactly simulated in them. This can be of great practical convenience.

(v) Atomic absorption is advantageous when the available spectral apparatus is only of moderate resolving power. It is easier to separate interfering neighbouring lines in absorption than in emission. The spectral selectivity is given by the sharpness of the line in the hollow-cathode lamp and not by the monochromator.

(vi) When the analysis has to be entrusted to a technician it is advisable to use a low-temperature flame, with which more elements can be determined by absorption than by emission.

(vii) When concomitants interfere, it is easier in the absorption method to change over to a flame in which the interference is much weaker or even absent (compare chapter 9). This may not be so easy to do in the emission method because of the critical dependence of emission on flame temperature.

(viii) The spectral resolving power of the overall apparatus is essentially determined in the absorption method by the linewidth of the background source and not by the monochromator (see § 9.2.2). Therefore it is possible in favourable cases to separate closely adjoining lines of the same element, such as those of different isotopes – that is, to resolve the isotope shift. This offers a simple method of isotope analysis by atomic absorption.

(ix) Since excitation of the atoms is unimportant, even undesirable, in the absorption method, lower temperatures often suffice. The flame can therefore be replaced by a heated carbon tube (electric furnace), for example. Moreover, the residence time of the atoms is longer in the furnace than in the flame, permitting detection of smaller traces. In particularly favourable cases, as in the detection of mercury, room temperature is high enough.

The absorption method thus offers a welcome extension of the older emission method. Today, the two methods are about equally common. In our opinion, however, AAS will never altogether supersede AES and besides, the two methods can be easily accommodated in a single instrument so that the analyst may choose either method on its merits. This is also true for AFS.

1.6. Comparison of Flame Methods with Non-flame Methods

In modern instruments means are usually provided for replacing the flame as atomiser (possibly also as excitation source) by a non-flame atomiser (or excitation source) such as the electric furnace. Accordingly, we offer the following comparison in which we do not distinguish AES, AAS and AFS.

Advantages of the flame over the non-flame methods:

(i) Flames can be easily generated with good reproducibility. The outlay required for instrumentation is small.

(ii) Flames need little attention; in the electric carbon furnace, for example, the carbon has to be changed.

(iii) Together with a suitable means of sample introduction, such as pneumatic nebulisation, flames offer excellent reproducibility unequalled by any competitive plasma device.

(iv) Many types of flames burning many different fuels with different (combinations of) oxidants are available so that a suitable flame can be found for almost any application.

(v) Owing to their excellent precision, single-beam instruments are usually adequate, again reducing the cost.

These advantages of the flame methods are offset by the following advantages of the non-flame methods:

(i) There is often a lower background or blank reading (see Glossary) and hence better detection limits. On the other hand, with a furnace AAS may be disturbed by a blank signal due to absorption or scattering by unvaporised matrix constituents at high concentration.

(ii) Higher atomic concentrations due to the longer residence time of the atoms in the observed space (as in a furnace) result in better detection limits. The residence time can be varied and optimised by accessories such as 'gas-stop' or 'mini-flow'.

(iii) The above two points, especially the second, permit smaller sample consumption and hence microanalyses are easier to perform.

(iv) The protective inert-gas atmosphere in the plasma source hinders formation of involatile oxides of the analyte atoms and other unwanted reactions, for example with radicals present in a flame.

(v) Many furnaces allow programmed heating of the sample to be made in several temperature steps, for example, to evaporate the solvent, to ash the sample, and to dissociate it. The programme can be optimised for each sample. This is impossible with only a flame and a pneumatic nebuliser.

(vi) When a given element exists in the sample in different compounds, their boiling points will generally differ. By gradual heating, these compounds can in favourable cases be determined successively (MECA; see § 3.2).

(vii) Because the atomic lines are usually markedly narrower in low-pressure gas-discharge sources than in atmospheric-pressure flames, the discharge source is better for isotope analysis (see § 6.3).

9

Special References for Chapter 1

For further reading see the reviews below as well as the books listed in the general references.

Alkemade C Th J and Zeegers P J Th 1971 Excitation and de-excitation processes in flames, in *Spectrochemical Methods of Analysis* ed J D Winefordner (New York: Wiley Interscience) ch 1 pp 3–125

Aspila K I, Chakrabarti C L and Bratzel M P Jr 1972 Pyrolytic graphite-tube micro-furnace for trace analysis by atomic absorption spectrometry, *Anal. Chem.* **44** 1718–20

Boumans P W J M and de Boer F J Studies of flame and plasma torch emission for simultaneous multielement analysis – I and II, 1972 *Spectrochim. Acta* **27B** 391–414, and 1975 **30B** 309–34

Fassel V A 1971 Electrical 'flame' spectroscopy, Invited paper at *Colloq. Spectroscopicum Int. XVI, Heidelberg, 1971* (London: Adam Hilger) p 63

Gilbert P T Jr 1960 Analytical flame photometry: new developments, in *Symp. on Spectroscopy 1959, ASTM Spec. Tech. Publ.* no. 269 pp 73–156

Kirkbright G F 1971 The application of non-flame atom cells in atomic-absorption and atomic-fluorescence spectroscopy, *Analyst* **96** 609

L'vov B V 1972 Application de la spectroscopie atomique par absorption dans les recherches physiques et chimiques, *Revue du GAMS* **8** 3

Marinkovic M and Vickers T J 1971 Free atom production in a stabilized dc arc device for atomic absorption spectrometry, *Appl. Spectrosc.* **25** 319

Morrison G H and Talmi Y 1970 Microanalysis of solids by atomic absorption and emission spectroscopy using an r.f. furnace, *Anal. Chem.* **42** 809

Pinta M (ed) 1971 *Spectrométrie d'absorption atomique: I. Problèmes généraux; II. Application à l'analyse chimique* (Paris: Masson et ORSTOM)

Talmi Y and Morrison G H 1972 Induction furnace method in atomic spectrometry, *Anal. Chem.* **44** 1455–66

Wendt R H and Fassel V A 1966 Atomic absorption spectroscopy with induction-coupled plasmas, *Anal. Chem.* **38** 337–8

West C D and Hume D N 1964 Radiofrequency plasma emission spectrophotometer, *Anal. Chem.* **36** 412–5

West T S 1974 Atomic-fluorescence and atomic-absorption spectrometry for chemical analysis, *Analyst* **99** 886–99

Winefordner J D and Vickers T J 1974 Flame spectrometry, *Anal. Chem.* **46** 192R

2   History

2.1. Flame Emission Spectroscopy (FES)

2.1.1. Qualitative flame emission spectroscopy. Flame emission spectroscopy is the oldest branch of spectrochemistry. Isolated observations by various scientists on alcohol flames and candle flames coloured by metallic salts go back to the middle of the 18th century, for example, to Melvill (1756) (see special references at end of chapter) but no correct explanations were reached. The prism and spectroscope became available around 1800. With the spectroscope Wollaston (1802) found the yellow line of sodium in flames but thought it was a peculiarity of the flame itself. Fraunhofer (1817) discovered the same line in absorption in the spectrum of the sun but drew no conclusions. Talbot (1826, 1835) showed remarkable insight for his time; the principles of qualitative atomic spectral analysis revealed in his publications were basically sound though very briefly described. His comments, however, had no lasting influence on the science of his time.

It was the comprehensive research of Kirchhoff and Bunsen (1860) that first brought general recognition to analytical spectroscopy or spectrochemistry. At an earlier date Bunsen (1853–1859) had invented the burner named after him with its nearly colourless flame. The efficient spectroscope developed by both men, in conjunction with the Bunsen burner, made it a great deal easier than it had been to observe the spectra of metals in flames. In 1860 the two men succeeded in demonstrating that the spectral lines were due not to compounds but to the elements themselves, a conclusion that was widely acknowledged. With this new research tool they soon discovered the elements caesium and rubidium. Shortly afterwards other researchers discovered thallium, indium, gallium, and further elements by spectroscopy. The power of the method, then still qualitative, was thus proved.

2.1.2. Quantitative flame emission spectrography. A few years later, in 1873, the first attempts were made by Champion *et al* to develop qualitative flame emission spectroscopy into a quantitative analytical method. Later efforts were made by Beckmann and Waentig (1910) and by von Klemperer (1910) and an important contribution was the introduction of nebulisation by Gouy in 1876. But general recognition of the quantitative method did not come until much later when, in 1928, the Swedish plant physiologist Lundegårdh, with his colleagues, developed analytical flame spectrography for studying plant metabolism. He was also the first to try out the reference-element technique in flame spectrography, a technique that had been developed by Gerlach in 1925 for spectrographic analysis with arc and spark. Lundegårdh's methods were adopted and extended from 1936 to 1940 by other laboratories with similar equipment. But such laboratories were few since the use of photographic plates for routine analysis was rather cumbersome.

2.1.3. Flame emission spectrometry. Direct photoelectric spectrometry of flame lines (or bands) was tried by Lundegårdh (1936) and by Jansen and co-workers (1932–1935). An important step towards the simplification of quantitative flame spectroscopy was the realisation that the spectral apparatus could be replaced in favourable cases by combinations of relatively simple, cheap optical filters. The first experiments in this direction were made by Jansen *et al* in 1935. The work of Schuhknecht (1937) made this simplified method – filter flame spectrometry – a practical success. This work, together with Waibel's in 1935, led to the first commercial flame spectrometer, manufactured by Siemens and by Zeiss. Flame spectrometry could now be used in many agriculturally oriented laboratories in Germany. A further simplification came when Lange replaced the earlier (vacuum) phototube by the selenium barrier-layer cell.

In the United States a parallel development began somewhat later, in 1945, with the work of Barnes and co-workers. After that, however, flame spectrometry advanced vigorously. In 1946 the first filter flame spectrometer based on the reference-element technique was described by Berry *et al* and the instrument became commercially available. An important step for the further expansion of flame spectrometry was the introduction in 1951 of Gilbert's direct-injection burner in a flame attachment for the Beckman DU spectrometer. Before the wide application of this instrumentation in spectrochemistry the direct-injection prin- ciple had been described by Weichselbaum and Varney (1949), but their burner was not very successful. The Beckman DU with flame attachment made it possible, by relatively simple means, to measure the intensity of atomic lines or bands (in the ultraviolet as well as in the visible) quickly and accurately in hot flames burning with oxygen, to record spectra, and in particular to analyse combustible materials such as gasoline. In consequence the number of elements routinely determinable and the variety of materials amenable to analysis by flame spectrometry rose considerably, and the number of published applications has been growing year by year. These advances have also stimulated the optical and electronics industries in other countries to develop similar instruments. The number of different commercial instruments has exceeded 200, a number too great to survey.

2.1.4. Interferences. As the method expanded and the limits of detection and the accuracy improved, increasing numbers of critical papers on the possible errors or interferences of analytical flame spectrometry appeared. These did nothing to discourage analysts, because of the significant advantage in the speed of flame spectrometry over chemical methods. Instead, flame spectrometrists kept seeking effective, simple, fast methods of alleviating or correcting interferences. At first these methods were empirical but over the years, as a result of many basic physicochemical studies, a whole branch of science dealing with these phenomena developed, so that today we are well informed about the commoner interferences and can offer methods soundly based on theory for their elimination (see chapter 9).

In many cases these interferences can be used to solve analytical problems. For example an element or radical, such as phosphate, that is not directly detectable by flame emission spectrometry (that is, by thermal excitation) can be determined indirectly by measuring the interference on an added metallic element caused by it under controlled conditions. The well grounded knowledge of interferences also provides the means of linearising nonlinear analytical curves in many analytical applications (see chapter 8).

2.2. Atomic Absorption Spectroscopy (AAS)

2.2.1. Qualitative observations. Like flame emission spectroscopy, flame absorption spectroscopy originated long ago. Spectral absorption lines, the so-called 'reversed lines', were observed in the beginnings of spectroscopy. One observer was Wollaston (1802), who reported seven dark lines in the solar spectrum. Fraunhofer (1814 and later) found 590 absorption lines in the sun and also in part in terrestrial light sources, and measured their positions more accurately. The solar absorption lines came to be known as 'Fraunhofer lines'. These men did not yet understand the relation between the absorption lines seen in the sun or in a candle flame and their chemical composition. Nevertheless, Fraunhofer had noticed the coincidence in position of the yellow double emission line of sodium in the flame spectrum with the absorption D lines observed in the solar spectrum. Brewster (1835) reported further coincidences of emission with absorption lines. Talbot (1835) foreshadowed the correct theoretical explanation of the dark 'reversed lines' upon a bright, continuous background spectrum. Foucault (1849), by passing sunlight through an electric arc, was probably the first to generate absorption lines in the laboratory. But a decisive breakthrough in recognising the power of qualitative atomic absorption spectroscopy with the flame came only, as in the case of emission, with the work of Kirchhoff and Bunsen (1860). Kirchhoff studied many reversed lines under laboratory conditions and provided a correct theoretical interpretation.

2.2.2. Background light sources. In later years, apart from the research of astrophysicists, qualitative and quantitative atomic spectral analysis was carried out mainly by emission, because effective, sufficiently hot background light sources were not available then for the laboratory. Wood (1905) made use of a flame saturated with sodium as a background source and observed both the atomic absorption and the atomic fluorescence generated by absorption in an evacuated glass tube in which some sodium was warmed. In 1912, he repeated the experiment with mercury. The first measurements of the absorbance as a function of concentration for the case of mercury vapour at the 254 nm line go back to Malinowski (1914). At that time such studies were mainly limited to the easily vaporised elements. The first step towards better background sources was taken by Paschen (1916) with his hollow-cathode lamps, but in developing these lamps, the idea of using them for analytical flame atomic absorption spectroscopy did not occur to him. Incidentally, Rasetti (1930) used atomic absorption by cool atomic mercury vapour to suppress the exciting primary radiation of mercury in Raman spectrography.

2.2.3. Atomic absorption spectrometry. In the ensuing years quantitative atomic absorption spectrometry was restricted, again except for astrophysics, to the determination of the concentration of mercury in laboratory air (Ballard and Thornton 1941, van Suchtelen *et al* 1949, Zuehlke and Ballard 1950). For this application commercial instruments were developed and marketed. To extend the method to the laboratory determination of elements other than mercury, a sufficiently hot plasma for generating atomic vapour was required, together with efficient, selective background sources to produce the resonance emission to be absorbed for each element.

The decisive impetus for the extension of analytical atomic absorption spectrometry to the less easily vaporised elements came with the realisation that flames can be used for generating the atomic vapour (atomising the sample), and that hollow-cathode lamps or gas-discharge tubes can serve as background sources. This step was taken independently and almost simultaneously in 1955 by Walsh and by Alkemade and Milatz†. From that time interest in the method has risen steadily, and since 1959 commercial instruments for AAS have appeared in increasing numbers. The accumulated experience in flame emission spectroscopy has aided the development of atomic absorption instrumentation.

2.3. Atomic Fluorescence Spectroscopy (AFS)

To the first two analytical methods, FES and AAS, a third has now been added – analytical atomic fluorescence spectroscopy (AFS), promoted largely by Winefordner and his colleagues since 1964. A few fundamental studies of atomic fluorescence had been carried out previously in flames by Nichols and Howes (1923, 1924), Badger (1929) and Boers *et al* (1956). In 1961 Robinson detected a weak enhancement of the magnesium resonance-line intensity due to fluorescence in an oxyhydrogen flame irradiated by a hollow-cathode lamp containing the same element. Not until 1962 did Alkemade point out the analytical potential of the atomic fluorescence method in flames. The number of practical analyses carried out by this method is still small since there are few commercially available instruments that can be used for atomic fluorescence. But it is certainly possible that this method will gain practical importance as soon as reasonably priced background sources can be obtained that will meet the special demands of atomic fluorescence spectroscopy.

---

† Before their application as a spectrochemical tool, quantitative atomic absorption measurements with sodium vapour had occasionally been used in flame studies (see [160] in the general references).

14

2.4. Special Flames and Non-flame Methods

In concluding this historical survey it should be remarked that all three methods, FES, AAS and AFS, have gained substantially in popularity through the recent successful introduction of special flames and special flameless sources for atomising the analyte, or through the further development of such sources to make them applicable to practical analysis. We mention first the premixed, fuel-rich nitrous oxide–acetylene flame, credited to Amos and co-workers (1965, 1966); it strongly inhibits the formation of stable oxides in the flame. A non-flame source is the inductively coupled high-frequency plasma, first developed for analytical applications by Greenfield *et al* (1964). We note lastly the electric tube furnace, which now enjoys many embodiments. The groundwork for these was laid by L'vov (1959, 1961) and later by Massmann (1965).

Special References for Chapter 2

For an interesting and recent introduction to the historical development of spectrochemistry since Kirchhoff and Bunsen refer to the Chairman's Address by Dr A Walsh at the 5th International Conference on Atomic Spectroscopy in Melbourne, Australia in 1975 and published in 1975 in the *Proceedings of the Royal Australian Chemical Institute* **42** 297. For a survey of the development of AFS see also the paper by J D Winefordner in 1975 in *Chem. Tech.* p 123.

Alkemade C Th J 1963 Excitation and related phenomena in flames, in *Proc. Colloq. Spectroscopicum Int. X, College Park, Maryland, 1962* ed E R Lippincott and M Margoshes (Washington, DC: Spartan) pp 143–70

Alkemade C Th J and Milatz J M W Double-beam method of spectral selection with flames, 1955 *J. Opt. Soc. Am.* **45** 583–4; 1955 *Appl. Sci. Res.* **B4** 289–99

Amos M D and Thomas P E 1965 The determination of aluminium in aqueous solution by atomic absorption spectroscopy, *Anal. Chim. Acta* **32** 139–47

Amos M D and Willis J B 1966 Use of high-temperature pre-mixed flames in atomic absorption spectroscopy, *Spectrochim. Acta* **22** 1325–43

Badger R M 1929 Flammenfluoreszenz und die Auslöschung von Fluoreszenz im Gasgemische bei hohem Druck, *Z. Phys.* **55** 56–64

Ballard A E and Thornton C D W 1941 Photometric method for estimation of minute amounts of mercury, *Ind. Engng Chem., Anal. Edn* **13** 893

Barnes R B, Richardson D, Berry J W and Hood R L 1945 Flame photometry. A rapid analytical procedure, *Ind. Engng Chem., Anal. Edn* **17** 605–11

Beckmann E and Waentig P 1910 Photometrische Messungen an der gefärbten Bunsenflamme, *Z. Phys. Chem.* **68** 385–439

Berry J W, Chappell D G and Barnes R B 1946 Improved method of flame photometry, *Ind. Engng Chem., Anal. Edn* **18** 19–24

Boers A L, Alkemade C Th J and Smit J A 1956 The yield of resonance fluorescence of Na in a flame, *Physica* **22** 358–60

Brewster D Report on the recent progress of optics, in *Report of the First and Second Meeting of the British Association for the Advancement of Science, York 1831 and Oxford 1832* printed London (1835) pp 308–22

Bunsen R 1859 Löthrohrversuche, *Ann. Chem. Pharmacie.* **111** 257–76

Champion P, Pellet H and Grenier M 1873 De la spectrométrie, spectronatromètre, *C. R. Acad. Sci., Paris* **76** 707–11

Foucault L 1849 *L'Institut* p 45 (cited in G Kirchhoff and R Bunsen 1860 *Ann. Phys. Chem.* **110** 188)

Fraunhofer J 1817 *Ann. Phys. (Gilberts Ann.)* **56** (26) 264–313 (cited in R Mavrodineanu and H Boiteux 1965 *Flame Spectroscopy* (New York, London, Sydney: Wiley))

Gerlach W 1925 Zur Frage der richtigen Ausführung und Deutung der 'quantitativen Spektralanalyse', *Z. Anorg. Allg. Chem.* **142** 383–96

Gilbert P T Jr 1955 Burner structure for producing spectral flames, *US Patent Office* no. 2714833

Gouy M 1876 Photometric research on colored flames, *C. R. Acad. Sci., Paris* **83** 269–72

Greenfield S, Jones I Ll and Berry C T 1964 High-pressure plasmas as spectroscopic emission sources, *Analyst* **89** 713–20

Jansen W H, Heyes J and Richter C 1935 Die Anwendung der Spektralanalyse zur quantitativen Bestimmung von Alkalien und Erdalkalien V. Mitteilung: Die direkte photoelektrometrische Bestimmung der Alkalien und Erdalkalien, *Z. Phys. Chem.* (A) **174** 291–300

Kirchhoff G and Bunsen R 1860 Chemische Analyse durch Spektralbeobachtungen, *Pogg. Ann. Phys. Chem.* **110** 161–89

von Klemperer R L 1910 Über quantitative Spektralanalyse, *Diss. Dresden*

Lundegårdh H Die quantitative Spektralanalyse der Elemente, part I, Jena 1929; part II, Jena 1934

Lundegårdh H 1936 Investigations into the quantitative emission spectral analysis of inorganic elements in solutions, *Lantbruks-Högskol. Ann.* **3** 49–97

L'vov B V 1959 An investigation of atomic absorption spectra by complete vaporization of a substance in a graphite cell, *Inzh.-Fiz. Zh., Akad. Nauk Belorussk. SSR* **2** (2) 44–52 (see also 1959 *Inzh.-Fiz. Zh., Akad. Nauk Belorussk. SSR* **2** (11) 56)

L'vov B V 1961 Analytical use of atomic absorption spectra, *Spectrochim. Acta* **17** 761–70

Malinowski A V 1914 Untersuchungen über Resonanzstrahlung des Quecksilberdampfes, *Ann. Phys.* **44** 935

Massmann H 1966 Spurenanalyse mittels Atomabsorption in den Graphitrohrküvetten nach L'vov mit einem Mehrkanalspektrometer, *2. Int. Symp. 'Reinstoffe in Wissenschaft und Technik'*, 1965, Dresden (Berlin: Akademie Verlag) p 297

Melvill T 1756 Observations on light and colours, in *Essays and Observations, Physical and Literary* vol II section IV pp 31–6 (Edinburgh)

Nichols E L and Howes H L The photoluminescence of flames, I and II, 1923 *Phys. Rev.* **22** 425–31 and 1924 **23** 472–7

Paschen F 1916 Bohrs Heliumlinien, *Ann. Phys.* **50** 901

Rasetti F 1930 Über die Rotations-Ramanspektren von Stickstoff und Sauerstoff, *Z. Phys.* **61** 598

Robinson J W 1961 Mechanism of elemental spectral excitation in flame photometry, *Anal. Chim. Acta* **24** 254–62

Schuhknecht W 1937 Spektralanalytische Bestimmung von Kalium, *Ang. Chem.* **50** 299

van Suchtelen H, Warmoltz N and Wiggerink G L 1949 Ein Verfahren zur Bestimmung des Quecksilbergehaltes der Luft, *Philips Tech. Rundschau* **11** 94–100

Talbot H F 1826 Some experiments on coloured flames, *Brewster's J. Sci.* **5** 77

Talbot H F 1835 On the nature of light, *Phil. Mag.* **7** 113

Waibel F 1935 Die quantitative Flammenspektralanalyse, *Wiss. Veröffentlichg. Siemens-Werken* **14** 32–40

Walsh A 1955 The application of atomic absorption spectra to chemical analysis, *Spectrochim. Acta* **7** 108–17

Weichselbaum T E and Varney P L 1949 A new method of flame photometry, *Proc. Soc. Exp. Biol. Med.* **71** 570

Winefordner J D and Vickers T J 1964 Atomic fluorescence spectrometry as a means of chemical analysis, *Anal. Chem.* **36** 161–5

Wollaston W H 1802 A method of examining refractive and dispersive powers by prismatic reflection, *Phil. Trans. R. Soc.* **A92** 365

Wood R W 1905 The fluorescence of sodium vapour and the resonance radiation of electrons, *Phil. Mag.* **10** 513–25

Zuehlke C W and Ballard A E 1950 Photometric method for estimation of minute amounts of mercury, *Anal. Chem.* **22** 953

3 The Flame

3.1. Function of the Flame

In the emission method a medium of high temperature is needed for exciting atoms to emit light. The medium itself should emit as little light as possible (see chapter 7), so that the emission of the analyte may be readily observed. It must also be hot enough to evaporate quickly the liquid components of the sample entering as aerosol, to volatilise the residual solid particles, and to dissociate the molecules (often capable of little or no emission) to atoms. Moreover, the heat content of the medium must be great enough to prevent too much cooling by these endothermic processes. The simplest such medium is the non-luminous Bunsen flame, in which a fuel gas is burned more or less completely with air or oxygen. A flame spectrometer generates this flame, or a similar one (see § 3.3), and includes the means of introducing the sample into the flame, of measuring the emission from the flame, etc. This chapter is primarily concerned with the flame.

In atomic absorption and fluorescence spectrometry a flame is often employed for converting the analyte to atomic vapour. The flame needs only to be hot enough to evaporate the sample and to dissociate the molecular compounds as fully as possible. If the analyte tends to form an oxide, its dissociation can be enhanced by adjusting the gases to make the flame reducing. The background absorption and emission of the pure flame itself are usually negligible (see chapter 7) or can be eliminated instrumentally in the measurements.

The performance of the flame in atomic absorption spectrometry is characterised by the product of the (mean) atomic concentration of the analyte and the flame thickness (the absorption path length in the flame), for a given concentration of the solution. In flame emission spectrometry the excitation of metal atoms plays an additional and decisive role. If thermal equilibrium always prevailed in the observed part of the flame, specification of the temperature of this part would suffice to account for the atomic excitation, and there would be no need to consider the combustion processes in detail. But in some applications, for example the utilisation of chemiluminescence, the inner cone or primary combustion zone (see figure 3) is directly involved, while certain deviations from thermal equilibrium originating directly in the inner cone may appear in the region above the inner cone; therefore we shall also have to take the primary combustion processes into consideration. In this chapter we shall outline such knowledge of the flame as seems requisite for an understanding of its role in practical analysis. We limit this account to pure chemical flames into which no analytical substances are introduced and whose energy content is furnished only by a chemical reaction. For further information refer to [50, 58, 60, 61, 71, 72, 78, 635].

19

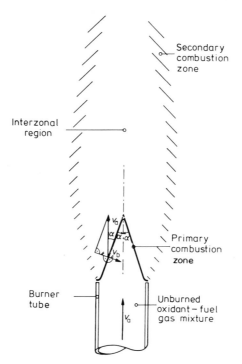

Figure 3. Schematic vertical section of a premixed laminar Bunsen flame. The magnitude of vector v_b, the burning velocity, is equal to the component of the velocity v_a of the unburned gas perpendicular to the primary combustion zone. See text for further explanation. (From R Herrmann and C Th J Alkemade 1960 *Flammenphotometrie* 2nd edn. Reproduced by permission of Springer-Verlag, Berlin and Heidelberg.)

3.2. Temperature of the Flame

The flames most often used in flame spectrometry flow out into the open air and hence burn at the constant pressure of the atmosphere. This fact governs the maximum attainable temperature for given fuel and oxidant gases. On the one hand, the flame gases greatly expand during combustion by virtue of Gay-Lussac's law and do work against the pressure, p, of the air. The energy thus consumed is no longer available for heating the gases. In terms of thermodynamics, it is not the change of internal energy (U) of the system but the change of enthalpy ($U + pV$, where V is the volume) that determines the temperature attained. On the other hand, the thermal expansion enhances the endothermic dissociation of flame molecules such as CO_2, H_2O and O_2. This energy too is unavailable for heating the gases. These effects limit the maximum temperature of chemical flames at 1 atmosphere pressure to about 5000 K. High temperature is favoured by a large heat of combustion per mole of fuel gas and high chemical stability (large dissociation energy) of the combustion products. The specific heat of the flame gases of course also plays a part; it depends on the number of degrees of freedom of the translational and internal

20

(rotational and vibrational) energy of the flame molecules. For a theoretical discussion of the attainable flame temperature refer to [61, 232].

In specifying a flame temperature one should distinguish the theoretical adiabatic temperature from the actual experimental temperature of the real flame. The calculation of the former is based on the assumption of ideal adiabatic conditions and of full thermodynamic equilibrium after the combustion process is completed, that is, the flame is regarded as homogeneous in every respect and heat losses are ignored. The theoretical flame temperature then depends only on the qualitative and quantitative composition and the initial temperature of the unburned gas mixture and on the pressure, which is assumed to be constant. It does not depend on the detailed course of the often complex combustion reactions. Theoretical adiabatic flame temperatures are given for some flames in table 1. Flames burning with air

Table 1 Theoretical flame temperatures[a] and maximum burning velocities

Gas mixture	Stoichiometric		Optimum[f]		Maximum burning velocity (cm s^{-1})[c]
	Mixing ratio[g]	Temperature (K)[b]	Mixing ratio[g]	Temperature (K)	
H_2–air	2·50	2380	2·35	2400[h]	300–440 [l][m]
C_3H_8–air	25	2267	24	2279[h]	39–43
C_2H_2–air	12·5	2540	9·8	$\begin{cases} 2606^{(h)} \\ 2570^{(i)} \end{cases}$	158–266[m]
H_2–O_2	0·5	3080	0·49	3085[h][i]	$\begin{cases} 1060\text{–}1400^{(m)} \\ 900^{(q)} \end{cases}$
C_3H_8–O_2	5·0	3094	4·4	3105[h][i]	370–390[o]
C_2H_2–O_2	2·50	3342	1·75	3430[h]	1100–2480[m]
C_2N_2–O_2	—	—	1·0	$\begin{cases} 4840^{(j)} \\ 4810\text{–}70^{(i)} \end{cases}$	140–270
H_2–N_2O[d]	1·0	2961	—	—	380–390
C_3H_8–N_2O[e]	10	2932	—	—	—
C_2H_2–N_2O	5·0	3150	2·8	$\begin{cases} \sim 3230^{(i)(n)} \\ 3255^{(k)} \end{cases}$	285[o]
CO–N_2O	1·0	2780[p]	0·6–0·7	$\sim 2890^{(p)}$	$\geqslant 40^{(p)}$

(a) These are calculated adiabatic flame temperatures at 1 atmosphere pressure, assuming that the gas mixture is initially at room temperature. (b) The temperatures listed are taken from [61] and/or [808]; where the values taken from these references (usually to the nearest 10 K) differ, the average is listed. (c) The range of values taken from [61, 91, 895]. (d) This flame, not used a great deal in practice, is described in [289, 367, 799, 900]. (e) A similar flame, not yet used in practice, is described in [250, 367]. (f) For the optimum gas mixture the theoretical temperature reaches its maximum. (g) The mole ratio of the oxidant gas (air, etc) to the fuel gas. (h) From [808]. (i) From [895]. (j) From [61, 492]. (k) Private communication from Dr P Th J Zeegers, Utrecht. (l) An accurate new determination can be found in [308, 316]. (m) A critical review of the data from the literature can be found in [180]. (n) The actual temperature of the fuel-rich, analytically useful nitrous oxide–acetylene flame is about 2900 K [735, 901]. (o) From [158]. (p) From [520, 522]; the temperature values are based on an initial temperature of 70 °C for the unburned gas mixture; the burning velocity also depends strongly on the rate of water supply. (q) From [317] for a stoichiometric gas mixture.

21

are seen to be a great deal cooler than those burning with pure oxygen or nitrous oxide. The inert nitrogen absorbs heat at the expense of the attainable flame temperature. The table also shows that flames of stoichiometric composition are not necessarily the hottest, a fact which is again related to the thermal dissociation of the stable combustion products at high temperature.

In real flames the ideal conditions assumed above (i.e. equilibrium and adiabatic combustion process) are often poorly or only approximately fulfilled and consequently, the measured temperature may depart considerably from the theoretical temperature, by 100 K or more (compare tables 1 and 2). When thermodynamic equilibrium does not prevail, the flame temperature cannot basically be defined in an unambiguous way. Different methods of measurement may then give different temperature values. Conversely, when temperatures measured in different ways agree, the flame can be assumed to be in a state of equilibrium [106, 367, 372, 492, 808].

Complete thermodynamic equilibrium requires that: (i) the spectral radiance at all wavelengths must obey Planck's radiation law for a cavity (*Hohlraum*); (ii) the velocity distribution of the particles must obey Maxwell's distribution law; (iii) the internal energy states of the particles (electronic excitation, vibration and rotation) must be occupied according to Boltzmann's formula (see chapter 6); and (iv) the ionisation and the dissociation of the molecules must correspond to the appropriate mass action laws (see chapter 5). The parameter T in these distribution and equilibrium laws must have one and the same value, and is simply defined as the temperature. In case this condition of overall equilibrium is not satisfied, the 'true' temperature is usually taken as the temperature value that corresponds to the velocity distribution.

These requirements for general equilibrium can be satisfied exactly only if the system is isolated from the outside world and given enough time for the separate kinds of equilibria to be established, that is, for all relaxation effects to die out. But in a flame these conditions are obviously not strictly met. The flame radiates energy to the surroundings, loses heat to the burner, and exchanges material with the surroundings by diffusion and convective mixing in the boundary zone, etc. Thus there is a net transport of energy and mass throughout the flame. Moreover, only a limited time is available for equilibration; at a height of 1 cm only about 1 millisecond has elapsed since the flame gases passed the combustion zone (§ 3.4.1), where the state of the system departs considerably from equilibrium.

If the relaxation times of the separate equilibria are short enough relative to all the transport times, it is still meaningful to speak of a local state of equilibrium characterised by a local temperature. It can be shown that this condition prevails for most (but not all) processes in the common flames, so that the (local) flame temperature can be specified with a sufficient degree of accuracy [162].

There are many methods for measuring flame temperature. Any quantity that depends on temperature could in principle be used as the basis for a temperature measurement. Optical methods such as the line-reversal method or the two-line method (see also chapter 6) are often preferred to the use of thermocouples or the like, provided that the excitation of the spectral lines corresponds to equilibrium.

For further discussions refer to [43, 45, 61, 78, 372, 492, 799, 808]. Experimental temperatures of various flames are reviewed in [78, 895].

There would be little point in listing the experimental temperatures in table 1 for comparison with the theoretical values. The experimental values depend strongly on the particular experimental conditions, which are often inadequately known or specified. Even when the mixing ratio of the fuel and oxidant gases is fixed exactly, the experimental temperature often depends sensitively on the burner design (which determines among other things the mode of outflow and degree of preheating of the gas mixture), the location of measurement (see figure 5), the nebulisation of the liquid (see also § 3.5.3), and other factors. Besides, unknown systematic errors can significantly affect the measurement, as in flames having a marked lateral temperature gradient or which are cooled by the insertion of a thermocouple. We refer to the critical discussions in [740–744, 799, 804–806, 842].

Most of the flames used in flame spectrometry have (actual) temperatures between 2000 and 3000 K [895]. Even the high measured temperatures of about 2900–2950 K for the fuel-rich nitrous oxide–acetylene flame, frequently used for analysis [735, 741–744, 901], and of 2950 K for the premixed stoichiometric oxyhydrogen flame [648] lie distinctly below their theoretical maximum values†. Such high temperatures are favourable for the excitation of atomic lines of high excitation energy (see chapter 6) and for the volatilisation and dissociation of samples that are hard to convert to atomic vapour (see chapters 4 and 5). Dissociation of stable metal oxides, however, depends not only on the temperature but also on the reducing conditions of the flame gas. With respect to the latter, the fuel-rich nitrous oxide–acetylene and nitric oxide–acetylene flames are more favourable than the fuel-rich premixed oxyhydrogen flame [594, 648] (see § 5.2.3). It is interesting to note that there are also analytical applications in which a very cool flame is best; for example, by adding nitrogen to the air, an oxygen–hydrogen–nitrogen flame of 1600 K can be obtained, as was shown in [279]. A hydrogen diffusion flame with added nitrogen has been used for determining non-metals such as sulphur and phosphorus [287, 290]. This flame burns with the oxygen that diffuses in from the surrounding air (see § 3.3) and has a temperature of less than 1000 K [290]. In this case the analytical line or band is excited by chemiluminescence (see chapter 6). Many hydrogen diffusion flames with added argon, used in atomic fluorescence spectrometry, have temperatures of about 1500–2100 K when water is nebulised [799]. Temperatures of 600–1700 K have also been reported for the hydrogen–argon and hydrogen–nitrogen diffusion flames [159, 547, 655–657]. Since combustion takes place at the margins of these flames, the measured temperatures are higher at the margin than in the centre [159]. All of these (low) temperature values have been measured with a thermocouple. An air–hydrogen flame chilled by a cold chimney has also been used for determining phosphorus and sulphur by utilising chemiluminescence [391].

---

† A temperature of 3000 K has been recently measured just above the primary combustion zone in a fuel-rich nitrous oxide–acetylene flame ($N_2O : C_2H_2 = 1·8$) [545]. When a shielding gas such as nitrogen or argon was used the temperature fell rapidly with increasing height of measurement [545].

The cool hydrogen diffusion flame has recently found special application in what is called *molecular emission cavity analysis* (MECA) [202, 204]. The procedure is as follows. A small cavity containing the solid, liquid or gaseous sample is introduced into the flame in line with the photodetector. The heating of the cavity by the flame results in the evaporation of the sample, and the emission of, for example, S_2 or HPO molecules inside the cavity is then measured to determine the sulphur or phosphorus content. Flame radicals seem to play a role in the molecular emission but the nature of this role has not yet been clarified. After ignition of the flame the S_2 emission pulse reaches a maximum at a time which depends on the kind of sulphur anion present in the sample. These anions can be discerned in mixtures by time resolution of the pulsed emission signal [552].

3.3. Classification of Flames

In flame spectroscopy, flames can be classified from various points of view. Firstly, a flame can be specified according to the kind of oxidant and fuel. Normally a fuel gas is employed, but in some cases a liquid organic fuel [188, 189, 279] or a combustible powder [747, 858] has been used for analysis. Nebulisation of an organic solvent can supply additional fuel to a conventional gas flame. The most common oxidants are normal air, pure oxygen and nitrous oxide. Mixtures of these gases have also been recommended as oxidants for optimising the temperature [250, 344, 358]. Sometimes a suitably selected inert gas such as nitrogen or argon is added to lower the flame temperature or the burning velocity (see § 3.4.1) [279, 332] or to raise the efficiency of fluorescence (see § 6.8) [514, 689]. The most common gas mixtures are listed in table 1, which also contains some less common mixtures such as oxygen–propane, nitrous oxide–propane and cyanogen. Recently the nitrous oxide–methylacetylene–propadiene (MAPP) flame has been used successfully in atomic absorption spectrometry [620]. This flame reaches a high temperature, about 3000 K, with a relatively low rise-velocity of the flame gases.

Besides the qualitative nature of the gas mixture, the quantitative mixing ratio is very important for characterising the flame. We distinguish sooting flames (only with organic fuels), fuel-rich flames, stoichiometric flames, and fuel-lean flames, depending on the fuel/oxygen ratio. Fuel-rich or reducing flames have proved especially good for determining metals that have a strong tendency to form oxides (see chapter 5). For example, the reducing conditions in an oxyacetylene flame are most satisfactory when the C/O ratio is about unity; they are only slightly enhanced at a higher C/O ratio giving rise to a sooty flame [75, 753].

Secondly, flames can be classified according to the manner of mixing the fuel gas with the oxidant. We distinguish the following kinds:

(i) Flames in which the inflowing fuel gas and air (or oxygen or nitrous oxide) are well mixed before reaching the combustion zone are called (fully) *premixed* flames. Since premixed flames used in flame spectrometry generally, but not necessarily, exhibit laminar flow, they are often also called *laminar* flames (see Glossary and § 3.4.4).

(ii) Flames in which the fuel gas and oxygen (or air or nitrous oxide) are first mixed in the flame are called *unpremixed* flames. We include flames in which the fuel is premixed with only a part of the consumed air or oxygen [322, 879]. Since the unpremixed flames used in flame spectrometry generally exhibit strong turbulence – which helps the mixing of fuel and oxygen – they are often also called *turbulent* flames (see Glossary).

With premixed flames the flow of gas mixture from the burner ports is usually kept as laminar as possible to give a stable, noiseless flame that mixes very little with the surrounding cold air. The liquid sample is mixed with the air stream by means of a chamber-type nebuliser (see Glossary), which normally forms a separate part of the instrument. The air–aerosol mixture is then mixed with the fuel gas and carried upward through the burner tube in laminar flow.

Nebuliser–burner combinations operate in a different way. In these the gases are not premixed. The unpremixed turbulent flame burns immediately above the concentric fuel and oxygen nozzles, while the sample is sprayed directly into the flame from a third liquid nozzle positioned generally at the centre. Such a device is called a *direct-injection burner* (see Glossary). It is often used for burning highly explosive gas mixtures (oxyhydrogen, oxyacetylene, etc) that have a high burning velocity (see table 1) and consequently a high risk of flashback (see § 3.4.1). It is also used for directly analysing combustible solvents such as gasoline. The flame of a direct-injection burner is strongly turbulent owing to the high speed with which the oxygen must emerge from the very narrow annular oxygen nozzle in order to nebulise the sample liquid.

With some care, gas mixtures such as oxyacetylene or acetylene with oxygen-enriched air, which have a high burning velocity, can also be safely burned after premixing [332, 341]. The hot nitrous oxide flames, owing to their relatively low burning velocity (see table 1), can be premixed without difficulty. Conversely, unpremixed diffusion flames† of low temperature (see § 3.2) are occasionally used in which hydrogen, with or without an inert gas such as nitrogen or argon, issues from a burner and mixes with the oxygen of the ambient air [290, 304, 655–658, 798, 799, 922]. These flames provide reducing conditions favourable for determining easily oxidised elements.

In this chapter we shall not discuss the less common flame variants, such as those in which direct injection is combined with complete premixing or, conversely, a chamber-type nebuliser is used without premixing the fuel gas and air [839], or those in which a turbulent flame is surrounded by a sheathing flame to stabilise it. For a comparison of laminar and turbulent flames in atomic absorption spectroscopy, refer to [790].

Flames can be classified, thirdly, according to their geometric form, which depends on the arrangement of the discharge orifice(s) in the burner head. There

---

† In the literature of analytical spectroscopy the terms 'turbulent flame' and 'diffusion flame' are often used interchangeably. In combustion research, an unpremixed flame is merely called a diffusion flame [61, 71]. This term does not, however, imply that the gas and air mix only by molecular diffusion; turbulent mixing is often dominant.

are conical, cylindrical, bulbous, planar (produced by a slot burner), and split flames (see Glossary).

It should be mentioned, finally, that a low-pressure oxyacetylene flame has been studied for possible applications in atomic absorption spectroscopy [213]. The analyte (iron) was added to the gas stream as a gaseous compound. Since it is homogeneous and versatile and can be calibrated absolutely, that is, without recurrence to reference samples, this flame might offer definite advantages. Emission measurements on aluminium have shown that fuel-rich nitrous oxide–acetylene and oxyacetylene flames below 100 Torr are especially advantageous for reducing metal oxides [820].

3.4. The Structure of Premixed Laminar Flames

We next consider the premixed laminar flames, which are the simpler of the two kinds and show an easily recognisable structure. The simplest example is the non-luminous Bunsen flame in which three zones can be distinguished, as shown in figure 3.

3.4.1. The primary combustion zone. The primary combustion zone, or inner zone, is sometimes also called the *inner cone* since it has a conical form when the burner port is round, as in the Bunsen burner. This combustion zone is about $0 \cdot 01$ to $0 \cdot 1$ mm thick at atmospheric pressure with laminar flow, and the gas takes about $10\,\mu s$ to pass through it. In air–acetylene flames, this zone is visible by its strong blue-green light, ascribed to the radicals C_2 and CH (see chapter 7). The inner zone of a premixed nitrous oxide–acetylene flame is blue-white [895]. The inner zone and the reactions occurring in it, which are generally not in thermodynamic equilibrium, can be important for analytical flame spectroscopy. For one thing, many atomic lines are much more strongly excited by suprathermal chemiluminescence in this zone than in the flame above it. This effect has given rise to a special branch of flame analysis, namely, chemiluminescence flame spectrometry (see § 6.7). Also, certain radicals such as H, OH and O and ions such as H_3O^+ are formed in excess in this zone. These neutral radicals are important for propagating the combustion and, on account of their relatively long life, also play a part in the higher regions of the flame, which are more commonly utilised in analytical flame spectroscopy.

Modern experimental methods such as mass spectrometry, electron spin resonance, and electrical probe measurements, along with optical spectroscopy, have been used to deepen our insight into the combustion reactions. It turns out that the primary combustion zone in oxygen-rich hydrocarbon flames consists of two regions. When the gas mixture enters the first region, saturated hydrocarbons are attacked by OH radicals and yield CO and H_2O; in the second region the CO is further oxidised to CO_2.

Although many details are still uncertain, it is now clear that the combustion is based on several simultaneous chain reactions in which short-lived fragments such as C_2H and CH may be formed, often in excited states. In both concentration and state of excitation these radicals considerably exceed equilibrium [176]. The reactions also generate other radicals such as OH, O and H in large excess. The following are probably typical reaction steps in the combustion of acetylene [60, 392]:

$$C_2H + O \rightarrow CH^* + CO \qquad (1a)$$

and

$$CH + O_2 \rightarrow CO + OH^*, \qquad (1b)$$

where CH^* and OH^* represent excited radicals that can radiate their excitation energy as a photon. These are among the processes accounting for the characteristic chemiluminescent radiation of the combustion zone. The concentrations of the molecular fragments and radicals are highly localised, so that the inner cone is very inhomogeneous. A review of chemiluminescent reactions in the primary combustion zone of the oxyacetylene flame can be found in [176]; combustion processes in general are discussed more fully in [50, 54, 91, 503].

It has been known for a long time that the combustion zone of hydrocarbon flames is strongly ionised. Studies with the ion mass spectrometer have in fact revealed many kinds of positive ions in this zone. Most of them are formed secondarily by charge exchange or by chemical reactions with primary ions that arise directly as the product of an energetic chemical reaction; this is called *chemi-ionisation*. Most frequently mentioned in the literature as primary ions are HCO^+ and $C_3H_3{}^+$. The reactions that produce them (e.g. in an acetylene flame) are thought to be the following [252, 254, 345, 348, 392, 549, 830, 846]:

$$CH \text{ (or } CH^*) + O \rightarrow HCO^+ + e^- \qquad (2)$$

and

$$CH^* + C_2H_2 \rightarrow C_3H_3{}^+ + e^-. \qquad (3)$$

In both cases the extent of ion formation depends on the concentration of (excited) CH radicals. Recent measurements and estimates of the energy needed for forming HCO^+ [311, 418, 709] show that in flames reaction (2) with CH in the ground state is energetically allowed and probably more important than the corresponding reaction with CH^* [345]. The following reaction has been proposed as a variant of reaction (2) [846]:

$$CH + O_2 \rightarrow CO_2H^+ + e^-. \qquad (2a)$$

From these primary ions, H_3O^+ can be formed as a secondary ion by the following reactions [830, 846]:

$$HCO^+ + H_2O \rightarrow H_3O^+ + CO \qquad (4)$$

and

$$CO_2H^+ + H_2O \rightarrow H_3O^+ + CO_2. \qquad (5)$$

The H_3O^+ ion recombines relatively slowly [532] according to the reaction [252, 827]

$$H_3O^+ + e^- \rightarrow H_2O + H \ (or \rightarrow OH + H_2). \tag{6}$$

Consequently it persists, unlike the other ions, in appreciable concentration in the flame gases above the combustion zone, where it may influence the ionisation of metals that are determined by flame spectrometry (see § 5.3.2).

Although negative ions such as OH^- are detected by mass spectrometry in the pure flame gases, these appear to play a minor part above the combustion zone in flames at atmospheric pressure [50, 252–254, 549]. The primary negative ion in carbon-containing flames might be C_2^- [254].

The participation of the CH radical in the formation of ions illustrates why strong ionisation, of the order of 10^{12} ions/cm^3, is confined to the combustion zone of hydrocarbon flames. In a premixed air–hydrogen flame at atmospheric pressure, the principal ion H_3O^+ is present at a concentration of $10^9 \, cm^{-3}$ at most [549] (compare also [345, 425, 426, 539]). The NO^+ ion has also been found in this flame but its concentration of about $10^7 \, cm^{-3}$ corresponds to thermodynamic equilibrium and may be entirely ignored in analytical flame spectroscopy.

Free electrons that are formed in an unequilibrated chemi-ionisation process or that have collided with suprathermally excited radicals or molecules may initially possess a relatively high kinetic energy [205, 223, 320]. Consequently, the 'electron temperature' that describes the velocity distribution of these 'hot' electrons exceeds the temperature of the flame gas. Ionising collisions of these electrons with neutral molecules or radicals might contribute to the excessive ionisation that is found in hydrocarbon flames [320]. The occurrence of elevated electron temperatures in the combustion zone of hydrocarbon low-pressure flames is demonstrated by electrical probe measurements [222, 223, 253]. According to a critical re-evaluation of these measurements the electron temperature is unlikely to exceed the flame temperature by more than a few hundred degrees. On the other hand, electron temperatures in the combustion zone of an oxyacetylene torch flame at 1 atm pressure were reported to exceed the gas temperature by more than 1000 degrees [320]. The question of the true magnitude of the electron temperature and the possible influence of 'hot' electrons has not yet been settled. For a review of experimental results we refer to [222], and for a theoretical discussion to [205]. A general discussion of flame ionisation, including experimental methods and data, can be found in [61, 71, 78, 419, 831].

It should be pointed out that experimental results obtained under special conditions, as in flames at low pressure or low temperature, atomic diffusion flames or shock waves, may not be directly applicable to flames of the sort commonly used in flame spectroscopy.

We now return to the phenomenological description of the combustion zone. The front of this zone tends to propagate towards the unburned gas mixture with a characteristic velocity, the *burning velocity*, which depends on the thermal conductivity of the gas mixture. Through thermal conduction the inflowing cold gas mixture is preheated to the ignition temperature. The burning velocity is also governed by the diffusion of radicals such as H out of the combustion zone towards the incoming gas. For a given gas mixture at a given pressure the burning velocity depends on the composition, the initial temperature, the humidity, the flow conditions, etc. Table 1 lists the maximum burning velocities for various gas mixtures. A detailed treatment of burning velocity can be found in [91].

For the flame to be ignited at all, the relative concentration of the fuel in the gas mixture must exceed a minimum of the order of 5 per cent by volume and remain below a maximum ranging from 8 to 94 per cent. These ignition limits are usually narrower for flames burning with air than for those with pure oxygen [635].

For the (conical) flame front shown in figure 3, a stable state is established in which at each point the component of the incoming gas velocity v_a normal to the combustion front is balanced against the burning velocity v_b, which is always normal to the combustion front. This balancing of velocities (marked with two strokes) is shown for an arbitrary point on the flame front in the figure. The greater v_a is relatively to v_b, the sharper will be the apical angle 2α and the stiffer the flame – that is, the less it will flicker except for extreme values of v_a. The ratio v_a/v_b is usually made to exceed 5 [550].

In figure 3 let us assume for simplicity that the inflowing air–gas mixture has the same velocity, of absolute value v_a, across the entire cross section of the burner, and that the burning velocity has a constant value v_b along the entire combustion front. Then, within the limits of stability (see below),

$$\sin \alpha = v_b/v_a. \tag{3.1}$$

Hence with a round burner port the primary combustion zone must take the form of a cone with apical angle 2α. Departures from the above assumptions result in a blunting of the inner cone.

The equation shows that as the efflux velocity v_a rises, the cone grows sharper, until the flame blows off the burner when the velocity becomes too great. But if the efflux velocity is less than the burning velocity ($v_b/v_a > 1$), there is again no stable state and the flame flashes back into the burner tube. With large burner ports, about 1 cm or more in diameter, the flashback may take the form of a noisy but generally harmless explosion. Flashback is hindered by covering the port with a metallic or porcelain burner cap having many small holes, as in the Méker burner (see Glossary), or by installing a screen inside the burner tube. The heat loss to the perforated material, when the holes are sufficiently small and the thermal conductivity of the material sufficiently high, is so large that the flame cannot burn inside the holes [91]. The maximum hole diameter that will avoid flashback for a given gas mixture, wall material and hole length has been tabulated in [72]. Other data on the risk of flashback for a slot burner with premixed oxyacetylene and nitrous oxide–acetylene can be found in [158, 341].

The stability of the flame front and hence of the flame itself is governed, incidentally, not by the ratio v_b/v_a alone as indicated by the simplified argument above, but also by other, secondary effects such as may arise at the upper edge of the burner port. The practical limits of stability have been investigated for various burner forms and various gas mixtures [72].

3.4.2. The interzonal region. Immediately following the primary combustion zone is the so-called *interzonal region* (see figure 3). It is usually the site for flame spectrometric analysis in premixed flames. This zone is surrounded by the secondary combustion zone (or outer zone; see § 3.4.3) and is sometimes also called the *reaction-free zone* or *interconal zone*. Unlike the sharply defined primary combustion zone, the interzonal region is often of appreciable extent and is more or less homogeneous in composition and temperature. The homogeneity, dimensions, etc, depend on the gas mixture, the mode of gas flow, the burner arrangement and the like. These properties give the flame an advantage over other excitation sources for spectrochemical analysis.

Observation of the interzonal region is often hindered by the surrounding secondary combustion zone. By suitable means, however, the latter can be lifted, as in the so-called *split flame* (see Glossary). The interzonal region of an air–hydrogen flame, when no solution is nebulised, emits practically no light and so looks dark. Even in an air–acetylene flame the background emission of the interzonal region is comparatively weak. In fuel-rich, non-sooting oxyacetylene and nitrous oxide–acetylene flames the interzonal region is brightly luminous and may extend several centimetres above the primary combustion zone. In the fuel-rich nitrous oxide–acetylene flame the interzonal region has a distinctive violet-red colour and is called the *red feather*, while in stoichiometric flames this region is scarcely discernible. The origin and spectral distribution of the flame background emission will be discussed in chapter 7.

After passing the primary combustion zone, the flame gases tend to rise vertically, as can be shown by photography with particles injected into the flame. The average rise-velocity of the gases is usually between about 1 and $10 \, \mathrm{m \, s^{-1}}$ [160, 620, 732, 924] and it can be greater still in unpremixed oxygen flames [784]. It is determined by the flow speed of the cold gas mixture and by the expansion of the gases after combustion. This expansion is mainly thermal (Gay-Lussac's law) and in the case of hydrocarbons is often also due to the increase in the number of moles in the combustion reaction. The rise-velocity is further increased by the buoyancy of the hot gases in the surrounding air but is again diminished by friction in the boundary zone. Owing to this rise of the gases, the vertical axis of the flame also represents a time scale in the sense that the time since passing the primary combustion zone increases with the distance from the burner. This is a matter of practical importance when certain slow reactions affect the emission or concentration of the analyte. A high rise-velocity is generally disadvantageous since it shortens the residence time of the analyte in the region under observation.

The (average) rise-velocity of the gases can be found, for example, by photographing incandescent carbon, aluminium or iron particles introduced into the flame and entrained with it. If the camera is made to rotate about a vertical axis, the plate shows streaks whose slope is a measure of the rise-velocity [160, 620]. It can also be found with a stationary camera and a rotating chopper that periodically interrupts the luminous streaks [455, 571, 924]. In another method the time of travel of the luminous particles between two horizontal diaphragm slits positioned one above the other is measured with a photodetector and an oscilloscope [730].

In atomic absorption spectroscopy one often tries to achieve a long absorption path in the flame. A slot burner (see Glossary) may be used to stretch the interzonal region as far as possible in one horizontal direction. One can also use a long horizontal absorption tube into which the burnt flame gases are forced to flow.

Since local equilibrium prevails on the whole in this region, its state can be characterised in general by specifying its temperature or temperature distribution. This will determine, for a given instrument design, nebuliser, flame gas composition, etc, which elements can be detected in emission and which spectral lines of an element are intense enough to be used for analysis (see chapter 6). In the absorption method also, the temperature can strongly affect the detection limits of refractory elements, since the temperature, along with other factors, governs the dissociation of the analyte into absorbing atoms (see § 5.2).

That the energy distribution of the particles as well as the chemical composition of the main components of the gas are in equilibrium can be inferred from various observations. Agreement of the temperature values measured in different ways with each other and with the theoretically calculated temperature (with allowances for heat losses, etc) is strong evidence for equilibrium [456, 804]. Theoretically, the equilibrium distribution of translational, rotational and vibrational energies should be reached quickly in the usual flames [162, 176]. Even in the interzonal region of fuel-rich oxyacetylene and nitrous oxide–acetylene flames a state of equilibrium may be assumed immediately above the primary combustion zone [61, 735; for a contrasting viewpoint see 753]. The presence in this zone of C_2, CN and NH radicals, otherwise found only in the primary combustion zone, is not necessarily contrary to chemical equilibrium. Equilibrium concentrations of CN and C_2 may become high enough to be noticeable at N_2O/C_2H_2 flow ratios below 2, and soot may form when this ratio drops below 1·6 [735, 745].

The following departures from thermodynamic equilibrium may be found in the interzonal region. The spectral radiance (see Glossary) in the flame departs in general from complete thermodynamic equilibrium. The equilibrium spectral radiance, given by Planck's radiation law, corresponds to that of a black body or cavity at the flame temperature. Even in sooty flames the spectral radiance is considerably below the Planck value but this departure is usually only of secondary importance for the general state of the flame [176]. Incidentally, the flame would be of no use at all for analytical emission or absorption spectroscopy if it actually behaved like a black body, characterised by an absorption factor equal to 1 at all wavelengths [165] (see also chapter 6). The concentrations of the electrons in hydrocarbon flames and of the radicals O, H and OH in air–hydrogen and air–acetylene flames also depart appreciably from their equilibrium values above the primary combustion zone (see below). Although these excess concentrations are still relatively small (of the order of 1 per cent or less of the total of all flame molecules [503, 808]), these departures can be important in flame spectroscopy (see chapters 5, 6 and 7).

Molecules and radicals such as N_2, NO and OH have relatively high ionisation energies, about 10 eV or more. According to the Saha equilibrium law (see § 5.3) the electron concentration in the interzonal region of a pure gaseous flame containing no soot or dust particles or added metal should then be very low, of the order of 10^7 cm^{-3}, corresponding to a partial pressure of about 10^{-11} atm. But considerably higher concentrations are found in pure hydrocarbon flames [251, 253, 455, 537, 653]. In hydrogen flames and carbon monoxide flames the electron concentration is much lower, although it may still sometimes exceed the equilibrium value [455, 537, 549]. The concentration of these excess electrons diminishes with increasing distance from the combustion zone. We recall from § 3.4.1 that the combustion zone is strongly ionised in hydrocarbon flames but not in hydrogen and carbon monoxide flames [61]. These facts make it clear that in hydrocarbon flames we are dealing with ions (H_3O^+) and electrons that are formed in great excess in the combustion zone and recombine only slowly in the rising flame gases [71]. In fuel-rich nitrous oxide–acetylene flames, an appreciable fraction of the electrons may be attached to CN, forming CN^- ions [735].

Certain radicals such as H, OH and O, which are formed in great excess in the combustion zone, may still be present in considerable excess above equilibrium in the interzonal region [50, 61, 239, 337, 455, 489, 526, 681, 832, 924–926]. This is particularly true of the cooler hydrogen flames, where the equilibrium concentrations are low. However, the excess decreases with increasing distance from the combustion zone owing to slow reactions in which the radicals recombine to form stable molecules, for example:

$$H + H + X \rightarrow H_2 + X \tag{7}$$

$$H + OH + X \rightarrow H_2O + X \tag{8}$$

$$O + CO + X \rightarrow CO_2 + X \tag{9}$$

$$\left.\begin{array}{l} H + O_2 + X \rightarrow HO_2 + X \\ HO_2 + H \rightarrow H_2 + O_2. \end{array}\right\} \tag{10}$$

Here X is any flame molecule which, acting as a third body, takes up the energy released in the recombination reaction so as to stabilise the recombination product. Such reactions are slow because they depend on three-body collisions, which are relatively infrequent. Equilibrium is attained when the concentrations of the reactants have declined so far that the recombination reaction and the reverse dissociation reaction balance each other.

The recombination reactions above do not proceed independently of one another because the radical concentrations are linked through rapid binary exchange reactions such as

$$H + H_2O \rightleftharpoons H_2 + OH \tag{11}$$

$$O + H_2 \rightleftharpoons OH + H \tag{12}$$

$$H + O_2 \rightleftharpoons OH + O. \tag{13}$$

The forward and backward reaction steps are balanced, so that a partial equilibrium is attained between the concentrations of the reactants and the reaction products. As a consequence of the partial equilibrium of reaction (11), for example, the ratio of the (excess) concentrations of H and OH is linked to that of the major components H_2 and H_2O. The concentrations of these major components are in general affected only slightly by the deviations of the radical concentrations from equilibrium because the radicals make up only a small fraction of the total flame gas.

The slow recombination of the flame radicals entails a gradual release of recombination energy, which often causes an initial increase in temperature with height in the flame [804, 924]. For further consideration of these radical reactions refer to [78, 91, 176, 503].

The equilibrium composition of the gas mixture in the interzonal region can be calculated for a given flame temperature from the composition of the unburned gas mixture by means of thermodynamics [61]. The temperature-dependent equilibrium constants of the participating chemical reactions enter as parameters into these calculations, which, today, are generally done by computer [272, 710]. Table 2 lists the calculated flame gas composition for a few typical flames (see also [61, 75, 239, 272, 402, 494, 503, 735, 745, 804, 808, 924]). The actual concentrations of the O, H and OH radicals may be appreciably higher in the cooler flames and just above the combustion zone owing to the relaxation effects (see above). In fuel-rich, planar flames (without gas shield) the actual free oxygen concentration may also rise considerably when additional oxygen from the ambient air takes part in the combustion. This has to be allowed for in calculating the degree of dissociation of metal oxides in such flames.

In premixed flames with a chamber-type nebuliser the composition, temperature and shape of the flame are generally little influenced by the introduction of a nebulised aqueous solution [78, 160, 165, 185, 343, 349, 735, 804, 808]. In unpremixed flames this effect is usually much greater (see § 3.5.3).

3.4.3. The secondary combustion zone. When, as is usual, the flame burns in the open air the interzonal region is externally bounded by a secondary combustion zone or outer zone. This takes the form of a cone (called the *outer cone*) in a Bunsen flame, but on a round Méker burner it takes the form of a cylinder. Compared to the primary combustion zone, this zone is much less distinct and more diffuse. Through molecular or turbulent diffusion, oxygen (and nitrogen) penetrate into the flame from the surrounding air, oxidising the carbon monoxide and hydrogen present in the boundary of the interzonal region to carbon dioxide (with weak emission of blue-violet light) and water. The OH emission also rises noticeably in this zone (see figure 4). The outer cone is more distinct when the primary combustion is incomplete because of an insufficient supply of air to the burner. Under these conditions it can even happen that the temperature at the edge of the outer cone, owing to the heat of the secondary combustion, is higher than on the axis of the flame (see figure 5). The outer cone has a stabilising effect on the flame but its radiation creates an extra background that has to be heeded in practical analysis, especially in working close to the limits of detection. This background can be eliminated by using a split flame (see Glossary).

3.4.4. The laminarity of the flame; flame emission noise. In the premixed flames under consideration an effort is commonly made to achieve *laminar flow*

Table 2 Calculated equilibrium compositions of some flames[†]

Flame	Mixing ratio[‡]	Assumed experimental temperature (K)§	References	Number density (cm^{-3})‖									
				H_2O	H_2	O_2	N_2	CO_2	CO	H	OH	O	NO
H_2–air	2·25	2018	[924]	1·3 (18)	7·2 (16)	1·5 (14)	2·6 (18)			2·5 (14)	1·9 (15)	1·8 (13)	6·2 (15)
C_2H_2–air	9·5	2398	[924, 925]	3·1 (17)	4·5 (16)	1·9 (15)	2·3 (18)	3·3 (17)	3·0 (17)	1·8 (15)	7·8 (15)	6·4 (14)	3·3 (15)
H_2–O_2	0·5	2650	[685]	2·4 (18)	1·8 (17)	6·6 (16)				3·3 (16)	1·1 (17)	1·2 (16)	
H_2–O_2	0·25	2640	[685]	1·6 (18)	1·1 (18)	7·6 (14)				7·8 (16)	2·8 (16)	1·3 (15)	
C_2H_2–O_2	0·65	2840	[685]	9·2 (17)	3·4 (17)	1·2 (16)		3·1 (17)	8·2 (17)	8·8 (16)	7·1 (16)	1·1 (16)	
C_2H_2–N_2O	3·0	2900	[735, 901]	2·1 (17)	1·5 (17)	5·0 (15)	1·2 (18)	1·3 (17)	6·9 (17)	7·0 (16)	3·1 (16)	8·8 (15)	8·3 (15)
C_2H_2–N_2O	2·04¶	2920	[735]	1·3 (16)	4·0 (17)	3·2 (12)	9·9 (17)	4·2 (15)	9·7 (17)	1·2 (17)	1·3 (15)	2·4 (14)	2·0 (14)

† From computer calculations by Dr P Th J Zeegers, Utrecht, using the thermodynamic data from [18].
‡ The mole ratio of the oxidant gas (air, etc) to the fuel gas.
§ The experimentally determined (mean) temperature of the actual flames is listed according to the references in the fourth column.
¶ When the N_2O/C_2H_2 mixing ratio falls below 2·0, carbon compounds such as HCN, C_2H_2 and CN can no longer be neglected [735].
‖ The number in parentheses is the power of ten by which the listed number should be multiplied. Thus, 3·2 (14) means $3\cdot2 \times 10^{14}$.

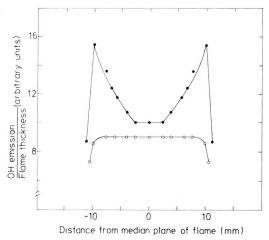

Figure 4. Intensity of the OH emission at 306 nm in a premixed fuel-rich air–acetylene flame, corrected for the change of flame thickness along the line of sight, plotted as a function of the distance from the central plane in the flame. The upper curve is for a flame burning in the open atmosphere, the lower curve for a flame surrounded by a nitrogen sheath. (From [925]. Reproduced by permission of The Combustion Institute, Pittsburgh, Pennsylvania.)

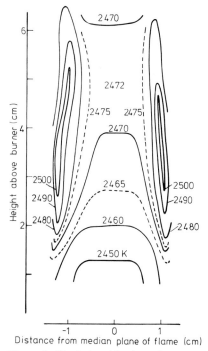

Figure 5. Vertical and horizontal temperature distribution (in K) in a fuel-rich air–acetylene flame on a rectangular Méker burner. (From [804, 808]. Reproduced by permission of Marcel Dekker Inc, New York.)

within the flame. It will then burn quietly and stably, and measurements of emission, absorption or fluorescence will show minimal fluctuation [97, 225, 399, 578, 790]. Another advantage of laminar flames is the limited lateral spread of the analyte vapour [367, 804]. Moreover, a strongly turbulent flame has the disadvantage of being cooled by mixing with the surrounding air. In fluorescence spectrometry this mixing lowers the fluorescence efficiency in flames diluted with argon instead of nitrogen [225, 689, 799]. If the flame is to be laminar, incidentally, the outflowing gas–air mixture must also be laminar.

Whether flow in a tube or in a flame is laminar or turbulent depends on the Reynolds Number Re (having dimension unity):

$$Re \equiv vd\rho/\eta, \tag{3.2}$$

where v is the mean flow velocity, d the diameter of the tube (or flame), and ρ the density and η the viscosity of the gas or liquid. For $Re < 2300$ the flow is mainly laminar and for $Re > 3200$ it is turbulent; between these values it is partly laminar, partly weakly turbulent. It is evident from the formula that the rise-velocity may not exceed a certain limit if the flame is to remain laminar. Also, for laminar flow, the burner tube or the holes in the burner head must have a certain minimum length L_m, related to the Reynolds Number by:

$$L_m = 0.03Re\, d, \tag{3.3}$$

where d is again the diameter of the tube or the holes.

In addition to this mathematical distinction, a turbulent flame can be recognised by its greater agitation and by the noise it makes. Owing to the irregular motions in the combustion zone, the flame front looks more diffuse and hence thicker. In premixed flames such turbulence should be avoided if possible. Laminar flow can be recognised, for example, by the straight, parallel streaks of light that appear on a photograph when carbon particles are entrained with the flame gases [732, 804]. Schlieren or shadow photography also readily distinguishes laminar and turbulent flames [61, 732, 790, 799].

Even in flames that are otherwise quite laminar, more or less regular vortices may form in the boundary of the outer zone and at a greater height these will pass over into a strong, disordered turbulence. These vortices have been clearly revealed by schlieren photography in the boundary of a premixed laminar flame burning with air [160]. This turbulence finally mixes the flame gases so thoroughly with the surrounding air that the flame comes to an end. Laminarity can be improved not only by proper burner design and flow conditions but also by means of an additional, analyte-free shielding flame [732, 804].

Flame emission noise can contribute to fluctuations in the photocurrent, a contribution which is often dominant over the shot noise of the photoelectrons in the low-frequency range. This holds true for the flame background as well as for the analyte emission. The flame emission noise may be caused by fluctuations in the temperature, the composition, or the shape of the flame [171]. The shot noise is inevitably associated with the statistical nature of the photoelectric emission process. The root-mean-square value of the flame emission noise increases proportionally with the mean photocurrent, whereas in the case of shot noise it increases proportionally with the square root of the photocurrent [171]. The two noise components are also distinguished by their noise spectra. Any fluctuating quantity can be formally represented as a superposition of an infinite series of sinusoidal components with different frequencies f. The noise spectrum describes the statistical distribution of these components over frequency, and when we are dealing, for example, with a fluctuating electric current i, it is expressed in $A^2 Hz^{-1}$ [76]. More precisely, the noise spectrum represents the mean square value of the current fluctuations, $\overline{(\Delta i)^2}$, per hertz bandwidth, as a function of f. Whereas shot noise has a flat noise spectrum (so-called white noise), the noise spectrum of the flame emission usually exhibits a strong increase as f decreases to zero (so-called flicker noise or $1/f$ noise).

Noise spectra over a wide frequency range have been reported for the emission of premixed and unpremixed, sheathed and unsheathed flames with and without metal vapour [171, 207, 445, 649, 841]†. In a premixed nitrous oxide–acetylene flame, flicker noise appeared dominant over shot noise for $f < 0.1$ Hz [649]. Because these components have different dependencies on photocurrent (see above), the frequency below which flicker noise becomes dominant over shot noise still depends on the photocurrent, that is, on the signal level. This frequency value is thus not to be considered as a characteristic of the flame alone. Such a characteristic, independent of the signal level, is provided by the value of the flame-noise spectrum at $f = 1$ Hz for example, divided by the square photocurrent. In addition, rather sharp peaks may occur in the flame-noise spectrum between about 10 and 100 Hz. These are associated with the presence of vortices in the flame boundary as discussed above [171]. Acoustical resonances in the burner ports may cause noise peaks in the kilohertz range [841].

3.5. The Structure of Unpremixed Turbulent Flames

It is hard to describe the unpremixed flame above a direct-injection burner because the strong turbulence obscures its structure (see § 3.3). We shall do no more here than to point out a few typical differences between unpremixed flames and laminar premixed flames.

3.5.1. The primary combustion zone. Even when the fuel gas (usually hydrogen or acetylene) and the oxygen (or air or nitrous oxide; see [799]) issue separately from the burner, a primary combustion zone can still be basically distinguished, but it is diffuse and spreads out because of the turbulence, more so in air–hydrogen flames than in oxyhydrogen flames. In the air–hydrogen flame this zone is not perceptible to the eye and its structure is hard to ascertain. The primary combustion zone of the nitrous oxide–hydrogen flame, on the other hand, is distinctly yellow [799].

The course of the combustion processes differs somewhat from that in premixed flames for the following reasons. For one thing, the fuel and oxygen, which are

---

† Similar noise spectra for the emission of hollow-cathode lamps have been reported in [310, 578].

supplied separately, mix less completely and less uniformly above the burner port than in premixed flames. However, the turbulence promotes the ignition of the cold incoming gas mixture. The burning velocity is hence a good bit higher than in premixed flames but it may vary sharply from point to point, depending on how thorough the local mixing is.

Normally the fuel gas (e.g. acetylene) emerges from the outer gas port and oxygen issues at high speed from an inner port, the ports being concentric. The aspirated liquid flows from the capillary tube in the centre of the oxygen port and is nebulised by the oxygen jet. This jet, which is loaded with mist droplets, has a conical shape and is quite stiff towards the bottom because of its high speed. The conical expansion of the jet diminishes its speed. The fuel gas, which is drawn inward by suction at the margin of the oxygen jet, mixes increasingly with the oxygen as it rises. This gradual mixing produces a diffuse, axially elongated primary combustion zone. Directly above the outer rim of the fuel port a part of the fuel diffuses into the surrounding air, creating a small diffusion flame called the *pilot flame*.

The turbulence makes the flame front irregular, ragged, and very erratic, characteristics which account for the loud hissing noise of the flame. The combustion process incessantly changes direction and speed, giving rise to shock waves. Moreover, in unpremixed, very turbulent flames, pockets of unburned gas mixture can be driven deep into the flame, where they are heated from all sides. The small explosions that ensue emit very sharp pressure pulses and the resulting noise, especially with gas mixtures of very high burning velocity such as oxyhydrogen and oxyacetylene, is very loud; it grows even louder either when the liquid is nebulised (owing to the explosive evaporation of the drops) or when the fuel/oxygen ratio is badly adjusted and the flame lifts from the burner. Instruments using these flames therefore need special soundproofing. The pressure pulses may further disrupt the liquid droplets already sprayed into the flame (see chapter 4), raising the concentration of metal atoms and thus intensifying the analytical lines.

In diffusion flames (in the stricter sense; see § 3.3) only the fuel gas issues from the burner, often with an inert gas such as nitrogen or argon, but with no oxidant. Consequently, combustion takes place only in the 'secondary' combustion zone where the fuel, usually hydrogen, mixes with the surrounding air; there is no 'primary' combustion zone.

3.5.2. The interzonal region and the secondary combustion zone. The properties of the interzonal region of the flame are affected by rapid mixing and dilution with the surrounding air. This makes the flame bulkier, club-shaped and rather short; a secondary combustion zone (outer zone) is not easily recognisable. The mixing with air, which starts at the bottom of the flame, lowers the attainable flame temperature and also makes the flame very inhomogeneous in rise-velocity, temperature, and analyte content [381, 727, 784]. In atomic spectroscopy the inhomogeneity in reducing conditions is of special interest in locating the best spot for observing elements that tend to form stable oxides. The turbulence in this part of the flame also affects the fluctuation of the analyte signal [784].

Owing to the mixing with the surrounding air, the turbulent flames are much harder to characterise and to understand than the more nearly homogeneous laminar flames. Thus, a change in the flow rate of hydrogen or the addition of acetone to the solvent does not affect the combustion and combustion products alone – it also alters the turbulence and hence the degree of mixing of the fuel and oxygen as well as the degree of entrainment of air. These effects react in turn upon the mean temperature and the shape of the flame [727, 784].

In direct-injection burners, the fuel and oxygen issue from the burner in separate, parallel, nearly vertical filaments. At low fuel flow there is a large difference between the velocities of fuel and oxygen, leading to strong turbulence in the boundary zone and hence, even in the bottom of the flame, to good mixing and optimum combustion [315]. As the fuel flow increases, the velocity difference and the turbulence diminish and the level in the flame where combustion is complete rises.

The hydrogen–argon diffusion flame (see § 3.3), which has to draw all of its oxygen from the surrounding air, may be conceived as a limiting case of the unpremixed hydrogen–air flame; these two flame types become more and more similar as the hydrogen/argon and hydrogen/air ratios are respectively raised [799]. Although the two flames are somewhat similar in form and temperature profile, the former is usually distinctly cooler [799]. In both flames the combustion zone is poorly defined, extending over a height interval of several centimetres.

The analytical utility of the unpremixed flames thus depends strongly on the location of measurement within the flame. For flame spectrometry it is therefore advisable to isolate carefully, by means of a diaphragm or the entrance slit of the monochromator, the part of the flame to be observed. This circumstance, among others, explains why different authors have sometimes made discordant observations on the same type of flame.

Their diffuse structure is no obstacle to the analytical use of unpremixed flames, which have certain practical advantages. It is sufficient that for a given experimental set-up the flame should be reproducible and the relative fluctuations of the signal sufficiently small. These requirements are often adequately met in practice.

That an interzonal region should exist implies that some degree of equilibrium prevails in it. In unpremixed oxyhydrogen and oxyacetylene flames there is a zone a few centimetres above the burner tip where the metal emission conforms fairly well to Boltzmann's formula (see § 6.4) [372, 381, 727, 784]. The temperature derived from the rotational and vibrational energy distribution in this zone also agrees well with the temperature found by the line-reversal method [372]. On the other hand, a distinct departure from equilibrium has been observed in the unpremixed air–hydrogen flame and the hydrogen–argon diffusion flame to a height of 4 cm [372]. Metal lines with excitation energy above 3 eV exhibit here a suprathermal emission that may be accounted for by chemiluminescence (see § 6.7.3), evidence that these flames have a (poorly defined) combustion zone several centimetres in vertical extent (see above). Unpremixed air–hydrogen flames are usually adjusted to be very fuel-rich and consequently the secondary combustion plays an important role [859].

Because of the higher flow velocity and the limited length of unpremixed flames, the residence time of the analyte vapour in these flames is usually shorter than in laminar flames with a chamber-type nebuliser. Hence slow relaxation or volatilisation processes may in certain cases cause greater departures from equilibrium or greater interference effects in unpremixed flames, as in the depression of a calcium signal by phosphate [175].

3.5.3. The influence of nebulisation on the flame properties.
Temperatures measured in unpremixed flames above the combustion zone turn out to be considerably lower than the theoretically calculated adiabatic temperature (see § 3.2). This does not necessarily refute equilibrium since the secondary entrainment of air and the direct injection of liquid cause much cooling. To exclude the effect of the liquid, the reversal temperature of a dry oxyhydrogen flame has been measured by the yellow line of sodium introduced as sodium oleate powder. The highest measured temperature appeared to be 2800 K, which is certainly short of the theoretical value of 3080 K (see table 1) [381], but if the measured temperature is corrected by the method described in [784] for the radial temperature gradient, a maximum of about 2900 K may be expected on the axis of the flame, closer to the theoretical value. Other methods of measuring the flame temperature without nebulising liquid are based on the intensity ratios of rotational lines or vibrational bands of flame radicals such as OH [349, 372, 494] or on the use of a thermocouple [799].

The direct injection of water into an unpremixed flame commonly cools it by a great deal, as much as 600 K at the rather high flow rate of 6 ml min^{-1} [190, 294, 306, 315, 349, 372, 381, 623, 727, 799, 910]. Energy is consumed in evaporating the water and heating the vapour to the temperature of the flame. Moreover, the water molecules dissociate partly, especially in hot flames, consuming additional energy [623]. This last factor is important in the very hot oxycyanogen flame; the aspiration rate (see Glossary) should therefore be kept low, although this again reduces the supply of analyte to the flame. It can be calculated, for example, that the introduction of 10 g water per mole of oxygen cools this flame by 1500 K, but an equal supply of acetic acid cools it a great deal less [492]. A theoretical estimate of these cooling effects has been found to agree with experiment [567]. At less than 1 ml min^{-1} of injected water the cooling is usually less than 100 K in oxyacetylene and oxyhydrogen flames [372, 910].

The temperature T of the unpremixed oxyhydrogen and oxyacetylene flames of a Beckman burner can be given approximately by an empirical formula as a function of the flow rate Q of water and the height h above the tip of the inner cone [910]. For oxyhydrogen this formula is:

$$T = (T_0 - ch^2)/(1 + aQ + bQ^2) \quad \text{(in K).} \tag{3.4}$$

The coefficients a, b, c and T_0 are positive and constant for a given flame and have been tabulated in [910]. Since the mist droplets nebulised in the flame do not all completely evaporate, Q should be taken as the effective flow rate (see chapter 4).

40

The denominator of the above expression agrees in form with the theoretical dependence of flame temperature on water flow rate [192].

The flow rate of water from a direct-injection nebuliser does not greatly affect the width of an unpremixed flame or its turbulence [315, 726, 909]. But injection of an alcohol, often used as a solvent, can double the height and width of the flame as well as the height of the inner zone [293, 315, 623, 725, 909], and also alters the turbulence. In hydrogen flames, hydrocarbon radicals formed from the alcohol change the emission spectrum of the inner zone [315, 727, 859].

In analytical spectrometry with an unpremixed flame, the background (see Glossary) must not be measured on the empty flame because the injected water affects the temperature and hence the flame background emission. Instead, the pure solvent used in the sample, or the blank solution (see Glossary), must be nebulised when the background reading is taken.

4 Pneumatic Nebulisation, Desolvation, and Volatilisation of the Analyte

4.1. Introduction

In a flame spectrometer the sample, usually in the form of a solution, passes through several phases: (i) it is nebulised, carried as an aerosol into the flame and evaporated; (ii) the liberated molecules are in part dissociated and the resulting free atoms are in part ionised; and (iii) the atomic or molecular vapour is excited to emission of light, or the atomic vapour is exposed to radiation from a background source to permit absorption or fluorescence measurement. The principles of these phases will be discussed in turn in chapters 4, 5 and 6.

The sample should be introduced by the simplest means possible in a reproducible and constant manner and with minimum wastage. A pneumatic nebuliser is often employed, in which a jet of compressed gas (e.g. air or oxygen) aspirates the sample liquid and nebulises it as a fine mist or liquid aerosol. On being nebulised, the sample solution is reduced to small droplets, which are carried into the flame either directly (in the direct-injection burner) or after passing through a spray chamber (in the chamber-type nebuliser), where the larger droplets separate. The liquid components of the droplets evaporate partly first in the spray chamber and the ducts leading to the burner, and finally in the flame. We shall call this process *desolvation*. After desolvation, salt crystals or oxide particles remain as a dry aerosol, which on further heating in the flame passes over into atomic or molecular vapour, a process we shall call *volatilisation*. The physical basis of aspiration, nebulisation and desolvation of the drops as well as the subsequent volatilisation of the solid (or molten) analyte particles will be discussed in this chapter.

The transport of the sample into the flame and the subsequent desolvation and volatilisation would present few problems in practice if all of the aspirated sample solution were to appear as vapour in the flame without altering flame properties such as temperature. Unfortunately there are various losses of sample and, particularly in direct-injection burners, the nebulisation and desolvation may react upon the flame. On the one hand, in chamber-type nebulisers much is lost in the spray chamber and the ducts to the burner by separation of the larger drops, which because of centrifugal forces hit the walls. On the other hand, particularly in direct-injection burners, losses may result because the larger mist droplets are incompletely desolvated in the flame owing to their brief residence time. These losses diminish the emission, absorption or fluorescence signal in the flame and worsen the detection limits. Furthermore, differences in the physical properties (such as surface tension) or in the matrix, among individual samples or between sample and reference solutions, may lead to variable losses of analyte. Such differences may be due to the presence of concomitants in varying concentrations and may cause transport, desolvation or volatilisation interferences (see chapter 9). Unless these interferences are corrected for, systematic errors may affect the analytical result. The magnitude of these losses is governed largely by the drop-size distribution of the mist. The factors affecting it will therefore be discussed here. Since the influence of these factors upon the nebulisation and the subsequent desolvation and volatilisation is very complex, and since many instrumental factors such as the form of the nebuliser and the size of the spray chamber enter too, some parts of the following discussion can only be qualitative or semi-quantitative.

Attempts have been made to derive a general theoretical expression for the spatial distribution of free analyte atoms in premixed flames with a slotted burner [603, 604] as well as with a circular burner [568]. These expressions are in general based on the vapour transport equation, the delay in sample vaporisation being taken into account. The transport equation includes entrainment of the vapour with the moving flame gas as well as atomic diffusion (see also [38, 216a, 804]). In [603, 604] the horizontal and vertical components of the gas velocity were considered. In [568] the size distribution of the droplets entering the flame was explicitly taken into account in the calculation of the desolvation and volatilisation rates†. Complicated expressions are obtained even when the mathematical model is much simplified. The restrictions of the idealised models should be kept in mind when attempts are made to apply the results in realistic situations.

4.2. Aspiration

In a pneumatic nebuliser the sample liquid is aspirated from the sample container through a capillary tube, usually by the suction (Venturi action) caused by the compressed-air or oxygen jet. The capillary tube may be either concentric with the air jet (in concentric nebulisers) or placed at right angles to it (in angular nebulisers). The high velocity of the jet creates a reduced pressure by virtue of Bernoulli's law at the outlet of the liquid capillary. The resulting suction pressure is opposed by the hydrostatic head of the liquid column lifted in the capillary. Another factor that reduces the effective suction pressure is the pressure drop needed to accelerate the liquid when it enters the narrow capillary. Yet another factor is the pressure drop due to threshold effects experienced by the liquid as it enters or leaves the capillary. These additional pressure drops depend, among other things, on the density and surface tension of the liquid. In most cases the suction

---

† Because of non-uniform, partial evaporation of the droplets before they enter the flame, the actual salt concentration of the droplets leaving the burner may not be uniform. This complication was not considered in [568].

pressure of the gas jet due to the Bernoulli effect (10 to over 100 cm of water) considerably exceeds these opposed pressure drops [41, 160, 209, 262, 349, 432, 438, 894, 907]. The aspiration rate, that is, the quantity of liquid aspirated per unit time, usually expressed in ml min^{-1}, is then little affected by the height of the liquid level in the sample container or by differences of density or surface tension among samples; otherwise the signal may change as the level falls or may be affected by the differences in physical properties of the liquid, thus giving rise to errors in the analytical result.

The surface tension of the liquid at the outlet of the capillary may also reduce the effective suction pressure [91, 721]. This effect might be expected to be more pronounced with concentric nebulisers, from which the liquid emerges more or less laminarly, than it is with angular nebulisers. Experimentally, however, no effect of surface tension has been found in concentric nebulisers [907].

The effective suction pressure for a given air flow cannot be found exactly by theory, since the configuration of the instrumental parts is very critical; this is true of direct-injection burners particularly with respect to the difference in height between the outlet of the liquid capillary and the orifice for the nebulising gas [432]. To be sure, a semi-empirical relation can be set up between the aspiration rate and the air pressure in a concentric nebuliser [574] (compare also [907]). It appears that a certain minimum pressure is needed if the liquid is to be aspirated at all.

According to the Hagen–Poiseuille law the aspiration rate should be directly proportional both to the effective pressure drop along the capillary tube and to the fourth power of the capillary diameter, and inversely proportional to the viscosity. This holds if laminar flow exists in the capillary, implying a sufficiently low Reynolds Number (compare § 3.4.4). For ordinary nebulisers in flame spectrometry this condition is satisfied [306, 309, 574, 907] but an exception must perhaps be made for organic solvents, for which the Reynolds Number may reach 1000 [907]. The dependence on viscosity has been fairly well confirmed in many cases by experiments with various solvents, in particular for concentric nebulisers [209, 262, 306, 315, 349, 574, 721, 725, 864, 907, 932]. The theoretical dependence on capillary diameter has also been experimentally verified [209, 907]. In summary, the Hagen–Poiseuille law is more accurately applicable in flame spectrometry when the aspiration rate is lower and the viscosity higher [907]. It should be noted, however, that in direct-injection nebulisers strong heating can occur, which may affect the viscosity and the precise adjustment of the (metallic) capillary tube.

It must be emphasised that the concentration of the analyte in the flame is not necessarily proportional to the aspiration rate [160, 210, 315, 324, 329, 623, 721, 727, 824, 853, 894, 909]. This is true even when all other adjustments of the flame spectrometer are kept fixed while the aspiration rate is controlled by means of a motor-driven syringe. One reason is that the fraction of sample lost is not constant but increases with increasing aspiration rate owing to deterioration of the nebulising action [432]. Another reason is that a rise in aspiration rate can lower the flame temperature, especially in direct-injection burners, substantially affecting the desolvation, dissociation and excitation of the sample. It is therefore often found that with rising aspiration rate the analyte emission passes through a maximum or reaches a plateau [329, 349, 727, 824, 894].

The presence of small air bubbles in the solution can slow the aspiration rate and make it erratic [773]. This difficulty can be avoided by de-aerating or boiling the solvent.

It has been observed in one case that the presence of an impact bead in the compressed-air jet induced a high-frequency pulsation in the aspiration rate [918].

4.3. Nebulisation

4.3.1. The nebulisation process. What fraction of the aspirated sample solution ultimately becomes available as metal vapour in the flame depends to a great extent on the drop-size distribution in the nebulisation process or, more exactly, on the fraction of liquid which, on nebulisation, is contained in droplets below a certain diameter. The larger drops, comprising a comparatively large fraction of the sample, are lost. Concerning the nebulisation process itself, little is yet known quantitatively despite its great importance in technology and industry. The process can be described qualitatively as follows: the liquid emerging from the suction capillary is shredded into liquid tatters, which contract into droplets owing to surface tension and are thrown into oscillation. Through the action of the powerful air jet or by striking an impact surface or the like, these drops can shatter into still smaller droplets [73, 78]. The disruption is enhanced by the strong turbulence and sometimes by shock waves or ultrasonic vibrations in the air jet. These nebulisation processes can be clearly demonstrated by photography [78, 261, 628].

A drop in an air jet is acted upon by both tangential and radial forces. The tangential forces are due to friction with the flowing air and the radial forces to the impingement of air normally upon the drop. The impact pressure resulting from the radial forces is $\rho_a v^2$, where ρ_a is the density of air in $g\,cm^{-3}$ and v is the relative velocity of the air with respect to the drop. These two forces oppose those that tend to hold the drop together, namely, viscosity and surface tension. At sufficiently high relative speeds the viscous forces can be neglected and the drop begins to burst when the impact pressure becomes at least three times the surface tension pressure $2\sigma/r$, where σ is the surface tension and r is the radius of the drop [450, 573]. Under these assumptions the critical radius r^* is given by the equality:

$$\rho_a v^2 = 6\sigma/r^*. \tag{4.1}$$

The greatest droplet diameter possible, $2r^*$, can thus be calculated and for pneumatic nebulisers in flame spectroscopy is of the order of 10 μm. But more important than this numerical result is the theoretically expected dependence of the critical diameter on the impact pressure and surface tension, as expressed by the equation.

However, the conditions prevailing in a pneumatic nebuliser depart considerably from those for which the above simple equation for the critical diameter $2r^*$ holds exactly. For example, the compressed-air jet is very inhomogeneous. Impact surfaces and the like are used for attaining the greatest possible velocity gradients and discontinuities in the air jet, so that with respect to the droplets (by virtue of their inertia) the relative velocity of the air stream will be kept high. Hence, it is hardly possible to calculate in advance the maximum drop diameter, much less the drop-size distribution. Empirical formulae have to be used, but these apply only to given apparatus and specific conditions. It is much better to measure the drop-size distribution directly (see below). A thorough treatment of the processes and methods of nebulisation can be found in [41, 55, 56, 62, 73, 78].

The energy absorbed (because of surface tension) in increasing the total liquid surface during nebulisation is usually negligible compared to the energy content of the air–mist system. In one case it was calculated that the surface work was equivalent to cooling the water drops by about 0·1 °C [160] and it has also been calculated [41] that, under the usual conditions, a power of 1 W is available for nebulisation in the compressed-air jet. This is a thousand times the power needed for nebulising water to a droplet diameter of 10 μm at an aspiration rate of 2 ml min^{-1}.

As a result of nebulisation the sample enters the flame not smoothly but in a 'dropwise' manner and consequently the emission or absorption of light in the flame is not steady but shows random fluctuation. This is in a way comparable to the shot effect in amplifier tubes or photomultiplier tubes, which is due to the random emission of individual electrons with discrete electric charge. For a mist of uniform drop size, the relative magnitude of the fluctuation is inversely proportional to $\sqrt{N}$, where N is the average number of aerosol particles passing through the observation volume per unit time. The fluctuation thus depends on the fineness of nebulisation. Since the former is governed by the laws of statistics, it will persist even when all other instrumental conditions (such as the air flow) are held strictly constant. Its effect can be diminished, however, by making the mist finer (by raising N) or by extending the time of measurement.

This drop-effect fluctuation has components not only at low frequencies, about 0–1 Hz, which cause a fairly slow random variation in the signal, but its so-called noise spectrum, that is, the distribution of noise components with respect to frequency (see also § 3.4.4), may in fact extend far beyond 1000 Hz. This might have consequences in the absorption and fluorescence methods when the background source radiation is modulated and the (interfering) thermal radiation of the analyte is not, the signal being measured by a synchronous AC meter. The components of the drop-effect fluctuation in the thermal radiation within a certain frequency band bracketing the modulation frequency then contribute to the fluctuations in the reading. The few noise measurements available show, however, that neither in direct-injection burners nor in laminar flames with a chamber-type nebuliser do the droplet statistics have more than a minor effect on the signal fluctuation [171, 784].

4.3.2. The drop-size distribution. There are more than 40 methods for determining a drop-size distribution, among which the cascade-impactor method and the microscopic method are easy to use [78, 160, 300, 446, 556, 637, 713]. (For a more thorough treatment of the theory and experimental methods see [55, 234, 576].) In the microscopic method the droplets strike a glass plate coated with magnesia. In this coating each drop makes a hole whose diameter corresponds to that of the drop and is measured under the microscope. Drop diameters below even 10 μm can be measured in this way.

Many capture and impact methods have the disadvantage that the drops may partly evaporate and the smallest of them may not be caught at all. In studies of nebulisers for flame spectroscopy, other methods have therefore also been applied. For example, a concentrated solution of methylene blue was nebulised and the solid particles left after evaporation of the solvent were caught and analysed microscopically according to size [894]. This analysis, together with the known concentration of the solution, gave the drop-size distribution. In another experiment, molten polyethylene glycol was nebulised and the size distribution of the captured solid particles was determined [536]. Dilute India ink has also been nebulised to make the droplets near 1 μm plainly visible [918]. Such methods have the drawback that the special test liquids may nebulise in a different way from the usual analytical solutions.

Drop sizes found with pneumatic nebulisers are mostly between about 1 and 50 μm. Although in many cases most of the drops are smaller than 10 μm [824] or even 2 μm [874] in diameter, the few larger drops, because of their greater salt content, are nevertheless relatively important (see § 4.6). The drop-volume distribution is therefore more significant for the nebulisation efficiency than the diameter distribution.

From the size distribution the weighted average value of the drop diameter D_0 can be calculated as a characteristic parameter for a given nebuliser. D_0 can be defined as the diameter of a drop whose volume to surface ratio equals that of the total aerosol. (It is also true that the mass of all drops of diameter less than D_0 is about half the total mass.)

We offer no empirical formulae or graphs of actual size distributions†, which will vary with the nebuliser type and with the arrangement of orifices, etc. In direct-injection burners, for example, a narrower liquid capillary makes for a smaller mean drop diameter and a more uniform mist [295]. Moreover, unknown systematic errors, depending on the particular method of measurement, may mar the comparison of data from different authors. In chamber-type nebulisers the drop-size distribution is greatly altered by (partial) desolvation in the spray chamber and by losses on the walls, effects which in turn depend strongly and often in an unknown way on the experimental conditions. The drop-size distribution in any given case can be found only by measurement. The *primary* drop-size distribution, measured directly at the nebuliser nozzle, should here be sharply distinguished from the *secondary* size distribution of the droplets reaching the flame (see also § 4.4). With direct-injection burners it is also doubtful that there is enough time for the size distribution that is measured without a flame to be fully developed before the mist actually enters the flame [300]. We shall therefore limit the discussion to

---

† The empirical dispersion formula often quoted in the literature (see [664]) for concentric nebulisers applies only to subsonic velocities and aqueous solutions. In practice, however, such nebulisers are best operated at about sonic velocities [432].

a more qualitative description of factors affecting the (mean) droplet diameter with the usual nebulisers.

4.3.3. The effects of gas flow, surface tension and viscosity.

In qualitative agreement with equation (4.1) for the critical radius r^*, measurements have shown that the characteristic diameter D_0 falls to about 10 μm when the gas velocity reaches sonic velocity [295, 664]; any further increase of velocity seems to offer little gain. Equation (4.1) contains the product $\rho_a v^2$ (where v is the relative velocity of the air with respect to the liquid droplet and ρ_a is the density of the air) as the factor determining the maximum drop size. This gives rise to the question of how the (absolute) velocity v_a and density ρ_a of the air are related to the absolute air pressure p at the entrance of the nozzle. In theory this relation can be derived for ideal conditions such as adiabatic expansion and no internal friction or wall friction. At comparatively low pressures, with $p - p_0 \ll p_0$ (where p_0 is the atmospheric pressure), v_a^2 is approximately proportional to $p - p_0$ and ρ_a remains practically constant, as has been verified by flow measurements with concentric nebulisers [574]. In the super-sonic range ($p > 1 \cdot 9 \, p_0$), $\rho_a v_a^2$ is proportional to p [160]. However, secondary effects make the application of these relations to practical cases rather dubious. Discrepancies may thus arise with concentric nebulisers because it is mainly the margin of the compressed-air jet that comes into contact with the liquid [574].

The expected effect of surface tension on the dispersion of the mist has been confirmed by direct measurements of drop size with concentric nebulisers and various liquids [664]. Attempts have therefore been made to increase the mist dispersion and hence the flame emission by adding wetting agents, such as Triton X-100, which greatly reduce the surface tension in small concentrations. Although a favourable effect on the flame emission has been reported in one case for a chamber-type nebuliser [293], the effect of wetting agents seems on the whole to be rather small [210, 318, 349, 640, 722, 881, 917]. This apparently unexpected result can be explained by the fact that the surface tension needs a certain length of time to reach equilibrium [160, 880, 881]. For wetting agents of high molecular weight and at low concentration, this time may be too long compared to the duration of the nebulisation process.

In an extended series of experiments with a direct-injection burner in which aliphatic acids such as acetic acid were added to the solution while the aspiration rate was held constant, a clear correlation was found between the atomic emission and the surface tension [881]. Evidently, the equilibrium time for the surface tension is short enough with these additives. More generally, this should be true of organic solvents, especially alcohols, of which the surface tension is only about one-third that of water [160, 315, 716, 725, 881]. In such studies, however, the experimental evidence is confused by other factors such as the rate of desolvation, the alteration of temperature and the thickness of the flame which also affect the observed flame emission. These circumstances may account for the apparently contradictory results of different authors on the role of surface tension with alcohols [339, 717, 725].

When the flow rate of the air is low and that of the liquid is high, D_0 can also be influenced by the viscosity of the liquid [664]. The viscosity affects the turbulence in the emerging jet of liquid [295] and in consequence probably affects the nebulisation.

4.3.4. The recombination of droplets. In contrast with the effect of nebulisation, recombination (or coagulation) in the spray chamber and elsewhere forms larger droplets out of smaller ones, an effect which coarsens the drop-size distribution and increases the deposition loss on the walls. Recombination has to be considered when there is a fair probability that two drops will meet and merge during their time of travel to the flame. (For a fuller theoretical treatment see [55, 91].) Recombination of water droplets with the usual nebulisers may be negligible up to travel times of 0.5 s [529]. In a concentric nebuliser, recombination has been observed to increase the mean drop diameter appreciably only when the air flow rate (in $cm^3 s^{-1}$) is less than about 5000 times the liquid flow rate (in $cm^3 s^{-1}$) [73, 664]. This condition may exist in the usual compressed-air nebuliser; in general, recombination is expected to depend on the initial drop-size distribution also.

To estimate these effects quantitatively, twin-nebuliser experiments have been performed in which an aqueous sodium solution and a methanol–water mixture were nebulised separately through two parallel, identical nebuliser units and then combined in a mixing chamber and conveyed jointly into the flame [91, 721, 724]. Mutual coagulation of droplets from the two aerosols increased the drop loss. Consequently, when the two solutions were nebulised together the sodium emission was less than when the methanol solution was not nebulised. The depressing effect increased with methanol concentration; the sodium emission fell to 88 per cent of its original value when the methanol concentration was raised to 91 per cent. The coagulation effect could be represented as a function of time by a mathematical expression from which a coagulation rate constant could be derived. This constant specifies the stability of the mist and depends on the physicochemical properties of the solvent, especially on the length of the non-polar group of an alcohol molecule. Other authors, however, have found that when a calcium and a phosphate solution are fed separately through twin nebulisers into the same burner, the phosphate does not depress the calcium emission. Recombination of aerosol particles was apparently unimportant in this case [356].

4.4. Desolvation

4.4.1. Desolvation before entry into the flame. For direct-injection burners the drop-size distribution measured when the flame is not lit is the same as the distribution of drops entering the flame, apart from possible post-nebulisation effects by shocks produced in the turbulent flame (see chapter 3). This is true only if the drop-size distribution is fully developed in the very short time before the mist enters the actual flame. It is then only the evaporation of the solvent and the volatilisation of the solid particles that determine how much of the sample can be made available quickly enough for excitation or absorption.

Conditions are different in chamber-type nebulisers. Here the deposition of the larger drops in the spray chamber and burner conduits influences the secondary drop-size distribution and the quantity of solution entering the flame. Also important are the effects of post-nebulisation at impact surfaces, recombination processes (see § 4.3.4) and finally the substantial effect of desolvation in the spray chamber and burner on the size of the droplets. We shall discuss here the desolvation process in the spray chamber.

It can be shown theoretically and experimentally that desolvation in the spray chamber does indeed have an important effect on the size distribution of the drops reaching the flame [160, 185, 209, 262, 306, 318, 439, 706]. This is demonstrated by experiments with spray chambers that can be heated to different temperatures or with spray chambers of various sizes, by spraying with preheated dry air or with air saturated with moisture, by using volatile solvents, and by other means. The effect of the desolvation on the drop-size distribution and hence on the analytical signal is hard to predict but the magnitude of the effect and the influence of the various factors can be judged from the following theoretical considerations.

In the theory, we consider a stationary spherical droplet of radius r in still air. The size of the drop is assumed to be large compared to the mean free path of air molecules but not so large that convection becomes involved. Let p_∞ be the spatially homogeneous vapour pressure far from the drop and let the vapour pressure at the surface of the drop be equal to the saturated vapour pressure p_s. We assume that the desolvation rate is governed only by diffusion of the vapour through the air (the diffusion coefficient being D), that the partial pressure of the vapour is everywhere relatively low, and that the release of vapour molecules at the surface of the drop in the desolvation process is rapid enough. The change of mass m of the drop with time t is then given by the Langmuir equation [114, 187, 575, 665, 890]

$$-\frac{dm}{dt} = \frac{4\pi r M D}{RT}(p_s - p_\infty),\qquad(4.2)$$

where M is the molecular weight of the vapour, R is the gas constant and T is the absolute temperature. It should be noted that the mass evaporated per unit time is proportional not to the surface $4\pi r^2$ but to the radius r. This is an essential consequence of our assumption that the rate of desolvation is determined only by diffusive transport of the vapour away from the drop.

The above equation yields the rate of decrease of surface F:

$$-\frac{dF}{dt} = \frac{8\pi M D}{\rho RT}(p_s - p_\infty),\qquad(4.3)$$

where ρ is the density of the liquid. This equation states that at constant temperature and for $p_\infty \approx 0$ the rate of surface contraction $-dF/dt$ is constant. For water droplets at room temperature in dry air, it amounts to about 10^{-4} cm^2 s^{-1}, so that a drop with a diameter of 50 μm is completely desolvated in about 1 second [62, 823]. This is roughly the residence time of the droplet in the spray chamber, so

51

that all droplets below 50 μm at the start, in a chamber-type nebuliser, might be supposed to have lost their solvent entirely before entering the flame. The following factors, however, upset this conclusion: (i) the droplets cool on desolvation – sometimes as much as 14 °C, reducing the desolvation rate by a factor of 2·5 [160]; (ii) the air in the spray chamber grows moister from the desolvation of the many nebulised droplets, so that p_∞ approaches p_s; (iii) as the drop shrinks, the changing radius and increasing salt concentration affect the saturation pressure p_s [270]; (iv) the desolvating drops are not quiescent but are moving with respect to the air and therefore they desolvate faster; and (v) if the solvent consists of two or more liquids, such as glycol and water, their relative concentrations change during desolvation and give rise to changes in $-dF/dt$ [380].

The consumption of energy by desolvation appreciably cools the nebulised liquid and the air stream [160] and hence the spray chamber and the connections to the burner. Thus heat is exchanged with the surroundings, and the energy consumed is partly recovered.

When a given total mass of liquid is dispersed over a larger number of finer droplets, the total quantity of solvent evaporated per unit time increases according to equation (4.2). An upper limit to the desolvation, however, is imposed by the saturation of the air. At a consumption rate of 10 l min^{-1} of dry air, no more than about 0·1 g min^{-1} of water can evaporate. If the flame heats the burner and the air passing through it, there can be further desolvation of the mist here [553].

The residence time and hence the desolvation of the drops in the spray chamber increase with the size of the chamber, and losses due to impact on the walls diminish. An increase in volume of the spray chamber from 0·5 to 2 litres, for example, appreciably raises the rate of entry of the sample into the flame [439], but if the volume is again increased by another factor of 4, there is little added gain, partly because losses on the walls have dwindled and partly because the approach to saturation of the air hinders further evaporation of the drops.

According to the Langmuir equation, the rate of desolvation increases with increasing saturated vapour pressure p_s. Low-boiling solvents such as ethanol (78 °C) or acetone (56 °C) are therefore helpful in delivering more analyte to the flame [185, 723].

4.4.2. Desolvation in the flame. When the nebulised mist – already partly evaporated in a chamber-type nebuliser – reaches the flame, the droplets are heated to boiling and the (remaining) solvent evaporates. In the following discussion of desolvation in the flame, we pay particular attention to the question whether there is enough time for complete desolvation.

In contrast to the desolvation of a water droplet at room temperature (see § 4.4.1), which is governed by diffusion of the vapour, the rate of desolvation of a drop with heat of vaporisation L at its boiling point T_b in a flame of temperature T_f (above T_b) is mainly governed by heat transport between the flame gas and the drop. Under the same postulates as before, we may assume that the transport of heat occurs only by heat conduction, the thermal conductivity λ being assumed constant. Part of the transported heat is used for bringing the vapour formed at T_b to the flame temperature T_f; the specific heat c_p (at constant pressure) of the vapour is thus involved. The drop then loses mass at a rate given by the equation [114, 429, 890]

$$-\frac{dm}{dt} = \frac{4\pi r\lambda}{c_p} \log_e\left(1 + \frac{(T_f - T_b)c_p}{L}\right),$$ (4.4)

where r is the drop radius. As before (equation (4.2)), the desolvation rate is proportional not to the surface but to the radius of the drop and consequently the surface shrinks at a constant rate dF/dt, provided at least that all other parameters can be assumed constant (see the experiments in [273]).

The boiling droplet is surrounded by a thin layer of solvent vapour and, outside this layer, by a mixture of solvent vapour and flame gas. The value to be inserted for λ in equation (4.4) depends on the ratio of the thermal conductivities of the solvent vapour and the flame gas. For water droplets boiling in an atmosphere of mainly nitrogen gas, the value of λ for nitrogen should be inserted, since it is the smaller one [273, 444]. This holds because the heat conduction through the flame gas is then the dominant factor limiting the rate of heat transport.

In calculating the desolvation rate, $\log_e(1 + x)$ may be approximated by x when $x \ll 1$. If this is done in equation (4.4), with $x \equiv c_p(T_f - T_b)/L$, it can be seen that c_p drops out. This is understandable since the condition $c_p(T_f - T_b)/L \ll 1$ means that the heat needed to bring the vapour to the temperature of the flame is small compared to the heat of vaporisation. However, for drops of water in a flame, the two quantities of heat are of the same order of magnitude and the approximation is not valid.

Since the residence time of the aerosol particles in the flame depends on the rise-velocity, it is possible to calculate at what height in the flame a drop of given initial surface area will be completely desolvated. From equation (4.4) it easily follows that the time t needed for complete desolvation of a drop having initial surface F_0 is given by [165]

$$t = \left(\frac{\rho F_0 c_p}{8\pi\lambda}\right) \Big/ \log_e\left(1 + \frac{(T_f - T_b)c_p}{L}\right),$$ (4.5)

where ρ is the density of the liquid. The relation between t and the height h at which the drop is just desolvated is given by

$$h = v_f t,$$ (4.6)

where v_f is the rise-velocity of the flame gases.

The proportionality between t and F_0 has been experimentally verified for drops with a uniform diameter above 40 μm injected singly into the flame by means of a special drop generator [446]. Theoretical calculations of the absolute desolvation rate in flames are given for various solvents in [273]. For water, as an example, the calculations agree with experiments in an air–acetylene flame [273]. Here the desolvation rate is determined mainly by the thermal conductivity of the flame gas itself and not by that of the surrounding water vapour. The theoretical value of this rate for cyclohexane appears to rise only fourfold when the gas temperature rises from 500 to 3000 K, an extreme case.

53

Some illustrations will next be given. For water droplets with an initial diameter of 1 μm in an air–acetylene flame rising at about 10 m s^{-1}, the minimum height for complete desolvation is calculated to be 0·03 mm. At a height of 3 mm even drops with an initial diameter of 10 μm will be fully desolvated; hence, in a laminar flame with a chamber-type nebuliser practically all drops are desolvated at the usual height of observation, 1 to 3 cm above the base of the flame. (Recall that with such a nebuliser part of the solvent has already evaporated in the spray chamber.) For still larger drops, such as may be expected from a direct-injection burner, the short residence time in the turbulent flame with its high discharge speed may not be enough to desolvate all the drops below the usual observation height of 1 to 3 cm. Analyte may thus be lost (see § 4.6.3). However, the turbulence of the gases may raise the desolvation rate above that in a laminar flame. It would be hard to evaluate the conditions in a turbulent flame quantitatively.

In an oxyhydrogen flame on a direct-injection burner, experiments with suitable dark-field illumination have shown that part of the nebulised water is not fully desolvated in the flame [381, 727] (see figure 6), but when a 90 per cent acetone solution is nebulised, desolvation is apparently complete [381]. The heat of vaporisation and surface tension of acetone are considerably less than those of water; these properties have a favourable effect on the desolvation rate and the drop size. The heat of combustion of acetone may also contribute (for comparison,

(a)   (b)

Figure 6. (a) Spray from a direct-injection burner with the flame unlit (dark-field illumination). (b) Spray from the same direct-injection burner with the flame burning. (From [727]. Reproduced by permission of the Wissenschaftliche Verlagsgesellschaft, Stuttgart.)

see below). Similar observations of scattering and reflection in fluorescence spectroscopy have shown the presence of undesolvated liquid drops in flames on direct-injection burners but not with chamber-type nebulisers [547, 671, 856]; in the former the drops affect the blank reading (see chapter 9). These effects can be avoided by using an organic solvent or by preheating the mist in a hot chamber, in which desolvation is complete [856]. Systematic studies have shown that with a direct-injection burner the desolvation of the water droplets increases with observation height, as expected [687]. It has also been found that the degree of completeness of desolvation depends on the bore of the liquid capillary and the flow velocity of the nebulising gas, both of which influence the drop-size distribution. Other experimental methods for demonstrating the (in)completeness of desolvation in the flame are surveyed in [166] (see also § 4.6.3).

When a combustible solvent is nebulised, the desolvation is accelerated by the heat of combustion of the vapour. Here the rate of mass loss of a drop in the flame is given by the expanded equation [890]

$$-\frac{dm}{dt} = \frac{4\pi r \lambda}{c_p} \log_e \left(1 + \frac{(T_f - T_b)c_p}{L} + \frac{Q\xi}{L} \right), \qquad (4.7)$$

where Q is the heat of combustion per unit mass of the reacting mixture and ξ is the ratio of the actual oxygen concentration in the flame to the oxygen concentration required for complete combustion of the combustible solvent.

The contribution of convection to the desolvation, which may be significant especially in turbulent flames, can be represented simply by adding to the above equation for the rate of mass loss a factor $(1 + 0.276 Re^{1/2} Pr^{1/3})$, where Re is the Reynolds Number (see § 3.4.4) and Pr is the Prandtl Number, equal to $\eta c_p/\lambda$, where η is the viscosity [890]. For ideal gases and water vapour Pr is approximately unity. For calculation of the relative droplet velocity see [216a].

The energy required to evaporate the liquid and to heat the vapour in a premixed flame with a chamber-type nebuliser is generally only a small fraction of the heat content of the flame gases [78, 165, 804]; only a portion of the aspirated liquid reaches the flame, and the drops are already partly desolvated in the spray chamber. This may not be true of flames on a direct-injection burner (see § 3.5.3); here, however, there may not be enough time for all of the aspirated liquid to be completely desolvated in the flame. The associated decrease in flame temperature has been discussed in § 3.5.3.

The desolvation process may be complicated by explosive disintegration due to crust formation by precipitation at high solute concentrations [196]. The simple equations above will then no longer hold.

4.5. Volatilisation of the Dry Aerosol

When a sample solution is nebulised, desolvation leaves a dry aerosol consisting of solid or molten particles. The formation of this aerosol and its conversion into vapour (*volatilisation*) are often very complex and on the whole poorly understood. The composition, size and state of the dry aerosol particles may depend on many

55

factors. Among these are: the qualitative and quantitative composition of the sample solution, including the solvent and concomitants; the (original) size distribution of the mist droplets; the flame temperature and the reducing or oxidising conditions in the flame; and the residence time of the particles in the flame, in other words, the observation height. During their brief period of heating in the flame, the particles pass rapidly and irreversibly through several phases and states that are hard to describe with the usual laws of chemical equilibrium.

The dimensions of the (solid) aerosol particles are easily calculated from the specified (original) drop sizes and concentration of the solution. The dimensions can also be found directly by capturing the particles in the flame and measuring them by electron microscopy [529] or by light scattering [553, 856]. The chemical composition and structure of the aerosol particles can be determined by x-ray diffraction or electron microdiffraction [97, 184, 760, 814, 815].

In analytical flame spectroscopy the following consequences of a limited rate of volatilisation are frequently encountered. Incomplete volatilisation means loss of analyte capable of emitting or absorbing radiation and impairs the detection limit – an example of this is the impossibility of determining aluminium in the cooler flames. Incomplete volatilisation also makes the analytical curve convex (see chapter 8), and incandescent, unvaporised solid aerosol particles may create a continuous background emission or, in the fluorescence method, an undesirable light scattering. Interferences can arise, finally, when the volatilisation of the analyte is influenced by concomitants. Such interferences may be due to the formation of involatile compounds between the analyte (e.g. calcium) and a concomitant (e.g. aluminium). Interference can also arise when the analyte (e.g. sodium) is dispersed or entrapped in a less volatile matrix (e.g. of CaO). Despite the many investigations that have been devoted to this problem, we shall unfortunately have to settle for a semi-quantitative or merely qualitative and often downright hypothetical presentation.

In the description of the volatilisation process we must distinguish the following three cases.

(i) When a drop of a sodium chloride solution is completely desolvated it leaves a sodium chloride particle which is heated to the melting point (1075 K) and then to the boiling point (~ 1700 K). In this case the flame temperature is usually higher than the boiling point, and so the particles or molten drops of salt are certainly not stable. The rate of volatilisation can presumably be represented by an equation such as equation (4.4) for a boiling water droplet. According to this equation the rate dm/dt is controlled by heat transport to the boiling droplet. However, when the diameter of the particles is less than the thickness of the surrounding diffusion layer – which is of the order of the mean free path of the vapour molecules – equation (4.4) no longer holds [196, 444]. The time needed for complete volatilisation is then no longer proportional to the initial particle surface F_0 (compare equation (4.5)), but is proportional to the initial radius r [487].

(ii) If the flame temperature is between the melting point and the boiling point of the aerosol particle, volatilisation from the liquid phase might be expressed by a Langmuir diffusion equation such as equation (4.2) for a water drop evaporating in air. An example of this case would be the evaporation of Al_2O_3 particles (melting point 2320 K, boiling point 3250 K) in the hotter flames. However, equation (4.2) only applies to a particle whose diameter is large compared to the free path of the molecules. Because the dry aerosol particles at normal solution concentrations are smaller, this equation should be modified according to [665, 929, 937]. The saturation pressure p_s of the vapour should here be taken as that corresponding to the surface temperature T_p of the molten particle, which in general is lower than the flame temperature T_f. The difference $T_f - T_p$ is given by (see [75]):

$$T_f - T_p = LMDp_s/\lambda RT, \tag{4.8}$$

where the other symbols have the same meaning as in equations (4.2) and (4.4). For a molten drop of Al_2O_3, we have, approximately [75]:

$$T_f - T_p \approx 3 \cdot 6 p_s, \tag{4.9}$$

where the temperature is in kelvin and p_s is in Torr. For less volatile substances with $p_s \lesssim 1$ Torr, T_p can usually be taken to be equal to T_f.

From the Langmuir equation (which is only approximate), it can be calculated that at a flame temperature of 3100 K a time of about 5×10^{-4} s is required for the volatilisation of a molten drop of Al_2O_3 formed from a drop of solution having an initial diameter of 20 μm and an aluminium concentration of 0·12 per cent [75]. During this time, according to equation (4.6), the Al_2O_3 drop will have risen about 5 mm.

(iii) For many metal oxides the melting point is higher than the flame temperature. A particle of MgO (melting at 3070 K and subliming at about 3040 K) in a flame would pass directly from the solid to the vapour phase [228]. As in the other two cases, this process is irreversible if, as is usual, the partial pressure of the oxide vapour in the flame after complete volatilisation remains below the saturated vapour pressure at the temperature of the flame. Therefore, if there is enough time available, the sample in this case too will vaporise completely.

For certain elements such as rhenium the partial pressure of atomic vapour after complete volatilisation may reach or even exceed saturation [389]. In such a case, account may then have to be taken of the condensation of the vapour which may also play a part in flame spectroscopy in the case of some metal oxides [502]. Calculations show that, at flame temperatures below 2000 K and at solution concentrations greater than 1 mol l^{-1}, chromium may be present partly as solid Cr_2O_3 in equilibrium [243]. Thermodynamic calculations of the equilibrium fractions of aluminium and silicon that are present as solid oxides in the cooler flames, with a temperature below 2500 K, are given in [272] and [285].

The theoretical question of the rate of volatilisation or the time needed for complete volatilisation is hard to answer quantitatively. It is only certain that, other things being equal, this time increases with the size of the aerosol particle. Experimentally, an unambiguous relation has been established between the (mean) particle diameter and the loss of free atoms of magnesium by forming involatile aluminates in the solid phase [824]. Quantitative calculations are thwarted because the particle is not necessarily spherical. A complex solution may even produce porous particles if the different components evaporate at widely different rates.

The question, which thermodynamical quantity determines the volatility of a solid particle of given dimensions at a given flame temperature, is still open to speculation. Both the surface energy [713] and the free energy of formation [154, 619] of the crystal have been proposed as determinative. Other authors consider the decomposition temperature, the melting point or the boiling point to be indicative of the volatility of a compound (e.g. [625, 647, 798, 815]). It is interesting that, on the basis of the free energy of formation, solid Al_2O_3 ought to be much more stable than MgO [619] although the melting point of Al_2O_3 (2320 K) is a great deal lower than that of MgO (3070 K). The existence of more than one decomposition product complicates matters. These considerations on the rate of volatilisation are further complicated by the possibility of explosive disintegration when the solution consists of several components with widely different boiling points or when, for example, it contains perchloric acid [196, 381]. Decrepitation is likely whenever the solvent becomes trapped in a melt derived from the solute (e.g. CrO) during the final phase of desolvation [529]. Chemical reactions at the surface of the solid aerosol particle with radicals in the flame may aid the volatilisation; as an example the liberation of Ca and Mg atoms could be partly accounted for by the reaction of the solid oxide particles with flame radicals (e.g. H) [410, 678, 750]. Such reactions, however, do not seem effective for Cr_2O_3 particles [243]. The formation of free Sn atoms from solid SnO or $SnCl_2$ particles in cool hydrogen flames could be due to similar reactions [290].

Finally, the validity of the above equations for the rate of volatilisation may also be impaired when the heat or vapour transport between the aerosol particle and the flame is aided by convection. This could be significant especially for the heavier particles, which, because of their inertia, possess a substantial velocity relative to the flame gases [929].

The energy consumed in volatilisation is totally negligible in comparison to the heat content of the flame in every case of interest in flame spectroscopy [78, 165]. This is true even when the concentration of the solution is of the order of 1 mol l^{-1}.

The nature of the chemical compound composing the solid or molten aerosol particles in the flame may not always be as well defined as it is for a sodium chloride solution. The fact that a metal is present as chloride in a solution does not guarantee that a chloride particle remains after desolvation. The chlorides of chromium and aluminium, for example, form hydrated molecules in aqueous solution, which on dehydration in the hot flame may yield involatile oxides instead of chlorides [315, 754]. For magnesium chloride the following chemical processes may be significant [410]. The hexahydrate is first converted in the solid phase at 430 K into $MgCl_2.H_2O(s)$ with loss of water and then may undergo two further reactions at the temperature of the flame:

$$MgCl_2.H_2O(s) \rightarrow MgCl_2(s) + H_2O(g) \qquad (1)$$

or

$$MgCl_2.H_2O(s) \rightarrow MgO(s) + 2HCl(g). \qquad (2)$$

The boiling point of $MgCl_2$ is 1685 K, so that even in the cooler flames reaction (1), unlike reaction (2), easily puts the magnesium into the vapour phase. For calcium and strontium reaction (1) may dominate [760]. In more complex salts such as nitrates, carbonates and phosphates, the heat of the flame can cause chemical changes prior to volatilisation, yielding oxides, pyrophosphates, etc [75, 410, 625, 760]. The formation of solid oxides such as those of chromium may also depend on the reducing conditions in the flame [434, 513]. It can be supposed that magnesium nitrate will change at 594 K into solid, involatile MgO, but if the nitrate particles are heated very quickly they may melt first. Volatilisation would then take place directly (and faster) from the melt [410]. In the hot, fuel-rich nitrous oxide–acetylene flame molybdenum oxides may be partly reduced by hydrocarbons and hydrogen to carbide and/or metal in the condensed phase [757]. Since molybdenum and its carbide are less volatile than its oxides, this reduction may lead to incomplete volatilisation even in hot flames.

Many of these mechanisms, however, are purely hypothetical and require closer study if they are to be clarified. Electron microdiffraction analysis of particles captured from the flame could be helpful. Vaporisation profiles measured on single droplets injected by a drop generator into the flame have also proved to be useful in studies of volatilisation rates [196]. Besides, various kinds of chemical interactions such as hydrolysis, complex formation, ion pairing, precipitation and surface adsorption may affect the fraction of analyte vapour released in the flame [844]. Still more complex are the conditions prevailing when several concomitants are present in the nebulised sample. The nature of the solid aerosol particles formed from it depends on many factors that are often obscurely inter-related. Volatilisation may proceed here in several stages and by different, concurrent mechanisms.

The form in which the analyte is released from the condensed phase – whether as atom, salt or (hydr)oxide – is irrelevant to its degree of dissociation in the flame provided that chemical equilibrium between the primary products of volatilisation and the constituents of the flame is quickly established in the gas phase (see also § 5.2.4).

In § 4.6.4 we shall discuss some typical examples and experimental methods for determining in practice whether volatilisation is complete.

4.6. The Efficiency of Nebulisation and the Fractions Desolvated and Volatilised

4.6.1. Definitions. In practical work the main interest does not lie in the detailed processes of nebulisation and evaporation but in the fraction of the aspirated analyte that is finally delivered into the flame. We shall next discuss the proper definition and measurement of this fraction and show what values may be expected in practice. In concluding and summarising this chapter we shall review the factors whose interplay can, in an often complex manner, improve or impair the efficiency.

We define the *efficiency of nebulisation* ε_n as the fraction of the analyte aspirated per second, for example of sodium in sodium chloride solution, that actually enters the flame, per second, in any form – mist, dry aerosol or vapour. The *fraction desolvated* β_s is defined as the amount of analyte present in the observed volume of the flame as a dry aerosol of, for example, solid or molten NaCl particles, or as vapour, divided by the total quantity of analyte present in any form in the same volume. Only those nebulised mist droplets that actually get into the flame and are also fully desolvated in time contribute to the product $\varepsilon_n\beta_s$. It is immaterial to the definition of β_s whether the sodium in the vapour phase occurs as free atoms, molecules or ions. It is also immaterial whether, after desolvation, the residual solid or molten aerosol particles are actually converted to vapour, that is, volatilised. Thus the product $\varepsilon_n\beta_s$ represents a property of the flame spectrometer that, although it depends on the nature of the solvent, does not depend on the nature and concentration of the solutes. We limit the discussion here to the commonest case in which the nebulisation and desolvation are practically unaffected by the solutes. Their concentration should therefore not be excessive [929]. The volatilisation of course depends greatly upon the solutes (see § 4.5).

In chamber-type nebulisers, since the desolvation is practically complete shortly after entry into the flame, β_s may be set equal to 1 (see § 4.4.2), but here ε_n is usually considerably less than 1 (see § 4.6.2). In direct-injection burners $\varepsilon_n = 1$ if the flame is taken to include the cooler boundary zone into which some of the droplets pass. However, desolvation is usually incomplete and depends on the height of observation in the flame. Hence the fraction desolvated, as defined here, depends on the chosen height of observation, the fineness of nebulisation and the solvent.

The efficiency ε_n often implicitly defined in the literature refers in many cases not to the fraction of aspirated analyte but to the fraction of aspirated liquid entering the flame (see [203] for example). Such a definition is unrealistic for chamber-type nebulisers. Some of the liquid reaches the flame as pure vapour (without analyte), while the rest, which reaches the flame as mist, may be more concentrated than the aspirated solution owing to partial evaporation of the solvent. Even when the definition of efficiency does refer to the fraction of analyte, its measurement is, in the literature, often restricted to the spray chamber of the nebuliser; that is, only the fraction of aspirated analyte that leaves the spray chamber is measured, regardless of possible further losses in the ducts or in the burner. These limitations or approximations must be kept in mind when comparing values of ε_n reported in the literature.

The effect of volatilisation on the quantity of analyte in the form of vapour in the flame is represented separately by the *fraction volatilised* β_v, which is defined as the amount of analyte present as vapour in the observation volume divided by the total amount of analyte present in this volume either as (solid or molten) aerosol or as vapour. Here again it is immaterial whether the analyte exists in the vapour phase as free atoms, molecules or ions (compare the definition of β_s). In § 5.1 we shall define the *fraction atomised* β_a, accounting finally for the effect of dissociation on the atom concentration in the vapour phase, which is all-important for atomic spectroscopy.

By combining these concepts we can observe that the fraction of analyte aspirated per second that actually passes as free atoms through the total (horizontal) cross section of the flame, per second, at the height of observation is given by the product $\varepsilon_n \beta_s \beta_v \beta_a$. This product is called the *efficiency of atomisation* ε_a. In inhomogeneous flames the fractions β_s, β_v and β_a may depend not only on the height in the flame but also on the distance from the flame axis. It is then more appropriate to speak of the *local* fraction desolvated, etc. The amount of analyte found within a finite observation volume in the state under consideration is then related to the average value of the local fraction over this volume.

4.6.2. The efficiency of nebulisation in chamber-type nebulisers. The efficiency of nebulisation ε_n in a chamber-type nebuliser can be determined by the following experimental methods. In one such method, for example, we measure the aspiration rate (in ml min^{-1}) at the nebuliser and the rate (in ml min^{-1}) at which liquid condenses on the spray chamber walls, by using a graduated cylinder in which the condensed liquid is collected [160, 371, 732, 857, 894]. But since the efficiency must be referred to the transport of salt or metal and not liquid, the difference between the two rates must be corrected for the intervening partial evaporation of the aspirated, nebulised, recondensed liquid. The extent of this evaporation is easily found by determining the metal concentration in the recondensed liquid with the flame spectrometer itself. In one instance the concentration was found to have risen about 10 per cent in the draining liquid [160]. Some salt is also lost in the tubing and in the burner, as can be discovered when we dismantle a burner or connecting tube that has been used for a long time and is coated with salt. The loss can be determined, if important, by flame spectrometry after rinsing these parts [218, 857, 894]. Losses can also occur when mist droplets leaving the burner pass into the cool boundary zone instead of into the flame. Similar losses can also result from outward diffusion of metal vapour in a narrow flame. Consequently the method outlined above, although easy to carry out and the one most often used, may generally give too high a value for ε_n.

For a more satisfactory determination of the efficiency of nebulisation, the product of total flame cross section O, (mean) rise-velocity of the flame gases v_f, and (mean) number density of metal atoms n in the flame should be found experimentally and compared with the product of aspiration rate F_l and concentration of the solution c [160, 371]. The following equation then gives ε_n:

$$\varepsilon_n = \frac{10^3 \times n O v_f}{F_l c N_A}, \tag{4.10}$$

where c is in mol l^{-1} and the other quantities are in CGS units; n is the number of metal atoms per cm^3 in the flame and N_A is Avogadro's constant. (See § 3.4.2 regarding the measurement of v_f.) This equation assumes that desolvation and volatilisation of the dry aerosol are complete ($\beta_s = \beta_v = 1$) and that the element occurs only as free atoms in the vapour phase ($\beta_a = 1$). The first two conditions are easily met, certainly in chamber-type nebulisers, provided that the element chosen for measurement is readily volatilised and occurs at low concentration. In most flames the third condition is adequately fulfilled for metals like copper, zinc and sodium [366, 367]. The ionisation of sodium in the flame can be easily suppressed, if necessary, by adding caesium (see § 5.3). An absolute measurement of n is most conveniently made by atomic absorption (see § 6.5.2); the flame temperature then only needs to be known approximately if at all. These methods, and those outlined below, have the drawback that they require a great deal of instrumentation.

With the same assumptions as in equation (4.10), the efficiency ε_n can also be found by measuring the flow rate F_u of the unburned gas mixture (in $cm^3 s^{-1}$) and calculating the number of moles κ of burned gases that are formed per mole of the unburned gas mixture. The following equation is used [367, 371, 913]:

$$\varepsilon_n = \frac{10^3 \times nF_u\kappa}{F_l c N_A}\left(\frac{T_f}{298}\right),$$ (4.11)

where the flame temperature T_f (in K) has to be known only roughly and the other quantities are the same as in equation (4.10). (Note that the symbol ε in [367] and ε_n in equation (4.11) have different meanings.) The factor $T_f/298$ represents the thermal expansion of the burned gas mixture. In equation (4.11) the contribution of the evaporated solvent to the expansion of the flame gases is ignored (see [367]). The equation becomes inaccurate when an appreciable but unknown quantity of air from the surroundings mixes with the flame gases.

For the direct determination of ε_n trapping methods have been used in which the aerosol leaving the nebuliser or burner is caught with a filter or other device. The quantity of metal salt in the trapped aerosol is determined (suitably by flame spectrometry) and compared with the known quantity of salt aspirated during the same time interval. The accuracy of this method depends on the completeness of capture. Electrostatic precipitation of the aerosol has proved satisfactory for this but requires special apparatus [278]. The aerosol carried in a gas stream can also be trapped almost completely by mixing it with steam, which condenses on the particles, and then washing it out with hot water [515]. Under special conditions, ε_n can also be determined elegantly and directly by means of a radioactive tracer such as ^{60}Co [853]. During nebulisation of the radioactive solution the aerosol is caught in a filter above the burner (without flame) and the radioactivity is measured.

In chamber-type nebulisers the experimental values of the nebulisation efficiency of water usually range from about 1 to 6 per cent [209, 322, 371, 439, 443, 732, 852, 853] but for some nebulisers values of ε_n between 10 and 30 per cent have been reported [432, 438, 443]. In view of the limitations mentioned at the beginning of this section, however, most published values must be regarded as only approximate and are probably generally too high.

If an average value of 10 per cent is assumed for the efficiency ε_n, we find from equation (4.10) or (4.11) a total number density of about 3×10^{14} cm^{-3} for a metallic element in the usual flames with a chamber-type nebuliser at a solution concentration of 1 mol l^{-1} [97, 166, 371]. This total density is the sum of atomic, molecular and ionic densities of the element in the vapour phase, complete desolvation and volatilisation being assumed.

The efficiency of nebulisation shows a dependence on the following instrumental factors or experimental conditions. The efficiency rises when, other things being equal, the aspiration rate is diminished, as by using a longer or narrower liquid capillary (see e.g. [824, 852, 853, 894]). This follows from the observation, reported in § 4.2, that the analyte signal generally rises less than proportionally with the aspiration rate. Comparison of the values of ε_n of different flame spectrometers should be made as far as possible at (approximately) equal aspiration rates. Other characteristics of the nebuliser, such as compressed-air speed, size and temperature of the spray chamber, outlet bore of the liquid capillary, and presence of impact surfaces (see e.g. [852, 918]), influence the fineness of nebulisation as well as the recombination and desolvation of the droplets and hence the efficiency of nebulisation. Heating the chamber or irradiating the mist in the chamber with an infrared lamp can raise the efficiency considerably, by a factor of 10 for example [218, 782, 852, 857]. The nature of the solvent also enters into all these effects, especially its viscosity, surface tension and saturated vapour pressure, and to a lesser degree its density. An efficiency of 4·5 per cent measured for water was found to rise threefold when 91 per cent methanol was nebulised [185, 293, 724, 853]. An efficient pneumatic nebuliser has been described that gave an efficiency of nebulisation of about 85 per cent for methanol solutions at an aspiration rate of about 2 ml min^{-1} [218]. Incidentally, losses outside the spray chamber in the connecting tubes and the burner may have a considerable effect on the actual value of the efficiency. Also, some of the analyte may get into the cooler boundary zone of the flame and must then be considered lost, especially when the emission method is used [160]. Molecular diffusion of the analyte vapour into the boundary zone is hindered because the diffusion coefficient is significantly smaller in this cooler region than inside the flame [804]. This effect noticeably restricts the progressive loss of emitting or absorbing analyte vapour with increasing height in a (laminar) flame.

The efficiency per se is, to be sure, not always the sole criterion of performance of a nebuliser in practical flame spectroscopy. The quantity of analyte present in a given height interval or volume of the flame may be more important than the through-put of analyte, that is, the quantity passing through a cross section of the flame per second. Thus, an increase of efficiency achieved by increasing the air flow may be (over)balanced by the resulting increase of rise-velocity and the shortened residence time of the particles in the observed height interval [432, 438] (compare also the influence of F_u on n in equation (4.11) for given values of ε_n, F_l, etc).

4.6.3. The fraction desolvated with direct-injection burners. Although all liquid aspirated by a direct-injection burner gets into the flame, there may be considerable losses because the mist droplets are not completely desolvated in the flame or are hurled out into the cool boundary zone. These lost drops are often visible to the naked eye. The fraction desolvated β_s is then less than 100 per cent and depends strongly on the height of observation since the desolvation progresses with the residence time of the mist particle in the flame.

For a direct-injection burner, β_s can be measured by the following experimental methods:

(i) By atomic emission or absorption spectrometry the absolute atomic concentration at a given height is determined for an element such as copper, zinc or sodium which occurs in the flame in the vapour phase only as atoms or for which the degree of ionisation or dissociation (see chapter 5) can be calculated. Here it is assumed that the salt crystals volatilise completely. Since the turbulent flame under consideration is quite inhomogeneous, the concentration is measured as a function of the distance from the flame axis. This function is then integrated over the total cross section of the flame and the result is substituted for nO in equation (4.10). This equation, in which v_f is the average rise-velocity of the flame gases, then yields a value for the fraction β_s.

(ii) The atomic concentration in the flame is measured by atomic absorption spectrometry in relative units for a suitable element (as above) at a known aspiration rate $(F_l)_1$ [909]. With the help of a device to control the sample flow, the measurement is repeated after the aspiration rate has been reduced to such a low value $(F_l)_2$ (all other factors being held constant) that the fraction β_s, by virtue of the much smaller drop size, may be taken to be 100 per cent. This condition is judged to have been effectively reached when a further reduction in the aspiration rate causes a proportional reduction in the atomic concentration in the flame. It should be recalled that β_s, when it is not 100 per cent, depends markedly on the aspiration rate, and so the atomic concentration in the flame will not be proportional to the aspiration rate (compare § 4.2). The ratio of the two atomic concentrations multiplied by the ratio $(F_l)_2/(F_l)_1$ yields the value of β_s at the given observation height and at the aspiration rate $(F_l)_1$. Since the controlled change of aspiration rate in this method may affect the relative distribution of the element over the cross section of the flame, a correction must be applied to the value obtained [909]. The slight change in flame temperature that may also result does not need to be corrected for, to a first approximation, when the atomic concentration is measured by absorption on a resonance line rather than by emission. It is curious that values of β_s found by the first method [785] appear to be systematically lower than those found by the second method under essentially the same conditions [909]. Ionisation could account for this discrepancy to the extent of a factor of 2 at most [909] but it may also be due to the loss of drops that fly out of the sides of the flame and that are compensated for in the first but not in the second method [909]. The latter would then give values that are too high.

(iii) Finally, attempts have been made to find the relative number of desolvated droplets in the turbulent flame, and hence β_s, by light-scattering measurements on the nebulised mist with and without the flame burning [609, 687]. Although there are certain difficulties in the interpretation of these scattering measurements (compare [166]), the values of β_s thus found appear to agree rather well with the true values. For the interpretation of scattering measurements in general refer to [553, 671].

Experimental values of β_s range from 10 to 100 per cent for a turbulent oxyhydrogen flame on a direct-injection burner at an aspiration rate of about 0·5 to

2 ml min^{-1} of water [409, 687, 784, 785, 909], while for the corresponding oxy-acetylene flame β_s is somewhat lower [909]. In both flames the values were found to depend on the height of observation, the composition of the flame gases, the aspiration rate and the capillary bore. For example, β_s was found to double when the height increased from 1 to 5 cm above the tip of a direct-injection burner [687]; a dependence on height was also found, as expected, in the cooler turbulent air–hydrogen flame [225]. Complete desolvation, for which $\beta_s = 1$, can be inferred indirectly if the atomic concentration integrated over the total cross section of the flame does not change with height [381]. This holds true only when the average rise-velocity does not change significantly with height (compare [785]) and when the element is easily volatilised and appears in the vapour phase only as atoms.

By using organic solvents β_s can be considerably improved: solutions of 90 per cent acetone or 80 per cent methanol gave values of 100 per cent for β_s in an oxyhydrogen flame even at low heights [381, 909]. No scattered light from undesolvated mist particles, which interferes in the fluorescence method, was observed when alcohols instead of water were nebulised into the turbulent flame [856]. The low surface tension and heat of vaporisation and also perhaps the heat released by the combustion of these solvents account for this gain. Moreover, the considerations discussed in § 4.6.2 for chamber-type nebulisers, concerning the influence of the properties of the solvent on β_s, apply to direct-injection burners also.

If β_s is near 100 per cent, equation (4.10) gives an average total concentration of metallic element of about 3×10^{15} cm^{-3} in the usual unpremixed flames on a direct-injection burner when the solution concentration is 1 mol l^{-1} [97, 166, 784]. This total concentration is the sum of the atomic, molecular and ionic concentrations of the element in the vapour phase, complete volatilisation being assumed. The expected total concentration is thus about an order of magnitude greater than in the usual flames with a chamber-type nebuliser (see § 4.6.2).

4.6.4. The fraction volatilised. The fraction volatilised β_v is much harder to determine quantitatively than ε_n or β_s. Even the simple question whether the solid or molten aerosol particles are completely volatilised at a given height of observation in the flame presents difficulties. We recall that β_v depends not only on the instrumental conditions but also on the qualitative and quantitative composition of the nebulised solution (see § 4.5). For example, when elements such as nickel, titanium and zirconium form bonds or complexes with oxygen-containing constituents in the solution, involatile metal oxides can form, hindering the liberation of atoms in the flame. This may explain why the atomic absorption signal in such cases is lower than when the element is present in solution as a metallocene or a fluoro complex, containing no oxygen [264, 770][†]. Hence β_v must be studied separately in each individual case. Owing to severe theoretical uncertainties, this quantity can only be found reliably by empirical means.

---

† This view differs from the hypothesis in [770], according to which this loss of signal is to be explained by processes in the *gas* phase.

In this section we present some experimental methods and results that can provide information on the extent of volatilisation. In summary we shall list all the factors that can influence the volatilisation. Most of the following experimental methods for determining the extent of volatilisation are easy to apply in practice.

(i) If the analytical signal from a simple metal salt solution does not depend on the nature of the anion, it is fairly certain that volatilisation is complete. The converse is not necessarily true since a change in the signal could also be caused by incomplete dissociation of the salt molecules in the flame (see § 5.2.2).

(ii) In the case of a composite solution consisting, say, of calcium and aluminium salts, if the height dependence of the calcium signal is the same as for a simple calcium solution, volatilisation interference by aluminium is unlikely.

(iii) If the concentrations of both free atoms and molecules of an element increase with height, volatilisation may be incomplete in the lower part of the flame. One should make certain, however, that desolvation of the mist droplets is complete (see § 4.6.2 and § 4.6.3).

(iv) If the total concentration of an element in the vapour phase is less than that calculated from the efficiency of nebulisation, the fraction desolvated, the aspiration rate, the solution concentration, etc (see also § 4.6.2 and § 4.6.3), then volatilisation is incomplete. If the element is present partly as free molecules (e.g. SrOH and SrO), the total concentration can be found by measuring the absolute atomic concentration and calculating the degree of dissociation and if necessary the degree of ionisation (see chapter 5) [367, 930]. But this method requires a fair amount of instrumentation and fairly accurate knowledge of the temperature and composition of the flame. The value of $\varepsilon_n \beta_s$ does not have to be known if the total concentration of the considered element in the vapour phase is compared with the atomic concentration of another element such as copper or sodium which is easily volatilised and present essentially as free atoms in the flame [455, 520]. The concentrations of the two elements compared in the flame should be referred, of course, to equal molar solution concentrations.

(v) A continuous background emission at high salt concentrations can indicate incomplete volatilisation.

(vi) Incomplete volatilisation (or desolvation) can be tested by trapping salt or oxide particles in the flame and observing them through an electron microscope [529].

(vii) Unvolatilised particles are revealed by Rayleigh scattering when the flame is illuminated from the side [671, 673]. The fraction scattered by undesolvated mist can be found by nebulising the pure solvent also; this fraction is then subtracted from the total scattering [856]. However, scattering may also result from the formation of large molecular aggregates by participation with flame molecules [856].

(viii) When volatilisation is incomplete, the relative loss of emitting or absorbing atomic vapour increases with solution concentration owing to the increasing size of the solid aerosol particles (see § 4.5). This gives rise to an analytical curve that is concave towards the concentration axis at higher concentrations (see chapter 8). Thus incomplete volatilisation can be inferred from the shape of the analytical curve after other sources of curvature (see chapter 8) have been allowed for [41].

The so-called halving procedure is a special case of this curvature method. The relative decrease in the intensity of metal emission is measured when the salt concentration in the solution is cut to half [343, 567]. When volatilisation of the crystals is incomplete, the intensity decrease will be less than 50 per cent, other factors that may affect this decrease being disregarded (compare chapter 8). This is due to the fact that at the lower concentration the crystals are initially smaller and need less time to volatilise, so that the *relative* loss as unvolatilised material is also smaller.

A scintillation spectrometer has also been used for finding the dependence of volatilisation on the mean initial diameter d_{max} of the solid aerosol particles in an oxyhydrogen flame. This method is based on the observation of the individual flashes emitted by the vapour clouds that are released from each particle [713]. The initial diameter can be easily controlled by varying the salt concentration. For example, the relative pulse height for sodium sulphate was found to be proportional to d_{max}^2 for $d_{max} > 0.1\ \mu$m and to d_{max}^3 for $d_{max} < 0.1\ \mu$m. Evidently volatilisation is complete in the latter case only, since the mass of a particle and the number of sodium atoms released from it are proportional to d_{max}^3. It should be noted that the proportionality to d_{max}^2 (i.e. to the surface of the particle) prevailing when volatilisation is incomplete does not conform to equation (4.4), according to which dm/dt is proportional to the radius r. In this experiment it was also found that the volatilisation rates of NaI, NaBr and NaCl diminish in that order and are all smaller than that of sodium sulphate.

A few typical experimental results will be presented next, giving a general idea of the extent of volatilisation in practical flame spectroscopy.

The alkali halides can generally be considered to be fully volatilised in flames. This has been demonstrated by method (i), for example, for potassium in an oxyhydrogen flame on a direct-injection burner [714]. It is also true for most of the other salts of the alkali metals and of copper, silver, zinc and cadmium, among others, whose boiling points do not exceed the flame temperature [97]. Only at high salt concentrations might there be doubt that volatilisation of the alkali chlorides is complete in turbulent flames [343, 349, 625]. The halving method (see above) showed that potassium chloride in an oxyhydrogen flame volatilises completely within 2 per cent up to a metal concentration of $1\ \mathrm{g\,l}^{-1}$ [343, 567]. The results were still more favourable in the hotter oxyacetylene flame but less so in the cooler air–hydrogen flame. That sodium chloride and other alkali halides volatilise completely in various premixed flames has been substantiated by many authors.

The compounds of the alkaline-earth metals are generally less volatile. However, the alkaline-earth halides may be expected to be fully volatilised in premixed air–acetylene flames if the solution concentration is not too high [367, 371, 557, 930]; results obtained by method (iv) confirm this. At high concentrations, however, a convex curvature may be found in the analytical curves of the molecular bands, which are practically free from self-absorption (see chapter 8) [455, 464]. In composite solutions and for phosphates, sulphates, silicates, etc, volatilisation may be considerably retarded. Calcium chloride and magnesium chloride have been shown by the halving method to volatilise fully in an oxyhydrogen flame on a direct-injection burner up to solution concentrations of 100 mg l^{-1} of Ca and 1 g l^{-1} of Mg [343, 567].

The well known involatility of oxides of uranium, vanadium and aluminium can be demonstrated by method (v) [525]. The involatility of aluminium in an air–acetylene flame is shown also by the negative result obtained by method (iv) [856]. The recently measured high dissociation energy of AlO is probably also partly responsible for this (see table 3). It is interesting that metallic aluminium is much more volatile than molybdenum, yet its *oxide* is much less volatile. The fact that molybdenum, but not aluminium, can be detected by atomic absorption in the air–acetylene flame emphasises the role of oxide formation in volatilisation [291]. The convexity of the analytical curve of molybdenum in the air–acetylene flame but not in the nitrous oxide–acetylene flame shows by method (viii) that even molybdenum is not necessarily fully volatilised in the cooler flame [894]. Aluminium and vanadium display convex curvature even in the hot nitrous oxide–acetylene flame [894].

In concluding this discussion of volatilisation, we summarise the chief factors that may have an adverse effect on it:

(i) High solution concentration of the analyte or matrix or both.

(ii) Tendency (and opportunity) of the analyte to form involatile compounds with oxygen, phosphoric acid, etc.

(iii) Use of a direct-injection burner, which gives a coarser aerosol than a chamber-type nebuliser.

(iv) Coarsening of the aerosol due to an increase in the aspiration rate or other cause.

(v) Low flame temperature.

(vi) An oxygen-rich adjustment of the flame, unless the metal oxide is more volatile than the metal carbide or the metal itself.

(vii) Low height of observation or high rise-velocity, either of which means little time for volatilisation.

5   Dissociation and Ionisation of the Analyte in the Flame

5.1. Introduction

In the preceding chapter we discussed the conversion of the sample in the flame into vapour. In this state the analyte can exist in different forms of free atoms, molecules, radicals or ions. Molecules or radicals are formed when the metal atoms combine with other particles from the flame gas or the concomitants, for example, CaO, CaOH, KCl. By giving off an electron a neutral atom or molecule becomes a positive ion, such as K^+, $SrOH^+$. This formation of molecules and ions is of importance for atomic emission, absorption and fluorescence spectroscopy because it reduces the concentration of the free analyte atoms capable of emission or absorption and thereby worsens the detection limit. However, if the resulting molecules themselves emit radiation, sometimes their spectral bands can be utilised in emission analysis. This is possible for instance with the CaOH radical, but not with KCl or LiOH. Some non-metallic elements, among them Cl and F, can be detected in flame spectroscopy only by their molecular bands, such as the InCl and SrF bands [406, 407, 436]. Because the majority of molecules (for exceptions see § 6.2) practically do not absorb in the usual concentration range in flames, absorption spectroscopy is restricted to atomic or ionic lines. Certain ions, for example those of the alkalis, do not emit or absorb in flames, while others, such as Sr^+, can emit or absorb radiation but occur in the cooler flames only in relatively low concentrations.

The formation of molecules and ions is of practical importance in flame spectroscopy not only because of the possible deterioration of the detection limit. What fraction of the analyte exists in the flame as molecules or ions may depend on the adjustment of the flame or on the presence of certain concomitants in the solution which can lead to interferences in quantitative atomic emission as well as in atomic absorption and fluorescence analysis (compare chapter 9). Furthermore, if the fraction of analyte that is bound in the flame in the form of molecules is appreciable, changes in temperature and flame composition can have a marked effect in AAS and AFS too [370]. Finally, molecular bands of concomitants may interfere with the blank measurements in the emission method and, to a much lesser degree, in the absorption method also.

For these reasons in this chapter we shall deal in some detail with molecular dissociation and ionisation of the analyte. We shall restrict the discussion to processes that proceed entirely in the vapour phase and disregard possible

69

interaction of the analyte vapour with solid aerosol particles. For simplicity's sake the volatilisation of these particles will be taken here to be complete.

Of particular interest in this discussion is the *fraction atomised* β_a. It is the ratio of the amount of analyte vapour present in the observed flame volume in the form of neutral or ionised free atoms to the total amount of analyte vapour present in the same volume as free atoms, ions or molecules†. Analyte losses due to incomplete desolvation or volatilisation are thus not taken into account in this definition (compare also § 4.6.1)‡.

For further study of the topics in this chapter we refer to [38, 59, 61, 71, 107, 831, 832].

5.2. Molecular Dissociation

5.2.1. The law of mass action. If for instance an NaCl solution is nebulised, because of incomplete dissociation of the NaCl molecules only a part of the sodium appears in the flame in the form of free atoms. The remaining part, sometimes quite minute, stays bound in the compound form as chloride and therefore does not contribute to the atomic emission or absorption signal. A similar situation exists if the analyte, say Ba, forms molecular compounds, for example BaO, with constituents of the flame gases. The *degree of dissociation* α is defined as:

$$\alpha = \frac{[\text{Na}]}{[\text{Na}] + [\text{NaCl}]} \qquad (5.1a)$$

or

$$\alpha = \frac{[\text{Ba}]}{[\text{Ba}] + [\text{BaO}]}, \qquad (5.1b)$$

where [Na] and [Ba] are the number of free Na and Ba atoms respectively per cm^3 in the flame. We shall now first investigate the factors on which α depends and its significance in flame spectroscopy. In the simplest case, namely when the metal in the flame only forms one molecular compound and no ions, α equals the fraction atomised (see § 5.1).

If chemical equilibrium is established in the flame the kinetic processes resulting in dissociation or molecule formation do not have to be considered in detail in discussing the degree of dissociation. (The process of dissociation of alkali halides through collisions is described in detail in [325].) It then suffices to formulate the *law of mass action*, which can generally be derived from statistical mechanics [38, 53, 78, 95]:

$$K_d(T) = \frac{[\text{M}][\text{X}]}{[\text{MX}]}, \qquad (5.2)$$

---

† A similar definition has been adopted in [367, 371, 913].
‡ These losses and those in the spray chamber are included in the definition of efficiency of atomisation ε_a (see § 4.6.1).

where M denotes a metal atom and X the other partner (e.g. Cl or O) in the molecule MX. The dissociation constant $K_d(T)$ only depends on the temperature T and on the kind, but not the concentration, of the reactants according to the equation:

$$\log_{10} K_d(T) = 20 \cdot 274 + \tfrac{3}{2} \log_{10} (M_M M_X / M_{MX})$$

$$+ \log_{10} (Z_M Z_X / Z_{MX}) + \tfrac{3}{2} \log_{10} T$$

$$- 5040 E_d / T. \tag{5.3}$$

Here K_d is expressed in cm^{-3}, T in K and the dissociation energy E_d in eV. M and Z are the molecular weight and the internal partition function of the considered partner respectively (see equation (6.4)). The partition function may depend somewhat on the temperature and can be calculated from the structure of the atom or molecule [64, 78, 229]. Values of atomic partition functions as functions of T are given in [38, 368].

The temperature dependence of $K_d(T)$ is mainly expressed by the last term in equation (5.3). The higher the value of T and the smaller the value of E_d, the larger the value of K_d and the more complete the dissociation. These dependences are connected with the endoergic† character of the dissociation reaction $MX \rightarrow M + X$. Typical values of E_d are $4 \cdot 24$ eV for NaCl and $5 \cdot 3$ eV for BaO [520, 521, 523], while for $E_d \simeq 5$ eV a change in the flame temperature of about 250 K leads to a change of K_d by a factor of 10. The energy that would be consumed for complete dissociation of the metal compounds even under the most unfavourable conditions (direct-injection burner: 1 molar solution concentration, $E_d = 10$ eV) is only a small fraction (<1 per cent) of the heat content of the flame [165].

The consequences resulting from the mass action law for the degree of dissociation will now be considered. We shall distinguish the two cases in which the partner X in the molecule is first a constituent of the solution introduced into the flame (§ 5.2.2) and then a constituent of the flame gas itself (§ 5.2.3).

5.2.2. Molecular compounds containing constituents of the nebulised solution.

Complex salts such as sodium sulphate and organometallic molecules are not stable but are decomposed in flames; only the simpler, stable compounds like KCl and $BaCl_2$ can exist in the flame. If we assume that Na and Cl do not react in any other way in the flame and that the sample solution only contains NaCl, we obtain from equations (5.1a) and (5.2) the following relationship for the degree of dissociation α:

$$\frac{\alpha^2}{1 - \alpha} = \frac{K_d(T)}{[\overline{Na}]}, \tag{5.4}$$

where $[\overline{Na}] \equiv [Na] + [NaCl]$. From this, the following points arise:

---

† In an endoergic reaction proceeding from left to right, energy is absorbed from the system in which the reaction takes place. In an exoergic reaction there is a release of energy.

(i) The dissociation becomes more complete when the solution concentration and hence the total concentration in the flame $[\overline{Na}]$, being proportional to it, decreases. According to equation (5.4) a fall in $[\overline{Na}]$ is equivalent to a rise in K_d which means better dissociation, that is, a higher degree of dissociation α.

(ii) The dissociation becomes more complete with rising temperature because K_d increases rapidly with the temperature (see § 5.2.1).

(iii) The atomic Na emission is not necessarily the same for different Na salts at the same Na concentration in the solution (influence of E_d, which equals 4·24 eV for NaCl, 3·84 eV for NaBr, and 3·11 eV for NaI [13]).

(iv) Since the dissociation energies of the different metal chlorides and consequently the values of K_d also are not identical, at a given solution concentration α depends, furthermore, on the kind of metal. Thus under identical conditions the dissociation of NaCl ($E_d = 4·24$ eV) is more complete than that of KCl ($E_d = 4·40$ eV).

(v) The presence in the solution of other chlorine compounds raises the concentration of free chlorine in the flame. By the law of mass action (equation (5.2) with $[X] = [Cl]$) this must lower the degree of dissociation of NaCl and with it the atomic Na concentration. If the concentrations of the other chlorine compounds greatly exceed the Na concentration, then $[Cl]$ and also α are practically independent of the Na salt concentration.

Because of the relatively high concentration of hydrogen in the commonly used flames and the non-negligible dissociation energy of HCl, we must expect to find the greater part of the chlorine in the flame bound in the form of HCl. The concentration of free chlorine is then lowered and the dissociation of the sodium chloride is stepped up. These and similar bonds of the hydrogen with the anions of alkali salts enhance the latter's dissociation. Therefore in the usual concentration range the dissociation of the alkali halides may in effect be considered complete. In 1877 Gouy [395] found that the emission of sodium in the flame is practically the same for all Na halides and other Na salts. Calculations and measurements have shown that the formation on NaCl in hydrogen-containing flames is negligible even with HCl solutions up to 1 M [160, 398, 567, 726]. This even holds for flames on direct-injection burners in which the chlorine content corresponding to a 1 M HCl solution concentration is markedly greater because of the higher efficiency of nebulisation (§ 4.6.3). But with a 1 M solution of HCl, the LiCl concentration in an oxyhydrogen flame on a direct-injection burner may be about one-half of the atomic concentration [343, 567]. Here the strong tendency to form LiOH (see § 5.2.3) has, however, a buffering effect [398].

The formation of (mono-)halides of the alkaline earths can be directly established by spectroscopy because their molecules emit band spectra in the flame. Excess halogen (in vapour form as CCl_4, Br_2, I_2) was found to increase the intensity of the molecular bands and to reduce the intensity of the atomic emission [934]. The intensity increase of the monohalide bands plotted against chlorine supply shows a maximum, presumably because at higher chlorine concentration $CaCl_2$ for instance is more readily formed than CaCl. Besides, a noticeable fraction of Ca and Ba may also be present in the form of CaOHCl and BaOHCl molecules in the air–hydrogen flame when a mole fraction of 10^{-3} per cent of chlorine is added to the flame [768].

Generally, it can be said that in flame spectroscopy a lowering of the atomic alkali and alkaline-earth signals due to the formation of halogen salts will only become manifest when the solution also contains halogen compounds in high concentrations and when the flame temperature is not too high. Additions of HCl, KCl, etc, may lead to interferences for other reasons also, a problem that will be discussed in chapter 9. The effects of 1 M HCl solutions, often found in air–propane, air–acetylene, oxyhydrogen and oxyacetylene flames, cannot therefore always be definitely attributed to a shift in the dissociation equilibrium alone. A more careful analysis would be required to separate the dissociation and the other effects. Different methods for this are described in [166].

The formation of dimers, such as Na_2, can be disregarded in flames. The low dissociation energy of such compounds ($E_d = 0.75$ eV for Na_2) and the relatively small metal concentration in the flame by virtue of equation (5.4) explain why α is practically equal to unity here.

5.2.3. Molecular compounds containing constituents of the flame gas.

Molecular compounds can also be formed through combination of the analyte with constituents of the flame gas, yielding for example LiOH, CaO, CaOH, and CuH among others. In non-sooting carbon-rich nitrous oxide–acetylene flames even compounds such as NaCN and SiC_2 may be detected [735].

In chemical equilibrium the degree of dissociation defined by equation (5.1b) is again given by the law of mass action (equation (5.2)):

$$\frac{\alpha}{1-\alpha} = \frac{K_d(T)}{[X]},$$

(5.5)

where [X] is the concentration of the free oxygen radicals, etc. As the total analyte content in the flame is much less than the total oxygen content, the value of [X], and by equation (5.5) α also, are here taken to be independent of the analyte concentration†. The dissociation of a given compound then depends only on the temperature and composition of the flame. The favourable effect on α of the increase in K_d with increasing T may be counteracted by a simultaneous rise in the free radical concentration [X] (e.g. [O]) [272, 735]. Under the usual flame conditions dissociation seems to be practically complete if the dissociation energy E_d is less than about 3 eV (e.g. CuH with $E_d = 2.8$ eV). However, compound formation can still be considerable for $E_d > 4$ eV (compare also [647]). In non-sooting carbon-rich nitrous oxide–acetylene flames ($N_2O : C_2H_2 \approx 1.7$) containing CN and C_2, monoxide dissociation is fairly complete for dissociation energies below about 6 eV [735].

---

† Contrary to statements in the literature, this is so even if the concentration of the *free* oxygen radicals in strongly reducing flames is less than the metal concentration because the equilibrium value of [O] by similar mass-action relations is again linked with the comparatively much higher concentrations of other flame molecules such as CO. This ensures a certain stabilisation of [O] against the removal of oxygen through metal oxide formation (compare [626, 735]).

Oxide formation in general is not significant with alkalis although even NaO_2 may incidentally occur as an intermediate compound in cool and markedly fuel-lean flames [527]. But alkali hydroxides not capable of emission or absorption are frequently formed. This tendency of the alkali elements increases approximately in the order Na, K, Rb, Cs, Li [398, 454, 488, 739, 800] (see table 3). Only in the hotter flames ($T > 2500$ K) with their correspondingly higher OH radical concentration may NaOH formation be detectable. Yet the formation of LiOH is considerable in most flames. In air–acetylene and in air–hydrogen flames even a loss of atomic Li content as high as about 90 per cent may be expected [455, 924]. But since the degree of dissociation is independent of the alkali content (see above) no curvature of the analytical curve due to hydroxide formation needs to be suspected; there is only a constant percentage loss of atoms capable of emission or absorption.

The alkaline earths are present in the flame predominantly as molecules. Ca and Sr mainly form monohydroxides and, to a lesser degree, also monoxides and dihydroxides; with Ba, monoxide is dominant. These molecules are identified by their band emissions (see spectrograms and wavelength tables in Appendixes 1 and 2). The degree of dissociation for BaO in two typical fuel-rich oxyhydrogen and oxyacetylene flames has been calculated to be about 1 per cent, compared with about 30 per cent for MgO and about 90 per cent for FeO [166]. (The value of the dissociation energy of BaO adopted in [166] is 0.3 eV too low [521].) The gas compositions of the flames in question have been computed from the flame temperatures and gas/oxygen ratios given in [685, 910] (compare table 2). The dissociation of the other alkaline-earth compounds is usually more complete. The atomic concentrations are generally still adequate for measurements on the atomic lines in normal flames.

The accuracy of the calculated degrees of dissociation of these elements very much depends on the accuracy of the available dissociation energy values. In the past different experimental methods often yielded inconsistent values of E_d. The values of E_d for some of the compounds important in flame spectroscopy are listed in table 3. Although several of these values are not quite definite and are reproduced here with reservations, nevertheless they can give an indication of the stability of the compounds.

It has been reported that chromium in hydrogen flames not only forms CrO but also forms CrO_2, CrO_3, $HCrO_3$ and/or H_2CrO_4 in comparable concentrations [243, 327]. (The formation of $HCrO_3$ presumed in [243] has not been confirmed by mass spectroscopy [327].) The value of E_d for CrO shown in table 3 is therefore not by itself decisive for the atomic fraction of Cr in the flame. This also applies to aluminium which exists in hydrogen flames chiefly as $Al(OH)_2$ and to a lesser degree as AlO [507, 661]. The principal compound of iron in such flames was found to be FeOH, together with smaller concentrations of FeO and $Fe(OH)_2$ [508]. An appreciable fraction of tungsten in stoichiometric nitrous oxide–acetylene flames proved to be bonded as H_2WO_4, WO_3 and WO_2 [735].

The elements Ti, Si, W, B, Zr, V, Sc, Ta, Re, U and the lanthanides, the oxides of which mainly have dissociation energies between about 6 and 8 eV, show a marked

Table 3 Dissociation energies E_d of metal oxides and metal monohydroxides

$MO \rightarrow M+O$

Metal	Mg	Ca	Sr	Ba	Sc	La	Ti	Zr	V	Cr
E_d (eV)	3·4–4·1	3·75†§	4·1†§	5·2–5·3‡§	6·6–7·0	8·2	6·7–7·2	7·8	6·4–6·6	4·7
Reference	5	775	229	229	60	5	5	5	5	229
	60	806	523	328	259	60	60	60	60	
	229		775	523		229	229	229	229	
	930			766						
				775						

$MO \rightarrow M+O$

Metal	Mo	Mn	Fe	Ni	B	Al	Si	Sn	Pb
E_d (eV)	5·0	4·2	4·1–4·3	3·8	8·0–8·3	4·6–6·3	8·3	5·4	3·9
Reference	60	5	60	229	5	5	229	5	60
	229	60	229		60	19		60	229
		229	930		229	60		229	355
		930				229			
						393			
						660			
						661			

$MOH \rightarrow M+OH$

Metal	Li	Na	K	Rb	Cs	Mg	Ca	Sr	Ba
E_d¶ (eV)	4·48–4·54	3·3–3·5	3·5–3·6	3·6–3·8	3·9–4·0	2·4	4·45	4·35	4·6–4·9
Reference	533	511	511	511	511	246	277	277	277
	928	278	533	533	533		523	523	766
		533					766	766	812

† Value based on rather arbitrary assumption of an electronic partition function for MO equal to 3 (compare [229, 775]).
‡ Value based on rather arbitrary assumption of an electronic partition function for MO equal to 6 (compare [229, 328, 775]).
§ The values $E_d = 4·3$, 4·2 and 5·7 respectively are recommended for CaO, SrO and BaO in [60].
¶ Some of the values listed may be affected by the uncertainty of the structure of the MOH molecule (linear or bent) in the electronic ground state (compare [482]). Dissociation energies that were obtained by assuming an *a priori* value for the excitation energy of the observed MOH bands in [404, 768] have not been included in the list (compare [482]).

Table prepared in collaboration with Dr P J Th Zeegers, University of Utrecht.

tendency to oxide formation (see also table 3)†. The majority of flames are not very suitable for producing atomic vapours of these elements but we have to bear in mind that the degree of dissociation of the oxides depends on the flame temperature and in particular also on the reducing conditions in the flame. The atomic metal concentration therefore can be greatly enhanced by strongly stepping up the fuel gas supply (H_2 or C_2H_2; compare table 2) or by spraying organic solvents [265, 298, 315, 330, 331, 334, 389, 635, 651, 728, 735, 791, 891]. Atomic emission spectra of V, Sc, Nb, Re, Ti, W and the lanthanides have in fact been observed in premixed fuel-rich oxyacetylene flames with admixtures of alcoholic solvents [282, 297, 332, 334, 341]; yet the same elements give no detectable atomic lines in oxyhydrogen or in stoichiometric oxyacetylene flames. Absorption of V and Sc atoms and others could be found in a strongly reducing oxyacetylene flame with alcohol admixture, but not in an air–hydrogen, air–acetylene or oxyhydrogen flame [282, 627]. The atomic concentration of these elements in a premixed nitrous oxide–acetylene flame, as expected, proved to be the more dependent on the mixture ratio, the larger the value of E_d [336, 626, 735, 745]. It also varies greatly with the height in the flame because the favourable reducing conditions only prevail within a small region immediately above the burner [265, 336, 787, 788]. As a result of the gradual entrainment of secondary oxygen from the surrounding air the reduction becomes weaker with increasing height. By means of a chimney or an inert gas (noble gas or nitrogen) shield [735] the undesirable effect of oxygen entrainment can be repressed. In view of the considerable spatial variation of the reducing conditions the volume of the flame region observed or the cross section of the radiation beam in AAS should be chosen small enough to ensure the best possible results [296, 336, 381].

A linear relationship has been found between the reciprocal atomic concentration $[M]^{-1}$ of the alkaline earths, silicon, boron, chromium or titanium, and the reciprocal radical concentration $[CN]^{-1}$, in their variations with the height of observation over a range of 1–2 cm in the red feather of a fuel-rich nitrous oxide–acetylene flame [819]. In this height range the flame temperature was practically constant. This linear relationship was explained by an assumed formation by M of one predominant compound involving oxygen with an M/O ratio equal to unity under chemical equilibrium conditions‡. The fraction atomised could be derived from the plot of relative $[M]^{-1}$ against relative $[CN]^{-1}$.

At first, the non-premixed, luminous oxyacetylene flames with excess fuel and with alcoholic solvents were frequently employed in practical analysis for producing atomic vapour of elements whose oxides do not dissociate readily [389]. Later the fuel-rich nitrous oxide–acetylene flame became a very attractive flame for atomic absorption analysis mainly because it was found to be much more suitable for reducing the stable metal oxides than the fuel-rich nitrous oxide–hydrogen, oxyhydrogen or air–hydrogen flames [275, 367, 648, 900]. This remarkable difference

---

† Aluminium would have the same strong tendency if we are justified in accepting the high dissociation energy of 6·3 eV for AlO determined recently [661]. (For a critical review see also [393].)

‡ The cause of the variation of [CN] with height in the flame was not taken into consideration in [819].

76

in dissociation effect seems to be due not to temperature or to temperature only, but rather to the dissimilar reducing conditions in these flames, that is, the unequal free oxygen contents. A much smaller concentration of oxygen atoms is to be expected more in the carbon-containing flames than in the hydrogen flames because of the much higher dissociation energy of CO (11 eV) compared with that of OH (4·4 eV) [367]. This however does not account for the fact that the reducing conditions in the fuel-rich nitrous oxide–propane/butane flame are less favourable than those in the fuel-rich nitrous oxide–acetylene flame [250]. An interesting explanation is given in [367]: on increasing the C/O ratio in the unburnt gas mixture for creating optimum reducing conditions, a critical limit is reached above which soot particles begin to form. These particles only slightly improve the reduction of the metal oxides. According to [367] soot formation in acetylene flames sets in at a value of the C/O ratio about twice as high as that at which it begins in propane flames. Hence a smaller atomic oxygen concentration can be obtained without soot formation in acetylene flames than can be obtained in propane, propane/butane and similar flames. (It is maintained in [735] that the critical C/O ratio in the nitrous oxide–acetylene flame is even larger than that assumed in [367].)

Some other metals, in particular Cu, Ag, Tl and Zn, exist in the usual flames entirely or largely in the form of free atoms [405, 735, 739, 803, 826, 909] and therefore can be very satisfactorily analysed by atomic flame spectroscopy. They are characterised by the strength of their absorption lines and also by their insensitivity to changes in the gas/oxygen or gas/nitrous oxide ratios and by the spatial distribution of the atoms in the flame [803].

It can be deduced from measurements of the nebulisation efficiency ε_n and of the absolute atomic concentration in the flame, at known solution concentration, that these metals in fact occur practically only as free atoms (see § 4.6.2) [367, 371, 557, 732, 896, 898, 930; for a critical discussion of the results of earlier measurements see 361, 371, 733, 803, 898]. Complete volatilisation of the solid aerosol particles has to be presumed (see § 4.6.4). Complete atomisation can also be inferred indirectly from a comparison of the relative atomic concentrations of the elements in the flame at the same molar solution concentration.

The metals Cu, Na and Ag are often used as reference elements for determining the fraction atomised β_a (see § 5.1) of other elements by comparing their atomic concentrations with the Cu concentration, etc, in the flame at the same molar solution concentration [371, 520, 524, 803, 930]†. Again complete volatilisation is presumed. With Na as the reference element a possible loss of atoms due to ionisation must be prevented by addition of an electron donor, for instance Cs (see § 5.3.1).

The value of β_a obtained by the reference method or by the measurement of ε_n and the absolute atomic concentration are of importance in the discussion of the

---

† Doubts about the justification for choosing Cu as the reference element in work with non-uniform nitrous oxide–acetylene flames and a slot burner have been raised in [898]. A subsequent detailed study however revealed that Cu is fully atomised in a similar flame with an Ar gas shield over a wide range of N_2O/C_2H_2 flow ratios [735].

limits of detection in AAS, AES and AFS. As the dissociation energies and the temperature and the composition of the flame in general are not known exactly, these measurements often yield more reliable information than theoretical calculations (compare also [735]). Measurements of absolute atomic concentrations by AAS generally are subject to errors within a margin of a factor of about 2, and in some cases of one order of magnitude [366, 735, 896]. If the experimental conditions are well known, dissociation energies can be evaluated from the measurements of β_a [403, 520, 766, 833].

Experimental methods for measuring β_a or for just establishing incomplete dissociation are described in [166]. It has been shown recently that incomplete dissociation of the oxide can also be inferred from relative measurements of the ratio of the atomic line emission to the oxide band emission as a function of the C_2H_2/N_2O ratio at constant temperature [559]. Another indication of incomplete dissociation is the pronounced weakening of the atomic absorption signal on decreasing the C_2H_2 supply [275, 735].

5.2.4. Deviations from dissociation equilibrium. In the foregoing discussions the existence of chemical equilibrium has been presumed. For fuel-rich nitrous oxide–acetylene flames this presumption has been roughly checked by experiment [735]. In general, however, deviations from chemical equilibrium due to one or the other of the following two causes are possible. Firstly the constituents of the flame gas themselves may not be in equilibrium; and secondly the dissociation or recombination reaction for the metal compounds may proceed so slowly that because of the short residence time in the flame the dissociation equilibrium is not reached at all. Little is known with certainty of the rate of the dissociation reactions. Accurate measurements of the degree of dissociation of alkali and alkaline-earth compounds in flames however point to a quick establishment of the equilibrium. This conclusion is supported by the following general kinetic considerations.

Formation of metal oxides etc, and the reverse dissociation are mainly brought about by binary exchange reactions with flame molecules or flame radicals (e.g.: $M+H_2O \rightleftarrows MO+H_2$; see below). One metal atom M on average has about 10^8 collisions per second with, let us say, an H_2O molecule at a partial H_2O pressure of about 10^{-2} atm. According to equation (4.6) at least 10^4 such collisions will occur if the metal atom is carried up 1 mm with a rise-velocity of not more than 10^3 cm s^{-1}. Even if due to the occurrence of an energy threshold the number of effective collisions, that is, those initiating a reaction, were lower by a factor of 10^4 a state of approximate equilibrium would nevertheless be reached within a height range of a few millimetres. But if the equilibrium lies almost entirely on the side of the metal oxide and the metal is only present at first in the form of atoms [550], it may take much longer before not only the (major) oxide concentration, but also the (minor) atomic concentration approach their respective equilibrium values.

Few measurements of metal–atom oxidation reactions in the gas phase have so far been carried out. The rate of formation of an alkali (A) oxide, AO_2, as a result of the reaction $A+O_2+X \rightarrow AO_2+X$ in the presence of a third partner X (e.g. an N_2 molecule) has been investigated in flames and in a fast gas-flow reactor [258,

347, 608]. The results obtained for this reaction, as well as for the aluminium oxidation reaction $Al+O_2 \rightarrow AlO+O$ studied in a fast gas-flow reactor [346, 346a, 347] are not very exact, the error margin being almost one order of magnitude. Examination of the oxidation reaction $Fe+O_2 \rightarrow FeO+O$ in a fast gas-flow reactor at about 1600 K yielded more accurate figures [347]. Extrapolation of the measured rate coefficient to a flame temperature of 2500 K for the establishment of the FeO/Fe flame equilibrium, at an assumed O_2 partial pressure of only 10^{-2} atm, is expected to give a relaxation time of about 10^{-5} s. For a rise-velocity of 10^3 cm s^{-1} such a relaxation time would correspond to a height interval of only 0·1 mm (compare above).

It is a well known fact that the concentration of some flame radicals, not only in the primary combustion zone but also in the interzonal region above it, can considerably exceed that corresponding to the law of mass action (see § 3.4.1 and § 3.4.2). To take into account such deviations in the calculation of the metal atom to metal oxide or hydroxide ratio we shall have to consider in detail the specific reactions actually leading to the formation or to the dissociation of these compounds.

If there is a large excess of C atoms in the primary combustion zone, even the very stable metal oxides (MO) can here be dissociated to a high degree according to [296, 389]:

$$MO+C \rightarrow M+CO.$$

As the dissociation energy of CO (11 eV) is markedly higher than that of the metal oxides this is an exoergic reaction and therefore usually very effective. The results of studies on the roles of H, OH and O radicals in the dissociation reactions of Cu, Ca, Fe and Na compounds in the combustion zone of low-pressure H_2–O_2–N_2 and CH_4–O_2–N_2 flames are reported in [217].

In strongly reducing acetylene flames unusual radicals, such as C, CH, C_2, NH or CN, otherwise only found in the combustion zone, can also play a significant part in the dissociation of the stable metal oxides above this zone [274, 275, 408a, 542]. Yet few relevant quantitative data are available at present. (For a critical discussion of the possible participation of these radicals in the dissociation reactions in carbon-rich nitrous oxide–acetylene flames see [735].)

In normal air–hydrogen and air–acetylene flames there may be an appreciable excess of H, OH and O radicals even several centimetres above the combustion zone (see § 3.4.2). The effect of this departure from equilibrium on the formation of metal oxides or hydroxides depends on which of the following competing reactions is dominant:

(1) $M+H_2O \rightleftarrows MOH+H$,   (2) $M+OH+X \rightleftarrows MOH+X$,

(3) $M+H_2O \rightleftarrows MO+H_2$,   (4) $M+OH \rightleftarrows MO+H$,

(5) $M+O+X \rightleftarrows MO+X$,   (6) $M+CO_2 \rightleftarrows MO+CO$,

(7) $MOH+H_2O \rightleftarrows M(OH)_2+H$.

Here M denotes a metal atom and X the third partner (e.g. N_2) in ternary reactions [169, 403, 739, 826, 833]. An example of reaction (1) is the formation of LiOH. Because this reaction is in partial equilibrium†, too high a concentration of H above the combustion zone results in an LiOH concentration there below that corresponding to general chemical equilibrium. In other words, the excess H concentration shifts the partial equilibrium of reaction (1) to the left at the expense of LiOH. The concentration of H however falls with increasing height due to recombination (see § 3.4.2) and hence the ratio of molecular to atomic Li concentration increases at the same time. This explains the considerable decrease of the atomic Li concentration with increasing height of observation that is sometimes noted. Similar considerations apply in the case of the alkaline-earth monohydroxides, which are also formed by reaction (1).

Certain elements such as Cr, Sn and the alkaline earths, as well as NO and SO_2, when present in the flame in high concentrations, may act as catalysts of the recombination of excess H and OH radicals [243–245, 280, 832]. Thus the H and OH concentrations and consequently also the concentrations of all metal compounds depending on these radicals are altered. From this in turn interference of a concomitant, for example Sn, on an analyte, for example Li (see § 9.2), or convex curvature of the analytical curve for Cr etc may result in the high concentration range [245] (see § 8.2).

Alkaline-earth oxides can be formed in reaction (3) as well as in reaction (4); no radicals are directly involved in the first case and for both reactions partial equilibrium may be assumed. If furthermore the stable H_2 and H_2O molecules are taken practically to be present in equilibrium concentrations, the degree of dissociation of the alkaline-earth oxides will similarly correspond to its equilibrium value. This will be so even if reaction (4) alone is responsible for the oxide formation in the presence of excess O and OH radicals, for H and OH again are bound to H_2 and H_2O in partial equilibrium by the reaction $H + H_2O \rightleftarrows OH + H_2$ (see § 3.4.2). This rapid exchange reaction equilibrates the *ratio* of the H and OH concentrations and, because of the partial equilibrium for reaction (4), indirectly the *ratio* of the M and MO concentrations also. For magnesium, which is bonded in the flame mainly as monoxide, the decrease of the excess H and OH concentrations with height then has no consequences for the height dependence of the atomic metal concentration [502].

Deviations from metal oxide equilibrium due to excess O radicals would only have to be expected if MO molecules were mainly formed directly from $M + O + X$ according to reaction (5). But this is an unlikely process. Because of the release of the considerable amount of recombination energy a third partner (X) is needed for stabilisation of the reaction product against re-dissociation through absorption of (part of) this energy‡ (compare § 3.4.2). Such reactions which result from three-

---

† In *partial equilibrium* the forward and the reverse reactions proceed at the same rates, in spite of the deviation of the H concentration from general chemical equilibrium [176]. If such a deviation does exist, there cannot be partial equilibrium simultaneously for all the listed reactions (1) to (7).

‡ By the law of conservation of total linear momentum this energy cannot be converted into translational energy of the MO molecule alone.

body collisions are infrequent. For similar reasons reaction (2) is usually insignificant for hydroxide formation; only with relatively low dissociation energies (e.g. $E_d = 2.5$ eV for CuOH) may it become predominant [825, 832]. The relative importance of reactions (1) and (2) is treated semi-quantitatively for a number of elements in [217]. Reaction (7) yields alkaline-earth dihydroxides, which seem especially important in fuel-rich hydrogen flames.

The unexpectedly strong emission and absorption of the atomic Sn lines in the cool air–hydrogen flame has so far not been explained [256, 389, 391]. Even hydrogen–argon and hydrogen–nitrogen diffusion flames (see § 3.2) in AAS give better sensitivities at the Sn resonance line at 286.3 nm than do the hotter air–hydrogen or air–acetylene flames [756, 758]. This however is not so at the Sn *non*-resonance line at 284.0 nm, for which the relative increase in population of the lower level with increasing temperature (see § 6.5.2) has a compensating effect [756]. Yet the lowering of sensitivity at the resonance line in the hotter flames cannot be accounted for with the change in the relative populations of the atomic ground and neighbouring excited levels according to Boltzmann's distribution law.

As a result of the high dissociation energy of SnO (see table 3) only a small fraction of Sn should be present in the cool hydrogen flames as free atoms. It is also surprising that addition of alcoholic solvents practically does not affect the atomic Sn absorption in the fuel-rich air–hydrogen flame [859], although in general it considerably enhances the reduction of the stable oxides. The degree of dissociation of SnO here thus seems to exceed the equilibrium value greatly. A connection between this abnormal Sn atomisation and its catalytic effect (see above) on the recombination of excess free H radicals has been suggested in [756], but a satisfactory explanation is still lacking.

Finally it should be mentioned that with an excess of H radicals above the combustion zone the concentration of the alkali chlorides remains below the equilibrium value (compare § 5.2.2) [239, 826].

5.3. Ionisation

5.3.1. The Saha equation. In 1891 Arrhenius discovered that ionisation occurred in some flames and that this was increased by the introduction of metal salts into the flame. Of all the metals the alkalis have the greatest tendency to ionisation, resulting in the formation of a positive alkali ion and a free electron. Since the electron cloud of an alkali ion has some similarity to that of the noble gases the ions neither emit nor absorb radiation in the flame. On account of the loss of atoms capable of emitting radiation, ionisation of the alkalis has thus always an adverse effect in analytical flame spectroscopy; but a singly ionised atom of the alkaline earths, which have two valence electrons, can emit and absorb radiation. In the hot oxyacetylene and nitrous oxide–acetylene flames such ionic lines are sufficiently strong to be of use in spectrochemical analysis [614].

A measure of the ionisation of a metal, for example potassium, and the ensuing atomic concentration loss is the *degree of ionisation* which is defined as:

$$\gamma = \frac{[K^+]}{[K]+[K^+]},\tag{5.6}$$

where $[K^+]$ and $[K]$ are respectively the numbers of free K ions and K atoms per cm^3 in the flame. The total concentration $[K]+[K^+]$ in the flame is again proportional to the K concentration in the sample solution. With the stipulation of thermal equilibrium the degree of ionisation can be calculated from the *Saha equation* [53, 61, 71, 107]:

$$K_i(T) = \frac{[M^+][e^-]}{[M]},\tag{5.7}$$

where M is a metal atom and $[e^-]$ the number of free electrons per cm^3 in the flame. The ionisation constant $K_i(T)$ only depends on the temperature T and on the kind of element, but not on its concentration. Using statistical mechanics it can be shown that its value (in cm^{-3}) follows from T (in kelvin) and the ionisation energy E_i (in eV) according to:

$$\log_{10} K_i(T) = 15 \cdot 684 + \log_{10} (Z_i/Z_a) + \tfrac{3}{2} \log_{10} T - 5040 E_i/T,\tag{5.8}$$

where Z_i and Z_a are the internal partition functions of the ion and atom respectively (see equation (6.4)). For the alkalis, $Z_i = 1$ and $Z_a = 2$; and for the alkaline earths, $Z_i = 2$ and $Z_a = 1$. Partition functions for atoms and ions dependent on temperature are listed in [38, 368]. In chapter 6 table 4 contains ionisation energies for some metal atoms, while more comprehensive tables can be found in [13, 16, 19, 75, 78]. Ionisation constants for alkalis and alkaline earths at $T = 2000$ K and 2500 K have been computed in [78, 753]. The similarity between equation (5.8) and equation (5.3) for molecular dissociation is to be noted.

If, for example, only the reaction $K^+ + e^- \rightleftarrows K$ is assumed to take place in the flame the following relationship for γ results:

$$\frac{\gamma^2}{1-\gamma} = \frac{K_i(T)}{[\bar{K}]},\tag{5.9}$$

where $[\bar{K}] \equiv [K]+[K^+]$ is the total alkali concentration in cm^{-3}. This equation only holds if the flame is electrically neutral, that is, if at every point in the flame $[e^-] = [K^+]$, a condition that is generally well satisfied in the usual flames†. Double ionisation, that is, loss of two electrons by the atom, is not to be expected in flames.

Application of equations (5.8) and (5.9) will show that for $[\bar{K}] = 3 \times 10^{12}$ cm^{-3} (corresponding to a solution concentration of about $0 \cdot 5$ mg ml^{-1}; see § 4.6.2) and $T = 2500$ K about 50 per cent of potassium is ionised whereas under the same

---

† If the flame is placed between two electrodes with a potential difference of about 300 V between them the neutrality condition no longer seems to be fulfilled [641]. The degree of ionisation and consequently the atomic concentration are then changed.

conditions only 7 per cent of sodium is present in the flame as ions. This difference in the degrees of ionisation is due to the difference in the ionisation energies (see table 4). Calculations of the degree of ionisation of alkalis, alkaline earths and other elements at various temperatures and concentrations can be found in [61, 564, 895, 897, 919]. With the alkaline earths one must however also account for the formation of molecular ions (see below).

When dealing with molecular dissociation we spoke of the dependence of the degree of dissociation α on temperature and total concentration. Similar considerations apply to the dependence of the degree of ionisation γ on temperature etc, yet the temperature and concentration dependences of dissociation and ionisation have different effects because an increase in γ gives rise to a decrease in atomic emission or absorption. Therefore with rising temperature a progressive fall of the atomic concentration as a result of the ionisation is to be anticipated. This fall eventually may compensate, if not even exceed, the increase of the thermal excitation with rising temperature (see § 6.4). Also the percentage atomic concentration loss becomes greater with a lowering of the total concentration. The resulting dependence of the percentage atomic concentration loss on the solution concentration causes concave curvature of the analytical curve (see chapter 8). Finally it follows from the Saha equation that the addition of an electron donor element like caesium, which itself fairly readily gives off electrons into the flame and so enhances the value of $[e^-]$, will reduce the degree of ionisation of the analyte, for example potassium. In this way the atomic emission or absorption of the analysis element can be enhanced, but ionisation interference can also result (see chapter 9). All these effects are often observed in flame spectroscopy and many quantitative interpretations have been put forward in the course of time [160, 298, 350, 357, 453, 454, 471, 516, 564, 705, 771, 794, 795].

The ionisation effects in general are more pronounced in the hot nitrous oxide–acetylene flames [179, 564, 792]. Although the degree of ionisation depends not only on T and E_i but also on the total element concentration, elements with $E_i \geqslant 7 \cdot 5$ eV on the whole probably are not or only slightly ionised in these flames [564, 695]†. Such elements are for instance boron and germanium while noticeable fractions of others, such as aluminium, lanthanum and niobium (with $E_i < 7$ eV), are usually present in these flames as ions [564, 613, 614]. But this troublesome ionisation can be suppressed to a great extent by addition of an alkali as electron donor. To prevent interference by ionisation in the analysis of caesium the use of a cool air–hydrogen flame is advisable [772].

Various experimental methods for investigating the ionisation of an element in a flame are available. Measurements of this kind are not only instructive for analytical flame spectroscopy, but through comparison with the Saha law can also provide an insight into the kinetics of the ionisation processes (see § 5.3.2). In addition to optical methods microwave, high-frequency and electric probe methods are also used in flame investigations [71]. Moreover, ion mass

---

† Theoretically calculated values of the degree of ionisation published without indication of the metal solution concentration and the efficiency of atomisation are in general meaningless.

spectrometry conveys information about the nature and concentration of ionised atoms and molecules. Flame spectroscopic methods easy to apply in practice are reviewed in [166].

Whereas addition of an electron donor like caesium depresses, that of an element with strong electron affinity, for example chlorine, enhances the metal ionisation. When HCl is contained in the sprayed solution negative Cl^- ions are formed in the flame according to [828, 830]:

$$HCl + e^- \rightleftarrows Cl^- + H$$

Given a high chlorine concentration the free electron concentration can thereby be greatly reduced and by the Saha equation the degree of ionisation of the metal is then raised. The occurrence of Cl^- ions and the resulting effects have been thoroughly investigated by means of electric probes, ion mass spectrometers and flame spectrometers [247, 254, 338, 343, 539, 567, 679, 828, 934]. For magnesium the rise in ionisation as a result of chlorine addition could be directly deduced from the intensity increase of the ionic line [934]. A similar strengthening of the Ba ionic line on addition of Br_2 was found in an air–acetylene flame while microwave absorption measurements revealed a simultaneous decline of the electron concentration [753, 933]. Evidently here the ionising action of the halogen is stronger than its effect on the halide formation. This also seems to be the case for the alkalis with the exception of Li. It has also been established that, in agreement with theoretical predictions, the mutual ionisation interference of the alkalis and the concavity of their analytical curves (see above) are increased by addition of HCl.

A Saha equation analogous to that for positive metal ions can be formulated for the equilibrium concentration of the chlorine ion, $[Cl^-]$, with the electron affinity (here 3·8 eV [71]) replacing the ionisation energy. It follows from the above-mentioned Cl^- formation reaction, however, that the ratio $[Cl^-]/[e^-]$ will depart from the equilibrium value if the concentration of H radicals does not correspond to chemical equilibrium (see § 5.3.2). Besides, a partial equilibrium between the reactants and products of the above chemi-ionisation reaction exists in flames since the rates of the forward and backward reaction steps are fast enough to balance each other [247].

On the introduction of tungsten together with potassium into the flame other negative ions such as HWO_4^- and WO_2^- can be formed which again will lower the atomic potassium concentration by enhancing the potassium ionisation [510].

Not only the presence of chlorine in the flame but also other factors may cause the concentration of free electrons in the flame to be different from the concentration of positive metal ions although the flame remains electrically neutral. With the alkaline earths also molecular ions like $SrOH^+$ appear and these are in considerably greater quantities than Sr^+ [195, 453, 454, 505, 534, 548, 564, 776]. Furthermore, negative ions like O^- and OH^- can be formed through attachment of electrons to flame radicals. Such ions are actually found in flames (see § 3.4.1), but they only play a minor part in the metal ionisation [453, 454, 679, 776, 829] because the electron affinity of the flame molecules and radicals is usually rather small, about 1·8 eV for OH [61, 71]. In carbon-rich nitrous oxide–acetylene flames an appreciable fraction of the electrons may be bonded to CN and so produce CN^- ions [735].

Of greater importance in the metal ionisation are the positive ions and the electrons that are also found in the pure hydrocarbon flames, that is, without metal additions (§ 3.4.1 and § 3.4.2). The concentration of these so-called *flame electrons* or *natural flame electrons* above the combustion zone may exceed the concentration of the metal ions, notably when the salt concentration of the sprayed solution is sufficiently low. The existence of free flame electrons manifests itself in practical analyses through various effects, all due to a lowering of the ionisation of the analyte. At low metal solution concentrations, giving a metal ion concentration in the flame which is very small compared with the concentration of the natural flame electrons, the value to be inserted for $[e^-]$ in the Saha equation (5.7) is practically independent of the metal concentration. In this low concentration range the degree of metal ionisation is then also practically independent of the metal concentration and consequently the analytical curve is straight, that is, without the characteristic concave ionisation curvature. By this restraint of the ionisation at low metal concentrations the limit of detection is also improved and mutual ionisation interference, for instance between $K \rightleftharpoons Na$, is reduced. As the concentration of the flame electrons drops at greater flame heights through gradual recombination, their favourable influence becomes weaker and the metal ionisation increases. Consequently the atomic metal emission and absorption decreases with increasing height of observation and the shape of the analytical curve is also altered.

The concentration of the flame electrons cannot always be deduced directly from the observation of the above-mentioned effects on the atomic metal concentration. Firstly the electrons released from the metal atoms themselves may retroact on the ionisation of the flame gas. Secondly the Saha equation for the metal ionisation is not always satisfied (see § 5.3.2) and so a quantitative basis for calculation of the total electron concentration from the observed degree of ionisation is lacking. Hence direct electric measurements of the electron concentration are then required [71, 216, 512, 551].

In calculating the effect of ionisation on the atomic concentration the possible formation of molecules of the element must also be taken into account. The formation of LiOH for instance (see § 5.2.3) acts as a buffer in the ionisation interference on lithium. If ionisation would reduce the atomic Li concentration by about 50 per cent without molecule formation, this reduction would be only 13 per cent for the case when LiOH is formed at a LiOH/Li ratio of 20/1 [78]. Ionisation interference through concomitants will then also be less effective [564] (see chapter 9). Similar considerations apply to the alkaline earths, for which the formation of molecular ions complicates the calculations still further. In the case of such elements, calculations of atom losses through ionisation, published in the literature, therefore must often be viewed critically.

5.3.2. Departures from Saha equilibrium. Ionisation mechanisms. Discussions on the existence, above the primary combustion zone, of thermodynamical equilibrium for metal ionisation according to the Saha equation have recently been resumed. It has been found that Saha equilibrium is not established immediately after the alkali and alkaline-earth atoms have appeared in the flame [71]. The departures from the Saha equilibrium generally decrease with increasing height of observation and rising temperature, and they become the greater, the higher the ionisation energy of the atom.

There are two possible causes for these departures. On the one hand deviations of the radical concentrations from chemical equilibrium (see § 3.4.2) may have a direct effect on the metal ionisation; on the other hand a certain sluggishness in the ionisation process itself may induce a relaxation in the establishment of equilibrium, which means that the ionisation rate appears to be too low for immediate equilibration. The former, for instance, is the case with the alkaline earths, uranium and tin [421, 504, 509, 534, 776], whereas relaxation effects are of significance for the alkalis, gallium, indium, thallium and the alkaline earths in H_2 and CO flames at moderate temperatures [216, 421, 424, 455, 461, 512, 513, 530, 535, 538, 682, 776]. Under otherwise identical conditions the deviations for the alkalis seem to be smaller in air–acetylene flames [160, 454, 682]. Before dealing with these deviations we must look more closely at the various possible ionisation and recombination mechanisms [61, 71, 421, 524, 829, 830, 831].

Ionisation in a flame can be brought about by any one of the following three mechanisms: (i) by collision of the metal atom with another particle that has enough energy to ionise the metal atom; (ii) by transfer to the metal atom of a positive elementary charge from another particle that is already ionised, that is, reverse transition of an electron; and (iii) by chemi-ionisation, in which the ionisation is essentially associated with an exoergic chemical reaction.

As a result of the very low radiant energy density in flames compared with a black-body radiator at the same temperature (see § 3.4.2), photoionisation can be disregarded as a likely ionisation mechanism [455].

To (*i*). On account of the low partial pressure of electrons in a flame (at the most about 10^{-4} atm) *collisions with electrons* are practically of no significance for the ionisation mechanism, in contrast to the conditions in hot plasmas. This can also be deduced indirectly from experiments showing that the measured ionisation rate per atom is independent of the alkali content and therefore independent of the electron concentration [455, 535].

Collisions with flame molecules (for instance N_2) can lead to ionisation and in reverse to recombination according to:

$$Na + X^* \rightleftarrows Na^+ + e^- + X,$$

where X^* is a thermally excited flame molecule. Optical and electric measurements of the state of ionisation as a function of the height of observation and of the flame temperature in $CO-O_2-N_2$ and $H_2-O_2-N_2$ flames have shown that this process really is effective for the alkalis and for Ga, In and Tl [424, 455, 461, 512, 513, 530, 535, 830]. If the metal atom is not yet ionised when introduced into the flame, the limited rate of the process makes the gradual rise in ionisation with increasing height of observation plausible. The activation energy of the ionisation process measured in flames corresponds to the ionisation energy and hence the ionisation rate increases rapidly with rising temperature and falling ionisation energy. Furthermore the large effective cross section found for the collisional ionisation of the alkalis by N_2 (about 10^4 Å^2; see [71]) seems to indicate that in the above collision process thermal excitation of the alkali atoms also plays an important part [165, 351, 455, 461, 513, 535, 708]. Ar atoms, though lacking internal excitation energy, also seem to have a large effective cross section for the ionisation of alkali atoms, according to [535].

A summary of the methods of measuring ionisation rate constants for alkali atoms in flames and of the results obtained is given in [183] together with an approximate expression for this rate constant as a function of ionisation energy and temperature. The rate constant is defined here as the probability of an ionising collision per second and per alkali atom.

To (ii). Charge transfer can contribute appreciably to the ionisation of metal atoms in and immediately above the combustion zone if there is a considerable excess of natural flame ions in this region (see § 3.4.1). The following reactions for example are known to occur [252, 423, 826]:

$$Na + H_3O^+ \rightarrow Na^+ + H_2O + H$$

or

$$Na + C_3H_3^+ \rightarrow Na^+ + \text{other products.}$$

In these reactions the charge transfer is associated with a chemical dissociation. As these flame ions occur preferentially in hydrocarbon flames, the ionisation rate of the metal atom is higher in acetylene flames than in hydrogen flames (compare [50, 423, 682]). The $C_3H_3^+$ ion is to be found exclusively in the combustion zone and consequently can be effective in the metal ionisation only if the metal salt is already volatilised and dissociated in this zone; this is not necessarily the case however.

Other forms of charge transfer which are associated with a chemical reaction involving the metallic element, are for instance [423]:

$$MO + H_3O^+ \rightarrow M^+.H_2O + OH$$

$$MO + H_3O^+ \rightarrow MOH^+ + H_2O$$

$$M + H_3O^+ \rightarrow MOH^+ + H_2$$

$$MOH^+ + H \rightleftharpoons M^+ + H_2O.$$

In these expressions M represents a Cr, Pb, Mn, Cu or Fe atom. In the presence of excess H_3O^+ ions the first three reactions proceed mainly in the forward direction whereas the fourth takes place in the forward and reverse direction at equal rates so that a partial equilibrium is established (see also § 5.2.4). This last named, equilibrated reaction is also observed with Sn [504] and the alkaline earths (see below). Metals with high ionisation energies of 6·5 to 8 eV, otherwise hardly ionised in thermal equilibrium, such as Pb and Cr, can be excessively ionised in acetylene flames by these transfer processes [240, 423, 828]. With increasing distance from the combustion zone this excess ionisation gradually decreases as a result of fairly slow recombination processes until a Saha equilibrium is finally reached.

A charge transfer also may take place without molecular dissociation, as, for instance, in:

$$Na + SrOH^+ \rightarrow Na^+ + SrOH.$$

This process explains the catalysis of the Na ionisation through introduction of an alkaline-earth species [776] (see also below).

To (iii). In *chemi-ionisation processes* part of the required ionisation energy comes from a chemical reaction in which energy is released; examples are [345, 453, 534, 776]:

$$Sr + OH \rightarrow SrOH^+ + e^-$$

$$UO_2 + OH \rightarrow HUO_3^+ + e^-$$

or

$$SrO + H \rightarrow SrOH^+ + e^-$$

$$UO_3 + H \rightarrow HUO_3^+ + e^-.$$

In the cases of the first and third examples, Sr^+ is then formed from $SrOH^+$ in the equilibrated reaction (see above):

$$SrOH^+ + H \rightleftarrows Sr^+ + H_2O.$$

Reactions of these two types of examples may similarly be observed with $SnOH^+$ [506]. The rate of the ionisation reaction involving the metal oxide as reactant (third example) is believed to be higher than that of the reaction involving the metal atom (first example) [420]. Chemi-ionisation may also be associated with a dissociation of the metal oxide, for example,

$$CO + MO \rightarrow M^+ + e^- + CO_2,$$

where M is again a metal atom [295].

Considerably less activation energy is usually required to bring about these chemical ionisation processes compared with the physical collision processes described before. In the latter the full amount of atomic ionisation energy must be supplied. Since this greatly exceeds the average energy of a particle in the flame, chemi-ionisation processes lead to an appreciably higher rate of ionisation. Contrary to the opinion expressed previously [534, 776] it must however be considered as doubtful that the two indicated processes of formation of $SrOH^+$ are actually rapid enough to establish a partial equilibrium with the corresponding reverse recombination processes in flames [421].

Chemi-ionisation can also be connected with radical recombination reactions in which chemical recombination energy is used to ionise a metal atom present as a third partner. Thus there are indications of the occurrence of the following reaction [455, 461]:

$$K + CO + O \rightarrow K^+ + e^- + CO_2.$$

Recombination of H radicals on the other hand leads to metal ionisation not in a direct, but in an indirect manner when chlorine is added as a catalyst [422, 679, 828]:

$$\left.\begin{array}{l} Na+Cl \rightleftarrows Na^+ + Cl^- \\ Cl^- + H \rightleftarrows e^- + HCl \\ HCl + H \rightleftarrows H_2 + Cl \end{array}\right\}.$$

In effect these reactions together constitute ultimately a recombination of two H radicals to one H_2 molecule:

$$Na + H + H \rightleftarrows Na^+ + e^- + H_2.$$

From the participation of flame radicals in most chemi-ionisation reactions it becomes evident that departures from Saha equilibrium have to be suspected should the concentrations of the radicals themselves not correspond to chemical equilibrium (see § 3.4.2). This would hold even if these reactions proceed infinitely fast in both directions.

In conclusion we have to point out that the role played by the flame radicals, the limited rate of the ionisation process and the competition between chemi-ionisation and ionisation through physical collisions and charge transfer greatly complicate the theoretical interpretation of the metal ionisation in and immediately above the combustion zone. The situation in the simple hydrogen and carbon monoxide flames is rather comprehensible, but much has still to be explained for the hydrocarbon flames with their natural flame ions, which do not correspond to equilibrium. Complications also arise especially when a large amount of chlorine is added to the sample solution because chlorine as a catalyst speeds up ionisation, binds the electrons by forming negative Cl^- ions and combines with the metals to form chlorides. Besides, the described effects lead to a dependence of the atomic metal concentration on the height of observation. In practical analysis this has a bearing on the optimum height adjustment of the burner. Further away, some centimetres say, from the combustion zone, however, calculations based on the Saha equation can provide a fair indication of the state of metal ionisation in flames which are not too cool. These calculations are still rather difficult for non-premixed flames because the distribution of temperature and material here is far from uniform and because the degree of ionisation depends critically on these factors. For the alkaline earths and for tin the ionisation calculations are further complicated by the formation of molecules and molecular ions. Here the ionisation effects cannot be evaluated without taking into account the interaction with the molecule formation. On the other hand the ionic lines of the alkaline earths can help greatly in the experimental determination of the degree of ionisation. Finally, involatile aerosol particles of metal oxides (e.g. uranium dioxide and nickel oxide) can also play a part in the ionisation [531, 683].

The ionisation rate of sodium is enhanced when uranium is present in the nebulised sodium solution, but not if the sodium and uranium are separately introduced into the flame by means of a twin nebuliser [534]. This can probably be explained by volatilisation of Na *ions* directly from the solid phase after the sodium in this phase has given up one electron to the uranium oxide [534].

6 Emission, Absorption and Fluorescence of Radiation

6.1. Introduction

In the preceding chapters the injection of the sample solution into the flame and its transformation there into atomic or molecular vapour have been dealt with. Now we will discuss the capability of the free atoms and molecules to absorb and emit or re-emit optical radiation.

In § 6.2 we first discuss in general the origin and the types of emission and absorption spectra. The individual spectral lines are not infinitely narrow, but for various reasons are spectrally broadened, which has important consequences especially for the absorption of radiation. This broadening is discussed for emission and absorption lines together in § 6.3, firstly only for the case of optically thin vapour layers, that is, without self-absorption. Presented in § 6.4 is a quantitative treatment of the thermal radiation intensity of atomic lines, without regard to self-absorption. Then in § 6.5 we quantitatively consider the factors on which the absorption of radiation by atomic vapour depends and in particular elaborate on the relation between absorption strength and atomic concentration. Here the spectral profile of the absorption line as well as that of the line of the background light source have to be taken into account explicitly. This discussion is not restricted to optically thin layers and therefore applies to any atomic concentration, however high. There follows in § 6.6 a discussion of the effects of self-absorption and of self-reversal on the radiation intensity and on the spectral profile of resonance lines at high atomic concentrations. For atomic emission this is of practical importance not only in flames, but also in discharge lamps and similar sources. Section 6.6.3 briefly covers the basis of the line-reversal method for the optical measurement of flame temperatures. In § 6.7 we shall consider the various specific excitation mechanisms (both physical and chemical) that may occur in flames. This discussion is of special interest to the practical application of suprathermal chemilumines-cence. The physical principles of atomic fluorescence spectroscopy are dealt with separately in § 6.8.

Evidently only a short outline of the main concepts and the new developments in atomic and molecular physics can be presented here and for a further study we refer to the special literature [15, 37, 38, 51, 52, 58, 60, 64–66, 70, 78, 79, 82, 83, 89, 101, 104, 107, 116, 176, 219, 860].

6.2. Emission and Absorption Spectra

6.2.1. Atomic line spectra. We discuss here the origin of atomic line spectra on the example of the sodium atom which consists of a positive nucleus and eleven electrons (11 being the atomic number of sodium in the periodic table). The electrons are moving around the nucleus in definite orbits which are arranged in different shells. Under the usual flame conditions, ten of these electrons remain permanently in the closed (or full) inner shells (here the K and L shell). The eleventh, so-called valence electron moves in an orbit in the outermost, open shell (here the M shell). The latter electron alone is responsible for the optical radiation processes under discussion. To visualise these processes we shall use the Ruther-ford–Bohr atom model with which an excitation and in particular an absorption process is explained by the transition of the valence electron from its innermost orbit of lowest energy E_0 to an outer orbit of higher energy E_q. The atom as a whole is then said to be transferred from the ground state into a higher or excited state. The (stationary) states of an atom are quantised, which leads to discrete energy values $E_0, E_1, E_2 \ldots E_q \ldots$. The choice of the zero point of the energy scale is arbitrary; often the energy E_0 of the ground state is taken as zero so that E_1 is actually the excitation energy of the first excited state, and so on. As a consequence of the law of conservation of energy, the excitation energy must somehow be supplied from outside the atom, for instance as kinetic energy in a so-called *collision of the first kind* with another particle or as radiation energy in a *photo-excitation process*.

If now the valence electron falls back from an outer orbit of energy E_q into an unoccupied inner orbit of lower energy E_p, an energy amount $E_q - E_p$ is released. This liberated energy can be converted into kinetic energy in a so-called *collision of the second kind* with another particle, but there can also result an *emission of a photon* (quantum of light). The energy of the photon, $h\nu$, is governed by Bohr's frequency condition: $h\nu = E_q - E_p$, where h is Planck's constant or quantum of action, equal to $6 \cdot 62 \times 10^{-27}$ erg s, and ν is the frequency of the emitted radiation. Because the energy differences $E_q - E_p$ are characteristic for the atomic species, the frequencies and wavelengths $\lambda = c/\nu$ (c being the velocity of light) of the emitted radiation are similarly characteristic. This fact is the basis of *qualitative flame analysis* in which the species of the atoms present in the flame are identified from the wavelengths of the emitted radiation.

It thus depends on the arrangement of the energy levels of the atom (represented in an energy level diagram) which wavelengths are expected to appear in the spectrum under certain general selection rules for the allowed optical (dipole) transitions (see e.g. § 6.7.3). Wave mechanics explains why only certain electron orbits in the atom are possible and why certain optical transitions are forbidden, but this problem will not be discussed here. In figure 7 we only produce as an example the energy level diagram of the sodium atom, which is typical for the group of the alkali metals. In addition the excitation energies in eV for some lines of the most important elements are listed in table 4.

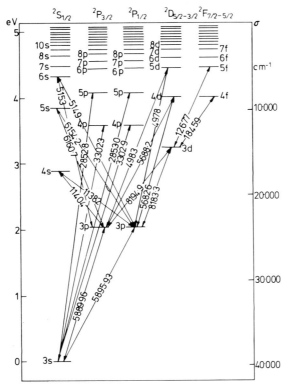

Figure 7. Energy level diagram of the sodium atom with some wavelengths (measured in air) indicated in ångström units. The splitting of the ^{2}D and ^{2}F terms is not shown in the figure because it is too small to be perceived. The top line of the diagram marks the ionisation limit. (Reproduced from R Herrmann and C Th J Alkemade 1960 *Flammenphotometrie* 2nd edn, by permission of Springer-Verlag, Berlin and Heidelberg.)

In the diagram each level indicated by a horizontal line represents a possible state or a collection of so-called degenerate states of the atom with a certain energy. The height of the lines in the diagram corresponds to the energy, the value of which relative to that of the ground level is indicated on the left-hand scale in electron volts ($1 \, \text{eV} = 1.60 \times 10^{-12}$ erg). Term values relative to the ionisation limit (the ionisation energy of the Na ground state amounts to $5.14 \, \text{eV}$) are expressed in wavenumbers on the right-hand scale. The arrows between the various levels mark the optically allowed transitions; this means that to every arrow in the diagram there corresponds a certain spectral line, the wavelength of which is given on the arrows in ångström units. We can see in figure 7 the well known pair (so-called doublet) of yellow Na D lines 5890/5896 Å† which are due to transitions from the lowest pair (doublet) of excited levels to the ground level. The D_1 and D_2 lines are

---

† In the following, wavelengths will be expressed in the now generally adopted unit of nanometre having the symbol nm ($1 \, \text{nm} = 10 \, \text{Å} = 10^{-9}$ m).

93

Table 4 Ionisation energies and spectral line characteristics of some metal atoms

Element	Ionisation energy†‡ (eV)	Wavelength† (nm)	Upper and lower level of transition		gf §	References §
			Symbols†	Energies†‡ (eV)		
Ag	7·57	328·07	$5^2P_{3/2}-5^2S_{1/2}$	3·78–0	0·9	284, 580, 646
Al	5·98	394·40	$4^2S_{1/2}-3^2P_{1/2}$	3·14–0	0·3	691
		309·27/8	$3^2D-3^2P_{3/2}$	4·01–0·014	1·0	690
Au	9·22	242·80	$6^2P_{3/2}-6^2S_{1/2}$	5·10–0	$0·1^5-0·5$	3, 75
Ba	5·21	553·55	$6^1P_1-6^1S_0$	2·24–0	1·3–1·6	267, 476, 579, 642, 690
Ba⁺	10·0	455·40	$6^2P_{3/2}-6^2S_{1/2}$	2·72–0	1·3	267, 690
Bi	7·29	306·77	$^4P_{1/2}-^4S_{3/2}$	4·04–0	0·3	597
Ca	6·11	422·67	$4^1P_1-4^1S_0$	2·93–0	1·4–1·8	267, 413, 476, 579, 595, 690, 802
Cd	8·99	228·80	$5^1P_1-5^1S_0$	5·42–0	1·2	597
Cr	6·76	357·87	$^7P_4-^7S_3$	3·46–0	2	3, 690
Cs	3·89	852·11	$6^2P_{3/2}-6^2S_{1/2}$	1·45–0	$1·4^5$	590
		894·35	$6^2P_{1/2}-6^2S_{1/2}$	1·39–0	0·7	590
Cu	7·72	324·76	$4^2P_{3/2}-4^2S_{1/2}$	3·82–0	0·9	284, 580, 662
Fe	7·9	248·33	$^5F_5-^5D_4$	4·99–0	3·9	4
K	4·34	766·49	$4^2P_{3/2}-4^2S_{1/2}$	1·62–0	1·3	1, 590
		769·90	$4^2P_{1/2}-4^2S_{1/2}$	1·61–0	$0·6^5$	1, 590
		404·41	$5^2P_{3/2}-4^2S_{1/2}$	3·06–0	0·03	449
Li	5·39	670·78	$2^2P-2^2S_{1/2}$	1·85–0	1·4–1·6	1, 821
		323·26	$3^2P-2^2S_{1/2}$	3·83–0	0·02	16
Mg	7·64	285·21	$3^1P_1-3^1S_0$	4·35–0	1·8	595, 802
Na	5·14	589·00	$3^2P_{3/2}-3^2S_{1/2}$ ⎱	2·10–0	1·3	1, 590, 638
		589·59	$3^2P_{1/2}-3^2S_{1/2}$ ⎰		$0·6^5$	1, 590, 638
		330·23	$4^2P_{3/2}-3^2S_{1/2}$	3·75–0	0·02	16
		819·48	$3^2D-3^2P_{3/2}$	3·62–2·10	9	3
Ni	7·63	232·00	$^3G_5-^3F_4$	5·34–0	0·9	3, 75
Pb	7·42	283·31	$^3P_1-^3P_0$	4·37–0	0·2	3, 597
Rb	4·18	780·02	$5^2P_{3/2}-5^2S_{1/2}$	1·59–0	$1·3^5$	590
		794·76	$5^2P_{1/2}-5^2S_{1/2}$	1·56–0	$0·6^5$	590
Sn	7·33	224·61	$^3D_1-^3P_0$	5·52–0	0·2	3
Sr	5·69	460·73	$5^1P_1-5^1S_0$	2·69–0	1·5–2·1	267, 476, 579, 595, 690
Tl	6·11	276·79	$6^2D_{3/2}-6^2P_{1/2}$	4·48–0	0·5	3
		377·57	$7^2S_{1/2}-6^2P_{1/2}$	3·28–0	$0·2^5\|$	3, 692
		535·05	$7^2S_{1/2}-6^2P_{3/2}$	3·28–0·97	0·5–0·9‖	3, 692
Yb	6·22	398·80	$^1P_1-^1S_0$	3·11–0	0·4	3
Zn	9·39	213·86	$4^1P_1-4^1S_0$	5·79–0	1·2–1·5	75, 595

† In general there is good agreement in the literature on the ionisation energy, energy level and wavelength data [16, 19, 20, 75, 78].
‡ All energy values are referred to the ground level.
§ The values of gf (g is the statistical weight of the lower level, f the oscillator strength for the absorption transition; see § 6.5.2) are taken from the references given in the last column. More comprehensive tables of the values of gf and f can be found in [2–4, 21, 22, 75, 284, 400, 690, 888]. Some of the data in [3] have been amended in later NBS publications [2, 4].
‖ According to [449] higher values have been found in a flame.
The values of gf have been compiled in collaboration with Dr P J Th Zeegers, University of Utrecht.

94

closely spaced in the spectrum because the energy difference of the excited doublet levels (which together form a term†) is relatively small (see also below).

The ultraviolet Na doublet 330·2/330·3 nm similarly arises from a transition to the ground level. The infrared doublet 818·3/819·5 nm on the other hand is due to a transition from a higher to a lower excited term, but not to the ground level. This doublet can also be detected in flames (see the spectral tables in Appendix 2). If the transition occurs to or from the ground level, the resulting line is called a *resonance line*, and in particular a *first resonance line* in case the transition occurs from the lowest excited term to the ground level. Hence the Na D_1 and D_2 lines are first resonance lines.

Basically, because all transitions in microphysics are reversible, to every allowed emission transition at wavelength λ there corresponds a possible reverse absorption transition at the same wavelength‡. The probability of a photon emission is proportional to the population of the upper energy level (see § 6.4), whereas the probability of photon absorption is proportional to the population of the lower energy level (see § 6.5). In a flame the population of the ground level for most atoms is by many orders of magnitude higher than that of the higher energy levels (see § 6.4) and thus resonance lines are mostly used for absorption and fluorescence measurements. Occasionally non-resonance lines are used in the case where atoms have a low excited level with an energy of about 1 eV or less above the ground level (see § 6.5).

Each energy level in figure 7 is characterised by three *quantum numbers*: the principal quantum number (n); the azimuthal (or orbital angular momentum) quantum number (l) and the inner (or total angular momentum) quantum number (j). These quantum numbers arise from the quantisation of the energy, the orbital angular momentum, and the spin angular momentum (due to the rotation of the valence electron about its own axis). The *principal quantum number* defines the shell in which the valence electron is located. It is inserted in front of the level symbol. In our example of the sodium atom $n = 3, 4, 5 \ldots$ since the first two inner shells (K and L shell with $n = 1$ and 2 respectively) are closed, that is, fully occupied

---

† In general a (*spectroscopic*) *term* comprises a *multiplet* of closely spaced levels that have the same atomic orbital angular momentum (L) and spin (S) but are distinguished by their total angular momentum (J) (see also below). The *term* (*value*) corresponds to the weighted average energy of the individual levels. Following common practice we use term values here to indicate energy, although strictly speaking they define the position of the multiplets in the spectrum by their wavenumbers $\sigma \equiv 1/\lambda$ (expressed e.g. in cm^{-1}, see right-hand scale of figure 7; 1 eV corresponds to 8066 cm^{-1}).

‡ At the flame temperature the excitation energy of the upper level is usually large compared with kT (k = Boltzmann's constant, T = absolute temperature). Therefore under thermal excitation conditions *stimulated* (or *induced*) *emission* is practically of no significance. The latter may become effective when the Na vapour in the flame is illuminated by a strong pulsed dye laser tuned to one of the resonance wavelengths of the Na atom (see § 6.8). For practical reasons here we do not consider *two-photon processes* in which the energy of each of the simultaneously absorbed photons is exactly equal to half the excitation energy. Such processes have been observed with Na, Cd and Zn atoms in a flame illuminated by a strong dye laser [168a, 354]. The probability of a two-photon process is proportional to the square of the intensity of the light source, and therefore does not need to be taken into account for conventional light sources.

by electrons. The *azimuthal quantum number* expresses the eccentricity of the elliptical electron orbit and can have the values $l = 0, 1, 2, 3 \ldots, (n-1)$. The quantum number $l = 0$ corresponds to zero orbital angular momentum, that is maximum eccentricity, and the quantum number $l = n - 1$ $(n > 1)$ to a circular orbit. The values $l = 0, 1, 2, 3, \ldots$ are indicated by the letters s, p, d, f, ... respectively in the term symbol. The corresponding capital letters S, P, D, F, ... indicate the quantum number $L = 0, 1, 2, \ldots$ for the total orbital angular momentum of the whole electron cloud in the atom. Since the orbital angular momenta of all electrons in the closed shells add up to $L = 0$, and since alkali atoms only have one valence electron, the azimuthal quantum number L of an alkali atom as a whole is equal to l. Therefore the S, P, D, ... terms of these atoms correspond respectively to a valence electron in an s, p, d, ... orbit.

The *inner quantum number* describes the mode of coupling between the orbital angular momentum and the spin of the valence electron through the associated magnetic fields. When the sense of the orbital rotation is the same as that of the spin rotation, then $j = l + \frac{1}{2}$, and if it is the opposite, then $j = l - \frac{1}{2}$; when l equals zero, $j = \frac{1}{2}$. There are no other possibilities for the Na atom. The value of j is written as a subscript behind the level symbol. The two closely spaced levels of the lowest excited term of sodium, $3p_{3/2}$ and $3p_{1/2}$ (see figure 7) thus only differ in their value of j. To this disparity in the values of j corresponds a slight difference in the excitation energies and hence in the wavelengths. This explains the so-called *fine structure* (FS) of the alkali resonance lines which appear as doublets (see figure 7).

The spins of all electrons in the inner closed shells add up to a zero resultant spin angular momentum. With the alkali atoms the quantum number S for the total spin of the whole electron cloud thus equals that for a single electron, that is $S = \frac{1}{2}$. For alkaline-earth atoms having two valence electrons with antiparallel or parallel spins we obtain $S = 0$ and $S = 1$ respectively. The *multiplicity* $(2S + 1)$ is written as a superscript in front of the term symbol for the atom as a whole. The alkali terms are doublets with $2S + 1 = 2$; the alkaline-earth terms are either singlets $(2S + 1 = 1)$ or triplets $(2S + 1 = 3)$.

By vector addition of the resultant orbital and spin angular momenta we obtain the total angular momentum of the electron cloud. It is denoted by the quantum number J which is written as a subscript behind the atomic level symbol; for alkalis we have $J = j$.

A further term splitting and resulting line splitting, called *hyperfine structure* (HFS), is caused by the magnetic coupling of the spin and orbital motion of the electron with the nuclear spin. The $3S_{1/2}$ ground level of the Na atom for instance is split into two sublevels very close together and this leads to a further doubling of each of the two Na D lines (the splitting of each of the two 3P levels through the nuclear spin is considerably smaller and is neglected here). Because the wavelength difference of the components of the hyperfine structure here is only about 0·002 nm and the linewidth under flame conditions (see § 6.3) about 0.005 nm, this splitting is hardly noticeable [850]. However it may become manifest in alkali vapour discharge lamps [206]. For other lines, such as the Cu 324·7 nm and the In 410 nm lines, a distinct HFS splitting may be found in the flame [417, 861, 871]. This

96

splitting has consequences for the strength of the atomic absorption when a Cu or In line source is used (see § 6.5.3).

A splitting of the atomic lines that is of analytical interest can result from the *isotope shift* when several isotopes of the element are present. The ^{6}Li and ^{7}Li first resonance lines at 670·8 nm for example are separated by an interval of about 0·016 nm, which by coincidence is approximately equal to the fine structure splitting. Since this shift is of a magnitude comparable to the linewidth in the flame, the isotope components of the Li lines are sufficiently resolved to allow determination of the isotope ratios by flame atomic absorption spectroscopy [75, 923].

The general features of the spectra of all the alkali metals are very much alike because each of these elements has only one valence electron. The alkaline-earth metals on the other hand have two valence electrons; their spectra also resemble each other, but they are different from those of the alkali metals. The singly ionised alkaline-earth atom (with only one valence electron), in contrast to the ionised alkali atom, emits a line spectrum in the flame that is fairly similar to the spectrum of the neutral alkali atom. The singlet term diagram of a neutral alkaline-earth atom basically differs from the term diagram of an alkali atom (see for instance figure 7) in that the resonance lines of all the alkaline-earth atoms, such as the blue Ca line at 422·7 nm, are not doublets but single lines. The triplet terms of these atoms do not give resonance lines, as a transition from an excited triplet term to the singlet ground term is forbidden.

6.2.2. Molecular band spectra.

When an alkaline-earth metal is present in the flame, then not only a line spectrum of the free atoms but also a molecular spectrum is observed, which is emitted by the alkaline-earth oxides or hydroxides. This band spectrum is sometimes of greater analytical importance in flame emission spectroscopy than the line spectrum. There are also non-metallic elements, such as Cl, which do not emit any atomic lines in flames but which sometimes produce a useful molecular spectrum, as for instance that of InCl. The appearance and the origin of molecular spectra, which incidentally also occur in the flame background emission (see chapter 7), will be dealt with next. We give here only a short outline and for a fuller treatment we refer to the general literature listed in § 6.1.

As in the case of atoms, the frequencies of the radiation emitted by molecules or radicals (e.g. OH) are governed by the equation $h\nu = E_q - E_p$ where E_q is the internal energy of the excited molecule and E_p its internal energy after the emission of a photon. Again we have to satisfy selection rules which forbid certain radiative transitions. Like atoms, molecules have distinct excited states with discrete excitation energies which give rise to discrete frequencies in the spectrum characteristic for the kind of molecule. But there are many more energy levels in the molecule than in the atom and hence many more lines in the molecular spectrum. When the practical resolution of the spectrometer is insufficient or when the lines are broadened too much in the flame (see also § 6.3), the many closely spaced molecular lines cannot be separated. They then appear to be bunched together in broad bands, and a typical band spectrum is observed as distinct from the line spectrum of atoms. Sometimes, for example with CO_2, the band spectrum even appears as a quasi-continuum covering an extended wavelength range (see chapter 7). Some records of molecular spectra, with identifications, are reproduced in Appendix 1. (For metal compounds, HPO and S_2 see, for instance, spectrograms 5, 6, 8–18, 20, 21, 25, 32, 34, 36 and 40–44; for flame molecules and radicals see spectrograms 28, 30, 33, 36, 38 and 45–48.)

To account for the multitude of excited levels and the resulting complexity of molecular spectra, we have to consider the three, fairly independent forms of internal energy of a molecule:

(i) The *electronic excitation energy* corresponding to that of the atoms. In molecules also the electron can make certain quantum jumps between different orbits corresponding to different electronic levels.

(ii) The *vibrational energy* due to vibrations of the atomic nuclei in the molecule about their rest positions. The molecule behaves like a flexible or extensible structure. There are distinct vibrational states with discrete vibrational energies corresponding to different vibrational amplitudes. Molecules with more than two atoms have several vibrational modes.

(iii) The *rotational energy*. A diatomic molecule, for instance, which can be visualised as a dumb-bell, can rotate freely about an axis through the centre of gravity perpendicular to the molecule axis. The rotational states are also quantised according to wave mechanics and again only discrete rotational energy values can exist. But the energy differences between consecutive rotational states are usually much smaller than those between consecutive vibrational states, which in turn are generally a great deal less than those between two electronic states.

The possible discrete values of the total energy of the simplest, diatomic molecule are schematically illustrated in the level diagram of figure 8. In this diagram we show, for example, two electronic states which are designated by $^1\Sigma$ (ground state) and $^1\Pi$ (excited state; for the meaning of these symbols see the general literature referred to above). The electronic states may occur in combination with any of the vibrational states designated by the quantum number $v = 0, 1, 2\ldots$, and lastly any vibrational state may be associated with any rotational state having the quantum numbers $J = 0, 1, 2, 3 \ldots$†. When the electronic and vibrational states remain unchanged, the transitions between the rotational states ($J' \rightarrow J''$) usually manifest themselves only in the far-infrared or microwave region of the spectrum because the energy differences are very small. Radiations due to combined vibrational and rotational transitions ($v' \rightarrow v''$; $J' \rightarrow J''$), in the same electronic state, mostly belong to the near infrared. The molecular bands in the visible and ultraviolet regions of the spectrum sometimes utilised in analytical flame spectroscopy are associated, apart from rotational and vibrational transitions, with electronic transitions.

---

† J refers here to the nuclear rotation only. Because of a coupling of the electronic angular momenta with the nuclear rotation, this J is not a strict quantum number, in contrast to the quantum number for the *total* molecular angular momentum.

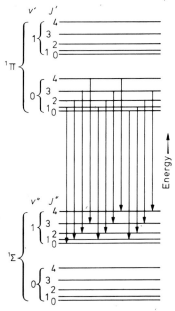

Figure 8. Level diagram of a diatomic molecule. The allowed rotational transitions for a given electronic transition $^1\Pi \to {}^1\Sigma$ and a given vibrational transition $v' = 0 \to v'' = 1$, shown as an example, together form a band. The separation of the vibrational and rotational levels has been exaggerated for the sake of clarity. (Reproduced from R Herrmann and C Th J Alkemade 1960 *Flammenphotometrie* 2nd edn, by permission of Springer-Verlag, Berlin and Heidelberg.)

In the example of figure 8 all molecular lines due to a transition ($^1\Pi \to {}^1\Sigma$; $v' \to v''$; $J' \to J''$) with given, constant values of v' and v'', but with varying values of J'' and J', together form a band. The J'', J' combinations have to satisfy certain selection rules, namely $\Delta J = 0$ or ± 1 (see figure 8). Since the energy differences corresponding to the various values of J are small, the lines of a band are very closely spaced. These lines usually crowd together (converge) towards a certain position in the spectrum – forming there a more or less sharp edge – the so-called *band head*. From the band head the overall intensity of the band falls off more or less gradually to the red or to the violet end of the spectrum. (The band is said to be *degraded* to the red or to the violet.) Every allowed combination $v' \to v''$ gives rise to a further band. The intensity distribution within a band is again determined by the temperature-dependent relative population of the associated rotational states (J').

In principle all (allowed) downward transitions in the level diagram of the molecule could also occur reversed in absorption, but again absorption transitions from the electronic ground state are much favoured on account of the comparatively high population of this state (see § 6.2.1). However the ground state is split up into many closely spaced rotational and vibrational states and therefore the population of each individual level is relatively low, as is consequently the intensity of the absorption lines starting from this level. It is found that the concentration of molecular compounds of metals introduced into the flame as trace elements is insufficient to produce a noticeable absorption spectrum. Absorption flame spectroscopy on the whole is therefore restricted to the use of the atomic or ionic metal lines. Yet for solutions of about 1 per cent concentration certain molecular bands of metal compounds, such as the CaOH band at 554 nm and the InO band at 273 nm, may show noticeable absorption [257, 416, 555]. For other alkaline-earth metals also, molecular absorptions were detected in the visible spectrum range. Absorption bands of indium halides were observed in the UV region when a solution containing 0.2 M indium and 2 N hydrogen halide was sprayed into an air–acetylene flame [416]. The LaO band at 441·8 nm, the YO band at 597·2 nm and the ScO band at 603·6 nm can also show up in absorption when a strong continuum background source is used [341]. In flame atomic absorption spectrometers equipped with long absorption tubes molecular absorption bands may cause spectral interferences. The absorption spectrum of SO_2 observed in a flame with the use of a long heated silica tube [359] and that of PO in a normal air–acetylene flame [359, 360] have been suggested as a means of S and P analyses. Furthermore molecules or radicals present under certain conditions as constituents of the combustion gases (e.g. OH [472, 892, 926] or C_2 and CN [341]) may give rise to appreciable flame background absorptions (see chapter 7). The presence of the latter two radicals in graphite furnaces was also detected by means of their absorption [401]. The emission and absorption band spectra of the flame gases themselves are discussed in more detail in chapter 7.

There has not always been full agreement in the older literature on the origin in the visible spectrum range of the alkaline-earth bands which are sometimes used in flame emission spectroscopy. Many of the alkaline-earth elements are present in the commonly used flames as monoxides or monohydroxides (see § 5.2.3). Whereas the ultraviolet, the blue and the infrared bands of Ca and Sr are doubtless oxide bands [82, 490], most of their bands in the visible must be due to their hydroxides. In the case of the Ca bands around 554, 602 and 623 nm, and of the Sr bands around 606, 647, 668 and 682 nm this was shown by measurements of their intensities in flames of constant temperature but with varying O and OH contents [238, 480, 481, 490] and by observation of the isotope shifts in their spectra when deuterium is used instead of hydrogen [227, 268, 376, 377, 475, 572]. The Ba bands in the visible are generally oxide bands, but at 487 nm and 512 nm have hydroxide bands superimposed on them [481, 483]. There is however no agreement yet on the assignment of some other alkaline-earth bands in flames. We have decided to list such disputed bands in the spectral tables in Appendix 2 as CaO(H), etc. In any case it is certain that most of the alkaline-earth bands are not due to polymer molecules such as Ca_2, Ca_2O, etc [455, 474, 490]. Band emissions of alkaline-earth–halogen compounds, such as CaF and SrF, have been observed under suitable conditions in oxyhydrogen flames with direct-injection burners and with a sufficient supply of halogens [406–408, 652].

6.2.3. Continuous spectra. Some elements (e.g. As, Al, Mo, V, Cr, Ni, Na, P, S, Cl) in a flame, in addition to a line or band spectrum, often give rise to a (usually weak) continuous spectrum (see Glossary). This manifests itself in an enhancement of the background radiation over an extended spectral range [391].

With elements such as Al, Cr and V the additional continuous radiation can be due to the thermal radiation of incompletely volatilised oxide particles, for example Al_2O_3 [654]. The temperature of these particles may sometimes be raised above the flame temperature, and the thermal radiation enhanced, through the release of chemical energy in the recombination of excess H and OH radicals on the particles' surfaces [242, 243, 525, 529]. For Cr_2O_3 particles temperatures up to 500 °C above the flame temperature have been established by measurement of their continuous radiation in hydrogen flames with an excess of such radicals [243].

A spectral continuum can also be produced by recombination processes in the gas phase, as for instance $K^+ + e^- \rightarrow K$ [52], $Na + OH \rightarrow NaOH$, $H + Cl \rightarrow HCl$, $Cl + Cl \rightarrow Cl_2$ or $SO + O \rightarrow SO_2$ [161, 391, 491, 681, 693, 826]. The proportionality between the intensity of the continuous radiation and the atomic metal concentration as well as the greater prominence of metal continua in cooler air–hydrogen flames are characteristic for the recombination processes between a neutral metal atom and a flame radical [161, 386, 881]. The significance of ion recombination continua in flames, incidentally, is still questionable [161].

In general the emission of continuous spectra in recombination processes can be explained as follows. In the formation of a neutral K atom from $K^+ + e^-$, or of an NaOH, HCl or Cl_2 molecule, energy is released that is made up of the discrete ionisation or dissociation energy and the initial translational energy, continuously distributed, of the reaction partners. This energy – referred to the centre of gravity of the reacting particles – can be radiated as a light quantum. On account of the continuous distribution of the translational energy, the frequencies and wavelengths of this radiation – within certain limits – will also then be continuously distributed. In these processes also a discrete part of the recombination energy can be converted into internal energy of the formed particle. Fundamentally such continuous spectra can appear in absorption too – so-called *photoionisation* or *photodissociation continua*. Photodissociation continua of Na halides, SO_3, and SO have been observed in graphite-tube furnaces between 200 and 400 nm [401, 632]. The long-wavelength limits of the SO_3 and SO absorption continua appear to correspond to the dissociation energies of $SO_3 \rightarrow SO_2 + O$ and $SO \rightarrow S + O$ respectively [632].

These continuous spectra of course have to be distinguished from those appearing to be continuous because of instrumental shortcomings, such as the lack of resolution of molecular bands or the scatter of radiation inside the monochromator. The true continuous spectra of the flame gases are discussed in chapter 7.

6.3. Spectral Line Broadening

When a spectrum, for instance that of sodium, is viewed in a spectroscope, the spectral lines are seen to be not infinitely narrow, but show a broadening depending on the width of the entrance slit. If the slit width is made very small, there still remains a line broadening because of diffraction. Apart from this apparent broadening caused by instrumental factors [369], the true spectral profile of a so-called monochromatic line itself has a certain width, usually much less than the broadening due to the spectral instrument and only detectable with instruments of high resolving power. The origin of the broadening of the true line profile lies in the light source itself and in the Na atoms present therein. Here we will discuss in some detail the usual causes of line broadening in flames. As a consequence of the reversibility of all radiation processes, the following considerations apply equally for emission and absorption lines. To some extent they are also significant for the spectral lines emitted by gas-discharge lamps, etc.

We first only consider radiation sources and absorbing media that are optically thin for the wavelength of the spectral line considered. This means that, as will be pointed out in § 6.5 and § 6.6, the (self-)absorption is weak, a condition satisfied for resonance lines usually at low atomic concentrations only. The spectral line profile, that is, the shape of the graph of the intensity or absorption plotted against the wavelength difference with respect to the line centre, is then independent of the atomic concentration. With optically thick radiating or absorbing layers the line profile is flattened which leads to a further broadening of the spectral line emitted or absorbed by the layer as a whole (see § 6.5 and § 6.6). In other words, we limit the discussion here to the spectral distribution of the line as emitted or absorbed by the individual atoms in a given medium. By extrapolation of the measured line profile to zero solution concentration one can correct for the additional broadening due to self-absorption [159, 543, 544, 547]. Furthermore we shall not deal with the apparent broadening resulting from any hyperfine structure (see § 6.2.1) and refer for more details to [868].

As a measure of the spectral width we adopt the so-called *half-intensity width δλ* which defines the width of the wavelength interval, expressed in nm, within which the spectral intensity or the absorption factor falls to one-half of the peak value (see figure 9).

Every atomic line has basically a so-called *natural linewidth* because an atom, even if conceived as isolated, always remains only for a limited duration in its excited state; for after a certain (mean) period of time it loses its energy through emission of a photon. According to Heisenberg's uncertainty principle the limited optical lifetime of the excited state (typically of the order of 10^{-8} s) leads to an uncertainty of the energy E_q of the excited state. This again results in a smearing-out of the wavelength of the emitted or absorbed photon over a range of about 10^{-5} nm. In the case of a non-resonance line the atom, on emitting a photon, falls from a higher to a lower excited level; then the uncertainties of both excitation energies contribute to the natural linewidth. For flames and discharge lamps however natural line broadening is usually negligible compared with the broadening effects described below.

The mean lifetime of an excited state can be further shortened through quenching collisions with other particles in which the excitation energy is transferred (see § 6.7.2). The resulting *quenching broadening* may indeed exceed the effect of the natural line broadening, but in most cases it is in turn small compared with the

Lorentz broadening [200, 460, 466] which also results from collisions with neutral particles. Here however the radiating particle is not transferred to another energy level, but during the emission or absorption process only undergoes transient perturbations†. These perturbations manifest themselves in a smearing-out of the wavelength of the emitted or absorbed photons. This broadening depends on the type of the collision partners and their concentrations as well as on the temperature. In the special case of collisions between atoms of the same element we speak of *Holtsmark* or *resonance broadening*. Such collisions however are rare for metal atoms in the commonly used flames because, in spite of their comparatively large effective cross section, the partial metal vapour pressure is seldom much above 10^{-4} atm. Therefore resonance broadening may be disregarded in flame spectroscopy [200, 369, 711, 746, 850] and the same holds for the so-called *Stark broadening* which is caused by the variable electric fields arising from the encounter with electrons, ions, and molecules with an electric dipole moment (e.g. H_2O) [200]. Yet this broadening is significant in high-pressure discharge lamps.

Of the types of line broadening mentioned so far, the Lorentz broadening is mostly dominant in flames. In the classical theory [79] the profile of a spectral line broadened only by collisions is described by a dispersion formula according to which the spectral intensity or the absorption factor decreases with increasing distance $|\lambda - \lambda_0|$ from the line centre λ_0 proportionally to $1/(1 + 4(\lambda - \lambda_0)^2/(\delta\lambda)_L^2)$. Hence we would expect a symmetrical profile and, at greater wavelength distances ($\gg (\delta\lambda)_L$), a decrease approximately proportional to $1/(\lambda - \lambda_0)^2$. In fact lines with collision broadening may show an asymmetrical profile in disagreement with a dispersion formula. The asymmetry is particularly marked in the outer line wings [473]; at the line centre the profiles observed in flames are mostly almost symmetrical [198, 460, 473, 812, 868].

It has also been found that the centre of the line disturbed by collisions is somewhat displaced from the position of the undisturbed line centre. The ratio of this *line shift* to the Lorentz half-intensity width and the direction of the shift, which is generally towards the longer wavelengths, depend on the kind of interaction between the radiating atom and its collision partner. For the first resonance lines of the alkaline-earth atoms in an air–acetylene, a nitrous oxide–acetylene and an oxygen–hydrogen–argon flame, ratios between about 0·25 and 0·45 have been established [460, 496, 602, 606]; for most of the investigated resonance lines of other atoms the ratio lies between about 0·5 and 0·1 [75, 868, 870, 871]. These results seem to deviate from the classical Lorentz theory but can be explained by the more refined Lindholm–Foley theory‡ [39]. Sometimes however the classical

---

† Two different theoretical approaches have been advanced to account for the line broadening through such adiabatic interactions with neighbouring particles, namely the impact theory and the so-called quasi-static theory [39, 104]. Under the conditions prevailing in flame spectroscopy the application of the former theory is justified [453, 460].

‡ The Lindholm theory with consideration of the van der Waals forces alone cannot fully account for the experimental results. The results of theory and experiment are in better agreement if the interaction between the radiating atom and the collision partner is described by a Lennard-Jones potential [198, 199]. The collision broadening has been calculated for various types of interaction potentials in [878]. The temperature dependence of the line shift is discussed in [75].

103

theory is adequate, as in dealing with the self-absorption effect (see § 6.6.1). The line shift may have a distinct effect in the absorption method [364, 369, 606, 871] because, as a consequence of the lower gas pressure in the discharge lamp, the shift is much less for the emission line in the lamp than for the absorption line in the flame (see § 6.5.2 and figure 12).

Besides the Lorentz broadening the *Doppler broadening* can also play an important part. It is due to the Doppler effect according to which the observed wavelength is increased or decreased according to whether the emitting (or absorbing) atom is moving away from the observer or moving towards him. The random thermal motions of the atoms in the flame therefore lead to a smearing out of the line with a Gaussian intensity distribution and a half-intensity width given by

$$\frac{(\delta\lambda)_D}{\lambda_0} = 7 \cdot 16 \times 10^{-7} \left(\frac{T}{M}\right)^{1/2}, \tag{6.1}$$

where λ_0 is the wavelength of the line centre, T the absolute temperature in K, and M the mass of the radiating atom in atomic mass units. From this a Doppler width of $4 \cdot 5 \times 10^{-3}$ nm is calculated for the first resonance doublet of sodium in a flame of temperature $T = 2500$ K. (The commonly used expression (6.1) only holds if the mean free path of the radiating particle in the gas is large compared with the wavelength [848], a condition that is fairly well fulfilled in flame spectroscopy.)

The ratio of the half-intensity widths due to Lorentz and to Doppler broadening, $(\delta\lambda)_L/(\delta\lambda)_D$, can be derived from the so-called *a-parameter* which plays an important role in the absorption and self-absorption theory and which can be expressed to a good approximation as:

$$a = \sqrt{\log_e 2}(\delta\lambda)_L/(\delta\lambda)_D = 0 \cdot 84(\delta\lambda)_L/(\delta\lambda)_D. \tag{6.2}$$

This parameter does not vary greatly with temperature, but depends on the composition and the pressure of the gas. For the first resonance lines of the alkali and alkaline-earth atoms in flames at 1 atm it is of the order of unity. Since resonance broadening (see above) is practically of no significance in flames, the collisional broadening and with it a can be taken to be independent of the metal concentration.

The a-parameter can be determined by experiment either directly from spectral-line profile measurements in emission or absorption, for example by means of a Fabry–Pérot interferometer [159, 198, 364, 369, 458–460, 473, 543, 544, 547, 812, 868, 869], or indirectly from curve of growth measurements [448, 449, 455, 460, 609, 850, 931] (see § 6.5 and § 6.6). Alternatively it may be derived by comparing the absorption measurements with a spectral continuum source and with a line source [363, 366, 598, 601]. It can also be calculated from an absorption measurement at the line centre obtained with a narrow-line source for a known absolute atomic concentration in the absorbing flame [732]. Recently a-parameters have also been derived from the analytical curves in atomic fluorescence spectroscopy with a continuum source [846]. Experimental values of a in oxygen–carbon monoxide–nitrogen, oxygen–hydrogen–argon, air–acetylene, nitrous oxide–acetylene and oxyacetylene flames for some alkali, alkaline-earth, Al, Ag, Mg, In,

Ga, Cu, Zn, and Mn lines are listed in [364, 543, 544, 601, 846, 868, 896], while values of a for the Ca line in hydrogen–argon and hydrogen–nitrogen diffusion flames have been reported in [547]. The experimental errors are difficult to assess; sometimes they could be rather large (about 50 per cent). Most of the established values of a lie between about 0·3 and 2. Measurements of the Lorentz broadening of atomic lines in heated graphite and similar furnaces with nitrogen or argon used as fill gas are described in [75]. Purely theoretical evaluation of a must still be considered as unreliable, but also values of a measured in flames not uniform in temperature and/or composition have limited physical significance as a depends on these factors.

When Doppler and Lorentz broadening occur simultaneously, the line profile can no longer be described by a simple mathematical function. In the classical theory the line profile is described by an integral expression, the so-called *Voigt function* [79]. This expression involves the convolution of a Gaussian function (associated with the Doppler broadening) with a Lorentzian function. The half-intensity width of the Voigt profile is however not simply given by the sum of $(\delta\lambda)_L$ and $(\delta\lambda)_D$ [887]. Since the Gaussian profile falls off more steeply than the Lorentzian profile with increasing distance from the line centre, the outer line wings of the Voigt profile approach a Lorentzian profile. The pure Lorentzian profile can be determined mathematically by deconvolution of the measured total profile with the Doppler profile, which is known at the flame temperature [263, 292].

The various (semi-)classical theories of collision broadening (in combination with Doppler broadening) only provide approximations and can only be applied with certain restrictions. A generally valid exact quantum-mechanical theory has been given and the validity ranges of the classical, approximate theories have been clearly set out in [663]. In general the Voigt theory is to be considered as valid only in lowest-order approximation [878].

The Doppler broadening in hollow-cathode lamps and similar devices is less than in flames mainly because of the lower temperature. From equation (6.1), there results for the yellow Na doublet a half-intensity width of $1·5 \times 10^{-3}$ nm at about room temperature. As a consequence of the lower gas pressure the Lorentz broadening also is much smaller here than in flames and is in fact almost negligible. These observations do not apply to high-pressure gas-discharge lamps. The half-intensity widths found in hollow-cathode lamps correspond to 'Doppler temperatures' of about 350 to 600 K [235, 479, 497, 868, 870]. With modulated hollow-cathode lamps the spectral profile was found to depend on modulation frequency and pulse duration [698, 699]. If the noble-gas pressure in hollow-cathode lamps is too low, thermalisation of the sputtered metal atoms may not be complete and extra Doppler broadening will result [186, 412, 414].

6.4. Thermal Radiation Intensity of Atomic Lines (without Self-absorption)

In § 6.2 we discussed the connection between the atomic and molecular structure and the wavelengths of the lines and bands appearing in the spectrum. As quantitative emission spectroscopy is based on intensity measurements on lines and bands, we shall now also deal with the factors that determine these intensities, assuming thermal equilibrium in the flame. Non-thermal radiation due to chemi-luminescence and fluorescence will be treated separately in § 6.7 and § 6.8.

To simplify the discussion we again take as an example a sodium salt that has been introduced into the flame. Let n be the number of free Na atoms per cm^3 in the flame, L the thickness of the flame in the direction of observation, and O the area of the flame surface observed by the photodetector. There are then nLO Na atoms in the volume element LO of the flame, which is supposed to be homogeneous. The atoms collide with the molecules of the flame gases and in the collisions energy can be transferred from the flame molecules to the Na atoms. As a result Na atoms in the ground state go over from this state into any one of the excited states indicated in figure 7, while at the same time excited Na atoms after quenching collisions fall back into the ground state. If thermal equilibrium is assumed to be established in the flame (see also § 6.7.4), then statistical mechanics yields the following relationship between the number n^* of Na atoms per cm^3 in excited level q and the total number n of Na atoms per cm^3:

$$n^* = n \frac{g_q}{Z} \exp \left(\frac{-E_q}{kT} \right).$$ (6.3)

This is known as *Boltzmann's equation* in which E_q is the excitation energy (see table 4), k is Boltzmann's constant, T is absolute temperature, g_q is the statistical weight of the excited level, and Z is the partition function or state sum, defined as:

$$Z \equiv \sum_{n=0}^{n=\infty} g_n \exp \left(\frac{-E_n}{kT} \right),$$ (6.4)

where the summation is taken over all the possible energy levels, including the ground level ($n = 0$). The energies, as usual, are referred to the ground level. If the energy of the lowest excited level of the atom is large compared with kT, as in the Na atom, then Z can be satisfactorily approximated to the first term (g_0) in the summation. In this case the number of atoms in the ground level is practically the same as the total number n, whereas for all excited levels $n^* \ll n$. For atoms such as those of chromium the ground term contains a set of several closely spaced levels. The partition function is then temperature-dependent and the number of the atoms in the ground level could be appreciably smaller than the total number n (see also § 6.5.1). Tables of calculated partition functions for atoms and ions at different flame temperatures are given in [368] and the possible practical importance of the partition function for the choice of reference element (see Glossary) in the emission method is discussed in [194].

We are not going to elaborate here on the statistical weights. According to theory they are related to the quantised orientations of the atom relative to a (hypothetical) magnetic field. For an atomic level with a total electronic angular momentum quantum number J the relationship is $g = 2J + 1$; hence g is of the order of magnitude 1–10 (see the level diagram of figure 7).

106

For every optical transition from a higher to a lower level, including the ground level, there exists a characteristic, temperature- and concentration-independent quantity A, called the *Einstein transition probability* (for spontaneous emission). It indicates the probability of an excited atom per unit time falling into a lower level, for example, the ground level, under spontaneous emission of a photon. If there are $n*$ atoms/cm^3 in the excited state, in the mean a number $n*A$ of photons with energy $h\nu$ are emitted per unit time and per cm^3. Then the radiant flux Φ (see Glossary) emitted by the flame volume OL within a solid angle Ω (expressed in sr) and over the whole width of the spectral line of central frequency ν_0, is :

$$\Phi = Ah\nu_0 n \frac{g_q \exp\left(-E_q/kT\right)}{Z} OL \frac{\Omega}{4\pi}. \tag{6.5}$$

In the special case of a resonance line $E_q = h\nu_0$. The extremely small width of the spectral line ($\delta\lambda \ll \lambda_0$; see § 6.3) can be completely ignored in this equation which therefore expresses, in watts, the radiant flux emitted over the whole spectral width of the line. In the calculation of the radiant flux received by the photodetector, Ω is the solid angle of the radiation beam emerging from the flame and falling on the photodetector. The fraction of flux lost in the optical components (filters, lenses, etc) through absorption, scattering, etc, must then also be taken into account. With strong resonance lines some of the emitted photons are 'lost' through self-absorption in the flame, as will be described in § 6.6.1. It can be seen from equation (6.5) that the intensity ratio of the Na D_1 and D_2 lines, having practically equal values of A and E_q, must be the same as the ratio of their statistical weights, namely $1:2$.

The values of A of the atomic lines listed in table 4 can be simply derived from the corresponding values of gf by means of equation (6.22) quoted in § 6.6.1.

If the line intensity in a flame actually corresponds to equation (6.5) at the flame temperature T, we speak of *thermal radiation* [162, 176, 826, 851] (compare § 6.7.4). Thermal radiation must be clearly distinguished from black-body radiation which only occurs in perfect thermodynamical equilibrium in a cavity (see § 6.6.1).

With the excitation energy expressed in eV and the temperature in K, the exponential factor in equation (6.3) takes the form $10^{-5040E_q/T}$. Its value for the yellow Na line with $E_q = 2 \cdot 10$ eV at 2500 K is found to be $5 \cdot 8 \times 10^{-5}$ and for the Na line at 330 nm with $E_q = 3 \cdot 7$ eV it is 4×10^{-8}. These figures show that under otherwise identical conditions, such as temperature, etc, the intensities of these resonance lines diminish rapidly with decreasing wavelength. The Na 330 nm line is so much weaker than the yellow Na line for the further reason that its value of A is about a hundred times lower (compare equation (6.5)). The exponential factor decreases sharply with falling temperature also; at 2000 K for instance it is only 5×10^{-6} for the yellow Na line. Resonance lines arising from a transition from the lowest excited level to the ground level are usually the strongest lines in the spectrum (as for instance the yellow Na doublet). Such lines therefore are sometimes called *raies ultimes*, or *persistent lines*, because they are the last lines to disappear on continued dilution of the sample under analysis. Some first resonance lines of the most important elements together with their excitation energies are listed in table 4. The wavelengths of the other resonance lines are to be found in the tables of Appendix 2.

When two lines of the same atom result from thermal radiation, the flame temperature can be simply determined by measuring the ratios of the line intensities. For this purpose only the ratio of the parameters A, ν_0 and g_q as well as the difference of the values of E_q for the two lines have to be known; the quantities n and Z are not required. Appropriate corrections should be made to allow for the wavelength dependence of the spectrometer response and for possible self-absorption effects. The theory of this method and the error sources in work with non-uniform flames are discussed in [740–744].

If the flame temperature is known accurately (see § 6.6.3), the atomic concentration n can be found by means of equation (6.5) through an absolute measurement of Φ [366]. It is however easier to establish the relative atomic concentrations for two different elements from the intensity ratios of their respective lines.

The excitation energy E_q, the flame temperature T and the transition probability A are of course not the only important factors that govern the atomic line intensity for a given solution concentration. The intensity also depends on the volatilisation, dissociation and ionisation processes dealt with in the preceding chapters. The resonance line of uranium for instance, in spite of its low excitation energy of $1 \cdot 44$ eV, does not show up because the uranium molecules are difficult to dissociate in the usual flames.

The thermal intensities of individual molecular lines and the overall band intensities are described by equations similar to those given for atomic lines. Because of the complex structure of the molecular spectra the relationships here are however considerably more intricate. We are not going to discuss this problem; we will only point out that the intensity distribution within a band, in particular the position of the intensity maximum, is directly linked with the flame temperature. The relevant literature has been quoted in § 6.1.

6.5. The Strength of Atomic Absorption

6.5.1. Definitions. The Beer–Lambert law.
In the usual procedure of atomic absorption spectroscopy, 'monochromatic' radiation from a discharge lamp is passed through the flame. If the wavelength λ of this radiation is identical to that of a resonance line of the metal atoms present in the flame, some of the radiation is absorbed. The absorbed photons raise metal atoms to a higher energy level. The excited atoms mostly lose their energy again in quenching collisions with flame molecules and not through re-emission of a photon, so that the absorbed radiation energy is ultimately converted into heat (see § 6.7.2). Hence the intensity of the lamp emission is reduced after passage through the flame. We shall now investigate the quantitative relation between the metal atom concentration in the flame and the lamp intensity attenuation, which is the basis of quantitative atomic absorption spectroscopy.

Let I_t and I_0 be the intensities (see Glossary) of the lamp radiation after passage through the flame containing atomic vapour and through the blank flame respectively. The quantity $\tau \equiv (I_t/I_0)$ is called the *transmission factor* (or *transmittance*) of the flame containing atomic vapour, and the quantity $\alpha \equiv (1 - I_t/I_0)$ the *absorption factor* (or *absorptance*). The usual measure of the absorption is the *absorbance*, defined as $A \equiv \log_{10}(I_0/I_t)$. We assume that the thermal radiation of the metal vapour in the flame and the background radiation of the flame itself have been eliminated by suitable instrumentation or their effects allowed for by appropriate corrections.

To facilitate the mathematical treatment we assume a parallel light beam and a flame with rectangular cross section. The beam enters the narrower flame surface normally. The flame coloured by the metal vapour has, for every ray, the same thickness L in the direction of the light beam. Let n be the total concentration (number per cm^3) of the free metal atoms uniformly distributed in the flame. When the atomic concentration in the direction of the light beam varies with position x and is described by a function $n(x)$, then in the following equations nL is simply to be replaced by $\int_0^L n(x)\,dx$.

It is of fundamental importance for our further considerations that the emission line of the discharge lamp as well as the absorption line in the flame have finite, nonzero spectral widths, however small (usually of the order of 10^{-2} to 10^{-3} nm). Within these very narrow wavelength ranges the emission of the lamp and the absorption in the flame vary strongly and continuously with wavelength. For the time being we assume that the emission and the absorption lines have their peak value at the same central wavelength λ_0 (see § 6.3).

We now select some hypothetical infinitesimal interval λ to $(\lambda + d\lambda)$ inside the wavelength range of the emission line, so narrow that within it the absorption factor is practically constant and the intensity of the lamp is described by the product $(I_\lambda)_0\,d\lambda$. The quantity $(I_\lambda)_0$ is the wavelength-dependent spectral intensity of the (non-attenuated) lamp radiation; it is normalised per unit wavelength and is expressed for example in $erg\,s^{-1}\,nm^{-1}$ (see Glossary). By definition the relationship between spectral intensity and total intensity I_0 is $I_0 = \int (I_\lambda)_0\,d\lambda$, the integration being carried out over the full width of the emission line. If $(I_\lambda)_t$ is the spectral intensity of the transmitted radiation, then according to the *Beer–Lambert law*

$$(I_\lambda)_t = (I_\lambda)_0 \exp[-k(\lambda)L], \tag{6.6}$$

where $k(\lambda)$ is the *absorptivity* (often called absorption coefficient) at wavelength λ and is expressed for example in cm^{-1}. It increases proportionally with the concentration of the metal atoms in the lower energy level (usually the ground level) of the radiative transition. In thermal equilibrium this concentration is again given by Boltzmann's equation (6.3) and hence is proportional to the total concentration n of the free atoms in the flame.

For a resonance line therefore the absorptivity is proportional to the concentration n_0 in the ground level which, from equation (6.3), is $n_0 = n g_0/Z$. In the absence of low excited levels, we have $Z \approx g_0$ (see equation (6.4)) and thus $n_0 \approx n$. If Z is, however, appreciably larger than g_0, then n_0 is a (slightly) temperature-dependent fraction of n, as for instance in the case of Cr, V, Ni, Al and Tl [836]. For vanadium g_0/Z falls from 16 to 7 per cent when T is raised from 1500 to 6000 K [855]. This may have an adverse effect when hot plasmas are used in the absorption method [259].

If a non-resonance line is utilised for the absorption measurement, by Boltzmann's equation (6.3) the absorptivity depends exponentially on the excitation energy E_p of the lower energy level and on the temperature. For the non-resonance line of the iodine atom at 206·2 nm this excitation energy is 0·94 eV; even at $T = 3000$ K the relative population of the lower level is only $1\frac{1}{2}$ per cent [845] (data for Sn and Pb at $T = 2800$ K are to be found in [411]; see also [283] for some Sc, Ga, In and Tl non-resonance lines). Calculated relative populations of lower levels are tabulated in [686] for a number of elements at different temperatures and predicted relative absorptions are compared with experimental values for a number of lines, either resonance or non-resonance. Yet on occasion non-resonance lines with an excitation energy E_p of about 1 eV have successfully been used in absorption spectroscopy with hot flames (e.g. for Sn [179], Sc [283], Fe [213] and P [616, 845]). For the Al line at 309·3 nm the lower level lies about 0·014 eV above the ground level (see table 4). Since the statistical weight of this lower level is 4 and that of the ground level is 2, the population of the former is even larger than that of the ground level. Other factors on which the absorptivity also depends will be discussed in § 6.5.2 and § 6.5.3.

The wavelength dependence of the absorptivity $k(\lambda)$ entails a wavelength dependence of the absorption factor $\alpha(\lambda)$, defined here for an infinitesimal wavelength interval within the line profile, for according to the Beer–Lambert law

$$\alpha(\lambda) \equiv [(I_\lambda)_0 - (I_\lambda)_t]/(I_\lambda)_0 = 1 - \exp[-k(\lambda)L]. \tag{6.7}$$

The spectral profiles of the absorption line, $\alpha(\lambda)$, and of the lamp emission line before and after passing through the flame, $(I_\lambda)_0$ and $(I_\lambda)_t$, are illustrated in figure 9 for some typical cases. Also shown is the graph of the absorptivity $k(\lambda)$ as a function of λ, which reaches its maximum k_m at the central wavelength λ_0.

For an optically thin flame, that is for $k_m L \ll 1$, series expansion of the exponential factor gives $\alpha(\lambda) \simeq k(\lambda)L$. The spectral profiles of $\alpha(\lambda)$ and $k(\lambda)$ are then alike and the corresponding spectral widths are equal and independent of n because $k(\lambda)$ grows strictly proportionally to n. For an optically thick flame ($k_m L \gg 1$), in the line centre $\alpha(\lambda)$ approaches its limit value of 1. Its profile is then flattened and its spectral width increased; the latter increases further the greater the optical thickness of the flame, that is the larger the value of $k_m L$ or the higher the atomic concentration.

The spectral resolution of the instruments usually employed in flame spectroscopy in most cases is insufficient for absorption measurements in such an infinitesimal interval $d\lambda$ within the emission line profile. The measurements in fact take in the total intensities I_t and I_0 integrated over the full width of the emission line. Their ratio is given by:

$$\frac{I_t}{I_0} = \frac{\int (I_\lambda)_t \, d\lambda}{\int (I_\lambda)_0 \, d\lambda} = \frac{\int (I_\lambda)_0 \exp[-k(\lambda)L] \, d\lambda}{\int (I_\lambda)_0 \, d\lambda}. \tag{6.8}$$

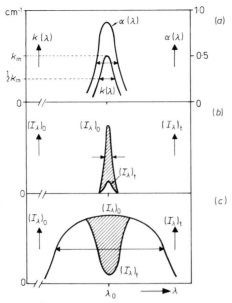

Figure 9. Spectral profiles of a strong absorption line in the flame (a) and of two typical emission lines of the background source (b and c). $\alpha(\lambda)$ and $k(\lambda)$ are the absorption factor and the absorptivity respectively, and $(I_\lambda)_0$ and $(I_\lambda)_t$ the spectral intensities of the lamp radiation before and after passage through the flame respectively. (b) illustrates a typical narrow emission line and (c) a typical broad emission line or the radiation from a continuum source used with a monochromator. The hatched areas represent the total absorbed radiation. The half-intensity widths are indicated by horizontal arrows.

The theoretical relationship between the measured quantity I_t/I_0 and the atomic concentration n in the flame, to which $k(\lambda)$ is proportional, is generally not easy to find from this equation. For such a derivation the wavelength-dependent functions $(I_\lambda)_0$ and $k(\lambda)$ have to be known, which depend on the conditions in the lamp and in the flame. In § 6.5.2 and § 6.5.3 we therefore shall only deal with two typical borderline cases which will suffice to demonstrate the most important factors determining the strength of the measured absorption.

6.5.2. Absorption of monochromatic radiation. We shall consider quantitatively the special case in which the half-intensity width of the emission line in the lamp is considerably smaller than that of the absorption line in the flame and in which the shift and hyperfine structure of the absorption line are negligible (see § 6.2.1 and § 6.3). If there is only Doppler broadening in the lamp, and no self-absorption (see § 6.6.1), the width of the lamp line at the most is about one-third of the absorption linewidth, which is equal to the width of the flame emission line (see figure 12). For the wavelength range in which the lamp emission line still has a perceptible spectral intensity $(I_\lambda)_0$ we are then justified in taking the absorptivity $k(\lambda)$ to be constant within an error margin of 10 per cent and equal to its peak value k_m. Equation (6.8) then simply becomes:

$$\frac{I_t}{I_0} = \frac{\exp\left(-k_m L\right)\int(I_\lambda)_0\, d\lambda}{\int(I_\lambda)_0\, d\lambda} = \exp\left(-k_m L\right). \tag{6.9}$$

A more quantitative discussion of the accuracy of this approximation is to be found in [902, 914]. Hence in the case of a narrow lamp line the absorbance A_1 is given by

$$A_1 \equiv \log_{10}(I_0/I_t) = (k_m L)\log_{10}e$$

$$= 0{\cdot}434 k_m L. \tag{6.10}$$

If the absorption line only exhibits Doppler broadening, the following relationship between the peak value $(k_m)_D$ of the absorptivity, the Doppler width $(\delta\lambda)_D$ and the concentration n_p of the atoms in the lower energy level of the absorption transition follows from theory [79]:

$$(k_m)_D = \frac{2(\pi\log_e 2)^{1/2}e^2\lambda_0^2 n_p f}{(\delta\lambda)_D mc^2}. \tag{6.11}$$

Here m and e are the mass and the absolute charge of the electron respectively, c is the velocity of light and $(\delta\lambda)_D$ the Doppler half-intensity width of the absorption line (see equation (6.1)). The concentration n_p is given by Boltzmann's equation (6.3) and for resonance lines equals the atomic concentration n_0 in the ground level. The constant f, called the *oscillator strength* (for absorption of radiation), is characteristic for the considered radiative transition and is proportional to the Einstein transition probability A for the reverse radiative transition† (see equation (6.22)). For strong absorption lines, such as K 766 nm, Na 589 nm, Cu 325 nm and Hg 185 nm, the quantity f (with dimension unity) is of the order of magnitude one. If λ and $(\delta\lambda)_D$ are expressed in cm and n_p in cm^{-3}, then equation (6.11) can also be written in the form:

$$(k_m)_D = 8{\cdot}4 \times 10^{-13}\frac{\lambda_0^2 n_p f}{(\delta\lambda)_D}\ (cm^{-1}). \tag{6.11a}$$

The f-values for some important atomic lines can be derived from the entries for gf in table 4, g being the statistical weight of the lower level of the radiative transition. Evidently the intensity of an absorption line is proportional to gf because according to Boltzmann's equation (6.3) the atomic concentration in the lower level (n_p) is proportional to the corresponding value of g.

For the Na resonance line at 589·0 nm with $f = 0{\cdot}65$ and $(\delta\lambda)_D = 4{\cdot}5 \times 10^{-3}$ nm $= 4{\cdot}5 \times 10^{-10}$ cm and a flame thickness $L = 10$ cm we find from equation (6.11a) that $(k_m)_D L = 4{\cdot}3 \times 10^{-11}\, n_0$. If, for instance, on nebulisation of a solution containing 1 mg Na per litre about 10^{10} free atoms per cm^3 are assumed to be in the flame (compare § 4.6.2), then according to equation (6.10) the (peak) absorbance A_1 is

---

† Note that the officially adopted symbol A stands for both the transition probability and the absorbance.

expected to be 0·19 and the (peak) absorption factor α_1 35 per cent. A solution concentration of 0·025 mg Na per litre would then give an absorption of 1 per cent. This special concentration value is called the *characteristic concentration* (see Glossary). Through lengthening the absorption path in the flame the characteristic concentration can be reduced further.

Under the stipulated conditions the absorbance is thus proportional to the atomic concentration in the flame. The percentage absorption factor $(1-I_t/I_0) \times 100 = (1-\exp(-k_m L)) \times 100$ on the other hand asymptotically approaches the limit of 100 per cent when n and with it k_m become very high. Only at lower concentrations is the absorption factor approximately proportional to k_m and hence also to n; we then have $\alpha_1 \times 100 \approx k_m L \times 100 = 230 \times A_1$ (see equation (6.10)). Tables have been compiled for the conversion of α to A in [92]. The concentration dependence of α and A is discussed further in § 8.4.

In the special case of a relatively narrow emission line the width and the spectral profile of the emission line do not enter into the relationship between absorbance and metal concentration.

When the absorption line in the flame also shows Lorentz broadening as well as Doppler broadening, then under otherwise identical conditions the maximum value k_m of the absorptivity is necessarily lowered. This holds because the integral of the absorptivity over the full linewidth, $\int k(\lambda) d\lambda$, is strictly independent of the line broadening (see equation (6.15) in § 6.5.3). An increase in the linewidth then entails a reduction in the maximum value k_m. Consequently, again under the assumption of a narrow lamp line, according to equation (6.10) the absorbance is also decreased.

Since with pure Doppler broadening the spectral line profile is very different from that with pure Lorentz broadening (see § 6.3), in general k_m cannot be derived from equation (6.11) for $(k_m)_D$ by simply substituting there $((\delta\lambda)_D + (\delta\lambda)_L)$ for $(\delta\lambda)_D$. By using the a-parameter (see equation (6.2)) the following approximation can be made with an error margin of about 10 per cent:

$$k_m = (k_m)_D \frac{(\delta\lambda)_D}{(\delta\lambda)_D + (\delta\lambda)_L} = \frac{(k_m)_D}{1+1\cdot2a} \qquad \text{if } 0 < a < 2, \qquad (6.12)$$

or

$$k_m = \frac{(k_m)_D}{a\sqrt{\pi}} = \frac{0\cdot56(k_m)_D}{a} \qquad \text{if } a > 2. \qquad (6.12a)$$

More detailed information on the ratio $k_m/(k_m)_D$ as a function of a can be found in [79, 182]. For the Na line at 589 nm with a value of a of approximately 0·5, measured in an air–acetylene flame [455], we obtain $k_m = 0\cdot62(k_m)_D$. The characteristic concentration for Na calculated before has now deteriorated by a factor of $1/0\cdot62$ (compare also [914]).

It is to be noted that the a-parameter does not depend on the atomic concentration n itself because resonance broadening can be neglected in flames (compare § 6.3). Hence the absorbance and n are also proportional in the presence of Lorentz broadening.

113

In the case of a resonance line, with $n_0 \approx n$, A_1 is altered by less than 10 per cent when there is a change in T of about 250 °C and nL remains constant. This constitutes a marked difference between the atomic absorption and the atomic emission methods. The possible temperature dependence of nL, of course, has the same effect in both methods if the same flame were used.

The temperature dependence of the absorbance A_1 for a narrow lamp line and a given total atomic concentration n and flame thickness L can be described by means of a combination of equations (6.10), (6.12), (6.11), (6.3) and (6.1). For resonance lines and with $n_0 \approx n$, A_1 depends on T only through the linewidth. If $a \ll 1$ it follows from equations (6.10), (6.11) and (6.1) that A_1 is proportional to $1/\sqrt{T}$. If, on the other hand, $a \gg 1$ (i.e. with predominant Lorentz broadening), according to equations (6.10), (6.12a), (6.11) and (6.2), A_1 is proportional to $(\delta\lambda)_L^{-1}$. By the classical Lorentz theory, at constant gas pressure $(\delta\lambda)_L$ should be proportional to $1/\sqrt{T}$[75, 79], but a more exact theory shows that the temperature dependence of the Lorentz width also depends on the particular kind of interaction between the atom and the collision partner. Although reliable experimental results are not available (compare [75]), it may be assumed in fact that $(\delta\lambda)_L$ does not vary much more with T than postulated by the classical theory.

For non-resonance lines the absorption factor changes with temperature not only because of the temperature dependence of the linewidth (see above), but also because the relative population of the lower energy level is affected by the temperature (see § 6.5.1). Since the population is rather sensitive to temperature changes, if the excitation energy of the lower level markedly exceeds kT, the temperature can be found from the ratio of the absorbances of a resonance and a non-resonance line of the same atom [233, 601, 605, 633].

It follows from equations (6.11), (6.1) and (6.3) that for a given element and a given flame the absorbances of the various lines (resonance and non-resonance) vary as $\lambda_0 f g_p \exp(-E_p/kT)$, where g_p and E_p are the statistical weight and the energy of the lower level respectively. The line of the background source is again taken to be narrow compared with the absorption line and only Doppler broadening is assumed in the latter. A more general theoretical expression is given in [601]. The theoretical relationship between the absorbance and the expression just given (apart from the rather insignificant factor λ_0) has been confirmed experimentally for several Co and Fe lines in [624]. Calculated and measured relative absorbance values for various Sc lines are compared in [283].

When the lamp line and the absorption line have comparable widths, the line profile of the former must be explicitly introduced into the calculation of the absorbance. Such calculations in general have to be numerical [732].

The width of the lamp line, if not too large, can also be approximately taken into account in the calculation by introducing, instead of k_m, an absorptivity $\bar{k}$ averaged over the width of the lamp line [499]. The value of $\bar{k}$ as a function of the ratio of the half-intensity widths of the emission and the absorption line can be evaluated roughly [379], but it has to be kept in mind that strictly speaking the ratio $\bar{k}/k_m$ is not independent of $k_m L$ and hence of the atomic concentration. The effect of the width of the lamp line on the absorption measurement has also been numerically

114

calculated with a computer [477]. The sometimes-adopted simplifications and the disregard of the possible line shift however often render such calculations rather unreliable.

When a spectral line source is used, the theoretical calculation of the absorbance may be complicated by the occurrence of a shift in the absorption line centre with respect to the emission line (see § 6.3 and [364, 602, 871, 898, 931]). Whether such a line shift actually has an appreciable effect can be deduced not only from line profile measurements but also indirectly by comparing the absorption measurements obtained with a line source and those taken with a continuum source [931].

The influence of the line shift can be readily visualised from figure 9 by supposing the centre of the absorption line (figure 9a) to be displaced relative to the centre of the (narrow) emission line (figure 9b). It is at once clear that a large reduction of the measured absorbance as a result of the line shift will only be found if the relative displacement is at least as great as the width of the absorption line. As this is not so however (see § 6.3 and figure 12), the resulting lowering of the absorbance is probably much less than 50 per cent. For lines with negligible hyperfine structure, such as the Ca resonance line, the effect of the finitely small linewidth of the hollow-cathode lamp (at low lamp current) and of the line shift on the absorbance in most cases is expected to be only of the order of 10 per cent [235, 602, 871] (see also figure 12). The effect of hyperfine structure will be discussed at the end of § 6.5.3.

Because the absorbance A_1 is dependent on the atomic concentration and also on the oscillator strength and on the a-parameter, any one of these quantities can be determined, absolutely or relatively, from absorption measurements if the other two are known. The possibilities of absolute concentration measurements are discussed in [75, 366, 602, 732, 898] and of measurements of the oscillator strength in [75, 593, 596, 599, 600]. For these purposes the absorption can also be measured with a continuum source (see § 6.5.3); in this case the absorption line-width does not have to be known, but the bandwidth of the monochromator used enters into the calculations. Through comparing absorption measurements with a line source and with a continuum source one can find the width of the absorption line and from it the a-parameter (see § 6.3).

6.5.3. Absorption of continuous radiation. After the special case of a comparatively narrow emission line from a spectral line source, dealt with in § 6.5.2, we shall now deal with the other extreme case – that of continuous radiation (see Glossary). The formulae for a continuous-radiation source also apply to a spectral line source when the emission line is excessively broadened by self-absorption (see § 6.6.1); in the case of a comparatively narrow absorption line such a line source effectively behaves as a continuum source. Thus the example of a K lamp emitting a line of a half-intensity width which was 2·5 times the width of the red K line in the flame has been reported [711]. A similar situation occurs when an auxiliary flame radiating a strongly self-absorbed resonance line is used as a background source [164].

We thus assume here generally that the spectral intensity $(I_\lambda)_0$ of the lamp (defined in § 6.5.1) in the wavelength range in which the metal vapour in the flame effectively absorbs, is practically independent of wavelength and equal to its value $I_{\lambda 0}$ at the absorption line centre. We also suppose that the spectral bandwidth of the monochromator used considerably exceeds this wavelength range, as it usually does. These conditions are more strictly formulated in [373, 601, 902].

In the special case now considered the intensity of the radiation absorbed in the flame is given by (compare equation (6.8)):

$$I_0 - I_t = \int (I_\lambda)_0 \{ 1 - \exp[-k(\lambda)L] \} \, d\lambda$$

$$= I_{\lambda 0} \int \{ 1 - \exp[-k(\lambda)L] \} \, d\lambda. \tag{6.13}$$

This quantity is proportional to the hatched area in figure 9(c). For an optically thin flame, that is for a low value of n, it approximates to (compare § 6.5.1):

$$I_0 - I_t \simeq I_{\lambda 0} L \int k(\lambda) \, d\lambda. \tag{6.14}$$

For the integral, classical dispersion theory gives

$$\int k(\lambda) \, d\lambda = \frac{\pi e^2}{mc^2} \lambda_0^2 f n_p, \tag{6.15}$$

where n_p is the atomic concentration in the lower energy level (usually the ground level) and the other symbols are those used in equation (6.11). For low concentration ranges one can therefore write:

$$I_0 - I_t \simeq \left(\frac{\pi e^2}{mc^2} \right) \lambda_0^2 I_{\lambda 0} L f n_p. \tag{6.16}$$

Thus in this range the absorption is again proportional to the product of n_p (or n) and L. In contrast to the previous special case the constant of proportionality here contains neither the half-intensity width nor the spectral profile of the absorption line and furthermore it is strictly independent of the flame temperature.

Let us now compare the absorption factor, that is the ratio of the total absorbed intensity, represented by the hatched area in figure 9(c), to the total, non-attenuated intensity represented by the total area under the $(I_\lambda)_0$ curve, with the corresponding ratio in figure 9(b). The absorption factor α evidently must be smaller for a continuous background radiation or a very broad emission line than for a relatively narrow emission line. Therefore in the former case the characteristic concentration (see Glossary) must be higher.

It is easy to evaluate this deterioration in the characteristic concentration. Division of the right-hand side of equation (6.16), valid for low concentration values, by I_0 then gives an expression for the absorption factor for continuous radiation: $\alpha_c \equiv (I_0 - I_t)/I_0$. If now $I_0/I_{\lambda 0} \equiv (\Delta\lambda)_{eq}$ is defined as the equivalent width of the emission profile in figure 9(c) (approximately equal to the half-intensity width) we get $I_0 = I_{\lambda_0}(\Delta\lambda)_{eq}$, and hence

$$\alpha_c = \frac{I_0 - I_t}{I_0} = \frac{\pi e^2}{mc^2} \frac{\lambda_0^2 n_p f L}{(\Delta\lambda)_{eq}}. \tag{6.17a}$$

For a relatively narrow emission line and an absorption line with Doppler broadening $(\delta\lambda)_D$ only, on the other hand, within the range of low concentrations it follows from equations (6.9) and (6.11), after series expansion of the exponential factor, that:

$$\alpha_l = \frac{I_0 - I_t}{I_0} = 2\left(\frac{\log_e 2}{\pi}\right)^{1/2} \frac{\pi e^2}{mc^2} \lambda_0^2 \frac{n_p f L}{(\delta\lambda)_D}. \tag{6.17b}$$

The ratio of the two expressions is

$$\frac{\alpha_c}{\alpha_l} = \frac{1}{2}\left(\frac{\pi}{\log_e 2}\right)^{1/2} \frac{(\delta\lambda)_D}{(\Delta\lambda)_{eq}} = 1 \cdot 06 \frac{(\delta\lambda)_D}{(\Delta\lambda)_{eq}}. \tag{6.18}$$

This ratio is a measure of the deterioration of the characteristic concentration that results from using a continuous background source instead of a comparatively narrow line source. With a continuum source $(\Delta\lambda)_{eq}$ is determined by the bandwidth of the chosen monochromator.

With continuously increasing values of n_p the absorption factor attains its limit value of 100 per cent earlier with the use of a narrow line source (figure 9b) rather than with the use of a continuous-radiation source (figure 9c). When in the latter case the absorption at the line centre approaches the value of 100 per cent, the absorption in the wings of the absorption line is still considerably below 100 per cent and continues to increase with rising concentration. The absorption line, represented by $\alpha(\lambda)$ in figure 9(a), then gains more in width than in height so that its half-intensity width increases†. With a continuous-radiation source the growth of the absorption still manifests itself in the absorption line wings and the total absorption increases further, although more slowly than in proportion with n. When, finally, the half-intensity width of the absorption line has become greater than the half-intensity width of the background radiation or the bandwidth of the monochromator, the total absorption approaches the limit value of 100 per cent (see also [164]).

Unlike the case of a narrow line source, with a continuum source there is no strict proportionality between absorbance and concentration. As for the absorption factor α_c, proportionality only prevails approximately in the range of low concentrations, that is, for small optical thicknesses (compare equation (6.16)). At higher concentrations the theoretical relationship between absorbance or absorption factor and concentration is very intricate and also dependent on the value of the a-parameter [164, 711]. The dependence of the total absorption on the atomic concentration is discussed further in § 6.6.1 and chapter 8.

---

† Note that the profile and the width of the absorptivity graph $k(\lambda)$ in figure 9(a) is strictly independent of the atomic concentration (see § 6.5.1).

In a given flame and with the use of a continuum source, the absorption factors of different lines, resonance or non-resonance, of the same atomic species vary as $(\lambda_0^2 f/(\Delta\lambda)_{eq})g_p \exp(-E_p/kT)$ within the range of low atomic concentrations. This follows from equations (6.17a) and (6.3); here the symbols g_p and E_p again denote the statistical weight and the energy of the lower level respectively. The relationship has been experimentally confirmed for a number of resonance lines of the iron atom [625]. The temperature of the flame and of the furnace etc, can again be determined from the ratio of the absorption factors of two lines with different values of E_p [233] (compare also § 6.5.2). In order to obtain reliable results the thermal emission of the atomic vapour should be eliminated in the absorption measurements, for example, by chopping the lamp radiation to allow AC measurements. This is especially important when, to find the temperature of an atomiser, a thermal background source such as a tungsten lamp is used, the temperature of which is comparable to that of the atomiser.

The theoretical considerations of § 6.5.2 and § 6.5.3 are only correct if certain simplifications are made, two of which will now be briefly outlined. Strictly speaking the refractive index should be taken into account in equations (6.11) and (6.15) for $k(\lambda)$. This index in fact is exactly unity at the line centre but due to the anomalous dispersion it may change rapidly around the centre. The deviation from unity is proportional to the atomic concentration. Calculations however show that under the usual conditions of flame spectroscopy these deviations are negligible [850].

So far we have dealt with single, isolated emission or absorption lines. Actually, due to (hyper-)fine structure and isotope effects, the atomic lines may be split into a number of components lying close together (see § 6.2.1). Because the oscillator strengths f of the various components and the concentrations of the isotopes are not all the same, the given absorption equations must be modified by introducing summations when the components are not resolved by the spectrometer. Such summations are fairly simple provided that the components do not overlap in the spectrum at all or that the overlap is complete [914]. In the case of partial overlapping considerable theoretical complications arise, the solution of which is dealt with in [198, 850, 861, 898]. These effects are in most cases fairly small and practically do not create much difficulty because, being constant, they do not affect the reproducibility of atomic absorption analysis. They may however possibly necessitate corrections in the theoretical equations [602, 898]. The theoretical aspect of the isotope effect in atomic absorption measurements is more fully discussed in [75].

6.6. Self-absorption and Self-reversal. The Line-reversal Method

6.6.1. Self-absorption. The absorption of radiation is of course an essential feature of atomic absorption spectroscopy but absorption also plays a part in atomic emission spectroscopy as an interfering side-effect on the intensity of strong resonance lines in high atomic concentration ranges. For there is a certain

118

probability that the photons produced inside the flame, for instance those of the first Na resonance doublet, will be re-absorbed in the flame by other Na atoms in the ground state before they can reach the photodetector. These latter atoms thereby become excited but since an excited atom loses its excitation energy in the usual flame conditions, normally through a radiationless transition in a quenching collision, the re-absorbed photons are lost, that is, ultimately converted into heat (see also the final part of § 6.6.2 on radiation diffusion). This loss is called *self-absorption*. Its probability increases with rising Na concentration and with extending flame thickness, and its effect is an attenuation of the radiation finally emerging from the flame, especially in the line centre for which the absorptivity is greatest (see § 6.5). Consequently the line profile is flattened near the centre and the half-intensity width of the emission line is enlarged. In the higher concentration ranges the total radiation intensity is then no longer proportional to n, but with continuing concentration rise approaches proportionality to $\sqrt{n}$ (see figure 15). In the intermediate range the relationship with n depends on the a-parameter of the line. Evidently equation (6.3) then no longer holds for the thermal radiation intensity and should be corrected for the radiation loss through self-absorption.

In metal-vapour discharge lamps also, self-absorption can occur in resonance lines when the metal-vapour pressure is sufficiently high. In this case, because of the much lower total gas pressure, the self-absorption will be accompanied by radiation diffusion (see also the final part of § 6.6.2). Here similarly it restricts the rise in the lamp radiation at the line centre and leads to a greater half-intensity width when the vapour pressure in the lamp is raised through a change in the operating conditions [840] (see figure 12). These effects are significant for the optimum setting of the lamp current [438].

The spectral distribution of the photons emitted by the atoms inside the flame or the lamp must be taken into account for a more detailed understanding of the self-absorption effect. This effect can be evaluated by considering consecutive thin flame zones at increasing distances from the flame surface. The fraction of the radiation originating from a particular zone that is lost through self-absorption in the following zones can then be calculated as a function of the wavelength from the absorptivity $k(\lambda)$. Integration over all the zones and over the entire spectral width of the emission line then yields the resulting total radiation [162].

For a thermally excited line and for a flame with uniform temperature distribution the same result can be obtained more directly by applying Kirchhoff's law, according to which in thermal equilibrium the absorption factor $\alpha(\lambda)$ of the radiating medium and its thermal spectral intensity are linked at any wavelength. Let Φ_λ be the spectral radiant flux (expressed e.g. in W nm^{-1}; see Glossary) emitted from a surface area O per unit wavelength in a solid angle Ω at right angles to the surface. Also let $B_\lambda^P(T)$ be Planck's spectral radiance for a black body (expressed e.g. in W cm^{-2} sr^{-1} nm^{-1}; see Glossary), which to good approximation is given by Wien's law as a function of wavelength and temperature in the form

$$B_\lambda^P(T) \simeq 2hc^2\lambda^{-5} \exp\left[-(hc/\lambda kT)\right]. \qquad (6.19a)$$

The approximation is fully justified for optical radiation at flame temperatures because here $(hc/\lambda kT) \gg 1$. From Kirchhoff's law we have that

$$\Phi_\lambda = \alpha(\lambda)B_\lambda^P(T)O\Omega, \tag{6.19b}$$

where $\alpha(\lambda)$ and $B_\lambda^P(T)$ are defined by equations (6.7) and (6.19a) respectively and T is the flame temperature.

Integration over all wavelengths and application of equation (6.7) then gives the total radiant flux of the emission line (expressed in W) as

$$\Phi = \int \Phi_\lambda \, d\lambda = B_{\lambda_0}^P(T)O\Omega \int \{1 - \exp[-k(\lambda)L]\} \, d\lambda. \tag{6.20}$$

For this integration Planck's spectral radiance is taken to be virtually constant over the entire linewidth and equal to its value at the central wavelength λ_0.

As can be inferred from equation (6.19b), the spectral profile of the emission line is the same as the profile of the corresponding absorption line, that is, $\alpha(\lambda)$. In particular the total radiant flux of the emission line is proportional to the area under the $\alpha(\lambda)$ curve in figure 9(a). All that was said in § 6.5.3 about the profile and the half-intensity width of the absorption line is therefore directly applicable to the (thermal) emission line. Because the integral

$$\int \{1 - \exp[-k(\lambda)L]\} \, d\lambda$$

appears in equation (6.13) as well as in equation (6.20) the total intensity of the emission line and the total absorption with a continuum background source as functions of (nfL) behave in the same way. The absorptivity $k(\lambda)$ contains the product (nf) as a factor. The curves representing the total intensity (with self-absorption taken into account) and the total absorption (with a continuum source) as functions of (nfL) are called *theoretical curves of growth*. Their shape depends on the a-parameter and is similar to that of the experimental analytical curves if corrections are applied for certain other curvature effects (see figure 15 and chapter 8). The dependence of this shape on the $k(\lambda)$ profile is discussed in [495a].

With the help of the exposition above the self-absorption effect in resonance lines can now be described in more detail. If the exponential factor in equation (6.20) is again expanded in a series for low concentration values, the following approximate result is obtained by the use of equation (6.15):

$$\Phi = B_{\lambda_0}^P(T)\left(\frac{\pi e^2}{mc^2}\right)\lambda_0^2 n_0 fLO\Omega. \tag{6.21}$$

In this range the self-absorption is evidently still negligible and Φ is proportional to n_0. The spectral profile of the emission line is then similar in shape to the $k(\lambda)$ profile and the half-intensity width is independent of n_0 since for every value of λ, $k(\lambda)$ is strictly proportional to n_0 (see also figure 10).

The asymptotic expression for Φ in the form of equation (6.21) must be identical with that for the thermal radiant intensity Φ (without self-absorption) in equation (6.5). If these two expressions are equated, the well known relation between the oscillator strength f for absorption and the Einstein transition probability A for

120

emission of a resonance line is found by means of equation (6.19a), with $c/\lambda = \nu$,

$$\frac{f}{A} = \left(\frac{mc}{8\pi^2 e^2}\right)\left(\frac{g_q}{g_0}\right)\lambda_0^2 = 1\cdot 51\left(\frac{g_q}{g_0}\right)\lambda_0^2, \qquad (6.22)$$

with λ_0 expressed in cm and A in s^{-1}. It can be seen from this equation that the values of f of the Na doublet lines are in the same ratio as the corresponding values of g_q, namely $2:1$, since the values of A are practically equal. As a consequence of the connection between the f-value and the absorptivity $k(\lambda)$ (compare equations (6.11) and (6.15)) the doublet line that appears as the stronger emission line, having the greater value of g_q, is generally expected to show a greater self-absorption effect also. Thus if there is strong self-absorption in a Na discharge lamp the intensity ratio of the doublet lines is found to be less than $2:1$, the value anticipated in the absence of self-absorption [438, 617, 783].

Attempts have even been made to determine solution concentrations by flame emission spectroscopy by measuring the intensity ratio of two lines of the analyte with different f-values [774]. The disparity in the f-values causes a difference in the self-absorption effect in the two lines, which depends on the concentration to be determined. This, like the absorption method, is practically independent of the flame temperature if the temperature dependence of the dissociation of the metal compounds in the flame, etc, is disregarded.

From equations (6.19b) and (6.20) we observe that for sufficiently high metal concentration in the flame (above about 10^{12} cm^{-3} Na atoms for the Na D doublet with a flame thickness of about 1 cm) the flame appears optically thick at the line centre (see § 6.5.1). The wavelength-dependent absorption factor $\alpha(\lambda)$ near $\lambda = \lambda_0$ then attains its limit value of unity and the spectral radiant flux Φ_λ of the thermal radiation by equation (6.19b) approaches the Planck value $B_{\lambda_0}^P(T)O\Omega$, which fundamentally it cannot exceed. Under this limitation the spectral radiant flux at the line centre hardly increases any further with rising concentration; however it does continue to increase in the line wings where the flame is still optically thin. The emission line profile thus only spreads laterally and its half-intensity width grows with rising concentration†. This is illustrated in figure 10, whereas figure 11 shows the changes of line profiles, determined experimentally, with solution concentration. More recent interferometric measurements of the half-intensity width of the Ca emission resonance line in air–acetylene and nitrous oxide–acetylene flames as a function of the solution concentration and of the flame thickness are described in [544].

With a further rise in the concentration the wavelength range within which the emission line virtually appears optically 'black' (i.e. $\alpha(\lambda) \approx 1$) thus continues to expand. Hence the total intensity can increase without limitation as the concentration increases. This rise is not proportional to n but within a restricted concentration range varies proportionally to n^p, with $p < 1$. Above a sufficiently high concentration (for the Na D doublet about 10^{13} cm^{-3} atoms for $L = 1$ cm) p stays constant and equal to $\frac{1}{2}$; in the intermediate range p may even be smaller than $\frac{1}{2}$, but

---

† The term 'concentration broadening' is sometimes used in the literature [369].

121

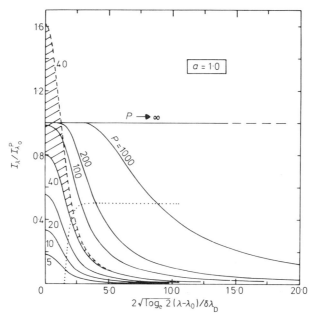

Figure 10. Computer plot of spectral profiles of a resonance emission line with a-parameter $= 1{\cdot}0$ for different atomic concentrations, showing the effect of self-absorption on the spectral linewidth. Only the right half of the profiles, which are symmetrical about the line centre λ_0, is shown. The ratio of the spectral intensity, I_λ, to the spectral intensity for a black body, $I_{\lambda_0}^P$, at the flame temperature is plotted against $2\sqrt{\log_e 2}$ times the wavelength distance from the line centre, divided by the Doppler width, $(\delta\lambda)_D$. In the ground state the atomic concentration n_0 is proportional to the value of P (in cm^{-2} s) indicated for each curve. ($P \equiv (\sqrt{\log_e 2}/\pi c)\lambda_0^2 n_0 f L(\delta\lambda)_D$, where L is the flame thickness and the other quantities are as in equation (6.11a).) If P tends to infinity, the profile asymptotically approaches the horizontal line marked $P \to \infty$ and the flame emission assumes black-body radiation characteristics. The dotted curve is the locus of the points on the profiles corresponding to half peak intensity. Note that the half-intensity width is concentration dependent and that the centre section of the profiles flattens out at high atomic concentrations. The broken curve shows the profile one would expect for $P = 40$ were there no self-absorption. The hatched area between the broken curve and the full curve for $P = 40$ represents half of the loss in total line intensity due to self-absorption in this particular example. (Reproduced from [495].)

only when the a-parameter is less than about 1 (see figure 15). The a-parameter, as well as the absolute value of n per unit of solution concentration, can be determined experimentally from the relative dependence of the total intensity (or total absorption; see above) on the metal concentration in the nebulised solution [160, 200, 448, 453, 455, 488, 610, 850, 916]. Under certain conditions this concentration may be taken to be proportional to the atomic concentration in the flame. We shall again consider the shape of these curves of growth when dealing with the analytical curves in § 8.3. For a further discussion of the theory see [79, 495a].

According to equation (6.20), the temperature dependence of the intensity of a resonance line with self-absorption is primarily determined by the Planck factor

122

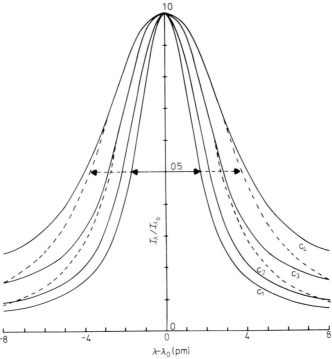

Figure 11. Experimental curves showing the normalised spectral intensity, I_λ/I_{λ_0}, of the 460·7 nm Sr resonance line as a function of wavelength distance from the line centre, λ_0 (1 pm = 10^{-3} nm). Curves labelled $c_1, \ldots, c_4$ correspond to 32, 128, 256 and 512 mg% Sr in the nebulised solution respectively. A shielded air–acetylene flame with a uniform temperature of 2275 K was used; the spectral line profiles were obtained with the aid of a Fabry–Pérot interferometer with negligible instrumental broadening. The broken curves show the profiles of the line wings for c_3 and c_4 corrected for spectral overlap of successive orders of the interferometer. Note the increase in the half-intensity width and in the flattening of the curves with rising concentration due to self-absorption. The profile for c_1 is practically free from self-absorption broadening. (The measurements were carried out by Dr B J Jansen at the Fysisch Laboratorium, University of Utrecht, The Netherlands [495].)

$B_{\lambda_0}^{\mathrm{P}}(T)$ which is expressed by equation (6.19a). The integral in equation (6.20) – which is dependent on the a-parameter – is controlled by T to a small degree only. The temperature dependence therefore is practically the same at strong and at weak self-absorption, that is, at high and low atomic concentrations. In each case its main cause is the change in the excited-level population with temperature (see equation (6.3)).

A comparison of equations (6.13) and (6.20) reveals that under otherwise identical conditions the absolute absorption signal $(\Phi_0 - \Phi_t)$ in work with a continuum source exceeds the absolute thermal emission signal Φ by the factor $\Phi_{\lambda_0}/B_{\lambda_0}^{\mathrm{P}}(T_f)O\Omega$. Here T_f is the flame temperature while the intensity I_0 in equation (6.13) is replaced by the radiant flux Φ_0. The spectral radiant flux of the lamp radiation Φ_{λ_0} can be formally expressed in terms of the *radiance temperature* T_r as

$$\Phi_{\lambda_0} \equiv B_{\lambda_0}(T_r)O'\Omega', \tag{6.23}$$

where O' and Ω' are the cross section and solid angle respectively of the radiation beam incident on the flame. If the product $O\Omega$ in the emission measurement is assumed to have the same value as $O'\Omega'$ in the absorption measurement then, with regard to the absolute signal, the use of the absorption method is found to be equivalent to increasing the flame temperature in the emission method by a factor of T_r/T_f [164, 176]. A continuum source with $T_r \approx 6000$ K (e.g. a high-pressure xenon lamp [929]) offers a considerable gain in this respect because the exponential dependence of the Planck factor on the temperature, especially in the ultraviolet region, is very strong.

The gain in absolute signal $(\Phi_0 - \Phi_t)$ for a line source can be similarly described by a radiance temperature [164, 176], but here also the half-intensity width of the emission line must be taken into account. The radiance temperature in this case can be very high (for a Cd lamp for instance about 7000 K [712]) because in low-pressure gas-discharge lamps the excitation is not a thermal effect but is due to suprathermally accelerated electrons. A further comparison of the absolute signals in the emission, absorption and fluorescence methods is to be found in [167, 904].

It should be noted that the experimental curves of figure 11 show symmetrical line profiles. Evidently the asymmetry, theoretically possible in lines with collisional (or Lorentz) broadening, does not show itself here (compare § 6.3). The Sr line shown in figure 11 had an a-parameter of $1 \cdot 3$ so that the collisional broadening was definitely not negligible compared with the Doppler broadening [812] (compare equation (6.2)). Moreover, even at the highest metal concentrations practically attainable the width of the emission line seems to be less than $0 \cdot 1$ nm (see figure 11). The line profiles thus encompass such a small wavelength interval that they cannot be resolved with the monochromators usually employed in flame spectroscopy. Only the total intensity represented by the area under the profile curve is measured in flame emission spectroscopy.

6.6.2. Self-reversal. In our treatment of self-absorption so far the temperature of the flame part coloured by the metal vapour, or the excitation of the upper energy level throughout the metal-vapour discharge lamp, were taken to be uniform, but with the common flames a colder outer zone which also contains metal vapour has to be taken into account. Similarly one must consider the possibility of the excitation strength in gas-discharge lamps considerably diminishing from the interior towards the window. At high Na concentration for instance the self-absorption of the yellow Na light can become so strong that at the line centre practically all the rays emitted by the hotter flame core or by the inner discharge column of the lamp are absorbed in the outer zone and are thereby lost. The absorbed radiation is in fact replaced by the radiation from the outer zone itself, but as the excitation there is less strong, this replacement is insufficient to compensate fully for the loss of radiation from the source centre. Hence a dip takes the place of the original peak in the spectral distribution curve of the emergent emission line because at the line

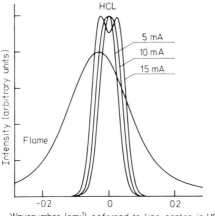

Figure 12. Spectral emission profiles of the atomic Ca line at 423 nm in a hollow-cathode lamp (HCL – Perkin Elmer Intensitron lamp) and in a nitrous oxide–acetylene flame with a-parameter value of 0·70. At higher lamp currents the increased line broadening due to self-absorption and even a self-reversal dip become manifest (the lamp profiles are normalised to the same peak height). The flame profile was measured in the absence of self-absorption and the line showed a red shift of 0·032 cm^{-1} ($+1$ cm^{-1} corresponds to $\Delta\lambda = -0\cdot016$ nm). (Reproduced from [364] by permission of Dr L de Galan.)

centre the absorption and with it the radiation loss is greatest (see figure 12). This phenomenon, in which the original maximum of the line profile so to speak is reversed, is called *self-reversal*†. Similarly to self-absorption it is normally only to be expected with strong resonance lines.

The appearance of a dip at the wavelength λ_0 of the line centre flanked by two maxima can be easily explained by means of a simple model. Suppose there are two zones in the flame – an inner core with uniform temperature distribution which emits a self-absorbed resonance line, and a cooler outer zone with negligible emission. Let the atomic concentrations and thicknesses of the two zones, as well as the Doppler and Lorentz widths of the resonance line in them, be the same. Then the profiles of the absorption factors $\alpha(\lambda)$ for the two zones are also equal. Furthermore, according to Kirchhoff's law (see § 6.6.1) the profile of the self-absorbed resonance line emitted by the inner core is similar in shape to that of the absorption-factor profile $\alpha(\lambda)$. The spectral profile of the resonance line resulting after passage of the core radiation through the absorbing outer layer is now

---

† In the literature the smaller Doppler width in the cooler outer zone is sometimes stated to be the main cause of self-reversal, often with a reference to the rather small width of the reversal dip (see also figure 12). The greater width of the emission profile of the radiation from the core of the flame or lamp however is not to be accounted for by the Doppler effect alone, but chiefly by the broadening effect of the strong self-absorption in this core (compare figure 11 and the pertinent discussion in § 6.6.1). Computer calculations of the line profile have shown that also with the *same* Doppler width in the core and in the outer zone there can be a reversal at the line centre [379, 731]. (See also the following paragraph.)

125

described – apart from a constant of proportionality – by the function $\alpha(\lambda)(1 - \alpha(\lambda))$. By differentiating this expression with respect to λ and equating the derivative to zero one finds that the line profile has three extrema if, at the line centre, $\alpha(\lambda_0)$ is larger than $0 \cdot 5$. It can be easily shown that the extremum at the line centre is a minimum symmetrically flanked by two maxima located at those wavelengths where $\alpha(\lambda)$ is exactly equal to $0 \cdot 5$.

The total intensity of an emission line with self-reversal is attenuated compared with that of a line without self-reversal. The intensity reduction becomes more pronounced with rising concentration so that the analytical curve is still more convex than for a flame of uniform temperature in which only self-absorption without self-reversal occurs (see chapter 8).

In absorption and fluorescence work self-reversal in the discharge lamp is especially troublesome when the absorption line in the flame is narrower than the emission line of the lamp, which may be the case at high lamp currents. The spectral band of the lamp radiation that is absorbed in the flame then comes mainly from the reversal minimum – the sensitivity is thereby reduced. Asymmetry of the two maxima of a self-reversed line can be observed when the line centres in the discharge volume and in the absorbing outer zone or absorbing flame are shifted with respect to each other [783, 840, 920] (see § 6.3).

The combined effect of self-absorption and self-reversal on the spectral profile of a vapour-discharge lamp is not easy to calculate. The characteristics of the lamp (excitation strength, temperature, vapour pressure, Lorentz and Doppler broadening) are often very far from uniform and not known exactly. Usually we endeavour to simulate the actual conditions by means of simple models and numerical (computer) calculations [226, 378, 379, 732, 783, 920]. (The earlier literature is reviewed in [281].) The lamp, for example, is thought of as being made up of two or three different adjacent zones each of which, in itself, is taken to be approximately homogeneous. But the zones differ from each other in excitation strength, vapour pressure, temperature, thickness, etc. The radiation contribution of each zone is calculated as a function of wavelength, the self-absorption and the additional absorption in the successive zones being taken into account. The contributions are then added up for each wavelength separately. Because of the large number of unknown parameters and the far-reaching simplifications, such calculations in most cases are only of a semi-quantitative nature. For more accurate investigations we usually have to resort to interferometric measurements [541, 868]. Similar model calculations have also been carried out for flames with a cooler outer zone [731, 842].

At high atomic concentrations self-reversal may also occur in light sources under uniform collisional–excitation conditions (e.g. in a flame of uniform temperature) if radiation diffusion plays a noticeable part. *Radiation diffusion* becomes effective when the probability of an excited atom being deactivated by resonance-photon emission is greater than the probability of deactivation by a quenching collision with a flame molecule. This is tantamount to saying that radiation diffusion will occur if the quantum efficiency of resonance fluorescence, Y, lies between about $\frac{1}{2}$ and its maximum value of 1 (see § 6.8, equation (6.24)). A photon generated inside

126

the source may then be re-absorbed and re-emitted (i.e. scattered) several times in succession, the more often the higher the atomic concentration and the greater Y. The photons jump from atom to atom before ultimately leaving the flame or being quenched. At each jump the wavelength of the photon may be changed (depending on whether the scattering is coherent or incoherent) and so may, due to the wavelength dependence of the absorptivity $k(\lambda)$ (see § 6.5.1), the probability of the photon being re-absorbed per unit path length. Complicated computer calculations are required to derive the spectral distribution of the photons escaping from the source. Such calculations have shown that under uniform collisional excitation conditions and for small quenching probabilities resonance lines may be self-reversed so that their profiles will show a dip at the line centre [326].

Self-reversal due to radiation diffusion, in a simple way, may be visualised by considering the fall in excited-level population close to the outer boundary of, for instance, an argon-diluted flame of uniform temperature. If Y is almost unity and if the atomic concentration is high, photoexcitation predominates over collisional excitation (see also § 6.7.4 and [176]). Near the flame boundary the spectral volume density of radiant energy at λ_0 (spectral energy per unit volume) and therefore the photoexcitation rate are lower than inside the flame. A uniform radiant energy density conforming to Planck's law would only exist if the flame were surrounded by a black-body radiator at the same temperature as the flame. The radiation at $\lambda \simeq \lambda_0$ from the black body towards the flame would compensate for the radiation emitted by the flame. Because of the absence of such a compensating black-body radiation, the radiant energy density is reduced. For a strongly self-absorbed line this decrease is more pronounced near the flame boundary than at its core; for, strong self-absorption implies that the flame is opaque for wavelengths near the line centre so that in this wavelength range the flame itself acts like a black-body radiator enveloping the flame core.

The resulting drop in the photoexcitation rate in the radial direction leads to a gradual decrease in the excited-level population with increasing distance from the flame axis. This effect on the line profile is the same *as if* the hot flame core were surrounded by a cooler outer layer.

An approximate analytical expression for the self-reversed line profile under uniform collisional–excitation conditions has been derived in [466]. The expression is believed to be valid for flames at 1 atm pressure when incoherent resonance scattering occurs and the value of Y is not too close to unity.

The usual disregard of the effect of radiation diffusion in the calculation of self-broadened and self-reversed spectral lines of low-pressure discharge sources, where Y is almost 1, is not justified.

6.6.3. The line-reversal method of temperature measurement.
With the help of the given equations the principle of the line-reversal method of flame temperature measurement can be easily explained [43, 58, 61, 72, 78, 176, 230, 740, 804, 808]. The method is based on the fact that the intensity of the thermal radiation emitted by the flame is exactly equal to the intensity of the radiation that would be absorbed by the same flame if a black-body radiator of the same temperature were placed

behind it. This is only so provided that the radiation in the two cases is observed under the same geometrical conditions, that is the same direction and same solid angle, and identical irradiated and emitting flame volume. This fundamental equality can easily be deduced from equations (6.13) and (6.20) if in equation (6.13) Φ_{λ_0} is substituted for I_{λ_0} and use is made of equation (6.23) for Φ_{λ_0} with T_r being put equal to the flame temperature T_f.

To measure the temperature, a black-body radiator is imaged in the flame coloured by an atomic metal vapour and the two sources together and under the same geometrical conditions are observed through a spectroscope. The radiation absorbed from the continuous spectrum of the background source within the wavelength interval of the resonance line is then exactly compensated by the thermal radiation from the flame within the same wavelength interval if the temperatures of the background source and the flame are equal. The presence in the flame of both emitting and absorbing metal vapour thus does not in any way alter the observed continuous spectrum. However, if the temperature of the background source is higher or lower than the flame temperature T_f, compensation fails to exist and the metal line stands out against the continuous background spectrum as an absorption or emission line, respectively. The unknown flame temperature is then found experimentally by adjusting the known temperature of the black-body radiator (or the radiance temperature of a suitably calibrated tungsten-ribbon lamp) so that the atomic line in the spectrum just disappears. This means that the line is just 'reversed' from a superimposed emission line into an absorption line.

It is necessary to be aware that it is not the translational temperature, describing the random thermal motions of the flame molecules, which is measured directly in this way. Actually we find the value of temperature T that has to be inserted in Boltzmann's equation (6.3) to render the population of the excited level in question correctly. We shall discuss in § 6.7.4 under what conditions the so-called *excitation temperature* determined here may deviate from the true translational temperature.

Thus the value of temperature derived from the line-reversal method is a measure of the *relative* population of the excited level (corresponding to the ratio n^*/n in equation (6.3)). It gives no information about the possible deviation of the absolute atomic concentration (n) from its chemical equilibrium value.

The accuracy and the possible error sources of this measuring procedure with (non-uniform) flames† have been discussed in [740, 804, 805, 808, 842]. A calibrated tungsten-ribbon lamp often serves as the temperature standard [765]. A well defined and reproducible auxiliary flame can also function as a secondary temperature standard as is described in [806, 811]. Accuracy and speed of line-reversal measurements can be considerably improved by resorting to photoelectric methods [321, 804, 808, 843].

The advantages of the line-reversal method are: the spectral response of the spectrometer does not have to be calibrated; the transition probability, etc, of the

---

† In [854] a method has been described to determine the temperature in the core of a non-uniform flame by applying the line-reversal method to a strongly self-reversed resonance line with resolved spectral profile.

128

resonance line plays no part in the measurement; and the flame is not disturbed as it would be if a thermocouple were used. The expenditure for the necessary additional equipment may be a drawback, but this can be avoided in the so-called *two-line method*, which is therefore more convenient to apply (see § 6.4 and [740]).

6.7. Excitation Mechanisms and Chemiluminescence

The expositions of § 6.4 depend on the assumption that thermal equilibrium has been established and that the population of the excited levels is given by Boltzmann's equation (6.3). This population is again connected with the intensity of the corresponding spectral line (see equation (6.5)). In equilibrium Boltzmann's equation holds irrespective of the particular kind of excitation mechanism(s) responsible for the transition of the atom to the excited level. Then it would hardly seem to serve any practical purpose to discuss here the possible, rather complex excitation mechanisms. In the flame, however, deviations from equilibrium can occur and affect the excitation of the analyte atoms or molecules. In order to elucidate their influence we in fact do have to deal in some detail with various possible excitation mechanisms, especially chemiluminescence. On account of the general reversibility of each particular process (so-called *principle of microscopic reversibility*) the discussion of excitation processes also has a bearing on the quenching of atomic fluorescence for quenching is the reverse process of collisional excitation. (The principle of the fluorescence method itself will be discussed in § 6.8.)

We will discuss successively in § 6.7.1: the excitation through absorption of photons produced in the flame itself, that is without background source; in § 6.7.2: the excitation through collisions in which translational energy and internal energy of the collision partner is transferred; in § 6.7.3: the excitation through chemical reactions, with (part of) the chemical energy being directly converted into excitation energy (chemiluminescence); and in § 6.7.4: the conditions, in general, required for thermal or non-thermal radiation to occur in flames. These excitation mechanisms are treated more extensively in [176].

6.7.1. Excitation through photon absorption (photoexcitation). As an excited atom can be deactivated through emission of a photon, so conversely can an atom be excited through the absorption of a photon with equal energy. The strength of this excitation depends on the radiation density in the flame which in turn in emission spectroscopy is governed by the concentration of the excited atoms. Therefore the probability of an atom being excited through absorption of a photon must be expected to increase with rising atomic concentration. If the atoms were excited mainly through photon absorption, then at low atomic concentrations in the flame the relative population of the excited level involved would be much lower than the Boltzmann population. This population would only be reached at atomic concentrations so high that the radiation density of the atomic line approaches the Planck value for a black-body radiator at flame temperature (see § 6.7.4).

Now it has generally been found in flame emission spectroscopy that the relative population of the excited level (n^*/n) is usually independent of the atomic concentration (n). This follows from the observed proportionality between radiation intensity and atomic concentration if self-absorption and other curvature effects (see chapter 8) are disregarded or corrected for. It has also been confirmed through temperature measurements by the line-reversal method (see § 6.6.3), for the line-reversal temperature, which is a measure of the relative population (n^*/n), proved to be the same for low and high metal concentrations [162, 804]. The different behaviour in special flames containing argon instead of nitrogen will be discussed in § 6.7.4. When deviations from a Boltzmann population are found at all, they mostly point to too *high* a population which can be accounted for by suprathermal chemiluminescence. Therefore the conclusion has to be drawn that besides the excitation by photon absorption also other, much more effective excitation mechanisms exist that determine the population of the excited level. The excitation of a metal line through photon absorption can become comparatively significant only if additional radiation energy of the corresponding wavelength is supplied to the flame from an intense background source. Then the population of the excited level may even greatly exceed the Boltzmann value [666]. This indeed is the main advantage of the fluorescence method.

6.7.2. Collisional excitation. Because of the relatively low concentration of free electrons excitation through electron impact, which plays such an important role for instance in the arc, need not be taken into consideration here [162, 170, 176]. Furthermore, it can be deduced from measurements of fluorescence quenching in metal-vapour cells and from excitation experiments in shock waves and atomic beams that under flame conditions collisional deactivation and excitation of metal atoms through transfer of translational energy alone is negligible [176, 582, 583]. Similar experiments and theoretical calculations have shown that on the other hand the participation of the internal energy of the flame molecules in the collisional excitation and deactivation processes is important [61, 70, 176, 197, 375, 484, 582, 583, 851]. Experiments with crossed molecular and atomic-alkali beams have yielded more detailed information on the relative shares of the vibrational and rotational energies as well as of the translational energy in the excitation process [520, 565] (see also [583]). Internal molecular energy and translational energy are in turn converted into each other through collisions between the flame molecules themselves. The conversion of the vibrational energy is facilitated by the fact that this energy is transferred in steps; the molecule in each collision goes over from one vibrational level to another neighbouring level and only exchanges a small part of its vibrational energy. In this indirect way therefore the excitation of the metal atoms ultimately depends on the distribution of the translational energy, that is on the translational temperature of the flame. A certain relaxation time is required for the adaptation of the vibrational-energy distribution to the translational temperature; yet experiments with shock waves have shown that at atmospheric pressure it is less than 10^{-4} s and hence is of no consequence in flame spectroscopy [484, 643, 851].

130

Since all processes are reversible it must also be assumed that excited metal atoms are mostly deactivated through conversion of their excitation energy into internal molecular and translational energy upon collisions with flame molecules. It is just the interplay of activating and deactivating (or quenching) collisions that results in a Boltzmann equilibrium population of the excited level, as described by equation (6.3). How this equilibrium is established and what deviations are possible in flames will be discussed more fully in § 6.7.4.

The noble-gas atoms, which only possess translational energy under flame conditions, have a very low efficiency in exciting and deactivating the metal atoms. This may lead to deviations from thermal radiation (see § 6.7.4) as well as to a high efficiency of fluorescence in Ar-diluted flames (see § 6.8).

Noble-gas atoms upon collision can readily transfer an excited atom into a neighbouring excited state if the energy exchange is only small [176, 582]. An example of this is the efficiency of noble-gas atoms in causing so-called mixing transitions between the Na and K doublet components in flames [468, 583]. The efficiency for such transitions for most flame molecules is also great and at least of the same order of magnitude as their efficiency for the quenching of the excited state [566, 582, 583, 586, 588, 589].

6.7.3. Chemiluminescence. In the emission method all the energy spent in the common flames for excitation ultimately comes from chemical reactions. The heat of combustion is distributed over the many degrees of freedom of the flame molecules and indirectly transferred to the particles to be excited in so-called inelastic collisions with these molecules (see § 6.7.2). In certain cases, however, a part of the energy released in a particular exothermic chemical reaction is directly converted into excitation energy. We then speak of *chemiluminescence* and for convenience also include those processes in which ion recombination directly contributes to the excitation. Chemiluminescence is to be expected in flames because the energy released in many reactions in the flame is of the same order of magnitude as the electronic excitation energy of the atoms and molecules (about 2 to 8 eV). The reaction energy does not necessarily have to equal the excitation energy exactly because a possible insufficiency or excess of energy can be easily compensated for by an additional exchange of kinetic energy. The possibility of a chemiluminescent reaction is, however, not only restricted by considerations of energy, but also by Wigner's rule which states that the total spin angular momentum of the particles involved must be the same before and after the reaction [70]. This rule, which does not always hold rigorously, is an analogy of the well known selection rule in atomic spectroscopy, which 'forbids' radiative transitions between states with different spins.

Deviations from thermal radiation due to chemiluminescence occur when the flame radicals taking part in the chemiluminescent reaction are not in chemical equilibrium (see also § 6.7.4). Chemiluminescence processes therefore are only of significance in flame spectroscopy when under certain circumstances deviations from chemical equilibrium are found. Such deviations are to be expected in the primary combustion zone in which the oxidation of the fuel gas takes place (see

131

§ 3.4.1). They may also be expected above this zone where the flame radicals H, OH and O, as a result of their slow recombination, may be present in excess concentrations (see § 3.4.2). If the concentrations of the reactants in a chemiluminescent reaction exceed the equilibrium values, fundamentally suprathermal radiation is to be expected.

Suprathermal chemiluminescence radiation is emitted not only by flame radicals such as CH, C_2 and OH in the combustion zone and even by OH above this zone (see § 3.4.1 and § 3.4.2), but it is also sometimes found in the emission lines and bands of elements which are introduced into the flame by the analyte solution. In the first case the chemiluminescent radiation enhances the flame background (see Glossary and the next chapter) and therefore may interfere with the measurement of the analysis line. In the second case the increased radiation intensity of the analyte can be utilised to improve the limit of detection (see Glossary) for the element concerned. This of course is possible only if the spectrum of the chemiluminescent radiation is specific for the element, that is, with line and band emission but not with chemiluminescent continuous radiation (see § 6.2.3). In this chapter we will only deal with the possibility of suprathermal chemiluminescent emission of atomic lines (e.g. of Sn, Tl and Fe) and molecular bands (e.g. of CuOH, InCl and HPO) emitted by the analyte. The literature on this topic is summarised in [176].

The conditions under which chemiluminescence effects are observed are also often favourable for strong reduction of the metal oxides (see § 5.2.4) [236, 296, 334, 389, 935]. By this reduction the concentration of the free atoms can be appreciably increased, especially with elements such as Sn, V and Nb which otherwise occur in flames practically only as stable oxides. This alone also enhances the atomic line intensities. In the literature the two effects, namely suprathermal excitation and strong reduction of oxides, are not always clearly differentiated. Experimental methods for separating these effects in practice are described in § 6.7.4.

There now follows a survey of the most important types of reactions that under certain conditions may lead to suprathermal excitation in flames. The interpretation of all the chemiluminescence effects can by no means be considered as final yet, in spite of extensive investigations carried out by different methods of measurement with variations of temperature, pressure, composition and height of observation in the flame.

Two fundamental types of chemiluminescence of atomic lines, both in regard to its occurrence in flames and to its origin, can be distinguished [176, 240, 826].

Suprathermal chemiluminescence of the first kind is only met with in lines with excitation energies of about 5 eV at most and is usually not particularly prominent. It is observed in and above the combustion zone of rather cool hydrogen and hydrocarbon flames, in the latter only provided that it is not masked by other chemiluminescence effects [201, 231, 680, 739, 826, 924, 926]. The suprathermal chemiluminescence of the first kind is due to recombination reactions of free radicals H and OH which are present in excess (see § 3.4.2) [70]. The recombination energy brings about the excitation of the metal atom, which does not play

132

a role in the reaction itself, but is only present as a non-active 'third body'. The reactions in question are:

$$H+H+M \rightarrow H_2+M^*$$

and

$$H+OH+M \rightarrow H_2O+M^*,$$

where M and M* are a non-excited and an excited metal atom respectively. Similar chemiluminescence can also be produced by recombination of CO with a free O atom, but this is generally of minor importance compared with the other reactions [455]. It is quite possible that the actual reaction takes place in two steps, as for instance:

$$\left.\begin{array}{l} H+H+X \rightarrow H_2^*+X \\ H_2^*+M \rightarrow H_2+M^* \end{array}\right\} \quad [826]$$

or

$$\left.\begin{array}{l} H+H+X \rightarrow H_2+X^* \\ X^*+M \rightarrow X+M^* \end{array}\right\} \quad [162]$$

where X is a stable flame molecule (N_2, H_2O) which stabilises the recombination of the radicals through (partial) energy transfer (see § 3.4.2).

A typical example of *suprathermal chemiluminescence of the second kind* is that found in ultraviolet lines with excitation energies of up to about 8 eV, with pre-mixed as well as non-premixed hydrocarbon flames, in or just above the primary combustion zone [58, 156, 157, 231, 240, 296, 826]. A very weak chemilumines-cence has been observed in the combustion zone of acetylene flames even in the Zn lines at 277 nm and 280 nm with excitation energies of about 8·5 eV, but not in the Hg line at 365 nm with an excitation energy of 8·85 eV. For lines with lower excitation energies this kind of chemiluminescence may possibly lead to a very considerable rise in the emission. There is however no such chemiluminescence in carbon monoxide or hydrogen flames, but the addition to the nebulised solution of hydrocarbon compounds, such as naphtha or methanol, or the addition of hydro-carbon compounds in the form of gases, such as acetylene, induces similar chemiluminescence effects also in non-premixed hydrogen flames [236, 381, 389]. In hydrocarbon flames an admixture of alcoholic constituents to the nebulised solution or an excess fuel-gas supply (giving a luminous flame) causes an additional enhancement of the atomic lines [296, 389]. Line-reversal measure-ments on Fe atoms in the combustion zone of a premixed acetylene flame however have shown that the chemiluminescent excitation as a function of the fuel-gas/oxi-dant mixture ratio passes through a maximum approximately at the stoichiometric ratio [240]. Therefore the further emission increase often resulting from excess fuel-gas supply has to be attributed to the favourable reducing conditions in luminous flames (see § 5.2.4).

The origin and the circumstances of the occurrence of this second kind of chemiluminescence indicate that it must be closely associated with the energetic

133

oxidation reactions and that fragments of hydrocarbon compounds play an essential part in it. Other observations have revealed certain relationships between these chemiluminescence effects on the one hand and the suprathermal ionisation and the chemiluminescent CH emission on the other hand [60, 61, 240, 295, 299, 301, 389 537]. Since the suprathermal ionisation itself is traced back to the CH radical (see § 3.4.1), attempts have been made to link the chemiluminescence of the second kind with the presence of these radicals. Accordingly the following chemiluminescent reactions or chain reactions have been suggested, listed here without further comment (compare also [176]):

$$CH + O + M \rightarrow CHO + M^*, \quad [381, 826]$$

$$CH + O(H) + M \rightarrow CO + H(H) + M^*, \quad [381]$$

$$CH + O \rightarrow CO^* + H \quad \text{with} \quad CO^* + M(O) \rightarrow CO(O) + M^*, \quad [156, 301]$$

$$CH + O + X \rightarrow CHO^* + X \quad \text{with} \quad CHO^* + M \rightarrow CO + H + M^*, \quad [212, 241]$$

$$CH + O + X \rightarrow CO + H + X^* \quad \text{with} \quad X^* + M \rightarrow X + M^*. \quad [231]$$

The second reaction (with CH and OH) has also been used to explain the remarkably strong radiation of the ultraviolet Sn lines in cool H_2 diffusion flames with alcohol addition [290]. The CO^* molecule, acting as an intermediary in the third reaction, may be either in the electronic ground state with high vibrational energy or in the excited metastable electronic state ($a^3\pi$) with moderate vibrational energy [156].

In all cases listed the energy necessary for excitation (and dissociation of the metal oxide) ultimately derives from the considerable reaction energy that is released in oxidation of the CH radical. The previously suggested chemiluminescent reaction [296, 389]:

$$C + MO \rightarrow CO + M^*$$

seems to be of significance only in luminous sooting flames in which the C line at 247·8 nm also appears.

There have also been attempts to account for the parallelism between chemiluminescence and chemi-ionisation by assuming the following luminescent ion-recombination processes:

$$M^+ + e^- + X \rightarrow M^* + X \quad [240]$$

or

$$H_3O^+ + e^- + M(O) \rightarrow H_2O + H(O) + M^*. \quad [389]$$

The excess of positive ions is again linked with the CH radical by chemi-ionisation (see § 3.4.1). Only further measurements and new thermodynamical data can reveal which reactions – and under what conditions – are of paramount importance in the commonly used flames. In particular, measurements of the intensity ratios of different chemiluminescent atomic lines could be instructive.

134

Finally the suprathermal chemiluminescence radiation in *molecular bands* has to be mentioned. Chemiluminescent excitation of the CuH and CuOH bands can occur in H_2–O_2–N_2 flames as a result of the following reaction [739]:

$$Cu + (O)H + X \rightarrow Cu(O)H^* + X,$$

and a contribution to the excitation of the diffuse SnOH bands at 470–510 nm could come from the reaction [244]:

$$SnO + H(+X) \rightarrow SnOH^*(+X).$$

Here part of the recombination energy is spent in the electronic excitation of the molecule.

More recent investigations have proved that the earlier supposition of a similar chemiluminescent emission of the alkaline-earth hydroxide bands was not correct [277].

Presumably suprathermal chemiluminescence also occurs in the bands of AlO at 484·2 nm and of BO_2 at 518·0 nm in the oxygen–hydrogen flame when naphtha is added to the solution [236]. The mechanism of the chemiluminescent excitation of similar compounds on the whole is still unknown and possibly may vary from case to case.

Of importance in the flame analysis of non-metals are the suprathermal chemiluminescent band emissions of compounds such as HPO, PO and S_2, found in cool hydrogen flames [60, 391, 636, 839] (see spectrograms 43 and 44 in Appendix 1). S_2 is thought to be electronically excited through the reaction:

$$(O)H + H + S_2 \rightarrow H_2(O) + S_2^*$$

and HPO through the reaction

$$PO + H + (X) \rightarrow HPO^* + (X)$$

or

$$PO + H_2 + OH \rightarrow HPO^* + H_2O.$$

The special method of chlorine determination by means of the InCl emission bands most probably utilises a chemiluminescence effect also (compare spectrogram 42). The enhancement of the band emission of NO, PO and CS in H_2 flames with acetylene addition similarly points to suprathermal chemiluminescence [386, 391].

6.7.4. General conditions for the occurrence of thermal and non-thermal radiation. In emission spectroscopy the radiation is said to be thermal if the relative population of the excited level corresponds to the Boltzmann equilibrium at the temperature T_f of the flame, so that equation (6.5) holds [162, 176]. This definition does not imply that for the radiation to be thermal also the degree of dissociation of the metal compounds or the degree of ionisation have to correspond to the equilibrium conditions (see § 5.2.1 and § 5.3.1). The criterion of thermal radiation therefore is only the *relative* population of the excited level, that is, the (n^*/n) ratio in equation (6.3).

Nothing can be deduced from the mere existence of thermal radiation in a given flame that will allow identification of the specific excitation mechanism(s) (collisional or chemical) operative in that flame. The only permissible conclusion is that the flame constituents playing a part in the activation and deactivation of the radiating atoms are themselves in physical as well as in chemical equilibrium. Physical equilibrium exists if the distribution of the translational and the internal energies over the various degrees of freedom of the flame molecules satisfies the equipartition law. Chemical equilibrium exists if the concentrations of all components (including the free radicals and the ions) fulfil the law of mass action and Saha's law (see § 5.2.1 and § 5.3.1). Under these conditions every separate activation process according to the *principle of detailed balancing* is in exact kinetic equilibrium with the reverse deactivation process. The mean number of excitation collisions per second with a collision partner of a particular type then equals the mean number of reverse quenching collisions per second with the same partner. If besides collisional excitation there is also excitation through a chemical reaction, as in chemiluminescence, this in itself does not necessarily make the radiation non-thermal, that is, affect the Boltzmann population. The excitation rate will of course be enhanced through this supervening chemical process, but at the same time the quenching rate will also increase in consequence of the reverse chemical process. If all particles taking part in the chemical excitation and the reverse chemical quenching process are in physical and chemical equilibrium, the mean rates of the two opposed processes will always be the same. In other words, the two processes will balance each other exactly. Hence the population of the excited level will not change and the radiation remains thermal.

Since absorption and emission of photons also contribute to the excitation and deactivation of the atoms, complete thermodynamic equilibrium furthermore requires detailed balancing for these two processes also. This is assured only if the spectral energy density at $\lambda \simeq \lambda_0$ in the flame corresponds to Planck's equilibrium value for a cavity radiator ('Hohlraum') at the flame temperature. In general this is not the case because the flame does not burn in such a cavity or furnace. The spectral energy density may only approach the Planck value in the interior of the flame and near the central wavelength of a resonance line at high atomic concentrations, that is, at strong self-absorption (compare figure 10; see also § 6.7.1).

Apart from possible suprathermal chemiluminescence effects, the population of the excited levels in the interzonal region of the commonly used flames is found to be in good agreement with Boltzmann's law. This can generally be deduced from temperature measurements by the line-reversal method [231, 453, 455, 784, 804, 808]. The temperature values thus found, when introduced into the appropriate equations (6.5) and (6.20), lead to a satisfactory description of the intensity of the atomic resonance lines. This holds not only in the higher concentration range where the spectral energy density at the central wavelength approaches the Planck value, but also at low concentrations for which the latter condition is not fulfilled. It therefore must be concluded that usually radiative processes are only of minor importance compared with other activation and deactivation processes (see § 6.7.1). The lack of detailed balancing for the photon absorption and emission processes then does not noticeably affect the population of the excited level.

This general statement does not hold for flames with a large excess of noble gas. The efficiency of the noble-gas atoms in activating and deactivating metal atoms is very low (see § 6.7.2). The deactivation of the levels through photon emission in such flames is then no longer insignificant compared with collisional quenching. The absence of detailed balancing for the radiative processes then manifests itself. It has in fact been established by the line-reversal method that in an oxygen–hydrogen flame diluted with argon instead of with nitrogen the populations of the excited levels of the yellow Na doublet at low Na concentrations remain below the equilibrium value (so-called *radiation depletion*) [464, 467]. The deviation proved to be equivalent to a lowering of the 'excitation temperature' by about 150 K. With rising Na concentration, that is with increasing radiant energy density, the population of this doublet, as a result of the increased photoexcitation, again comes near to the equilibrium value. The observed dependence of the excitation on the Na concentration conforms to theory [176, 466, 469]. The magnitude of the latter deviation increases with rising fluorescence efficiency which describes the relative probability of deactivation through photon emission (see § 6.8). Such deviations are therefore to be expected mainly with the special argon-diluted flames which are often used in fluorescence work for achieving a high fluorescence efficiency; in fluorescence measurements however possible deviations from thermal radiation are of no concern.

If for example in the forward chemiluminescent reaction step

$$H + OH + Tl \rightleftarrows H_2O + Tl^*$$

the concentrations of the H and OH radicals exceed their equilibrium values, the rate of this reaction will be higher than in the equilibrium state. This holds because this rate is proportional to the product of the H and OH concentrations. If on the other hand the concentration of H_2O on the right-hand side practically does not deviate from equilibrium (see § 3.4.2), the probability of quenching of an excited Tl^* atom through the backward reaction step will be the same as in the state of equilibrium. As a consequence the relative population of the excited level will rise above the Boltzmann value and the radiation intensity of the thallium line will now be suprathermal.

Whether certain deviations from chemical equilibrium lead to noticeable departures from the thermal radiation intensity for a particular atomic line with excitation energy E_q does not only depend on the absolute rate of the chemiluminescent excitation itself; the latter should always be considered in proportion to the rates of the other (thermally equilibrated) excitation processes. The thermal excitation rate increases with rising temperature T and decreasing excitation energy E_q roughly as $\propto \exp(-E_q/kT)$. With low values of E_q, as for instance in the case of the yellow Na doublet with $E_q = 2 \cdot 1$ eV, suprathermal chemiluminescence therefore is only expected to appear in rather cool flames where $T < 1800$ K

137

[680, 926]. In hot flames it is expected to become noticeable only in ultraviolet lines for which $E_q > 4$ eV [236]. In other cases a possible suprathermal chemiluminescence is masked by the thermal radiation.

The absolute strength of suprathermal chemiluminescence furthermore depends on the rate of the quenching collisions with other molecules which cause a radiationless transition of the chemically excited particle to a lower energy level (compare § 6.7.2). It is the interplay of excitation and quenching processes that governs the stationary (non-thermal) population of the excited level. The quenching rate is determined by the composition of the flame gas and varies for the different lines of a given atom. In particular the quenching effect of noble-gas atoms is much smaller than that of molecules (see § 6.7.2) and replacement of nitrogen molecules by argon or helium atoms must be expected to enhance the intensity of the suprathermal radiation [156, 201, 389].

The intensity of a suprathermal chemiluminescent resonance line can also be weakened through self-absorption at higher metal concentrations, as may be deduced from the convex curvature of the analytical curve for the suprathermal Be line at 234·9 nm [389]. However, in calculating the degree of self-absorption, the possibility of the Doppler broadening exceeding its equilibrium value given by equation (6.1) has to be taken into account [58, 162, 176]. Extra Doppler broadening could result when the excited atoms gain additional translational energy in the chemiluminescent reaction.

When under certain conditions (fuel-rich flame; alcoholic solvent) the emission of atomic lines, for example of V or Nb, is greatly intensified, suprathermal chemiluminescence does not necessarily have to be taken for the cause alone. Such an enhancement may also be due to strong reduction of the stable metal oxides (see § 5.2.3 and § 5.2.4). It may even be that the same chemical reaction (e.g. $C + MO \rightarrow CO + M^*$; see § 6.7.3) brings about a chemiluminescence effect as well as a reduction of the oxide.

The following experimental methods are available for separating the reduction and the chemiluminescence effects, and for actually establishing the suprathermal character of the radiation [166, 176].

By the line-reversal method one measures the excitation temperature, that is, the temperature value that has to be inserted in equation (6.5) for a correct description of the observed radiation intensity (see § 6.6.3) [240, 381]. If this value is higher than the flame temperature, the excitation must be suprathermal. In this way suprathermal radiation was found for the Ca 422·7 nm, Fe 372·0 nm and Cu 324·7 nm lines in a turbulent air–hydrogen flame at a height of a few centimetres [372]. The departure from thermal radiation became larger with increasing excitation energy E_q (see above regarding the effect of E_q). However, in the hotter oxygen–hydrogen flame the radiation of these lines proved to be thermal [372], in agreement with theoretical expectations (see above). Furthermore, the intensities of several lines of a given atom with different, known excitation energies, transition probabilities and statistical weights can be compared with one another for detecting departures from Boltzmann's equation [372, 389]. Thus an excitation temperature (see § 6.6.3) of 3000 K was established in the turbulent air–hydrogen flame just

138

mentioned for a series of Fe lines with values of E_q of about 4·2 eV [372]. The real flame temperature was about 2000 K.

The presence of non-thermal radiation of an atomic line can also be indirectly ascertained by measuring the line intensity as a function of the height in the flame, relative to its intensity at some reference height. Once the real flame temperature is known, also as a function of the height in the flame, departures from the Boltzmann population of the excited level can be detected by comparing the experimental intensity curve with the curve calculated from Boltzmann's equation (6.3) [859]. In this comparison a possible change in atomic concentration with height (e.g. because the degree of dissociation varies with height) can be allowed for after determination of the relative concentration change from atomic absorption measurements. In this indirect manner a non-thermal excitation of the Sn line at 380·1 nm and of the Cu line at 324·7 nm has been found in a turbulent air–hydrogen flame with the use of a 2-propanol–water mixture as solvent [859]. Similarly the non-thermal radiation of the blue K doublet at 404·4/404·7 nm has been established in premixed air–hydrogen flames [926].

The contribution of the reduction of the metal oxide or hydroxide to the enhancement of the resonance line emission can also be separated from that of the chemiluminescence effect by atomic absorption measurements [381, 434]. The non-thermal radiation of metal atoms in flames is dealt with in greater detail and more quantitatively in [176].

This exposition of the complicated and partly not yet fully understood excitation conditions in flames is not meant to imply that the flame is not a very useful analytical excitation source. Even if complete physical or chemical equilibrium is not established, the excitation conditions as such may be well reproducible with a given choice of solvent and flame adjustment. The deviations from equilibrium then have the same effect on the signals produced by the sample and the reference solution. Certain deviations from equilibrium may even be desirable because they offer some analytical advantages such as stronger reduction of the metal oxides and suprathermal chemiluminescence.

6.8. Atomic Fluorescence

As the newest branch of analytical flame spectroscopy, besides atomic absorption and chemiluminescence spectroscopy, the atomic fluorescence method (see Glossary) has been developed, in which the fundamental limitation of the thermal emission method can again be overcome. This limitation is connected with the exponential Boltzmann factor in equation (6.5) for the intensity of thermal radiation and consequently the thermal intensity decreases rapidly with increasing excitation energy E_q of the line. In atomic fluorescence spectroscopy a part of the primary radiation of an atomic or ionic (see [672]) resonance line, produced for instance in a metal-vapour lamp, is absorbed in the flame by atoms or ions of the same metal. The atoms thereby excited lose their energy either in quenching collisions with flame molecules (see § 6.7.2) or through re-emission of a secondary photon in a random direction (fluorescence)†. The intensity of the fluorescence radiation is evidently a measure of the concentration of the atoms in the flame and hence of the concentration of the nebulised solution. Calibration, as usual, is carried out with reference solutions of different but known concentrations.

On account of its great and stable spectral selectivity, atomic resonance fluorescence has also found application in atomic absorption spectroscopy as a means of spectral selection, the so-called 'resonance monochromator' [684].

We shall discuss briefly here the physical basis of the method. The theoretical aspects are dealt with more fully in [79, 89, 102, 116, 176] whereas a general introduction to the development and the application of atomic fluorescence spectroscopy with continuously operating as well as pulsed light sources is to be found in [102, 116, 669, 796, 881, 906, 908, 915].

The success of the fluorescence method is based on the fact that the intensity of the fluorescence radiation in a given solid angle can greatly exceed the intensity of the corresponding thermal radiation. Fundamentally, this is only possible if the energy density of the primary radiation beam, within the absorption linewidth, is much greater than the thermodynamic equilibrium value of a cavity radiator ('Hohlraum') at the flame temperature [167, 176] (see also § 6.7.1 and § 6.7.4). Since the radiance temperature of a continuum source or that of a discharge lamp at the wavelength of the atomic line centre can be much higher than the flame temperature (see § 6.6.1), this condition is easy to satisfy, especially if a tunable dye laser is used as an excitation source [215, 215a, 305, 353, 354, 397, 569, 570, 665a, 669, 670, 672, 673]. In the latter case even *saturation*‡ of the excited level may be reached [168a, 666, 696, 697, 795a]. When saturation is approached, the fluorescence intensity is hardly affected by variations in the laser intensity (for a critical discussion see [168a]) or in the fluorescence efficiency Y (see definition below).

The irradiance (expressed in $W\,cm^{-2}$) obtainable with a continuous-wave (cw) dye laser at the wavelength of the Na–D doublet is about 10^4 times greater than that obtainable with an Na line source focused under a solid angle of 0·05 sr in the flame [397, 911]. With the use of a cw dye laser with an output power of 100 mW and a bandwidth of 0·003 nm the fluorescence intensity of the Ba line at 553·5 nm appeared to exceed the thermal emission intensity by a factor of 3000 in a hydrogen–oxygen–argon flame [397]. Much higher peak ratios are expected to occur with pulsed dye lasers having an output power ranging from 1 to 100 kW, notwithstanding their larger bandwidths. With intense pulsed dye lasers frequency doubling can be applied to excite near-ultraviolet resonance lines. For a recent

---

† The term photoluminescence is more appropriate here than fluorescence, but the latter is more commonly used.

‡ Saturation sets in when the excitation by the source beam becomes so strong that the population of the upper level (n_1) of the resonance line approaches the value $n_0(g_1/g_0)$; here n_0 is the population of the ground level and g_1/g_0 is the ratio of the statistical weights of the upper and the ground levels (see § 6.4). The stimulated emission rate (see § 6.2.1) is then nearly equal to the photon absorption rate and dominates the spontaneous emission rate. At the same time the absorptivity is reduced ('bleaching effect').

140

discussion of the use of CW dye lasers in analytical AFS in comparison with pulsed lasers we refer to [219a, 397, 667a].

When the wavelengths of the absorbed primary and the fluorescence radiation are identical, we speak of *resonance fluorescence*. (Note that this term does not imply that the fluorescence line is necessarily a resonance line.) Basically, *non-resonance fluorescence* is also possible in which at least three energy levels of the atom are involved, as is shown diagrammatically in figure 13. The conditions illustrated in diagrams (a) and (b) are given for instance in thallium, which has a low excited level $(6^2P_{3/2})$ with an energy of 0.97 eV above the ground level. There also exists a second, higher excited level $(7^2S_{1/2})$ from which an optical transition to the ground level $(6^2P_{1/2})$ at 377.5 nm as well as a transition to the first excited level at 535.0 nm are allowed (see table 4). Examples of the conditions represented in (c) and (d) are those of the alkalis which possess two excited levels (doublets) close together (see § 6.2.1).

When the upper energy levels of the primary (exciting) line and of the observed fluorescence line are the same, we speak of *direct line fluorescence* (figure 13 (a and b)), and when these two levels are different, of *stepwise line fluorescence* (figure 13 (c and d)). In the latter case the atom excited by the primary radiation, before emission of the observed fluorescence line, first makes a radiationless or radiative

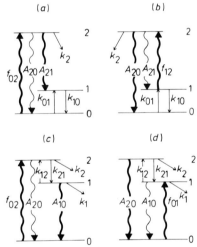

Figure 13. Schematic representation of the various types of non-resonance fluorescence with the use of a level diagram for an atom with ground level 0 and two excited levels 1 and 2. Wavy lines indicate radiative transitions; bold wavy lines with arrowheads pointing upwards represent atomic transitions due to absorption of the primary radiation; those with arrowheads pointing downwards represent observed fluorescence transitions; and straight lines with arrowheads denote radiationless transitions due to collisions with flame molecules. The following parameters characterise the probabilities of the various transitions between the numbered levels: f – oscillator strength; A – Einstein transition probability per second; k – probability per second of radiationless transition. (Reproduced from [167] by permission of the International Union of Pure and Applied Chemistry, Oxford, UK.)

transition to the other excited level. In the cases of diagrams (b) and (d) the frequency of the fluorescence line is higher than that of the primary line (so-called *anti-Stokes fluorescence*). The difference in the photon energy is made up by the thermal energy of the flame and here a temperature dependence is to be expected. Further information on classification of the various non-resonance fluorescence processes can be found in [676, 668].

Little use of non-resonance fluorescence has so far been made in practical analytical work although it sometimes has certain advantages over resonance fluorescence. There is no interference from scattering due to non-evaporated aerosol droplets in the flame or other causes if the wavelength of the primary line is filtered out in the output beam [354]. It may also be preferable to measure the fluorescence at a wavelength other than that of the primary line if the latter lies in the wavelength range of a strongly interfering background radiation [499]. Furthermore, in some cases the probability of emission of a secondary photon is greater for the non-resonant than for the resonant fluorescence line [607]. Observation of the fluorescence at a non-resonance line as in the case of figure 13(a) has the advantage that the fluorescence is not weakened by self-absorption and that with the use of a continuum source the analytical curve does not flatten out and become horizontal (see § 8.5) [167].

The possibilities of using non-resonance fluorescence for atoms having terms that show multiplet structure, such as Tl, Ga, Cr, Pb, Sc, V and the alkalis, have been discussed in [167, 208, 288, 354, 673–675, 838, 885]. References [469, 500, 588, 675] deal with the complicated theoretical calculations of the fluorescence intensities for atoms with several excited levels lying close together.

It may be noted that the flame temperature can be found from the ratio of the intensity of the fluorescence transition $2 \rightarrow 0$, with radiative excitation $1 \rightarrow 2$, to that of the fluorescence transition $2 \rightarrow 1$ with radiative excitation $0 \rightarrow 2$ (see figure 13) [167, 667, 668].

We shall now concern ourselves only with resonance fluorescence, which so far has been of greater practical importance, and we shall presume excitation by primary radiation from the ground level only.

The total intensity of the primary radiation absorbed in the flame as a function of the spectral characteristics of the primary light source and of the absorption line is again described by the equations derived for the atomic absorption method in § 6.5. Yet whereas in the absorption method the relative absorption or absorbance is relevant, it is the *absolute* amount of absorbed radiation that is of primordial importance in the fluorescence method. The half-intensity width of the lamp emission line therefore is much less critical here and the demands made in this respect on the performance of the discharge lamp are less stringent. For the same reason the bandwidth of the monochromator used with a continuum source does not have to be extremely small [286, 856]. In contrast to the situation in atomic absorption spectroscopy, various additional factors influence the magnitude of the fluorescence signal to be measured, namely the efficiency and the self-absorption of the fluorescence and the solid angle of acceptance of the optical detection system.

142

The *efficiency of fluorescence Y* is defined as the relative probability of an atom, excited by a primary photon, emitting a secondary photon. Let A be the Einstein transition probability per second of photon emission (see § 6.4) for the observed fluorescence line and k the probability per second of an excited atom losing its excitation energy through a quenching collision with a flame molecule. Then for the relative probability Y of an excited atom being deactivated through emission of a photon we obtain the *Stern–Volmer equation*:

$$Y = \frac{A}{k + A}, \qquad (6.24)$$

where A, as well as k, is expressed in s^{-1}. If several radiative transitions from the excited level being considered are allowed, then summation over the Einstein transition probabilities of all possible radiative transitions has to be substituted for A in the denominator of the equation.

In commonly used flames the efficiency Y, which is independent of the metal concentration, generally only amounts to about 1 to 10 per cent because the quenching through collisions with flame molecules like N_2 and CO_2 largely exceeds the deactivation through photon emission $(k \gg A)$ [214, 467, 469, 499, 502]. Incidentally, this is also a necessary condition for thermal radiation to occur in atomic emission spectroscopy (see § 6.7.4). However, when the nitrogen molecules are replaced by noble-gas atoms, for example of argon, then values of Y of up to more than 50 per cent can be expected in the stoichiometric hydrogen flame. The intensity of the fluorescent radiation is then enhanced by one order of magnitude. Similar considerations apply here as those put forward in dealing with the corresponding enhancement of chemiluminescent radiation in § 6.7.3.

The fluorescence efficiencies for the first resonance lines of Na, K, Rb, Li, Tl and Pb in the stoichiometric oxyhydrogen–argon flame are in the order: 75, 37, 33, 15, 33 and 22 per cent [464, 469, 470, 499, 502, 583, 585, 588, 589]. The stoichiometric or fuel-rich hydrogen flame is so satisfactory because the effective cross section of H_2O for quenching of the alkali first resonance lines is small compared with that of O_2 [499, 585–589]. Yet for the first resonance line of Sr for instance, H_2O has a comparatively large quenching cross section [462, 463]. This difference in the effect of H_2O has so far not been fully explained.

The improvement of the fluorescence signal on addition of argon instead of nitrogen has generally been confirmed in practice [615, 689, 838, 856], but in turbulent hydrogen flames this substitution is less effective because of the entrainment of air from the surroundings [689]. Nevertheless values of Y of about 20 to 30 per cent were found for the Mg 285·2 nm, Mn 279·8 nm and Fe 248·3 nm lines with the turbulent hydrogen–argon diffusion flame (see § 3.3) [689]. In a premixed air–acetylene flame Y is about 10 per cent for the Mg 285·2 nm line [931]. A summary of values of Y in different flames is to be found in [837] and a survey of quenching cross sections and their theoretical interpretation in [566, 582, 584]. An interpretation of the slow decrease of alkali quenching cross sections with rising temperature for diatomic quenchers has been presented in [583, 584].

The absolute radiant flux Φ_F emitted as fluorescence in a solid angle of Ω_F sr can be derived from equations (6.24) for Y and (6.16) for the absorbed radiation with a continuum source. Assuming low metal concentration so that self-absorption can be disregarded (see below), we obtain:

$$\Phi_F = \left(\frac{\pi e^2}{mc^2}\right)\lambda_0^2 Lf n_0 \Phi_{\lambda_0} Y\left(\frac{\Omega_F}{4\pi}\right). \tag{6.25}$$

L is the thickness of the flame in the direction of the primary radiation and Φ_{λ_0} is the primary spectral radiant flux (see Glossary) at λ_0, which can be calculated from the radiation characteristics of the continuum source and the parameters of the optical system. The other symbols have the same meaning as those used in equation (6.16).

For a line source equations (6.9), (6.11) and (6.24), with the assumption of a relatively narrow emission line and of an absorption line broadened only by the Doppler effect, lead to:

$$\Phi_F = 2\left(\frac{\log_e 2}{\pi}\right)^{1/2}\left(\frac{\pi e^2}{mc^2}\right)\lambda_0^2 Lf n_0 Y\left(\frac{\Phi_0}{\delta\lambda_D}\right)\left(\frac{\Omega_F}{4\pi}\right). \tag{6.26}$$

Here Φ_0 represents the primary radiant flux integrated over the total width of the lamp emission line. This equation follows from equation (6.9) if in the latter the exponential factor is approximated by $(1 - k_m L)$, where k_m is the peak value of the absorptivity $k(\lambda)$. This approximation only holds if $k_m L \ll 1$, that is for low metal concentrations when a resonance line is used.

By comparing equations (6.25) and (6.26) we can see that a continuum source and a line source (at low concentrations) yield about the same fluorescent flux if $\Phi_{\lambda_0} = \Phi_0/(\delta\lambda)_D$, that is if the radiant flux per unit wavelength of the continuum source is equal to the radiant flux of the line source divided by the width of the absorption line (see also [167, 904]).

If Y proves to be considerably less than unity (see equation (6.24) with $A \ll k$) then, for a given constant value of k, Y is proportional to A and therefore, according to equation (6.22), also proportional to f. Then by equations (6.25) and (6.26) the fluorescence intensity is proportional to f^2 [499], whereas the intensity of the thermal radiation and the strength of the absorption are proportional to A (or f). Furthermore, it follows from the two equations just mentioned that for low concentrations, apart from secondary effects (see § 8.2), the fluorescence signal is proportional to n_0 and consequently to the solution concentration. More elaborate equations for the fluorescence signal are to be found in [499, 607, 903]. The absolute values of the signals obtainable in the emission, absorption and fluorescence methods, under given flame and source conditions, are theoretically compared in [167, 904].

Self-absorption of the fluorescence radiation has not yet been considered here. The secondary photons, for instance those generated on the flame axis, have to travel through a certain pathlength in the flame in order to escape and some of them may hereby be lost through self-absorption. The fractional loss increases with

144

rising atomic concentration and extending flame thickness in the direction of observation. At high concentrations the self-absorption is detrimental in flattening the analytical curve or even turning it downwards. These effects are discussed more fully in § 8.5.

The theoretical treatment of absorption of primary photons and of re-emission and re-absorption of secondary photons in general is very intricate. When the fluorescence efficiency Y is close to unity, the emitted tertiary and higher-order photons have also to be taken into account. The radiation transport through the flame has then to be described by an integro-differential equation – a problem well known in astrophysics. An approximate solution, based on some simplifying assumptions but acceptable under flame conditions at 1 atm pressure, is set out in [167, 466]. Radiation-transport effects may become important in atomic vapour cells at low gas pressure (i.e. negligible quenching) and at high atomic densities (i.e. large probability of re-absorption of resonance radiation). The resonance radiation is then said to be trapped or imprisoned inside the atomic vapour cloud. As a consequence the intensity ratio of resonance line fluorescence and direct line fluorescence decreases as the atomic density increases [428].

The measurement of the fluorescence efficiency supplies information on the effectiveness of the molecules in deactivating an excited atom through impact. By means of the principle of detailed balancing (see § 6.7.4), from this the effectiveness of the molecules in exciting an atom from the ground level can be deduced. The resulting data are of great importance for atomic and molecular physics and have attracted the interest of theoretical physicists. References [176, 583] give an introduction to the problems involved and a survey of the experimental data; the more recent literature is reviewed in [566, 582].

If the resonance fluorescence is observed in a certain direction, for instance at right angles to the primary radiation beam, partial polarisation may, in principle, occur for certain radiative transitions (compare [79]). The degree of polarisation at sufficiently high gas pressure is lowered through depolarising collisions; this reduction however is counteracted by quenching collisions which also become more important at higher gas pressure. This explains the rather high degree of polarisation of the order of 10 per cent observed for the Cd line at 228·8 nm in an air–hydrogen flame at 1 atm pressure [271]. Other workers however found no significant polarisation of the fluorescent resonance lines of Tl, Cr, Ag in an air–acetylene flame [577, 671] and of the Rb resonance doublet at 780/795 nm in an oxyhydrogen–nitrogen flame [588]. On the other hand a high polarisation (and anisotropy) appeared in the radiation scattered by non-volatilised analyte particles. These dissimilar polarisation effects can be utilised for improving the fluorescence/scattering ratio, especially in turbulent flames [671].

Finally, molecular band fluorescence of alkaline-earth compounds and PO has also been observed in cool hydrogen flames with the use of a xenon lamp as background source [415, 478, 501]. The spectral intensity distribution of the fluorescence radiation was rather similar to the thermal emission spectrum.

7   Background Emission and Background Absorption of Flames

7.1. Introduction

Even without any sample or solvent introduced into the flame an emission or absorption signal which is caused by the empty flame itself may be observed. The origin of this signal is the so-called *background emission* or *background absorption of the flame*; its strength depends, apart from other factors, on the setting of the monochromator or the characteristics of the spectral filter. The spectrum of the flame background emission consists of rather sharply outlined bands of diatomic molecules or radicals, such as OH, on which diffuse bands of polyatomic molecules, for example CO_2, and continua due to recombination processes are superimposed. Spectra of the background emission of empty flames are reproduced in Appendix 1 (spectrograms 45–48). The absorption of empty (non-sooting) flames can be completely disregarded in the visible spectrum region, but in the ultraviolet and infrared it can be of importance in atomic absorption spectroscopy. Background fluorescence due to CH radicals near 431·5 nm has been observed in an oxy-acetylene flame with a tunable dye laser as excitation source [193]. NH, CN and OH fluorescence bands have been observed in an argon-shielded air–acetylene flame with a continuum source [351a].

This introduction deals with the general analytical significance of background emission and background absorption. In § 7.2 the origin and spectral distribution of the background emission of some frequently used flame types are described, as well as the factors controlling the intensity of this radiation. In § 7.3 there follows a brief explanation, based on the fundamental considerations in chapter 6, of the excitation mechanisms of the most prominent components of the flame background emission. Finally the spectral distribution and the strength of the flame background absorption are discussed in § 7.4. We will only consider here those aspects of flame background emission and absorption that are most essential for analytical flame spectroscopy; for a more comprehensive treatment we refer to the literature on this special subject [50, 58, 60, 61, 70, 72, 78, 82, 176, 715, 826, 827].

The *analytical importance* of the undesirable background emission is the resulting deterioration in the limit of detection and the reduction of the number of analytical lines or bands available for trace analysis. The ratio of the net intensity of the analytical line to the background intensity should be kept sufficiently high to make quantitative analysis possible. This can be done by narrowing the spectral bandpass of the monochromator or filter and by the appropriate choice of flame, fuel/oxidant mixture ratio, wavelength and height of observation in the flame. Even if all relevant conditions are kept as stable as possible, more or less rapid, spontaneous fluctuations of the flame background emission are unavoidable. These fluctuations reduce the precision of measurement and set a lower limit to the concentration level that can still be detected (limit of detection; see Glossary). Apart from these random fluctuations there may also be unnoticed changes in the (mean) intensity of the background emission due to variations in the measuring conditions or sample composition and as a consequence systematic errors may be introduced into the analytical results (see chapter 9). The background emission depends, for example, on the fuel/oxidant mixture ratio, which may not be strictly constant, and possibly it may be affected by the nature of the solvent and the rate of solution aspiration, or by the concomitants in the nebulised solution. Because of the risk of such interferences it is desirable to have a high ratio of net analyte intensity to background intensity.

Hence in practical flame spectroscopy we not only want the analytical lines to be excited as effectively as possible (high temperature; chemiluminescent excitation; strong reduction of the metal oxides; etc), but also the flame background at the wavelength of the analytical line, within the bandpass of the filter or monochromator, should be as weak as possible. In practice we often have to compromise in order to fulfil these two requirements and for this purpose the intensity ratio of the net analyte emission and the interfering background emission have to serve as criteria. The oxyacetylene flame for example is very hot and excites the introduced atoms strongly, yet for the Ag analysis at 338 nm or the Cs analysis at 852 nm the somewhat cooler oxyhydrogen flame, although providing weaker excitation, is to be preferred because its background emission is considerably lower and the analyte to background intensity ratio in the hydrogen flame is more favourable [302, 427].

The background emission can also interfere in absorption spectroscopy especially if it contributes appreciably to the photocurrent measured. It then causes curvature of the analytical curve (see § 8.4) and by its fluctuations lowers the precision of measurement. Even if the spectral lamp radiation is periodically interrupted ('chopped') and an AC amplifier tuned to the chopping frequency used, the inevitable shot-effect fluctuations caused by the (uninterrupted) flame background emission can introduce an inaccuracy in the absorption measurement. This holds particularly for instance in the white-luminous reducing acetylene flames with their strong background radiation [787, 788, 914].

Usually superimposed on the shot-effect fluctuations, especially with turbulent flames, is a flicker-noise component which is associated with the flame flicker. Its noise spectrum generally shows a fall-off with increasing frequency and the flicker noise is thus especially important in the low-frequency range (see § 3.4.4 where references to the literature are also given). The standard deviation of the flicker noise, unlike that of the shot-effect fluctuations, is proportional to the photocurrent itself and not to its square root value (see § 3.4.4).

148

Absorption by the empty flame in most cases can be disregarded in atomic absorption spectroscopy. Occasional moderate background absorption of the flame at the wavelength of the analytical line does not need to cause interference if a suitable measuring technique is used. But when the background absorption is very strong, interference can result from its fluctuations or from its dependence on the nebulised solvent or on the concomitants (see chapter 9). In the case of such dependences the zero reading for the absorption measurements should always be taken with nebulisation of a blank solution (see Glossary).

Among the various instrumental methods for correcting or automatically compensating the (quasi-)continuous background emission or absorption are: (i) the base-line method, that is recording the analyte plus background spectrum over a restricted wavelength interval including the analytical line; (ii) sequential or simultaneous measurement of the analyte plus background emission at the wavelength of the analytical line, and of the background emission alone at the same or at an adjacent wavelength respectively; similarly the background absorption can be corrected for by either using a nearby source line that is not absorbed by the analyte or by using the continuum emitted by an auxiliary deuterium arc within the bandpass of the monochromator [517]; (iii) intensity modulation of the spectral light source combined with AC detection for eliminating the DC background emission in absorption spectroscopy; (iv) wavelength modulation, that is a periodic scanning of the background plus analyte spectrum combined with an oscilloscope [430] or synchronous detection [807, 838a], and other techniques of derivative spectroscopy [809]; (v) modulation of the sample supply to the nebuliser (intermittent or alternating sample flow [434, 441, 442, 762]); (vi) Zeeman modulation methods utilising the wavelength shift of the atomic line in the source or in the absorbing vapour in a non-flame atomiser by a magnetic field [560, 560a, 921]; and (vii) the spectral stripping method in flame emission spectroscopy utilising a multichannel detector for recording a restricted wavelength interval including the analytical line, and a multichannel analyser [248]. A discussion of the advantages and disadvantages of some of these methods has been given in [862].

7.2. Background Emission

It is hardly possible to give a description of the background emission of flames that will be valid generally because the radiation is not only sensitive to temperature and varies with wavelength and the kind of gas mixture, but also depends on many other factors which include the gas/air or gas/oxygen mixture ratio, the gas purity, the burner type, the gas flow (laminar or turbulent), the sprayed solvent and the height of observation in the flame. The last two are particularly important for flames with direct-injection burners. The appearance of a recorded flame spectrum is also affected by the resolving power of the spectrometer. When large slit widths are used the detailed structure of the spectrum is smeared out and molecular bands look more like continua (quasi-continua).

In flame spectroscopy usually the part of the flame above the primary combustion zone is utilised for the analysis. The background radiation of this region emanates partly from the secondary combustion zone and partly from the interzonal region (see chapter 3). In the usual flames these two radiation components are difficult to separate and we shall therefore deal with them jointly. The use of a split flame (see Glossary) however makes it possible to eliminate the former component in actual analysis. Because of the use of chemiluminescence effects in practical flame spectroscopy (see § 6.7.3) we shall include in our discussion the radiation components that are characteristic for the primary combustion zone of hydrocarbon flames. Sometimes such radiation is also found above this zone in flames with a great excess of acetylene, and even in hydrogen flames with direct nebulisers when organic solvents are used (see spectrograms 2, 28 and 30 in Appendix 1) [293, 387, 562].

The spectra of *hydrogen flames* always show the marked OH bands with band heads at 281 nm, 306 nm (strong) and 343 nm (see spectrograms 27 and 31–33). The hotter oxyhydrogen flames also emit the Schumann–Runge O_2 bands between 250 nm and 400 nm, and a weak continuum in the blue and ultraviolet regions of the spectrum [648]. The continuum, which in the absence of impurities practically does not exist in the spectrum of air–hydrogen flames [924, 927], is attributed to the recombination of OH and H to H_2O (see § 6.2.3) [50, 58, 409, 826]. When on addition of nitrogen compounds free NO radicals exist in the flame, a green (quasi-)continuum can also result from the recombination of NO and O to NO_2 [826, 827] and NO bands appear in the ultraviolet (see spectrograms 45 and 46). The intensity of the O_2 bands can be appreciably reduced by feeding a slight fuel-gas excess into the hydrogen flame. The strong infrared vibration–rotation bands of H_2O at $0 \cdot 9$–$1 \cdot 1$ μm (see spectrogram 9) and at $1 \cdot 8$–$3 \cdot 0$ μm make these spectrum regions unsuitable for flame analysis; with oxyhydrogen flames weak wings of these bands extend into the red end of the visible spectrum. The background emission from the primary combustion zone does not substantially differ from the emission of other flame parts and therefore this zone does not manifest itself distinctly here.

Compared with other flames the hydrogen flames, including the recently used nitrous oxide–hydrogen flames [367], have the advantage of weak background emission. In the radiation from the region above the primary combustion zone of the nitrous oxide–hydrogen flame one finds bands of OH, NO (between 220 nm and 280 nm) and NH (at $336 \cdot 0$ nm and $337 \cdot 0$ nm) and further a (quasi-)continuum (from 350 nm to 600 nm) which has to be attributed partly to the recombination of NO and O [289, 900]. In the emission from the combustion zone a NH_2 band system additionally appears in the visible spectrum region [289].

Much more complex are the spectra of *hydrocarbon flames*, for example oxyacetylene flames, which not only produce the bands and the continua observed in hydrogen and carbon monoxide flames, but also emit radiations of hydrocarbon radicals (see spectrograms 11, 28 and 30). The radiation from the region above the primary combustion zone of an air–acetylene flame contains, besides the above-mentioned OH bands, another fairly strong component which with a spectrometer of only moderate resolving power appears as a quasi-continuum. Its spectrum extends from 600 nm to below 300 nm and gives the flame the well known blue colour similar to that of carbon monoxide flames. With flames burning with excess fuel in the free atmosphere this blue radiation and the ultraviolet OH bands are observed mainly in the emission from the outer zone of the flame in which the secondary combustion takes place [451, 542, 546, 924] (see figure 4). Also weak violet CN bands are found in air–acetylene flames, the intensities of which diminish with increasing distance from the combustion zone [201]. The oxyacetylene flames produce the O_2 bands referred to above, and some CO bands between 205 nm and 245 nm. In the infrared we note the existence, apart from the H_2O bands, of some strong vibration–rotation bands of CO_2 around $2 \cdot 8$ μm and $4 \cdot 4$ μm, and several CO bands at $2 \cdot 3$–$2 \cdot 8$ μm. If the primary oxygen supply is insufficient, the H_2O and CO_2 bands are mainly due to emission from the outer flame zone. The loss of energy ensuing from the strong radiation of these infrared bands causes a cooling of the rising flame gases [804].

The primary combustion zone of hydrocarbon flames is distinct from the other parts of the flame by its emission of mostly visible bands, namely CH bands at 387/9 nm and 432 nm and C_2 bands between 436 nm and 563 nm (the so-called Swan bands; see spectrogram 47). These bands may also be emitted by the region above the primary combustion zone with a very fuel-rich gas mixture (see spectrogram 48) [332]. Furthermore, the ultraviolet OH and CO radiations from the combustion zone are stronger than those from the zone above it [156, 157, 201]. In the cooler flames especially we may detect between 300 nm and 400 nm the Vaidya bands of the CHO radical. With an excess of fuel the atomic carbon line at $247 \cdot 8$ nm also appears (see spectrograms 30 and 45) [296, 387, 935]. If there is a considerable oxygen deficiency or if substances such as aromatic compounds are sprayed, the flame becomes white luminous as a result of the thermal continuous emission of unburnt soot particles [293].

In the spectrum of the primary combustion zone as well as of the red-violet interconal zone (see § 3.4.2) of fuel-rich nitrous oxide–acetylene flames we can detect CN, C_2, CH and NH bands between 300 and 700 nm [613, 694, 695] (see spectrogram 48). If the nitrous oxide–acetylene flames are fuel-lean or if they are premixed, their total background emission is generally much weaker than if they are fuel-rich or non-premixed and turbulent [650, 694]. The background emission is greatly suppressed by excluding the radiation from the secondary combustion zone through the use of a split flame (see Glossary) [447, 542, 546], yet the emission of the radicals mentioned persists if the flame is fuel-rich.

To complete this section we should mention that the very hot, rarely used *oxycyanogen flames* have a relatively weak background emission; in particular there is no OH emission unless the solvent in the nebulised solution contains hydrogen (see spectrograms 46 and 47 in Appendix 1) [384].

7.3. Excitation Mechanisms of Background Emission

The quantitative interpretation of the excitation of the flame background emission is often involved and in spite of diverse comprehensive investigations is not yet entirely clear. The background emission of the primary combustion zone especially and, to some degree, of the flame region above it also cannot be described in terms of thermal equilibrium (see chapters 3 and 6). Hence not only the concentrations of the emitting particles and the flame temperature, but also the various formation and excitation mechanisms have to be known. Moreover the probability of quenching collisions between the suprathermally excited particles and the flame molecules as well as the possibility of self-absorption must be taken into account (see § 6.7.3). An additional difficulty in dealing with the primary combustion zone arises from the very pronounced spatial variations of the gas composition [236, 269, 296]. Most of the investigations in this field were undertaken not directly for the purpose of explaining the background emission of the flames commonly used in analytical spectroscopy, but for understanding the combustion reactions or the structure of the radiating particles. Therefore these investigations were mostly carried out under special experimental conditions, for instance with flames of low temperature or burning at low pressure, with discharge tubes, explosions or shock waves. It is then difficult to apply the obtained experimental results in an interpretation of the phenomena observed in the usual analytical flames. Examples of combustion reactions that lead to the formation of excited CH and OH radicals were given in § 3.4.1. Here we shall discuss in some detail how the emission of the typical discrete OH bands at about 306 nm and the blue quasi-continuous radiation characteristic for carbon monoxide and hydrocarbon flames may arise (see also [176]).

Excited OH radicals can be formed in the primary combustion zone of, for instance, acetylene flames by the following chemiluminescence reaction [58, 70]:

$$CH + O_2 \rightarrow CO + OH^*.$$

Other possible reactions, also of importance for hydrogen flames, are [50, 58, 70, 269, 526]:

$$O + H \rightarrow OH^* \quad \text{and} \quad H + O_2 \rightarrow OH^* + O.$$

Here also intermediate excited states or complexes can occur. Furthermore, an OH radical can be raised from the ground state into the excited state when it is involved as the third partner in a recombination reaction between two other particles such as (see § 6.7.3 and [50, 70, 175, 526, 926])

$$OH + (O)H + H \rightarrow OH^* + H_2(O).$$

This reaction explains why the presence of H and OH radicals in hydrogen and acetylene flames in excess of their equilibrium concentration (see § 3.4.2) also leads to a suprathermal excitation of the OH bands above the primary combustion zone [201, 526, 924, 926]. Such suprathermal excitation is particularly conspicuous in cooler flames because here the thermal excitation through collisions with flame molecules is relatively weak (see § 6.7.4).

The origin as well as the true structure of the so-called blue 'continuum' radiation of carbon monoxide and hydrocarbon flames have been much debated. According to more recent opinions described in [50, 176] this is not simple recombination radiation. It is possible to show [307] that the flame radiation is not really a continuum, but has rotational fine structure which is smeared out at only moderate spectral resolution. The origin of the emission can be visualised in the following manner.

CO and O recombine in a three-body collision with a flame molecule as the third partner in the reaction:

$$^1CO + {}^3O + X \rightarrow {}^3CO_2^* + X,$$

where ^1CO denotes a CO molecule in the singlet ground state, ^3O an O atom in the triplet ground state and $^3CO_2^*$ a CO_2 molecule in an excited triplet state. The third partner X acts as a stabiliser of the newly formed CO_2 molecule (see § 3.4.2). This molecule then goes over, without radiating, into a neighbouring excited singlet state ($^1CO_2^*$) from which it can make an allowed radiative transition to the electronic singlet ground state

$$^1CO_2^* \rightarrow {}^1CO_2 + h\nu.$$

In this transition since the initial and the ground states are stable the frequency ν of the emitted photon can only have discrete values and the resulting spectrum will have a discrete structure. The expected complexity of the spectrum however may explain why at low resolution and high temperature the spectrum appears as a quasi-continuum. As a result of a quenching collision with a flame molecule X, the excited molecule $^1CO_2^*$ can also pass radiationless into the ground state

$$^1CO_2^* + X \rightarrow {}^1CO_2 + X,$$

while X will take up the released excitation energy. As all processes are basically reversible the reverse transition is also possible; it is described by the above reaction with the arrow pointing in the opposite direction. Excited $^1CO_2^*$ molecules can thus be formed in two ways: either through recombination of one CO molecule and one O atom; or through collisional excitation of a CO_2 molecule in the ground state by an X molecule. The spectral intensity I_λ (see Glossary) of the quasi-continuous radiation resulting from either process can therefore be expressed as the sum of the corresponding components [924, 927]:

$$I_\lambda = a_\lambda[CO][O] + c_\lambda \exp(-hc/k\lambda T)[CO_2].$$

Here a_λ and c_λ are coefficients dependent on wavelength and to some extent on temperature and overall composition of the flame, but not on the concentrations [O], [CO] and [CO_2]. The other constants have their customary meaning (see § 6.4). The first term on the right-hand side of the equation embodies the possibility of suprathermal excitation when O radicals are present in the flame in excess of their equilibrium concentration (see § 3.4.2). The second term represents the contribution of the thermal excitation of the CO_2 molecule through collisions with flame molecules. By means of this equation the visible and near-ultraviolet radiation from the region above the primary combustion zone of some air–acetylene, oxyacetylene, nitrous oxide–carbon monoxide and air–carbon monoxide flames could be calculated in absolute units with an error margin of a factor 2 [520, 924, 927]. For this calculation the concentrations of the partners involved, the temperature and the flame dimensions have to be known.

A similar mechanism seems to account for the emission of the (quasi-)continuous radiation resulting from the reaction $NO + O \rightarrow NO_2$ [176].

7.4. Background Absorption

The background absorption of the commonly used flames in the visible and near-ultraviolet spectrum regions is usually negligible. Thus, in a pure air–acetylene flame an absorption factor of only about 0·001 per cent was found at 589 nm and of about 0·1 per cent at 345 nm and 285 nm [160, 924]†. Appreciable absorption in flames containing OH radicals is observed at the resonance band at 306 nm which overlaps for instance the 306·8 Bi line [631, 648, 749, 801]. The strength of this absorption depends only on the OH concentration and the flame thickness, but not on the excitation conditions. Therefore the OH absorption can be utilised to determine the OH concentration in the flame, which may vary according to the position in the flame, the fuel/oxidant mixture ratio and the kind of nebulised solvent [58, 411, 526, 734, 925, 926]. The temperature dependence of the partition function of OH should be taken into account when relative OH concentrations are to be derived from measured OH absorbances in a wide range of flame temperatures [648]. Further, there is the possibility of appreciable absorption by C_2 radicals in acetylene flames and by CN radicals in nitrous oxide–acetylene flames; such radicals with excess fuel supply also exist above the combustion zone [341].

In the ultraviolet region the Schumann–Runge O_2 bands can also appear in absorption. Below about 250 nm a strong, apparently continuous absorption is found in air–acetylene and air–carbon monoxide flames, which with decreasing wavelength rapidly rises to high values [382]. In oxyhydrogen flames also, O_2 bands and an underlying weak continuum have been observed in absorption [648]. At the 213·9 nm Zn resonance line the background absorption can amount to as much as 40 per cent; if the flame is supplied with excess acetylene this absorption, to be attributed to the O_2 molecule [60], is somewhat less. At the same wavelength the hydrogen–nitrogen diffusion flame shows a background absorption of about 5 per cent [655–657]. Similarly appreciable absorption at the wavelength of the 185·0 nm Hg resonance line has been reported for pure acetylene and carbon monoxide flames, but not for oxyhydrogen flames [892]. Another example of background absorption is that at the wavelength of the 193·7 nm As line, strong in air–acetylene and air–hydrogen flames, but considerably weaker in the hydrogen–argon diffusion flame [728]. The nitrous oxide–acetylene flame with a nitrogen-gas shield is fairly transparent between 180 nm and 220 nm and can be used for the determination of non-metals such as As and Se which have absorption lines in this spectral region [540]. Below 170 nm CO_2 molecules can also contribute to the

---

† These low absorption factors were not measured directly but were derived from absolute measurements of the spectral intensity of the flame background radiation through application of Kirchhoff's law in thermal equilibrium (see § 6.6.1). Note that [924] contains a correction to one of the experimental values reported in [160].

154

background absorption of the flame [70]. Because of the strong atmospheric absorption, spectroscopic measurements are not possible at all below 170 nm without removing the air from the light path. In agreement with this strong background absorption is the finding that below 250 nm the absolute spectral radiance (see Glossary) of the background emission of an oxyacetylene flame is of the same order of magnitude as the Planck value for a black-body radiator at the temperature of the flame [385, 913]. This link between background absorption and background emission is stipulated by Kirchhoff's law (see § 6.6.1) when thermal equilibrium is established.

The use of the absorption method in the infrared region can again be complicated by strong H_2O and CO_2 absorption bands.

An (apparent) background absorption in flames with a large excess of acetylene can be caused through absorption by and scattering at soot particles [50]. Such absorption depends on the concentration and size of the particles and has a continuous spectral distribution. Absorption factors of 3 per cent due to soot particles have been observed in practical absorption analysis near the 422·7 nm Ca line [791].

8 Shape of Analytical Curves

8.1. Introduction

The measure (= relative intensity, absorption factor or absorbance) obtained from the meter reading, normally increases with rising concentration of the solution nebulised in the flame spectrometer. In the ideal case the measure is proportional to the concentration of the solution. The *analytical curve* which represents the *analytical (calibration) function* $x = g(c)$, where the measure x is a function of the solution concentration c (see Glossary), is then a straight line passing through the origin of the coordinate system. To determine this straight line we require only one reference point which can be found by means of a single reference solution. From now on, whenever we speak of the measure, we mean the *net* measure which is found by subtracting the blank measure from the total measure (see Glossary). The blank measure may, for example, originate from the background emission or background absorption of the flame, or in fluorescence spectroscopy from the scattering of the primary radiation at non-desolvated aerosol particles.

In practice unfortunately the analytical curves often show convex[†] or concave[†] curvature (see figures 14 and 15). Also combinations of these two types of curvature occur, giving rise to S-shaped (sigmoid) analytical curves. Convex analytical curves are particularly undesirable because they often render the concentration measurement less precise[‡]. In fluorescence spectroscopy the analytical curve may even have a turning point (inversion) or level out into a horizontal straight line (plateau); the measure then respectively decreases or stays constant as the concentration rises to high values (see figure 16). The analytically useful concentration range is restricted by such large departures from proportionality.

In this chapter we shall discuss the possible causes of curvature of the analytical curve. First in § 8.2 we consider the causes, common to FES, AAS and AFS, of deviation from proportionality between solution concentration and concentration of the optically active particles in the flame, mostly atoms. The dependence of the

---

[†] We define convex and concave curvature in the following way. If the slope, dx/dc, of the graph of the measure x as a function of the concentration c decreases with increasing concentration we shall speak of convex curvature, and in the opposite case of concave curvature. The curves can be visualised as the contours of convex and concave lenses respectively viewed from 'above', that is from the side of high values of x.

[‡] Convex curvature of the analytical curve however does not always render the analysis less precise; for, when the precision of reading the meter deflection is governed chiefly by fluctuations in the analyte transport (e.g. fluctuations in the nebulisation), the effect of these fluctuations on the precision of the reading is also mitigated by the convex curvature of the analytical curve.

emission, absorption and fluorescence signals on the atomic concentration in the flame is treated separately in § 8.3, § 8.4 and § 8.5 respectively.

Furthermore, departures from proportionality between solution concentration and meter reading may occur when the output current of the photodetector is not proportional to the radiant flux incident on the light-sensitive surface. This feature is often to be regarded as a saturation effect; in general it leads to curvature of the analytical curves. The effect is only to be expected for very high radiant flux. Furthermore, the deflection of the meter need not necessarily be strictly proportional to the input current; this again can lead to curvature of the analytical curve. Similar departures from proportionality are possible in the performance of amplifiers or, in AC methods, of rectifiers. With photon-counting techniques convex curvature of the analytical curve may be caused by a pile-up or overlap of the photomultiplier pulses in the electronic circuitry [485, 486, 786, 793]. The extent of this effect depends on the pulse width and pulse rate and can be reduced by cutting down the aperture of the optical system [786]. In general, with counting techniques, the upper limit of linear operation may be several orders of magnitude lower than with conventional current measurement techniques [486].

The derivative dx/dc is generally referred to as the *sensitivity* of the spectrometer (see Glossary). In the case of curvature of the analytical curve this sensitivity varies with the concentration c. Absolute and comparative statements on the sensitivity can only be made in absorption spectroscopy as x then unambiguously represents the absorbance or the absorption factor (see Glossary). In emission and fluorescence spectroscopy on the other hand x is measured relatively and expressed in arbitrary units (such as scale divisions) so that comparisons of different instruments are rather meaningless [362].

If in the emission or fluorescence methods the radiant intensity were measured in absolute units the sensitivity could be defined unambiguously for them also. Absolute radiation measurements and quantitative knowledge of all parameters governing the sample transport, the excitation, etc, would allow absolute concentration determination without reference solutions [366]. This would similarly be true for the absorption method if here the parameters determining the absorbance were also accurately known [366, 596]. Absolute calibration however requires full theoretical knowledge of the nature of the processes taking place in the flame spectrometer, but at the present time we do not possess this knowledge (compare [168]).

The degree of departure of the analytical curve from proportionality is sometimes stated in terms of the slope, $d(\log x)/d(\log c)$, of the graph of the measure x plotted on double-logarithmic scales against the concentration c (compare figure 15). The relationship between x and c in the vicinity of a point on the analytical curve at which this slope is p can be formulated as:

$$x = \text{constant} \times c^p.$$

When $p = 1$ then x is proportional to c and the analytical curve, plotted on a linear scale, is a straight line passing through the origin; if c changes by 1 per cent, so does x. For $p > 1$ the analytical curve is concave and for $p < 1$ is convex. Data for p (or an

158

equivalent quantity) as a function of concentration have been given by some authors for certain atomic lines and measuring conditions (e.g. [389, 700]).

It is often advisable to choose the quantity measured or to transform the coordinate system so that for the working range of concentration a linear relationship, that is a rectilinear analytical curve, results [518, 519]. This facilitates the numerical evaluation of the analytical results. A straight-line graph can often be obtained simply by plotting the analytical curve on double-logarithmic scales. Such representation also has further advantages, namely: the concentration range is easily extended to several decades; the concentration scale expands for decreasing concentration; percentage scale-reading errors are independent of the concentration; a given concentration ratio corresponds to a fixed distance along the log c axis [518, 519].

The shape and the position of the analytical curve often depend on certain parameters, for instance the height of observation in the flame or the concentration of a concomitant (see figure 14). This dependence can be expressed in either a two-dimensional graphical representation by plotting separate curves corresponding to various discrete parameter values (see figure 14) or otherwise can be illustrated in a three-dimensional representation [612, 729].

8.2. Dependence of Analyte Concentration in the Flame on Solution Concentration

The relation between the concentration of the solution and the concentration of the observed particles in the flame is affected by various factors and processes, such as nebulisation, desolvation, volatilisation, dissociation and ionisation (see chapters 4 and 5). If the influence of these factors itself depends on the concentration of the analyte, deviations from proportionality are expected to show in the analytical curve.

High salt concentration of about $10\,000\ \mathrm{mg\,l^{-1}}$ or more can change the physical properties of the sample solution, for instance the viscosity and vapour pressure, so much that the aspiration rate and the efficiency of nebulisation are altered, usually reduced (§ 4.6.2 and chapter 9). The *sample transport* to the flame thus becomes dependent on the salt concentration and curvature, usually convex, of the analytical curve in the range of very high concentrations results (see figure 15 and pertinent text). Such curvatures are observed with chamber-type nebulisers as well as with direct-injection burners [309].

Convex curvature of the analytical curve can also be due to incomplete and concentration-dependent *volatilisation* of the dry aerosol particles in the flame (see § 4.6.4). The time necessary for complete volatilisation of a particle depends on its size, therefore on the concentration of the solution, and the fineness of nebulisation, and furthermore on the composition of the particle and the flame temperature. Hence such curvatures are to be particularly expected with direct-injection burners in the range of high solution concentrations. Volatilisation losses with chamber-type nebulisers in general are only likely to occur at concentrations of

159

more than $1\,g\,l^{-1}$; examples were given in § 4.6.4. The analytical curve for the CaOH band at $622\cdot7$ nm in an air–acetylene flame with a chamber-type nebuliser was found to be rectilinear for concentrations up to more than $1\,g\,l^{-1}$ when a $CaCl_2$ solution was sprayed. But when a silicate was added convex curvature started at a concentration as low as about $10\,mg\,l^{-1}$ [779, 929]. (Calcium silicate is known to be involatile.) For similar reasons the analytical curve may become convex if involatile compounds, for instance of Ca and Al or Ca and P, are formed (in the presence of excess Al or P) from the constituents of the nebulised solution. The degree of these curvatures usually diminishes with increasing height of observation in the flame, that is, with lengthening the time available for volatilisation. Also, concave curvature of the analytical curve in the lower concentration range can come about when a certain quantity of the analyte, for example Ca, is lost for the analysis by becoming bound in the solid phase by a fixed quantity of a concomitant, for example Al [172, 865].

The analytical curve for a solution of more complex composition may even have a turning point (inversion) as a result of volatilisation effects. Thus the analytical curve for Ca in an air–propane flame on addition of a 1 N solution of H_2SO_4 and of a fixed quantity of Fe showed a peak at a molar Ca concentration of about twice the molar Fe concentration [777]. At this molar ratio the iron has an optimum protective effect against the depression of the calcium through sulphuric acid. At higher calcium concentrations the protective effect decreases rapidly and the strong depression of the calcium through the sulphuric acid sets in again. Consequently in a certain concentration range beyond the peak of the analytical curve the Ca emission falls off with increasing Ca concentration.

In principle convex curvature of the analytical curve can also originate from concentration-dependent dissociation of the metal compounds in the flame, for example Na_2 or KCl (see § 5.2). Yet compounds containing two or more metal atoms, such as Na_2 or Sr_2O_2, hardly ever occur in flames [929] (see § 5.2.2). A curvature effect due to a varying degree of *dissociation* of halides, for example KCl, is usually negligible because most of the chlorine is bonded to the hydrogen in the flame in the form of HCl. Only when chlorine is added to the solution in excess, for instance as hydrochloric acid, can the formation of metal chlorides become significant. But then the degree of dissociation is independent of the metal concentration and the analytical curve, though shifted, remains straight. For the same reason formation of metal oxides or hydroxides in the flame does not lead to curvature because the degree of dissociation of these compounds is also independent of the metal content (see § 5.2.3).

Certain elements, for instance Cr, in very high concentrations can have a catalytic effect on the recombination of excess free H and OH radicals in the flame (see § 5.2.4). When these elements form molecular compounds in the flame that are dependent on the H or the OH content the degree of dissociation is no longer constant at high element concentrations. If measurements are taken on an atomic line, then convex curvature may appear in the analytical curve [245].

Of greater practical importance is the frequently observed concave curvature of the analytical curve due to a concentration-dependent *ionisation* of the analyte (see

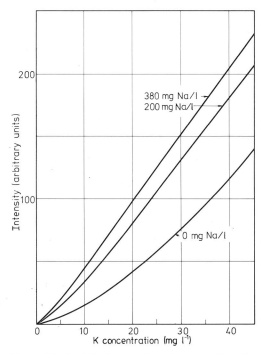

Figure 14. Analytical curves for potassium with sodium additions of different concentrations (after [794]).

figure 14). As pointed out before in § 5.3.1 this steepening of the analytical curve with increasing concentration can be explained by means of the Saha equation for ionisation in the following way: at low concentrations the analyte, for instance K, is ionised in the flame to such a high degree that a large part of it cannot contribute to the atomic signal. According to the Saha equation (5.7) the ionised K fraction and thus the relative loss in signal must be expected to diminish continuously with increasing K concentration. This means that in practice the atomic K concentration in the flame and hence the atomic emission, absorption and fluorescence signal must first rise more than in proportion to the solution concentration, which leads to the observed concavity of the analytical curve. At high concentrations the ionisation can be virtually neglected so that this type of curvature only manifests itself in the low concentration range and only with elements of low ionisation energies and/or with hot flames [174, 179, 322, 703, 711, 717]. Such concave curvature has also been found in atomic emission spectroscopy with hot inductively coupled argon plasmas [781]. As can be seen from figure 14 the concavity of the analytical curve for K is furthermore influenced by the amount of Na present besides K. Addition of Na, Cs, etc, not only enhances the K concentration (see chapter 9), but also straightens the analytical curve [179, 221, 427, 453, 567, 614, 778, 780, 891]. This may be attributed to the rise in the free electron concentration in the flame as a result of the addition of a partially ionised element. Through the electron concentration enhancement the ionisation of K is suppressed and the cause of the concave curvature is rendered less effective or eliminated completely. Such straightening of the analytical curve for low K concentrations is also possible without addition of a concomitant to the solution, namely in case the flame gases themselves produce enough free electrons (see § 5.3.1).

When the degree of ionisation γ is close to unity, that is at strong ionisation, according to equation (5.9) the atomic K concentration is approximately:

$$[K] \equiv (1-\gamma)[\bar{K}] = \text{constant} \times [\bar{K}]^2,$$

where $[\bar{K}]$ is the total element concentration which is proportional to the solution concentration. Therefore the atomic emission, absorption or fluorescence then increases with the square of the solution concentration. The results of calculations of the atomic concentration $[M]$ as a function of $[\bar{M}]$ in a flame of temperature $T = 2400$ K for M = Na, K, Rb, and Cs are graphically represented in [929]. The curvature of the analytical curves resulting from the ionisation effect is found to be dependent on the ionisation energy and on $[\bar{M}]$. The value of $d(\log [M])/d(\log [\bar{M}])$ (see § 8.1) varies between 1 and 2. The limiting value 2 corresponds to the given quadratic relationship: $[M] = \text{constant} \times [\bar{M}]^2$.

The opposite effect, namely strengthening of the concave curvature, is found after addition of hydrochloric acid [567], for through the formation of Cl^- ions in the flame the metal ionisation and the resulting curvature of the analytical curve are enhanced (see § 5.3.1).

The lowering of the degree of ionisation with increasing total analyte (or solution) concentration, on the other hand, leads to convexity of the analytical curve when an ionic line, for example the Sr^+ line at 407·8 nm, is used for analysis [298, 453, 781]. As the existence of a Saha equilibrium for $SrOH^+$ in flames must be considered doubtful according to recent investigations [421], the shape of the analytical curve for Sr^+ cannot be deduced from the Saha equation.

Finally, a concavity of the analytical curve originating from partial ionisation of the metal atoms must show itself in measurements not only on the atomic lines but also on the oxide or hydroxide bands of the element [322, 457]. This is concerned with the chemical-equilibrium relations between the metal atom and the metal oxide or hydroxide concentrations, according to which the molecular concentration [MO(H)] varies proportionally with the atomic concentration [M] (see § 5.2.3).

A particular case of concave curvature of the analytical curve is found in work with nitrous oxide–acetylene flames burning on a rectangular slot burner [884]. Consider an analyte that is almost completely volatilised at the height of observation in the concentration range of interest. The time required for complete volatilisation may however depend on the analyte concentration. The higher the concentration the shorter will be the time during which the free analyte atoms and molecules can diffuse sideways out from the median plane of the flame. Therefore with increasing concentration one expects a decrease in the percentage loss of free analyte atoms in the median plane that is connected with this *lateral diffusion*. Consequently the atomic concentration in this plane will increase more rapidly than in proportion to the solution concentration. When the analyte emission or absorption is measured along a line through the flame centre parallel to the length of the rectangular slot the analytical curve will be concave. This effect is closely related to the lateral diffusion interference discussed in § 9.2.

8.3. Dependence of the Emission Signal on the Atomic Concentration in the Flame

In emission spectroscopy departures from the proportionality between concentration of the radiating particles in the flame and radiation intensity can be caused by self-absorption which leads to convex curvature of the analytical curve of resonance lines in the range of higher solution concentrations (see figure 15). This curvature effect can be distinguished from the effect of a concentration-dependent efficiency of nebulisation and/or fraction volatilised (see § 8.2) by comparison with the analytical curve for an oxide or hydroxide band of the same element. Such molecular bands normally show no self-absorption in flames (see § 6.2). If the analytical curve for the band emission is found to be convex its curvature therefore must be attributed to a decrease in the efficiency of nebulisation or in the fraction volatilised. Consequently the analytical curve for the atomic emission can be corrected for any occurring variations of this efficiency, etc, by dividing the atomic emission by the corresponding band emission [455, 929]. This holds because

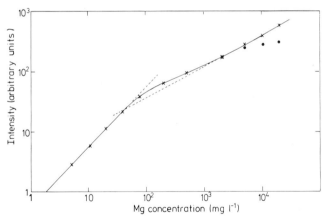

Figure 15. Analytical curve plotted on double-logarithmic scales for atomic emission of the 285·2 nm Mg resonance line in an air–acetylene flame at $T = 2410$ K. · · · · measured values (uncorrected); × × × measured values corrected for variations of efficiency of nebulisation and/or of fraction volatilised; —— curve calculated from theory for a-parameter $= 0·4$ (see § 6.3); – – – asymptotes of theoretical curve for low and high concentrations, corresponding to a linear and a parabolic relationship between emission and concentration respectively. The right-hand asymptote intersecting the theoretical curve at intermediate concentration is a characteristic feature for a-parameter values below 1. (Reproduced from [929] by permission of the American Chemical Society.)

variations in the nebulisation efficiency, etc, affect the atomic and molecular concentrations in the flame to the same extent.

As discussed in § 6.6.1 self-absorption, that is absorption of radiation by the same kind of atoms as those emitting it, causes a radiation loss that increases in percentage with increasing atomic concentration. Hence the atomic emission in the higher concentration range rises less than in proportion to the concentration. This is apparent in the double-logarithmic plot of figure 15 in the slope of the analytical curve which becomes less than 45°. It is a characteristic feature that with strong self-absorption the emission at high concentration increases with the square root of the concentration (see § 6.6.1). In a graph plotted on linear scales the analytical curve then assumes the shape of a parabola with a horizontal axis. In the double-logarithmic plot of figure 15 the analytical curve again becomes rectilinear for the concentration range from $1000 \, \text{mg} \, \text{Mg}/\text{l}$ upwards, but with a slope of $\tan^{-1}\frac{1}{2}$. Self-absorption and the ensuing convexity of the analytical curves are only to be expected for the resonance lines which result from transitions to the ground state. Therefore such curvatures are observed with the Na 589 nm, K 770 nm, Cs 852 nm, Li 671 nm, Mg 285 nm, Ca 423 nm, Sr 461 nm, Cu 237 nm, and Ag 328 nm lines, etc, but not with the Na line at 819 nm (compare figure 7 and table 4). The resonance lines of ions like Sr^+ can also exhibit perceptible self-absorption if the ionic concentration in the hot flames becomes high enough.

The foregoing statement that at high concentrations, which means at strong self-absorption, the emission increases as the square root of the concentration is only valid if the flame temperature is uniform over the whole cross section of the observed flame volume that is coloured by the metal atoms; the concentration of the emitting atoms need not be uniform however. Furthermore, absence of resonance broadening has to be assumed so that the a-parameter is independent of the metal concentration (see § 6.3).

If the temperature in the outer coloured flame zone is lower than at the centre, self-reversal will occur at high concentrations (see § 6.6.2). The emission then increases less than in proportion to the square root of the concentration [160, 389, 813]. This can be visualised as follows. With rising concentration the centre of the volume elements that effectively contribute to the emerging radiation continuously shifts towards the cooler outer zone for more and more of the radiation from the inner volume elements is lost through self-absorption. As the solution concentration increases, the mean temperature of the effectively radiating elements drops and renders the analytical curve even more convex. This additional effect of the self-reversal can in practice be easily confused with the transport effect mentioned in § 8.2.

From what has been described in § 6.6.1 it follows that the strength of the self-absorption and with it the position of the more or less marked knee in the double-logarithmic plot of the analytical curve (see figure 15) also depend on the flame thickness, on the a-parameter and on the oscillator strength f of the analytical line. In general self-absorption is stronger in thicker flames. The magnitude of the a-parameter in particular affects the shape of the intermediate section of the analytical curve where it changes from a linear to a parabolic shape. The difference

164

in the value of f is the reason why the curvature of the analytical curve due to self-absorption for the Na resonance line at 330 nm sets in at a much higher Na concentration than it does for the yellow Na doublet [431]. Moreover, the position of the knee in the analytical curve for a given flame and line corresponds to a value of the solution concentration that is the lower, the greater the number of free metal atoms that are delivered by the nebuliser per unit of solution concentration. Conversely, from the position of the knee in the curve, conclusions can be drawn about the efficiency of nebulisation.

It has been pointed out in § 6.7.3 that self-absorption can also occur with chemiluminescent resonance lines. Therefore convex analytical curves are also found when use is made of suprathermal chemiluminescence, for example for the resonance lines Bi 223·1 nm, Cd 228·8 nm and Be 234·9 nm [299, 389].

With flames in which there is an excess of noble gas instead of nitrogen, at low metal concentration the radiation intensity may be considerably different from the thermal radiation intensity (see § 6.7.4). This arises from the absence of equilibrium between excitation by photon absorption (self-absorption) and deactivation by photon emission. Since in argon-diluted flames the efficiency of resonance fluorescence (defined by equation (6.24)) is high this radiative non-equilibrium may have a noticeable effect on the excited-state population. But the radiation intensity inside the flame generally increases with rising atomic concentration and therefore the excitation of atoms through photon absorption becomes more important with rising concentration for resonance lines. Hence the departure from the thermal radiation intensity becomes smaller at higher atomic concentrations. This concentration dependence of the relative population of the excited state leads to concave curvature of the analytical curve in emission spectroscopy [176, 467, 469]. Flames of the type in question however are mostly used in fluorescence spectroscopy only and not in emission spectroscopy.

8.4. Dependence of the Absorption Signal on the Atomic Concentration in the Flame

In absorption spectroscopy the shape of the analytical curve depends on the choice of the quantity that is plotted as a function of the solution concentration: the absorbed intensity $(I_0 - I_t)$, the absorption factor $\alpha \equiv (I_0 - I_t)/I_0$, or (the usual choice) the absorbance $A \equiv \log_{10}(I_0/I_t)$ (see § 6.5). For low concentrations the analytical function is linear in all these cases, both with a line and a continuum background source. In the region of higher concentrations the analytical curve for α necessarily bends and approaches a horizontal asymptote when α approaches its maximum value of 100 per cent. The resulting strong convexity also depends on the spectral profile of the emission line of the background source. When a continuum source is used with a monochromator whose bandwidth is large compared with the width of the absorption line, α does not normally reach its saturation value. In this case actually the total absorption is measured (see § 6.5.3). On the basis of the theory described in § 6.6.1 it can be shown that then the analytical curve plotted

165

with α or $(I_0 - I_t)$ as ordinate is fundamentally similar in shape to the curve of growth in emission spectroscopy which depends on the a-parameter. For high solution concentrations c the same parabolic relationship $\alpha \propto \sqrt{c}$ is therefore to be expected as that in emission spectroscopy which is due to self-absorption (see § 8.3) [164]. Such a relationship has in fact been confirmed by experiment [333, 352, 609, 807].

If the bandwidth of the monochromator is of the same order of magnitude as the width of the absorption line (about 0·01 nm or less) the saturation value $\alpha = 100$ per cent can also be approached when working with a continuum source [624]. This will occur in fact if the solution concentration becomes so high that the width of the absorption line profile becomes noticeably greater than the bandwidth of the monochromator. It has to be remembered that the spectral profiles of a resonance line are the same in emission and absorption (see § 6.6.1). The width of the absorption line profile therefore increases with rising concentration in exactly the same way as shown for an emission line in figure 10. For further discussion of the subject we refer to [176, 369, 495a].

Under ideal conditions the absorbance A is strictly proportional to the atomic concentration n in the flame for any value of n (see § 6.5.2). For this reason the absorbance in general is preferred to the absorption factor as a measure of the solution concentration [363] although the use of A in practice also has some disadvantages [929]. The conditions are ideal when the background source (e.g. a single-mode narrow-band dye laser) is strictly monochromatic, that is when it emits only one spectral line the half-intensity width of which is very small compared with the width of the absorption line. The details of the emission line profile under these circumstances are of no consequence. Similarly the spectral profile and a possible shift of the absorption line then have no effect on the shape of the analytical curve although they do influence its slope. This however only holds with the (acceptable) proviso that there is no resonance broadening (see § 6.3). A further restricting condition for exact proportionality between absorbance and atomic concentration is the requirement that all light rays have to travel in the flame along paths of equal lengths (see below).

Departures from the discussed ideal proportionality between A and n arise when the spectral width of the lamp emission line is not small compared with that of the absorption line. Many investigations have been concerned with the theoretical and experimental aspects of the ensuing convex curvature of the analytical curve [109, 363, 364, 379, 477, 528, 711, 731, 761, 783, 867, 870, 871, 902][†] (see also the literature on the calculation of the emission line profile cited in § 6.6.2). The shape of the analytical curve then depends on the spectral profiles of both the lamp line and the absorption line and on their displacement relative to each other (shift)[‡].

[†] Similar problems exist in the spectrometry of liquid samples with non-monochromatic light sources [155].

[‡] In consequence of the far-reaching assumptions made in some theoretical computations for the purpose of simplification (pure Doppler or Lorentz profiles of both lines; no line shift; no hyperfine structure) these computations are often of little practical interest.

Because the half-intensity width and the entire profile of the lamp line usually vary with the lamp current – self-absorption and self-reversal increase with it – the shape of the analytical curve may depend on the operating conditions for the lamp [109, 364, 382, 783, 822, 867, 871]. Hence the convexity is often more pronounced at higher currents. In this case, the onset of the curvature of the analytical curve also depends on the value of the oscillator strength of the absorption line [438].

Consider, for example, a line source emitting radiation practically without self-absorption or self-reversal. The Doppler width of the lamp line is assumed to be about one-half of that of the absorption line. Let the a-parameter for the lamp line, because of the natural line broadening, be 0·01 and that for the absorption line, because of collision broadening, be 0·5. Calculation then gives for $A \leqslant 1·3$ (corresponding to $\alpha \leqslant 95$ per cent) a deviation from proportionality of less than 2 per cent [379].

Apart from the influence exerted by the lamp linewidth which we have considered so far, the absorbance may not be proportional to the atomic concentration n in the flame for the following reasons:

(i) Within the spectral bandpass of the monochromator or filter the lamp may also emit other lines of the analysis element, especially in the case of elements with complex spectra like Fe and Ni [365, 382]. If the other lines are not resonance lines they are virtually not attenuated by the flame and the (apparent) absorbance indicated by the meter will, for high values of n, tend to a constant limiting value $\log_{10}((I_0 + i_0)/i_0)$ where I_0 and i_0 are the non-attenuated intensities of the used resonance line and the interfering non-resonance line respectively. The curve of the (apparent) absorption factor $\alpha(\lambda)$, plotted as a function of n (contained in $k(\lambda)$), is then described by equation (6.7) modified through multiplication of the right-hand side by the constant factor $I_0/(I_0 + i_0)$ [751]. An example of this is the analytical curve for the Ni resonance line at 232·00 nm rendered strongly convex by the presence of a neighbouring unabsorbed line of the same element at 231·98 nm [260, 374].

The influence of non-resonance lines in the lamp radiation can be neatly avoided by using resonance fluorescence in which only the resonance lines are re-emitted as secondary radiation [764, 835, 875]. Such interferences can also be lessened by the use of both selective modulation [836] and a boosted hollow-cathode lamp with a pair of auxiliary electrodes for excitation of the atomic vapour [220, 260, 893]. Convex curvature can also be prevented through instrumental means, namely by a shift of the zero point so that the reading of the transmission factor of the flame becomes zero in the region of very high concentrations [639]. Also, if conditions permit, we can set the spectrometer to a narrower bandwidth in order to isolate, for instance in a Pb analysis, the interfering Pb line at 217·56 nm from the resonance line at 217·00 nm [789].

If the neighbouring line is also a resonance line of the analyte but has a different oscillator strength f, the analytical curve for the absorbance will again be convex, but saturation will not be reached. For low values of n the slope and position of the analytical curve are mainly governed by the line with the larger value of f, and for high n by that with the smaller value of f [711, 835]. Because the analytical curve for the line with the larger value of f is steeper than that for the other line, the gradual transition from one to the other section in the analytical curve results in a convex curvature. Similar to this interference by neighbouring resonance lines is the case of an analytical line with fine or hyperfine structure or with isotope shift [303, 365, 369, 379, 761, 867, 871]. The various components of the line usually have different absorptions because of their dissimilar values of f or dissimilar isotope concentrations. An example is the Cu doublet at 324·7/327·4 nm which, if not resolved by the spectrometer, gives a curved analytical curve [835, 876]. Moreover, the first component of this doublet has a partially resolved hyperfine structure in the flame† [364]. Sometimes, as with the Mn line at 403·1 nm [364] or the Ga line at 403·3 nm [871], the intrinsic half-intensity widths of the components may exceed their spectral separation in a flame but not in a low-pressure discharge lamp. The components of the absorption line in the flame then overlap† whereas the lamp emission line shows a resolved structure. This again complicates the relationship between the absorbance measured and the solution concentration when the spectrometer does not isolate the components of the emission line.

(ii) Within the spectral bandpass of the monochromator the lamp may emit lines of foreign elements that are not present in the flame. They can be lines of the noble gas used for the discharge [617] or lines of (metallic) impurities contained in the lamp, for example, a Cu impurity in a Zn lamp [382]. Such neighbouring lines, not attenuated by the flame, affect the analytical curve in a similar manner as described in (i). In particular their presence can make the analytical curve dependent on the bandwidth of the monochromator [438]. The emission of lines of the noble gas in the discharge lamp can be prevented by special instrumental arrangements [834] or alternatively their adverse effect on the measurement can be eliminated by means of intermittent or alternating nebulisation or aerosol supply [433, 441, 442].

(iii) If in addition to the chosen resonance line the lamp spectrum also includes a continuum the analytical curve will again be convex [611, 617] unless alternating or intermittent nebulisation or aerosol supply is used (see (ii)). The relative intensity of the continuum depends on the lamp current and the gas pressure [382]. That part of the continuum which lies within the narrow spectral interval of the absorption line is indeed attenuated in the flame, but usually this part is comparatively insignificant even with high metal concentration in the flame [173, 711, 871]. The effect of additional continuum radiation on the analytical curve is thus very similar to the effect of non-attenuated foreign lines (see (ii)).

(iv) Finally, a distortion of the analytical curve, easily remedied by instrumental means, can be caused if the thermal line or continuum radiation emitted by the flame itself affects the signal reading. Once more the analytical curve becomes convex [783], but this result can be obviated by appropriate defocusing and/or modulating the lamp radiation while using AC detection [173, 382, 787, 788].

---

† Clear distinction must be made between the spectral overlap of neighbouring lines of fine-structure and hyperfine-structure components in the observed spectrum (which may be due to insufficient resolving power of the monochromator) and the intrinsic overlap of these lines or components in the source (flame or lamp). The latter overlap would be observed even with a monochromator with infinite spectral resolving power.

The discussion above leads to the conclusion that a purely theoretical prediction of the shape of the analytical curves under the practical conditions met in atomic absorption spectroscopy is hardly feasible. The broadening and shift of the absorption line and the effect of self-absorption or self-reversal on the lamp line profile cannot be deduced from theory alone. However when the absorption and emission line profiles have been determined experimentally in a concrete case and if no complications arise (see above) the analytical curve can be computed with reasonable accuracy. Such computations have been carried out for some Al, Ca, Ga, In and Mn lines for an air–acetylene flame and several hollow-cathode lamps under different operating conditions [867, 871]. The calculated analytical curves could generally be described by the following expression in concentration c:

$$A = Sc - Bc^2,$$

where A is the absorbance and S and B are positive coefficients. This expression was found to give the deviation from linearity to an accuracy of 0·4 per cent for values of A up to 2. If the lamp current was held sufficiently low the calculated deviations from linearity for values of A up to 1 were found to be less than 8 per cent.

Previous calculations in which line shift and hyperfine-structure splitting had been disregarded led to the following approximate expression for the analytical curve [761]:

$$A = Sc/(1 + Dc),$$

in which the coefficient D is inversely proportional to the absorption linewidth. When this expression is expanded in a power series in c it is found to be equivalent to the expression given before up to second order in c, with B replaced by SD.

In practical atomic absorption spectroscopy of course we need not rely on theoretical calculations; we simply determine the analytical curve empirically.

8.5. Dependence of the Fluorescence Signal on the Atomic Concentration in the Flame

The theoretical appraisal of the shape of the analytical curves even in emission spectroscopy (with the problems of self-absorption and self-reversal) and in absorption spectroscopy (with the influence of the emission and absorption line profiles) is often difficult. However, the conditions in atomic fluorescence spectroscopy are even more intricate, for the fluorescence signal depends on the absorption of the primary radiation as well as on the self-absorption and self-reversal of the fluorescence radiation. All of these factors are in turn influenced in a complicated manner by the concentration and spatial distribution of the atoms in the flame. In this connection the dimensions of the irradiated and the observed flame volume, the a-parameter, the fluorescence efficiency Y, the spectral profiles of the lamp radiation and the absorption line in the flame and other circumstances play a part. In the theoretical calculation of the analytical curve very restrictive simplifying assumptions therefore have to be made. Hence the numerical results of the calculations cannot be directly applied to the majority of real flames and in practice the analytical curves must always be determined by measurement. Yet the calculations convey an understanding of the characteristic shape of the analytical curves and provide guidance for optimisation of the fluorescence method in practice. We shall not give here a quantitative treatment of the theory; for this we refer to the literature [102, 116, 167, 465, 670, 837, 905, 906, 929, 931]. Only the most important factors that affect the shape of the analytical curves will be qualitatively discussed and explained by means of a simple, idealised model.

The simplifications usually stipulated in a theoretical treatment of the subject are the assumptions of rectangular flame cross section with uniform atom and temperature distribution, and of a parallel beam of primary as well as of observed fluorescence radiation. The two beams are usually taken to be at right angles to each other. The spatial distribution of the radiant energy density of the incident primary beam is supposed to be uniform and the spectral bandwidth of this beam is either very large or very small compared with that of the atomic line in the flame. Also, the fluorescence efficiency and the a-parameter are considered to be spatially uniform. The further, usually tacitly implied condition that the spectral distribution of the re-emitted secondary photons be independent of the spectral distribution of the primary photons, normally seems to be satisfied under flame conditions [466]. In the calculations tertiary re-emission of photons absorbed from the fluorescence beam is usually disregarded as this would complicate the calculations. This neglect seems justifiable if the fluorescence efficiency is not more than 50 per cent. Evidently, in practice, some of these presumptions are only approximately or hardly at all fulfilled [167]. On the other hand similar conditions are often laid down in calculations of analytical curves in emission or absorption analysis. However, the consequences of such simplifications in fluorescence analysis are frequently more crucial than in other methods.

We shall examine here the usual case in which the fluorescence is measured on an atomic resonance line and in which the (mean) wavelengths of the primary exciting radiation and the fluorescence radiation are the same (so-called resonance fluorescence; see § 6.8). At low atomic concentrations (n) the absorbed primary radiant energy is proportional to n, with a continuum as well as a line source (see § 6.5). The spatial distribution of the excitation of the atomic vapour by the primary photons is then practically independent of n because the gradual attenuation of the primary radiation beam in its passage through the flame is still slight. The absorption of the secondary photons is also negligible at low concentrations. The fluorescence flux Φ_F is thus proportional to n and not dependent on the geometrical conditions. The analytical curve therefore is expected to follow initially a straight line through the origin – if we disregard the other causes of curvature discussed in § 8.2. This in fact is confirmed in practice [181, 279, 354, 394, 644, 838].

Proportionality no longer exists when the concentration rises so high that the flame seen in the direction of the primary and/or the fluorescence beam becomes optically thick. The product of absorptivity at the line centre, k_m, and flame

170

thickness is then no longer small compared with unity (see equation (6.10)). The absorbed primary radiant energy now increases less than in proportion to n and the percentage loss in fluorescence radiation through reabsorption is no longer negligible. These two facts combined lead to convex curvature of the analytical curve. The shape of the curve in the higher concentration range is strongly influenced by the geometrical conditions and it is distinctly different for a line and a continuum source [837]. Let us consider these questions more closely for the conditions illustrated in figure 16 (see the figure caption also). A flame with rectangular cross section of dimensions $(L+\Delta L)\times(l+\Delta l)$ is irradiated on one side over a width of l mm by a parallel uniform light beam of radiant flux Φ_p. The fluorescence beam of radiant flux Φ_F is observed over a width of L mm in a direction at right angles to the primary beam. Let the atoms be uniformly distributed over the entire flame cross section $(L+\Delta L)\times(l+\Delta l)$.

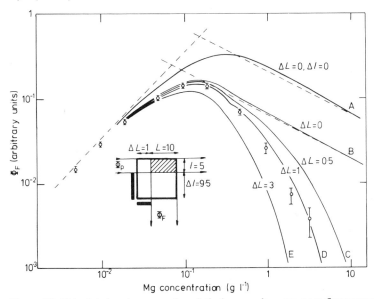

Figure 16. Calculated and measured analytical curves for resonance fluorescence of the 285·2 nm Mg line in an air–acetylene flame at $T=2410$ K with the line source emitting a relatively narrow line. Rectangular cross section $(L+\Delta L)\times(l+\Delta l)$ of the flame uniformly filled with atomic vapour is assumed. The hatched area $(L\times l)$ represents a rectangular section of irradiated flame volume observed by the spectrometer. The dimensions (in mm) are those of the actual experimental set-up. Φ_p: primary radiant flux incident from the left; Φ_F: observed radiant flux of resonance fluorescence plotted in arbitrary units as a function of solution concentration c; —— calculated curves for a-parameter $= 0·41$ (compare figure 15) if atomic concentration $n = 5·1\times10^{13}$ cm^{-3} with nebulisation of 1 mol l^{-1} Mg solution; L and l: for all curves equal to the actual experimental values; $\Delta l = 0$ for curve A, $\Delta l = 9·5$ mm for all other curves; ΔL: values (in mm) entered for each curve; – – – asymptotes for very low and very high concentrations; $\bigcirc$ measured values; $\underline{\mathrm{I}}$ error margins of fluorescence measurements. The positions of measured points should be compared with the calculated curve D. (After [931], reproduced by permission of Pergamon Press, Oxford, UK.)

171

The case of a line source. We first assume that a line source emitting a comparatively narrow line is used and that the entire cross section of the flame is irradiated and observed ($\Delta L = \Delta l = 0$). The amount of primary radiation absorbed in the flame is then given by the Beer–Lambert law as a function of n (see § 6.5.2). If n is very high, almost all the primary radiant flux Φ_p is absorbed and a part of it, defined by the fluorescence efficiency Y (see § 6.8), is converted into secondary photons. On their way out of the flame some of these photons are again absorbed by atoms in the ground state and their contributions to the outgoing fluorescence radiation are lost. Since within a small slice of the flame perpendicular to the direction of the primary beam the excitation of the atoms by the primary photons is uniform and the fluorescence is observed at right angles to the primary beam, this fluorescence loss is quite analogous to the energy loss of a thermal emission line due to self-absorption [167]. In the calculation of the self-absorption of an emission line the thermal excitation is similarly taken to be uniform along the line of observation. It follows from the theory of self-absorption (see § 6.6.1) that in the range of high concentrations (n) the fraction of generated photons that can leave the flame without being reabsorbed diminishes in proportion to $1/\sqrt{n}$. This explains the parabolic shape of the analytical curve in emission spectroscopy where the number of photons generated by the atomic vapour increases as n, but the number of photons actually leaving the flame only as $\sqrt{n}$. In fluorescence excitation with a line source the number of generated secondary photons for high values of n approaches a constant limiting value, controlled by $\Phi_p Y$, because all primary radiation is absorbed in the limit $n \to \infty$. The observed fluorescence radiation as a result of self-absorption must thus decrease in proportion to $1/\sqrt{n}$ for high values of n. Hence the analytical curve A in figure 16 with continuously rising high concentration goes down to zero so that a maximum results for an intermediate concentration value. The shape of the curve near the maximum depends on the a-parameter. As can be seen in figure 16, the right-hand-side asymptote in the double-logarithmic plot in fact has a slope of $\tan^{-1} -\frac{1}{2}$ corresponding to a $1/\sqrt{n}$ dependence.

This inversion of analytical curves obtained with line sources has been confirmed in practice, for instance for the resonance lines of Cd at 228·8 nm [394, 621], of Na at 589·0/589·6 nm [397, 467], of Pb at 217·0 nm and 283·3 nm [838] and for several Cu resonance lines [561, 797]. A maximum in the analytical curve has also been observed when a graphite-rod atomiser was used in combination with a line source [688]. Obviously this turning of the curve sets an upper limit to the useful concentration range in fluorescence analysis.

The instrumental set-up may also be such that the primary beam does not irradiate the flame over its whole width, so that between the fluorescent section and the spectrometer lies a part of the flame of thickness Δl that is not irradiated but absorbs the fluorescence radiation (see figure 16). This additional absorption of the fluorescence radiation from the irradiated section has an effect similar to that of self-reversal in emission spectroscopy (see § 6.6.2). The result is a steeper convex curvature of the analytical curve and a downward displacement of the descending asymptote, parallel to itself, for high concentrations (see curve B in figure 16, calculated for $\Delta L = 0$ and $\Delta l = 9 \cdot 5$ mm).

In another configuration only a part of the irradiated flame volume is observed by the spectrometer so that the primary beam then travels a distance ΔL in the flame before entering the observed volume (shown as the hatched area in figure 16). The primary radiation is then attenuated by pre-absorption, the more so the larger the product $n\Delta L$. Thus the convexity of the analytical curve especially beyond the maximum is enhanced still further (see curves C, D, E in figure 16, calculated for increasing values of ΔL). This effect can be visualised by imagining that with rising concentration the flame section in which most of the primary line radiation is absorbed moves more and more to the left and at the same time contracts so that it recedes from the observed volume.

The computed effects of pre-absorption of the primary radiation as well as of self-absorption and self-reversal of the fluorescence radiation have been confirmed for the $285 \cdot 2$ nm Mg resonance line by experiment [931] (see the positions of the measured points relative to the calculated curve D in figure 16). This work was carried out with an electrodeless Mg discharge lamp as line source, a special flame of rectangular cross section and with parallel light beams.

A point of interest to note is that at the maximum of the analytical curve the absorptivity k_m at the line centre is of the order of magnitude of 1 cm^{-1} while the dimensions of the flame cross section shown in figure 16 are of the order of magnitude of 1 cm [931]. Under these conditions analysis by atomic absorption spectroscopy is still possible. The corresponding absorbance is about $0 \cdot 5$ (see equation (6.10)). For work with a given flame the upper limit of the useful concentration range in AFS therefore is markedly lower than that in AAS.

If the fluorescence line is a non-resonance line and if its lower level lies sufficiently high above the ground level, no self-absorption or self-reversal will occur. The fluorescence signal with increasing concentration then tends to a constant, nonzero limiting value; the analytical curve has no descending branch, but ends in a straight horizontal line ('plateau'). Yet the curve may turn downwards even in this case if the effective primary radiant flux is attenuated by pre-absorption (see above). This probably explains the slight downward turn of the analytical curve for the non-resonance line of Pb at $405 \cdot 8$ nm which was excited through absorption of the $283 \cdot 3$ nm resonance line [838], an example of so-called direct line fluorescence (see § 6.8).

The case of a continuum source. The conditions are completely different when the resonance fluorescence is excited by means of a continuum source. Firstly, we again assume that $\Delta L = \Delta l = 0$ (see figure 16) so that there will be no pre-absorption or self-reversal. In the high concentration range the amount of the primary radiant energy that is absorbed, when integrated over the whole width of the absorption line, increases proportionally with $\sqrt{n}$ (see § 6.5.3). A constant fraction Y (defined as fluorescence efficiency) of this absorbed primary radiation is transformed within the flame into fluorescence radiation. As a result of self-absorption only a fraction of the fluorescence radiation proportional to $1/\sqrt{n}$ emerges from the flame (see above). The radiant flux reaching the spectrometer is thus proportional to $(Y\sqrt{n})\times (1/\sqrt{n})=$ constant. The analytical curve asymptotically approaches a horizontal straight line (plateau) of nonzero ordinate value in the range of which the signal no longer varies with concentration.

If the front zone of the flame nearest the observer is not irradiated (see the inset in figure 16 with $\Delta l \neq 0$), self-reversal again plays a role. The plateau of the analytical curve is displaced downwards parallel to itself and is also extended into the range of lower concentrations. As a consequence the upper limit of the analytically useful concentration range is lowered. There is a similar effect in the case of pre-absorption of the continuum radiation (see the inset in figure 16 with $\Delta L \neq 0$). Before ending in the plateau the analytical curve may pass through a maximum. The height of the maximum above the plateau and the shape of the curve in its vicinity generally depend on the a-parameter and on the degree of self-reversal or pre-absorption.

Plateaus in the analytical curves described are in fact found in practice when continuum sources are used [319, 856]. Analytical work is of course impossible in the corresponding concentration range.

The results of theoretical computations of analytical curves for work with continuum sources have been confirmed by experiments [847, 931] carried out under the conditions as illustrated in figure 16. When a dye laser with comparatively large spectral bandwidth was used the analytical curves were similar to those obtained with a continuum source, as expected [354].

With strong laser irradiation, saturation (see § 6.8) of the atomic transition may set in. Consequently the irradiated part of the atomic vapour becomes more transparent for the laser radiation as well as for the resonance-fluorescence radiation. The reduction in laser-beam absorption and/or self-absorption leads to an extension of the linear branch of the analytical curve in atomic fluorescence spectroscopy [570, 659, 677].

The discussed characteristic inversion or plateau of the analytical curve is also found when the fluorescence output of a 'resonance monochromator' is plotted as a function of the electric input power [683]. The atomic concentration in the vapour cell of this 'monochromator' increases monotonically with increasing electric heating of the cell.

For further theoretical discussion and explanation of the shape and the position of the analytical curves in fluorescence spectroscopy under different operating conditions we refer to the literature cited at the beginning of § 8.5.

9 Classification of Interferences of Concomitants

9.1. Introduction

We have seen that under constant instrumental conditions a relation exists between the optical signal and the analyte concentration which is represented in the simplest case by a straight analytical curve and in the more general case by a curved one. The position and form of the curve may, however, be affected by the nature and concentration of concomitants present in the solution. In this chapter we shall discuss the various ways in which concomitants such as salts of other metals, acids or organic constituents may give rise to what is called an interference on the measure. An interference does not cause an error in the analytical result (the sought concentration of the solution; see Glossary) if the flame analysis is done properly. This holds because flame analysis is a relative method and the reference solutions should normally contain the same concomitants as the sample in sufficiently similar concentrations. Accordingly, the interferences to be considered here, though affecting the measure, are not necessarily to be thought of as analytical errors. Our discussion will serve only as a basis for understanding the possible kinds of systematic errors and the means of eliminating them when the sample and reference solutions differ. The discussion bears also upon the intentional use of certain interference effects for improving the detection limit, for straightening the analytical curve or for suppressing unintended interferences of concomitants. In the present chapter we ignore the analytical significance of interferences and offer only a general classification and understanding of interferences as far as they are known.

For various reasons concomitants may be present in the sample solution readied for analysis, and their effect on the analyte signal may take many forms. The sample – for example, human blood serum to be analysed for potassium – may originally contain, along with the analyte (potassium), other elements such as sodium, calcium and magnesium in combination with various anions, as well as organic constituents such as proteins and surface-active materials such as bile acids. Substances such as these are called *concomitants in the* (original) *sample*. Errors will appear in an (improperly conducted) analysis only when the concomitants affecting the potassium measure are missing from the reference solutions or occur in variable, uncontrolled concentrations in the serum. It should be noted that the preparation of reference solutions that simulate the sample solutions in every respect is often laborious or practically impossible (e.g. because of the complex composition of biological samples).

Concomitants may also be introduced at fixed, known concentrations during chemical pretreatment of the sample solution, as in extraction, protein removal, or dilution. Concomitants of this kind will be called *additives*. The pretreatment is designed to avoid errors in the analytical result due to concomitants in the sample that occur in uncontrollable, varying concentrations. Moreover, an internal-reference element (e.g. lithium; see Glossary) is sometimes added to the solution in order to suppress errors due to variable instrumental factors such as unsteady nebulisation. For an element hard to excite thermally, such as tin, an alcohol may be used as an additive to enhance the emission by chemiluminescence (see § 6.7.3). In these last examples, the concomitants added to the sample solution cause no analytical error, for the blank and reference solutions of course receive the same additions.

In the following we assume that the instrumental factors, such as the height of observation and the emission of the background source in atomic absorption and fluorescence methods, remain constant. For simplicity we also assume constancy – not usually fully realised in practice – for the temperature and the level of the sample solution in the container from which it is aspirated by the nebuliser (see § 4.2). The background emission or absorption of the flame gases will enter into the discussion only when its spectral distribution or intensity depends on the composition of the nebulised solution.

The measure (see Glossary) is not always determined solely by the concentration of the analyte in the nebulised sample or reference solution. It has to be corrected for the *blank measure* that results when the blank solution (see Glossary) is nebulised. Ideally, the signal obtained by nebulising the blank solution – the radiant intensity or absorbance due to the flame gas itself and the concomitants – would be zero. But in the emission method this is seldom true, often because of inadequate spectral selection, and the blank measure must be subtracted from the total measure to obtain the net measure. The analytical result is finally obtained by comparing the net measures of the sample solution and reference solutions with the aid of the analytical curve. Here we shall discuss in a general way how concomitants may affect the blank signal and the net signal. Correspondingly, we shall divide interferences into blank interferences and analyte interferences.

This formal distinction between blank interferences and analyte interferences does not necessarily imply that correction for the blank measure is always made by a *separate* reading of the blank measure. In some flame atomic spectrometers this correction is performed automatically by instrumental means, particularly with respect to the continuous part of the background emission or absorption spectrum (see § 7.1). Since these automatic correction procedures are not always fully effective, a residual error in the analytical result due to blank interferences is still possible. A discussion of blank interferences is thus appropriate here too.

We speak of an *interference* when the blank measure or the net measure systematically departs from the value found on nebulising, respectively, the pure solvent or a simple solution containing only the analyte (at the same concentration and in the same chemical binding as in the sample). From this definition it follows that the effect of the nebulised solvent itself upon the measure is not to be called an interference. The discussion below is limited to the most common case in which the sample is pneumatically nebulised as a solution into the flame, with distilled water as solvent.

9.2. Classification of Interferences according to their Mechanisms

9.2.1. General. Table 5 offers a classification of the various kinds of interferences of concomitants according to their mechanisms. There are many problems in arriving at a suitable and unambiguous classification and nomenclature of the various interferences. Earlier investigators not only built their own flame spectrometers but also coined their own terms for the interferences that they observed. These various classifications and terminologies overlap considerably, so that even now a term such as 'excitation interference' can have quite different meanings.

Table 5   Types of interferences of concomitants

I Blank interferences

  Ia Cross-spectral interference
  Ib Background interferences
      Ib1 Interference on the flame background emission, absorption or scattering
      Ib2 Spectral overlap of emission or absorption by concomitants

II Interferences on the analyte signal

  IIa Non-specific interferences
      IIa1 Sample aspiration, nebulisation and desolvation interferences
      IIa2 Flame shape interference

  IIb Specific interferences
      IIb1 Solute volatilisation interference
      IIb2 Dissociation interference
      IIb3 Ionisation interference
      IIb4 Excitation and quenching interference
      IIb5 Lateral-diffusion interference

There is another difficulty in that interferences can be classified from entirely different points of view. For example, according to the type of concomitant, they can be divided into cation and anion interferences (see § 9.3.2), or, according to the part of the flame spectrometer in which the interference arises, into nebulisation interference, etc. If emphasis is on the mechanism, we have drop-size interference, ionisation interference, etc. Again, the nomenclature may express a certain judgment referring to interferences of the 'first' and 'second' kind, or to 'true' and 'apparent' interferences. In the current literature a subdivision into 'physical' and 'chemical' interferences is common. We feel, however, that the physical versus chemical dichotomy is less suitable; for example, interference from the volatilisation of the solid aerosol particles has both a physical aspect (e.g. sublimation temperature) and a chemical aspect (nature of the molecular bonding). Terminology found in the literature may also be based on the manner of correcting the interference: an example is 'multiplicative' and 'additive' interferences.

Our classification, like any other scheme, is not free from shortcomings, but we have tried to construct an unambiguous classification based on the mechanisms of the interferences, and to select the simplest and most obvious terms. There will be some inevitable overlap with earlier terms used in other senses.

We next give a brief, general explanation of table 5 for the emission method as well as for the absorption and fluorescence methods, starting with the subdivisions of class I (see also [237]).

9.2.2. Blank interferences. In *flame emission spectroscopy* the blank measure or blank interference can arise from two different causes.

Ia. Cross-spectral interference, due to instrumental inadequacy. Interfering radiation at wavelengths other than that of the analytical line or band, due to unwanted spectral transmission by the filter (filter leakage) or to stray light or limited resolution in the monochromator, etc, contributes to the blank measure. For example, measurement of the emission of the magnesium line at 285·21 nm may suffer interference from the closely adjacent sodium line at 285·28 nm when sodium is present at relatively high concentration.

Ib. Background emission, which spectrally overlaps the emission of the analyte, at least partly. Unlike Ia, it cannot in principle be entirely eliminated by improving the spectral resolution of the spectral apparatus, since it is also present at the wavelength of the analytical line or band. A concomitant can cause a background interference in two ways: *Ib1. Interference on the flame background emission,* the concomitant altering the continuous flame background emission or the superposed band emission of the flame gases. Such interference is especially likely with direct-injection burners and combustible constituents of the solution. *Ib2. Overlapping emission from the concomitant,* usually of quasi-continuous spectral distribution. Many elements such as molybdenum and barium display a nearly continuous emission or an extensive band structure covering a wide range of the spectrum, apart from their specific line emissions. Such an element can raise the background at the analytical wavelength. For example, the CaOH band at 554 nm interferes in measuring the barium emission line at 553·6 nm when calcium is present at relatively high concentration [755].

It must be emphasised that these last two interferences, Ib1 and Ib2, cannot be entirely relieved even with the best possible monochromator or interferometer. They are to be distinguished from the interference arising from the use of an instrument of poor spectral resolution such as a colour filter, which may be unable to separate adjacent lines of the analyte and the concomitant. Such interference is classed under Ia; it can be fully eliminated by improving the spectral resolution. The same is true of filter leakage, that is, residual transmission by a filter of a part of the spectrum where a concomitant element has a strong interfering line. However, when there exists an intrinsic spectral overlap (see § 8.4(i)) of the interfering and analytical lines in the source, the interference cannot be eliminated by better spectral resolution and is to be classed under Ib2.

In *flame atomic absorption spectroscopy* with a spectral line source (see Glossary), blank interferences would not be expected since the spectral resolution is effectively determined by the very narrow linewidth (usually about 0·001 nm) of the lamp emission. Adjacent absorption lines (with half-intensity widths between 0·001 and 0·01 nm) of different elements, such as the 285·21 nm Mg line and the 285·28 nm Na line, do not interfere here. Blank interference due to thermal emission of concomitants in the flame is easily excluded instrumentally, for example, by modulating the lamp emission and by applying AC detection.

Since neither molecular bands nor continua of metal compounds normally show. detectable absorption in the flame (see § 6.2), very low sodium concentrations can be readily determined in the presence of much calcium although the CaOH bands overlap the yellow absorption doublet of sodium [437]. Sometimes, however, molecular absorption due to high concentration of a concomitant, especially with a long absorption path, may cause a blank interference (see § 6.6.2) [341, 555]. The CaOH band at 554 nm gives rise to an interfering absorption at the barium resonance line at 553·6 nm when a solution containing 0·1 per cent or more of calcium is nebulised into an air–acetylene flame [257, 554, 555], but in the hotter and reducing nitrous oxide–acetylene flame this overlapping absorption is much slighter since calcium has less tendency to form molecules in this flame [257]. In an insufficiently hot graphite furnace, molecular absorption by chlorides of potassium, barium and aluminium, for example, is more readily observed [401].

There are other sources of blank interference in the absorption method. If the lamp emits resonance lines of metallic impurities (such as copper) that are not separated by the spectral apparatus from the analytical line (e.g. of zinc or selenium), there will be a cross-spectral interference by copper on the zinc or selenium determination [382, 438, 816]. Furthermore, in short-wavelength regions where the flame itself absorbs appreciably, a kind of background interference (type Ib1) may arise if this background absorption is influenced by concomitants.

A concomitant present in the flame as unvolatilised oxide or salt particles at high solution concentration will scatter a part of the source radiation. This causes an interference similar to that of overlapping absorption by a concomitant, analogous to interference of type Ib2 in emission. Radiation losses up to 5 per cent at 200 nm have been ascribed to this scattering effect [382]. Other reports on radiation loss due to scattering appear in [211, 340, 805]; the effect is more pronounced in the air–propane flame than in the hotter air–acetylene flame. However, doubt has been expressed, based on theoretical calculations of the scattering, that it is in every such case the chief cause of the observed loss of radiation, rather than molecular absorption [553, 555, 556].

Overlap of an emission line from the hollow-cathode lamp (e.g. the Cu line at 324·754 nm) by a closely adjacent absorption line (e.g. the Eu line at 324·753 nm) can also lead to blank interference of type Ib2 in the determination of the first element [178, 335] (see [592] for a survey of overlapping lines). The half-intensity width of absorption lines in flames is usually between 0·001 and 0·01 nm, accounting for this interference, which cannot be removed by better spectral resolution. A similar spectral line coincidence occurs with the resonance lines of the two lithium isotopes in a flame [75]. In non-flame atomisers with lower temperature and lower total gas pressure the chance of spectral overlap of isotopic lines is reduced, so that they are more suited to isotope determinations [414, 630].

With a continuum radiator there is greater likelihood of cross-spectral interference for other elements. It is easily shown by Kirchhoff's law (see § 6.6.1), according to which the thermal emission and the absorption factor of a metal vapour in a flame are inter-related, that the ratio of the absorption factors of two adjacent atomic lines equals the ratio of their thermal emission intensities. This holds since for two adjacent emission lines Planck's factor in equation (6.19b) is practically the same. In flame atomic absorption spectroscopy with a continuum radiator, therefore, the cross-spectral interference is as large as in the emission method. A multi-element line source also involves more risk of cross-spectral interference in absorption work [498] though less than a continuum radiator does.

In *flame fluorescence spectroscopy* blank interferences can arise in the same way as in absorption spectroscopy. Here too there is the possibility of line coincidence between analyte and concomitant. If the background source (e.g. a continuum radiator) emits not only at the wavelength of the analytical line but also at neighbouring wavelengths, adjacent interfering lines can also fluoresce. Interference from thermal emission of concomitants can be easily avoided, as before, by modulating the lamp radiation. In the use of resonance fluorescence in particular, scattering of the primary radiation by unvolatilised particles formed from concomitants can be troublesome [856]. Besides, a concomitant can alter the scattering on unevaporated mist droplets through its effect on desolvation and thus indirectly cause a blank interference of type Ib1. A systematic treatment of scattering on mist droplets or unvolatilised particles in turbulent and laminar flames, including its dependence on wavelength, scattering angle and polarisation, can be found in [671].

9.2.3. Non-specific interferences. Interferences of class II relate solely to the analytical line or band itself. We distinguish specific and non-specific interferences. A non-specific interference, group IIa, causes the same percentage change in the emission or absorbance of all lines or bands of any element at any concentration, if we exclude secondary effects. Non-specific interferences have the same effect in the emission, absorption and fluorescence methods under otherwise identical conditions, and so they will be discussed for all three methods jointly.

An interference affecting the aspiration rate, for example, will react in the same way, to a first approximation, on sodium and lithium emission or absorbance and hence belongs to group IIa (more specifically, type IIa1). This fact is, incidentally, utilised in the reference-element technique (see Glossary) for compensating potential errors due to non-specific interferences but a closer examination shows that, to a second approximation, a non-specific interference may not have exactly the same relative effect upon different elements or even upon the same element at different

180

concentrations. This can happen when the two lines under comparison (in emission or absorption) have analytical curves of different shape owing to differences in self-absorption, ionisation, or spectral width of the background source line (see chapter 8). The two lines may then show unequal percentage changes of signal when the solution transport rate changes by 1 per cent. These secondary effects disappear when we consider not the effect on the emission or absorbance as such but the effect on the so-called apparent concentration (see § 9.4 and Glossary). For simplicity we shall ignore these secondary effects and, although they may depend to a certain extent on element and concentration, we shall classify these interferences as non-specific, that is, under group IIa.

Accordingly, all interferences involving transport of the sample into the flame and desolvation will be called non-specific and classed as type IIa1. We recall that the coarser drops cannot always desolvate quickly enough, especially in the turbulent flame of a direct-injection burner (see § 4.6). Concomitants that alter the evaporation rate of the solvent thus indirectly affect the quantity of atomic vapour that is quickly released in the flame.

Similar non-specific effects are exerted upon the analyte signal by changes in the flame shape (type IIa2), especially in direct-injection burners, in which the flame shape can be strongly influenced by the nebulisation of liquid (see § 3.5.3). Concomitants that alter the nebulisation therefore also affect the flame shape. A concomitant such as an alcohol may further affect the burning velocity and hence the shapes of both the combustion zone and that part of the flame above it.

Referring to chapter 4, we should expect concomitants that affect the surface tension, density, viscosity and/or vapour pressure of the liquid to cause non-specific interference of type IIa1 in both chamber-type and direct-injection nebulisers. These factors were seen to affect the aspiration rate and the drop-size distribution of the mist and hence the rate of delivery of analyte to the flame. Non-specific interferences seem on the whole to be more conspicuous in chamber-type than in direct-injection nebulisers [343]. Since a change of aspiration rate also usually entails a change of drop-size distribution, etc, the emission or absorbance need not be proportional to the aspiration rate (see § 4.2). Consequently, such changes cannot be corrected for by merely dividing the measure by the aspiration rate. The effect of a variable aspiration rate can be eliminated by means of a device for mechanically controlling the flow of liquid into the nebuliser.

9.2.4. Specific interferences. Specific interferences, group IIb, unlike group IIa, in principle act in different degrees upon different elements; that is, the mechanism of the interference involves factors that are specific for the affected element and that may depend on its concentration and chemical bonding in the solution. Among the specific interferences we distinguish those that affect the number of free analyte atoms and molecules (in the ground state) made available in the observed part of the flame (types IIb1, IIb2, IIb3 and IIb5) and those that affect their excitation or, in the fluorescence method, the quenching of photoexcited atoms (type IIb4).

Specific interferences of the first category occur in both the emission method and in the atomic absorption and fluorescence methods, with the difference that in emission it is sometimes possible to utilise a molecular band of the analyte instead of the atomic line. However, in the absorption method there is a wider choice of flames for minimising these interferences, because little or no attention need be given here to cross-spectral interferences or deterioration of the excitation conditions.

Excitation interference is significant primarily in flame emission spectroscopy. It is an advantage of the absorption and fluorescence methods that they are basically immune to this interference, provided that a resonance line is employed. However, with a non-resonance line these methods are also subject (though much more weakly) to excitation interference that affects the population of the lower energy level.

9.2.5. Specific interferences involving the gas phase only.

We shall first describe in more detail the specific interferences that involve the gas phase only. As an example, HCl may affect the degree of dissociation of both KCl and LiCl and thereby influence the atomic concentrations of K and Li in the flame (type IIb2). But since KCl and LiCl have different dissociation energies, this *dissociation interference* is manifested to different degrees in the two cases. This kind of interference usually becomes appreciable only at higher concentrations of the concomitant and is more conspicuous in direct-injection than in chamber-type nebulisers because the latter deliver only a small fraction of the aspirated solution to the flame (see § 9.3.1) [343, 567]. Elements that occur in the flame partly as atoms and partly as oxides or hydroxides are especially subject to dissociation interference in direct-injection nebulisers when the flame temperature or flame composition is altered by addition of organic components (see § 5.2.3)†. Dissociation interferences usually depend little if at all on the concentration of the analyte and tend to diminish at higher flame temperatures. For the latter reason very hot flames are often used in atomic absorption spectroscopy [875]. Dissociation interferences on elements occurring partly as hydroxides in the flame can be caused by concomitants such as chromium, tin and the alkaline earths if present at very high concentrations, since they catalyse the recombination of excess H and OH radicals (see § 5.2.4). Therefore, the concentrations of H and OH radicals and hence the degree of dissociation of the hydroxide depend on the concentration of these concomitants [245].

Ionisation interference occurs when, for example, the addition of caesium

---

† In fuel-rich flames it may happen that the concentration of free oxygen atoms is commensurate with the concentration of a concomitant element. If the latter element has a strong affinity for oxygen, we might expect that the free oxygen concentration would be markedly reduced by it. This reduction would in turn affect the degree of dissociation of the analyte oxides or hydroxides, which would result in a dissociation interference. This so-called oxygen-competition effect, however, is doubtful because the atomic oxygen concentration is 'stabilised' by chemical-equilibrium relations with the relatively high and hence practically unalterable concentrations of other flame molecules (see the first footnote in § 5.2.3) [735, 753]. These molecules act as a 'pool' of oxygen, resupplying any loss of atomic oxygen due to metal-oxide formation. Oxygen-competition effects have been suggested [285, 626, 769] as an explanation of the enhancement of, for example, atomic V and Li signals by Ti and Si, respectively, in acetylene flames. This hypothesis, however, has been criticised in [735].

changes the degree of ionisation of potassium or sodium, since the caesium supplies the flame with electrons that suppress the ionisation. This raises the signal from the K or Na atoms. The interference varies among elements – in this case sodium and potassium – because of their different ionisation energies. It is characteristic of ionisation interference that it approaches saturation as the concentration of interferent rises. The opposite effect results from addition of a chlorine compound, which supplies free negative Cl^- ions to the flame and thus raises the degree of ionisation of the analyte (see § 5.3.1). Ionisation interferences equally affect the atomic lines and the oxide or hydroxide bands, if any, of the analyte, since the mass action law requires a fixed ratio of atomic to molecular concentration in a given flame. An ionic line, however, is affected in the opposite sense. The ionisation equilibrium also shifts whenever the flame temperature is changed by the addition of a combustible constituent [295]. Ionisation interferences are in general more pronounced in hotter flames and at lower concentrations of the analyte.

Since the excitation of atomic lines or molecular bands is usually characterised by a state of thermal equilibrium (see § 6.7.4), it suffers interference normally only through a change of flame temperature. *Excitation interference* due to temperature shift may be expected when the nebulised solution contains a combustible solvent such as alcohol or when the thermal and thermochemical properties of the nebulised liquid differ from those of water. Incombustible concomitants can also change the temperature, especially in direct-injection burners, if only by affecting the flow rate or desolvation of the aerosol in the flame. They thus alter the extent to which the flame is cooled upon nebulisation of the solution (see § 3.5.3). We must consider these effects specific (type IIb4), as the temperature dependence of line and band emission varies with the excitation energy.

A change of flame temperature naturally also changes the degree of dissociation and ionisation because the equilibria of these processes are temperature-dependent. Such temperature interferences also make themselves felt in flame absorption spectroscopy. The general temperature dependence of absorbance was treated in § 6.5.2. The flame background emission also varies with temperature and, therefore, a concomitant that changes the flame temperature may cause different types of interference; excitation interference (IIb4) is only one of these possible types.

In special cases, in a direct-injection burner, alcohol added to the sample solution can bring about a suprathermal excitation through chemiluminescence (see § 6.7.3). This excitation interference, deliberately employed to improve the detection limit, is similarly specific for the analyte and for the chosen spectral line.

Another kind of excitation interference is sometimes mentioned in the literature. It is suggested that when the interferent has an energy level with an excitation energy close to that of the analytical emission line, an interference may occur through resonant transfer of excitation energy. But if thermal equilibrium prevails, as it usually does, the population of any energy level is unalterably fixed by Boltzmann's law (equation (6.3)) – the temperature T being taken as constant – and this kind of interference is theoretically impossible (see also § 6.7.4). Moreover, the greatest partial pressure that a metallic interferent is likely to reach in the flame is so low that interference by this route, in the absence of equilibrium, is *a priori* very improbable [164, 176]. Conversely, in flame fluorescence spectroscopy it is unlikely that quenching collisions with atoms or molecules of a concomitant metallic element will cause direct interference on the fluorescence signal [796]. However, a specific *quenching interference* could arise indirectly, for example, on nebulising an alcohol solution, which would change the flame composition and consequently the fluorescence efficiency (see § 6.8).

Reports of an alleged excitation interference due to a transfer of radiant energy from a concomitant to the analyte must be viewed very critically. The excitation of an analytical line by the absorption of photons emitted by a concomitant whose frequency does not match that of the line exactly is a very improbable process.

9.2.6. Specific interferences involving the condensed phase. The specific interference of types IIb2, IIb3 and IIb4 described so far all operate in the gas phase. However, a more or less specific interference can also occur in the solid phase or in the melt when a concomitant influences the volatilisation of the aerosol particles containing the analyte after desolvation (type IIb1). There are two typical cases of a clearly specific *volatilisation interference* [163]. In one, the interferent (e.g. phosphorus or aluminium) forms an involatile compound with the analyte (e.g. calcium) in the solid aerosol particle (see § 4.5). This so-called chemical effect reduces the concentration of free atoms and molecules of calcium in the flame, and the calcium emission or absorption drops. Since elements (such as calcium, strontium and potassium, for example) differ greatly in their tendency to form involatile complexes with phosphorus, aluminium, etc, and since the complexes differ in volatility, this interference too must be regarded as specific (group IIb). This kind of interference depends strongly on the element and on its concentration and is found especially in direct-injection burners and at low flame temperatures. The interference characteristically already makes itself felt at a concentration of interferent as low as that of the analyte. With rising interferent concentration the depression of the analyte signal often approaches saturation. This behaviour is revealed as a bend point or 'knee' followed by a horizontal plateau in the interference curve (see Glossary), a plot of the analyte signal (calcium) as a function of the interferent concentration (phosphorus). Another example is the interference of iron on chromium in an air–acetylene flame, which has similarly been ascribed to the formation of an involatile compound $FeCr_2O_4$ [754, 759].

In the other typical case of a specific type IIb1 interference, an analyte such as aluminium when present in a solution without concomitants tends to form involatile oxide particles in the solid phase. This tendency may depend among other things on the nature of the aluminium salt in the nebulised solution. Thus, aluminium nitrate forms more oxide in the condensed phase and gives a weaker aluminium signal in the flame than does aluminium chloride, which is inherently more volatile. Now, the presence of other metal salts or acids in an aluminium salt solution can influence the volatility of the aerosol particles by affecting the chemical compounds that are formed from the analyte. For example, hydrofluoric acid forms aluminium fluoride, which has a lower sublimation temperature and aids the volatilisation of the aerosol particles; it brightens the emission lines and bands of aluminium, and to the

184

same degree [563]. An enhancement may also occur when the concomitant inhibits the reduction of a volatile metal oxide to an involatile metal or carbide in fuel-rich nitrous oxide–acetylene flames [757] (see also below). Such effects, which raise the volatility of the solid or molten particles, are opposite to the effect of phosphorus on the alkaline earths discussed above. The two effects must be clearly distinguished. Both kinds depend specifically on the nature and concentration of the analyte. Varieties intermediate between the two kinds described here are met with in practice; either the cation or the anion of the concomitant salt, or both together, can play a role. The state of affairs is often hard to clarify.

A less specific kind of volatilisation interference can arise when after desolvation the analyte is dispersed in a matrix of a concomitant present in large excess. Depending on whether the analyte alone or the interferent alone tends to be involatile, this dispersion effect may cause either enhancement or depression of the analytical signal. The *dispersion effect* can be utilised in alleviating, for example, the depression of calcium by phosphorus or the involatility of the analyte itself by adding an excess of a volatile compound such as sodium or potassium chloride [191, 881]. Such an effect may account for the enhancement of the calcium signal in the oxyhydrogen flame by an excess of a salt of an aliphatic acid [881]. An opposite effect is exerted by much calcium on the strontium signal in a direct-injection burner [720]; its depression is most probably due to occlusion of the strontium salt by calcium oxide in the aerosol particle, whose large size delays complete volatilisation. We should not expect these dispersion or matrix effects to be very specific, being physical rather than chemical in nature.

It has also been suggested that the interferences on calcium of a number of elements in the air–acetylene flame may be due to their influence on the reaction(s) of the desolvated solid particles with flame radicals [678]. These reactions are thought to control the liberation of calcium atoms from the solid phase (see also § 4.5).

Another kind of interference, most noticeable in the nitrous oxide–acetylene flame on a linear slot burner, was first reported in [558] and called *lateral-diffusion interference* (type IIb5). High-boiling acids and involatile salts in certain concentration ranges appeared to enhance the emission and absorption signals of lithium, calcium, barium and aluminium when the optical axis was parallel to the slot but not when it was perpendicular. Similar enhancements by various concomitants (such as Fe, Mo, Cr) were later reported for the same flame and burner ([352a, 602, 626, 754, 757]; for a more complete review see [882]). These effects are caused by a change in the transverse horizontal distribution (perpendicular to the slot) of the analyte in the flame. The analyte concentration appears to be raised in the median plane of the flame and lowered at the edges (see figure 17).

This interference was ascribed in [558] to an effect of the concomitants on the lateral diffusion of the spray droplets or solid particles in the flame. However, a later and more detailed study [882] points to a change in the lateral diffusion of the free analyte atoms and molecules *after* volatilisation. The concomitants are here thought to delay the complete volatilisation of the analyte. In the hot nitrous oxide–acetylene flame and even in the presence of concomitants the analyte is

185

believed to be volatilised practically completely before reaching the height of observation, but owing to the delay there is less time for the free analyte atoms or molecules to diffuse laterally out of the centre of the flame. This raises the analyte concentration in the median plane (through which absorption measurements are normally taken) at the cost of analyte concentration in the edges. The observations in figure 17 can be accounted for in this way. In cooler flames this enhancement may be overbalanced by a stronger (overall) depression due to incomplete volatilisation at the height of observation caused by the concomitants.

L'vov *et al* [602], who theoretically and experimentally studied the flow pattern of the aerosol particles and burnt gas above a slot burner, have recently arrived at a different explanation for the lateral-diffusion interference on Pb observed by them with excess aluminium nitrate. Their explanation comes close to that originally suggested in [558].

Lateral-diffusion interference depends on the rate of diffusion of analyte in the gas phase as well as on the rate of solute volatilisation. The diffusion rate will be specific for the analyte but is not likely to vary widely in a given flame gas mixture. Also the effect of the interferent on the volatilisation rate may be more or less specific (see above in this subsection). The extent of the interference may further depend sensitively upon the flow patterns of both the flame gas above the burner and the aerosol droplets in the flame. The dependence on aerosol flow pattern may

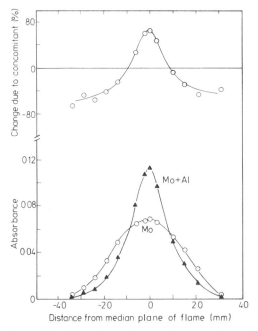

Figure 17. Effect of aluminium on the absorption profile of molybdenum measured in a direction perpendicular to the slot of the burner in an argon-shielded nitrous oxide–acetylene flame. (From [882]. Reproduced by permission of the American Chemical Society.)

186

explain why the enhancement of molybdenum by nickel and zinc disappeared when the solution aspiration rate was decreased [882]. In a nitrous oxide–acetylene flame on a circular slot burner, these enhancement effects were virtually absent [882].

More detailed experiments are needed to elucidate the precise nature of the enhancement effects typically seen in hot flames on linear slot burners, which cannot be ascribed to ionisation. The enhancement of calcium absorption or emission by phosphoric acid in such a flame has unambiguously been shown to be due to lateral diffusion and not to ionisation interference [883]. According to [757], however, the enhancement of molybdenum and vanadium absorption by added aluminium in the nitrous oxide–acetylene flame cannot be explained by a lateral-diffusion interference alone, for the enhancement was found with a slot burner parallel and perpendicular to the optical axis as well as with a Méker burner. This can be explained if we assume that addition of aluminium salts forming alumina accelerates the heat transfer to the particles and inhibits the reduction of the oxides [757]. Molybdenum oxides are known to be more easily volatilised than the corresponding carbide and metal, which have high melting points, around 2900 K (see also § 4.5). A favourable effect of alumina on the oxidation of Mo and Mo_2C particles has also been found in a hot electric furnace (see reference cited in [757]).

A mixed volatilisation and dissociation interference has been proposed to explain the fact that the atomic signal from analyte present in the solution as a metallocene (the organic ligand having no oxygen) is greater than that from the same analyte present as acetylacetonate (the metal being bonded directly to oxygen in the solution) [770]. The liberation of metal atoms from the condensed phase should be favoured with the metallocene. If, besides, chemical equilibrium between the atom and its oxide in the gas phase is assumed to be noticeably delayed (see also § 5.2.4), the atomic concentration is further enhanced. However, in the nitrous oxide–acetylene flame, where this type of interference was observed, no large departure from metal-oxide dissociation equilibrium is expected, according to a recent study [735].

9.2.7. Concluding remarks. The magnitude of the various interferences obviously depends not only on the nature and concentration of the interferent and analyte, but also on the kind of flame and nebuliser. Moreover, there are many other factors that make it impossible to transfer directly quantitative or even qualitative statements about the magnitude of an interference by concomitants from one instrument to another†. Thus, transport interference (type IIa1) also depends on the air consumption, the exact arrangement of the air and liquid nozzles, and the temperature of the air, the aspirated sample and the walls of the spray chamber, among other factors. Ionisation interference also depends on the height of observation in the flame and the rate at which the sample is supplied to the flame. Unless the

---

† Compare, for example, the disparate results reported for a Perkin–Elmer Model 303 and for an Instrumentation Laboratory Model 153 flame atomic absorption spectrometer regarding the interference of various iron and alkali salts on molybdenum at 313·3 nm, at equal solution concentrations and in similar, fuel-rich nitrous oxide–acetylene flames in the two instruments [591].

component parts of the instrument are standardised precisely by the manufacturer and the experimental conditions are specified exactly, the results obtained by one author cannot be taken over by another, even for the same type of instrument, without repeating the preliminary test experiments.

There is the further difficulty that when several concomitants are present in the same sample solution, their interferences on the analyte cannot usually be simply added or multiplied. For example, the presence of one concomitant (e.g. caesium) may suppress the interference of another concomitant (sodium) on the analyte (potassium) – a phenomenon that is put to good use in eliminating the interference of sodium on potassium. Still more complex is the interplay of concomitants in solid phase volatilisation interference (type IIb1). The organic compound EDTA is known to diminish the depression of magnesium by phosphorus, but if calcium is also present as a concomitant, EDTA actually increases this depression [440].

9.3. Classification of Interferences according to the Nature of the Concomitant

9.3.1. General. Up to this point we have classified and discussed interferences according to their mechanisms. However, since practical flame analysis is directly concerned not with mechanisms but with sample solutions of specified composition, we shall next offer a classification and a brief discussion of interferences according to the type of concomitant. For simplicity we group concomitants as cations, anions, salts, acids or organic substances present in the solution; a sharp differentiation, however, is not always possible.

There is generally a characteristic difference in their effects between concomitants that form only a minor component of the solution (e.g. metal salts) and those that may make up a fair fraction of the entire solution (e.g. added acids or alcohols). The number of molecules of a constituent of the nebulised solution per mole of solvent is 10 to 100 times greater than it is per mole of total gas mixture in the flame, depending on whether a direct-injection or a chamber-type nebuliser is used. Owing to this considerable dilution factor, only a major constituent of the solution can be expected to have any direct effect on the gross flame properties such as temperature, composition, background emission and shape. Changes in the flame properties can in turn bring out interferences of types Ib1, IIa2 and IIb. A minor constituent of the solution can influence these properties indirectly only in a direct-injection burner. This occurs when the constituent affects the physical properties of the solution in such a way as to alter the supply rate of solvent into the flame and the degree of flame cooling caused by it. The direct effect of concomitant metal salts is generally limited to changing the transport of analyte, altering the volatility of the dry aerosol particles, and participating in the equilibria among the metal atoms and ions or salt molecules in the flame. There is of course also the emission or absorption of radiation by these concomitants, which may cause a blank interference, but they have no effect on the excitation of an analytical emission line or the quenching of a fluorescence line.

188

9.3.2. Cation and anion interferences. A cation interference is one in which only the cation of the concomitant metal salt in the solution, and not its chemical bonding, is responsible for the interference. An example in flame emission spectrometry is the blank interference of types Ia and Ib2 caused by the emission of a metal line. Specific cation interferences may be of types IIb1 and IIb3 in the emission method as well as in the absorption and fluorescence methods. For instance, magnesium emission or absorption is suppressed by aluminium in the cooler flames through the formation of an involatile compound of these elements with participation of oxygen. The mutual depression of the rare earths, probably also due to effects in the solid phase, may be counted as a cation interference'[390]. The mutual ionisation interferences of the alkali and alkaline-earth metals are very well known and readily observed in the hotter flames. The latter kind of cation interference raises the concentration of the neutral metal species in the flame and thus the atomic signal but it weakens the signal from an ion line, such as that of Sr^+.

An anion interference may be recognised when the addition of different salts or acids of a given anion causes the same kind of interference on the measure [276, 849]. If the anion raises the background (see § 6.2.3) it causes a blank interference of type Ib2. Specific anion interferences of types IIb1, IIb2 and IIb3 result from the formation of involatile complexes of phosphorus or sulphur and alkaline-earth metals, for example, with participation of oxygen, or from the formation of metal chlorides or bromides in the gas phase. Also the effect of the free negative halogen ion on the metal ionisation in the flame may be counted as an anion interference. The effect of the anion on the volatilisation of particles formed from an aluminium salt solution was described in § 9.2.6. The enhancement of aluminium by hydrofluoric acid and ammonium fluoride, presumably due to the formation of the more volatile aluminium fluoride compound, may be an anion interference of type IIb1 [452, 563].

Certain anion effects cannot yet be satisfactorily explained. For example, cobaltous sulphate and cobaltous chloride solutions of equal molarity yield identical atomic Co signals in hydrogen-rich but not in carbon-rich reducing acetylene flames [735]†. This difference cannot be explained by volatilisation or oxide-dissociation effects.

9.3.3. Acid and salt interferences. A non-specific interference of type IIa1 arises when an acid, for example hydrochloric acid, alters the physical properties of the nebulised solution (viscosity, surface tension, density or vapour pressure). This changes the rate of transport of analyte into the flame and generally depresses the analytical signal. This interference is usually noticeable only with fairly high concentrations of interferent [298]. It is commonly found that the amount of interference per mole of acid decreases with rising acid concentration; the interference

---

† In [735] a distinction is made between 'hydrogen-rich' and 'carbon-rich' nitrous oxide–acetylene flames according as the N_2O/C_2H_2 ratio is between 2 and 3 or less than 2, respectively. In the hydrogen-rich flames there is an abundance of hydrogen molecules whereas the carbon-rich flames contain more reactant carbon than the nitrous oxide can oxidise to carbon monoxide.

curve (emission or absorbance plotted against interferent concentration; see Glossary) is then steeper at the low-concentration end [706]. In a direct-injection burner this non-specific interference may combine with a specific interference of type IIb2, IIb3 or IIb4 when a change in the transport and evaporation of the solvent in the flame affects the flame temperature, etc (compare § 9.2.5). If this alters the flame background emission too, then we have a blank interference of type İb1 as well. A specific acid interference of type IIb1 can occur if a change in the physical properties of the solution alters the fineness of nebulisation and hence the rate of volatilisation of analyte from the condensed phase (see § 4.5). The enhancement by perchloric acid, sometimes noted, is a special case. This acid perhaps causes small explosions in the mist droplets evaporating swiftly in the flame [381, 390], refining the aerosol particles and aiding the volatilisation of the analyte.

Salts as such can influence the physical properties of the solution provided that their concentration is high enough. Interferences like those due to acids may result; for example, an excess of copper salt depresses the absorbance of lead through its effect on the nebulisation [382]. A similar effect may be at work in the depression of the absorption signal of various metals by an excess of sodium chloride [729]. It has also been shown that potassium chloride, besides enhancing sodium and lithium emission by cation ionisation interference, impedes the transport of the analyte to the flame [357]. Unvolatilised salt particles may cause a background interference in emission, absorption and fluorescence by their incandescence or scattering of radiation. A typical interference arises when a concomitant salt such as sodium chloride present in excess in the desolvated aerosol creates a volatile solid matrix in which the analyte (e.g. calcium as oxide or phosphate) is dispersed and 'baked in' [191, 881]; on further heating and with partial volatilisation of the more volatile components, the large mixed crystals probably shatter, speeding the volatilisation of the analyte (see § 4.5).

Here again, we do not always have a satisfactory explanation. For example, alkali chlorides in relative excess depress the atomic emission and absorption of molybdenum at 313·3 nm in a reducing nitrous oxide–acetylene flame [591, 884], whereas alkali nitrates and sulphates markedly enhance the molybdenum absorption [591].

9.3.4. Interferences of organic substances. Interferences of concomitant organic substances, especially alcohols, are by far the most complex and manifold. They include practically all the types of interferences listed in table 5. One kind or another may predominate depending on whether the nebuliser is of the direct-injection or chamber type, whether the flame is rich or lean, and whether the analyte is volatile or involatile. We summarise below the commonest interferences for the simple case in which the sample solution contains only one organic concomitant along with the analyte salt.

An organic concomitant sometimes raises the background emission considerably, thus causing a blank interference (group Ib). It can change the background emission by taking part in the combustion, and it can bring out the C_2 bands and even the atomic carbon line if the flame is fuel-rich to start with. With much alcohol the

190

flame turns white from incandescent soot particles. Also, the background absorption of the flame can be raised by organic substances at wavelengths below 250 nm [787, 788]. On the other hand, the promotion of desolvation by alcohols or other such compounds helps the fluorescence method by diminishing the blank measure due to scattering.

By altering the physical properties of the nebulised solution, an alcohol often substantially raises the rate of transport of analyte into the flame (type IIa1). Even protein can raise the transport [177]. Alcohols lower the surface tension and raise the vapour pressure, thereby improving nebulisation and desolvation. Evidently these favourable effects outweigh the higher viscosity of alcohol–water mixtures, which by itself would be expected to lower the rate of transport and hence the analytical signal. It should also be kept in mind that the rate of recombination of the mist droplets may depend on the organic constituents (see § 4.3.4). The helpful effects mentioned above are welcomed as means of improving the detection limit. They are especially effective in chamber-type nebulisers, in which ordinarily only a small fraction of the aspirated solution reaches the flame. A five-fold improvement of salt transport is not exceptional. Incidentally, a non-specific interference of type IIa2 may arise, chiefly in direct-injection burners, when combustion of the organic substance changes the shape of the flame (see § 3.5.3) [293, 623].

All these interferences are usually accompanied by a specific interference on the analytical line or band, since organic concomitants such as protein tend to shift the flame temperature [185, 293, 295, 312, 623, 881] and bring about interferences of types IIb1 to IIb4. In chamber-type nebulisers, however, these interferences are usually unimportant compared to the transport interference. It is worth noting that participation of the organic substance in the combustion does not necessarily *increase* the temperature. It has been shown experimentally and theoretically that when acetone–water and propanol–water mixtures are nebulised, the flame may be either hotter or cooler than with pure water [160, 161, 185, 381, 727], depending on whether the unburned gas mixture (without the aerosol) is lean or rich in fuel. In the fuel-rich case the organic molecules are incompletely burned and the heat liberated by this combustion may not be enough to heat the products to the original temperature of the flame; the flame temperature is then depressed. In a direct-injection burner an added alcohol may also raise the temperature (above that prevailing when pure water is nebulised) by reducing the considerable cooling due to the flow of water into the flame, in consequence of the higher viscosity (see § 3.5.3) [295, 312]. Another favourable circumstance arises sometimes from the fact that the organic solvent (e.g. acetone) has a lower heat of vaporisation than water.

When a metal such as calcium or titanium occurs largely as oxide or hydroxide in the flame, the reducing effect of an added alcohol raises the atomic emission or absorption [372, 618, 791]. This is an interference of type IIb2.

The type of organic ligand (namely, whether with or without oxygen) to which a metal is bonded in the solution may also affect the concentration of metal atoms in the flame [770]; no satisfactory explanation yet exists – see the relevant discussion near the end of § 9.2.6. At all events the observed effect cannot be due to the influence of the organic ligand on the bulk properties of the spray or flame since it is present at the same low concentration as the analyte.

The volatilisation interference (type IIb1) of organic additives on an involatile analyte can be ascribed to the resulting finer mist droplets or the greater volatility of the resulting organometallic complex [295]. In general, volatilisation is known to be aided by better dispersion of the aerosol, so that the involatile constituents end up in smaller particles.

Organic additives sometimes enhance the analyte emission very considerably. The gallium emission at 417·2 nm is augmented 39-fold by acetone at 50 volume per cent (see also spectrogram 7 in Appendix 1); enhancement factors may exceed 100 in anhydrous organic solvents [295]. The great enhancement by alcohols of the emission lines of tin, among other elements, has been discussed in § 6.7.3. In these cases there is probably a suprathermal chemiluminescence, which must be classified as a specific excitation interference of type IIb4.

9.4. Experimental Identification of an Interference

It is not always easy in practice to decide which of the interference mechanisms classified in § 9.2 is operative, especially in working with simple means of spectral selection such as a filter instrument. A few suggestions for distinguishing them may be helpful.

Blank interferences (class I) are most easily recognised by means of blank solutions containing the interferents in various concentrations but not the analyte, even as a detectable impurity. The ranges of interferent concentration must of course correspond to those found in the actual samples.

If in the emission method an interference on the net measure (class II) is present, we may ask whether the effect is due to a change of flame temperature or to a change of transport, that is, a change in the concentration of analyte available at the height of observation in the flame. This can be decided by measuring the interference for analyte lines of different excitation energies. If the effect, expressed as a percentage change in apparent concentration, is the same for all lines of the atom, then we have at least no specific interference of type IIb4 but instead an interference acting on the atomic concentration in the flame. The *apparent concentration* is the concentration value derived, without correction for interference, from the emission measured in the presence of interferent by means of an analytical curve constructed from reference solutions lacking the interferent (see also Glossary). In this way the disturbing effect of secondary factors such as ionisation and self-absorption on the shape of the analytical curve is circumvented in identifying the interference on the analyte concentration. An excitation interference can also be distinguished from a transport interference, for example, by measuring the same atomic line in both emission and absorption, as is possible with some instruments.

Another way to test for a change in the available atomic concentration (group IIa) is to measure the change in the emission signal of the analyte at two concentrations chosen so that self-absorption is negligible at one and strong at the other.

192

At the latter concentration the interference effect will be percentually half as great, because when self-absorption is strong the intensity is proportional to the square root of the concentration (see § 6.6.1).

It may also be helpful to observe the interference with different analytes. In this way, for example, the non-specific effect of hydrochloric acid on the transport of sample solution can be distinguished from its specific effect on dissociation or ionisation. An ionisation interference on the alkaline earths can be easily recognised by simultaneous measurements on their atomic and ionic lines.

If an additional chamber-type nebuliser is available, a volatilisation interference (type IIb1) can be identified by a twin-nebuliser experiment, interferent and analyte being nebulised simultaneously but separately into the flame by two parallel nebulisers (see also Glossary). Some indication of the type of interference is often afforded by the shape of the interference curve (interference plotted as a function of interferent concentration; see Glossary), as discussed in § 9.2.6. A good many authors also recommend measuring the interference as a function of the observation height or of the horizontal distance from the axis of the flame (compare figure 17), to gain information about the nature of the interference.

Other such tests for differentiating interferences are possible; they have more than academic interest. When the mechanism of an interference is recognised, methods will come to mind for best eliminating it or perhaps putting it to use for such purposes as improving the detection limit or straightening the analytical curve.

Glossary

Commonly used terms and abbreviations peculiar to the field of analytical flame spectroscopy are defined or explained in this alphabetical glossary. The latest recommendations on nomenclature by the International Union of Pure and Applied Chemistry (IUPAC) are incorporated†. To help the reader identify the new terms with the earlier ones encountered in the older literature, some of the earlier terms are cross-referenced to the new ones. To coordinate the (often new) English terms with the German terms, the German equivalents are added in brackets after the English entry. Terms in small capitals refer to entries elsewhere in the Glossary.

---

† IUPAC, Division of Analytical Chemistry, Commission on Spectrochemical and Other Optical Procedures for Analysis: *Nomenclature, symbols, units and their usage in spectrochemical analysis: I. General atomic emission spectroscopy*, 1972 Pure Appl. Chem. **30** 653–79; *II. Data interpretation*, 1976 Pure Appl. Chem. **45** 99–103; *III. Analytical flame spectroscopy and associated non-flame procedures*, 1976 Pure Appl. Chem. **45** 105–23.

AAS. [*A-Methode.*] Abbreviation for ATOMIC ABSORPTION SPECTROSCOPY.

absorbance; internal absorbance; symbol: *A*. [*dekadische Extinktion*; *dekadisches Absorptionsmass.*] A quantity defined by $A = -\log_{10} \tau_i = \log_{10} (I_0/I)$, where τ_i is the internal TRANSMITTANCE, I_0 is the INTENSITY of the light incident upon a sample medium, and I is the intensity of the attenuated light transmitted by it. *A* refers to the transmission properties of the sample alone, excluding those of the apparatus (cell windows, etc).

absorption coefficient; linear absorption coefficient. [*dekadischer Extinktionsmodul*; *spektraler Absorptionskoeffizient.*] The ratio of ABSORBANCE to path length in the transmitting medium. Distinguish from: ABSORPTIVITY.

absorption factor; absorptance; symbol: *α*. [*Absorptionsgrad.*] The ratio of the RADIANT FLUX absorbed in the interior of an optically clear (non-scattering) medium to the radiant flux entering the medium. The *absorption factor*, which may include the properties of the cell windows, etc, is distinguished from the *internal absorption factor* α_i [*Reinabsorptionsgrad*], which excludes them. Compare: TRANSMITTANCE.

absorption spectroscopy. [*Absorptionsspektroskopie.*] See ATOMIC ABSORPTION SPECTROSCOPY.

absorptivity; Napierian absorptivity; symbol: *k*. [*Absorptionskoeffizient.*] The (internal) ABSORPTION FACTOR per unit path length, defined by $k = \lim_{l \to 0} \alpha/l$, where l is the path length. It is expressed for example in cm^{-1}. Distinguish from: ABSORPTION COEFFICIENT.

accuracy. [*Zuverlässigkeit* (literally, 'reliability'); *Genauigkeit*.] The quality of an ANALYTICAL RESULT or procedure as characterised by its total error, that is, the sum of the RANDOM ERRORS and the SYSTEMATIC ERRORS, all errors being added according to the rules for the propagation of error (see VARIANCE); the degree of agreement between an analytical result and its (known or assumed) true value. Compare: PRECISION.

addition technique; analyte addition technique. [*Zugabeverfahren*; also *Zumisch-verfahren, Aufstockungsverfahren, Additions-Verfahren*.] A technique used for obtaining the ANALYTICAL RESULT in which successive known quantities of the ANALYTE are added to aliquots of the SAMPLE SOLUTION. The net MEASURES obtained from these are plotted against the added concentrations; the intercept of the extrapolated plot with the negative concentration axis gives the concentration of the original sample solution, provided that the ANALYTICAL CURVE is linear down to low concentrations. The plot itself provides a test of the linearity and establishes the slope of the curve (the SENSITIVITY). The technique is useful when a complex (e.g. biological) MATRIX is ill suited for the preparation of REFERENCE SOLUTIONS and when unknown INTERFERENCES by this matrix may affect the sensitivity.

additive. [*Zusatz*; *Zugabe*.] A substance added to a sample or SAMPLE SOLUTION during preparation of the sample, usually at a constant concentration, to overcome INTERFERENCES or to facilitate analysis in other ways. A deliberately introduced CONCOMITANT.

aerosol. See DRY AEROSOL; MIST.

AES. [*E-Methode*.] Abbreviation for ATOMIC EMISSION SPECTROSCOPY.

AFS. [*F-Methode*.] Abbreviation for ATOMIC FLUORESCENCE SPECTROSCOPY.

analog(ue). [*analog*.] An adjective describing a READING (or a technique of measurement or an instrument that provides a reading) that is, at least ideally, a unique indication on a continuous scale of the quantity to be measured. Compare: DIGITAL.

analyte; analysis element. [*Analysenelement*.] The element whose concentration or quantity is to be found. Distinguish from: INTERFERENT; CONCOMITANT; ADDITIVE. See also QUANTITY OF ANALYTE.

analyte addition technique. See ADDITION TECHNIQUE.

analyte signal; net signal. [*Nutzsignal*.] A SIGNAL resulting from the presence of ANALYTE in the sample analysed. More particularly, the optical signal, namely the radiation emitted, absorbed, or fluoresced by the analyte within the SPECTRAL BANDWIDTH specified for the analysis. The magnitude of the analyte signal is the NET MEASURE. Compare: BLANK SIGNAL.

analytical curve; analytical function. [*Analysenkurve, Analysenfunktion*; *Bezugskurve, Bezugsfunktion*; (older term): *Eichkurve* (not recommended).] The graphical representation of the relation between the NET MEASURE and the ANALYTE concentration in the solution is generally called the *analtytical curve*; the relation described by this graph is generally called the *analytical function*. The *analytical calibration curve* (or *function*) is obtained when the measure is plotted on the ordinate axis as a function of the concentration on the abscissa.

196

This graph is established by a series of calibration measurements on REFERENCE SOLUTIONS having known composition; the reference solutions should be treated in the same way as the SAMPLE SOLUTIONS to be analysed. In the EVALUATION of actual sample solutions, the (unknown) concentration is derived from the measure through an inversion of the analytical calibration curve. The inverted calibration curve (or function), that is, the graph representing the concentration as a function of the measure, is therefore called the *analytical evaluation curve* (or *function*). In the simple case of an analysis for only one component, only the analytical calibration curve is usually plotted and the unknown sample concentration is read on the abscissa. The distinction between analytical calibration and analytical evaluation functions is less trivial in the case of multicomponent analyses when the measures for the individual components are interdependent because of various INTERFERENCES. The analytical curve was formerly often called *working curve*.

analytical flame spectroscopy. See FLAME SPECTROSCOPY.

analytical material. [*Ausgangsmaterial*; *Analysensubstanz*.] A substance to be subjected to chemical analysis for one or more constituents. See also ANALYTICAL SAMPLE; MATERIAL.

analytical result; result of analysis. [*Analysenergebnis, Messergebnis*.] The final result expressed as the CONCENTRATION or QUANTITY OF ANALYTE after all operations of preparation, measurement, and EVALUATION have been carried out, including the application of corrections such as those due to INTERFERENCE by CONCOMITANTS. Distinguish from: SIGNAL; MEASURE; READING.

analytical sample. [*Probe*; *Messobjekt, Messgegenstand*.] A representative portion of an ANALYTICAL MATERIAL submitted for analysis.

apparent concentration. [*scheinbare Konzentration*.] The CONCENTRATION of ANALYTE found when an INTERFERENCE exists that is not corrected or removed. The apparent concentration may be either greater (an ENHANCEMENT) or smaller (a DEPRESSION) than the true concentration.

aspiration rate; rate of solution (or **liquid**) **aspiration.** [*Ansaugmenge*; earlier: *Ansauggeschwindigkeit*.] The volume of liquid aspirated by the NEBULISER per unit of time.

atomic absorption spectroscopy; abbreviation: AAS. [*Atomabsorptionsspektroskopie, Atomabsorptionsmethode*; shorter terms: *Absorptionsmethode, A-Methode*.] A method in which the absorption by free ANALYTE atoms of a SPECTRAL LINE (or of a small part of a CONTINUUM) radiated from an external LIGHT SOURCE is employed for quantitative analysis. For this purpose the analyte, usually present in a SAMPLE SOLUTION, has to be converted to ATOMIC VAPOUR by DESOLVATION, VOLATILISATION, and DISSOCIATION, in a flame for example.

atomic emission spectroscopy; abbreviation: AES. [*Atomemissionsspektroskopie*; shorter: *E-Methode*.] A method in which the emission of an ATOMIC LINE of the ANALYTE in a suitable excitation source such as a flame is employed for quantitative analysis. More broadly, any kind of EMISSION SPECTROSCOPY employing atomic lines.

atomic fluorescence. See FLUORESCENCE.

atomic fluorescence spectroscopy; abbreviation: AFS. [*Atomfluoreszenzspektroskopie, Fluoreszenzspektroskopie*; shorter: *F-Methode.*] A method in which the FLUORESCENCE of an ATOMIC LINE of the ANALYTE, excited by a suitable BACKGROUND SOURCE, is employed for quantitative analysis. The analyte passes through the same processes as in ATOMIC ABSORPTION SPECTROSCOPY.

atomic line; atom line. [*Atomlinie.*] A SPECTRAL LINE emitted or absorbed by a free, neutral atom. Distinguish from: BAND LINE; IONIC LINE.

atomic vapour. [*Atomdampf, atomarer Dampf.*] A vapour of free atoms, in particular of an ANALYTE in a flame, formed after evaporation of the aerosol and DISSOCIATION of the compounds containing the analyte.

atomisation. [*Atomdampferzeugung.*] The generation of ATOMIC VAPOUR by an ATOMISER.

atomisation efficiency. See EFFICIENCY OF ATOMISATION.

atomiser. [*Atomdampferzeuger.*] Any device such as an electrically heated furnace, a CATHODIC SPUTTERING CHAMBER, or a flame, capable of converting the ANALYTE in a SAMPLE SOLUTION or solid sample into its ATOMIC VAPOUR. The term *atomiser* must no longer be used for NEBULISER.

atomiser–burner. See DIRECT-INJECTION BURNER.

background. [*Untergrund.*] See BLANK SIGNAL; FLAME BACKGROUND.

background source; light source. [*Hintergrundstrahler.*] The source of radiation used in AAS and AFS in the measurement of the absorption or FLUORESCENCE of the ANALYTE present as ATOMIC VAPOUR in, for example, the flame.

band. [*Bande.*] See SPECTRAL BAND.

band head. [*Bandenkopf.*] The sharp edge of a SPECTRAL BAND lying on either the long-wavelength or the short-wavelength side of the band.

band line. [*Bandenlinie*]. One of the many SPECTRAL LINES constituting a SPECTRAL BAND.

bandwidth. See SPECTRAL BANDWIDTH.

baseline; zero line. [*Null-Linie.*] In instruments with ANALOGUE recording, the trace of the BLANK READING recorded as a function of time or wavelength.

bias. [*Richtigkeit; systematischer Fehler; Verfälschung; Bias.*] The presence or magnitude of SYSTEMATIC ERROR in an ANALYTICAL RESULT. Distinguish from: (i) PRECISION, characterised by RANDOM ERROR, and (ii) ACCURACY, characterised by total error.

black body, black-body. [*schwarzer Körper.*] An ideal body or surface that completely absorbs electromagnetic radiation of all wavelengths.

blank background. See BLANK SIGNAL.

blank measure. [*Blindwert.*] The MEASURE of the BLANK SIGNAL; in finding it the DARK CURRENT, if observable, is excluded.

blank reading. [*Blindablesung.*] The READING corresponding to the BLANK SIGNAL. From it the BLANK MEASURE is inferred. See also BASELINE.

blank scatter; blank measure fluctuation. [*Blindwertschwankung(en).*] The statistical fluctuation or SCATTER of the BLANK MEASURE. Its numerical value, a STANDARD DEVIATION, can be expressed also in terms of the equivalent standard deviation of the concentration of the ANALYTE, although the blank measure contains no information about the analyte. The conditions of measurement, especially the RESPONSE TIME of the read-out system, must then be specified. The series of measurements that yields the blank scatter must be carefully planned so that all of the sources of scatter listed in the article NOISE except those directly related to the analyte may play their full part.

blank signal; blank background. [*Blindsignal*; *Blinduntergrund.*] A SIGNAL observed when a BLANK SOLUTION is introduced; specifically, the radiation emitted, absorbed, or scattered (within the SPECTRAL BANDWIDTH specified for the analysis) when the blank solution is nebulised into the flame or PLASMA in AES, AAS, or AFS, respectively. It may be part of a spectral CONTINUUM or consist of more or less unresolved molecular BANDS. Because nebulisation of a solution affects the temperature and composition of the flame, the blank background often differs measurably from the FLAME BACKGROUND, especially in FES. The term *blank background* (but not *blank signal*) may also refer to the entire spectrum that exists when the blank solution is nebulised. Compare: ANALYTE SIGNAL.

blank solution. [*Blindlösung.*] A solution that does not intentionally contain the ANALYTE but in other respects has as far as possible the same composition as the SAMPLE SOLUTION. In the simplest case it consists only of the SOLVENT and is then called a *solvent blank*.

Boltzmann's constant; symbol: k [*Boltzmann-Konstante.*] The ratio R/N_A, where R is the gas constant and N_A is Avogadro's constant. Its value is $1 \cdot 38 \times 10^{-16}$ erg K^{-1} or $1 \cdot 38 \times 10^{-23}$ J K^{-1}.

bracketing technique. [*Einschachtelungsverfahren.*] A method of obtaining the ANALYTICAL RESULT by graphical or numerical interpolation between the MEASURES of two REFERENCE SOLUTIONS, one having a slightly lower and the other a slightly higher ANALYTE concentration than the SAMPLE SOLUTION. The method is to be recommended especially when the instrument is equipped with means of stably suppressing the zero point (see ZERO SUPPRESSION).

buffer. See IONISATION BUFFER; SPECTROCHEMICAL BUFFER.

Bunsen burner. [*Bunsen-Brenner.*] A burner having a single large hole as outlet port for the gas mixture. Distinguish from: MÉKER BURNER; SLOT BURNER.

burner. [*Brenner.*] See BUNSEN BURNER; DIRECT-INJECTION BURNER; MÉKER BURNER; PREMIX BURNER; REVERSED DIRECT-INJECTION BURNER; SLOT BURNER.

burning velocity. [*Verbrennungsgeschwindigkeit.*] The velocity with which the front of the PRIMARY COMBUSTION ZONE would be propagated toward the unburned gas mixture if the latter were at rest (see § 3.4.1). Distinguish from: RISE VELOCITY.

calibration curve. See ANALYTICAL CURVE.

carrier gas. See HOLLOW-CATHODE LAMP.

cathodic sputtering chamber. [*Kathodenzerstäubungsgefäss.*] A HOLLOW-CATHODE LAMP employed not as a LIGHT SOURCE but as an ATOMISER for AAS. The sample is placed upon the cooled hollow cathode and atomised by sputtering. The chamber has a window at each end, permitting a beam of light from a separate BACKGROUND SOURCE to pass through the sample vapour.

chamber-type nebuliser. [*Vorkammerzerstäuber.*] A nebuliser, separated from the burner, that blows the MIST into a *spray chamber*, where the larger droplets collect upon the walls and drain away before the mist reaches the burner. Formerly called *indirect atomiser* [*Indirekt-Zerstäuber*]. Distinguish from: DIRECT-INJECTION BURNER.

characteristic concentration. [*charakteristische Konzentration.*] In AAS, the CONCENTRATION of an ANALYTE giving a net absorption of 1 per cent (or an ABSORBANCE of $0\cdot0044$) under specified conditions. It provides a (reciprocal) measure of SENSITIVITY at low concentrations. Distinguish from: DETECTION LIMIT.

chemiluminescence. [*Chemilumineszenz.*] Emission of radiation as a direct result of a chemical reaction leading to an excited atom or molecule (see § 6.7.3).

chopper. See MODULATION.

coefficient of variation. [*Variationskoeffizient.*] An earlier synonym of RELATIVE STANDARD DEVIATION, the term approved by the IUPAC.

combustion zones. See PRIMARY COMBUSTION ZONE; SECONDARY COMBUSTION ZONE.

concentration; symbol: *c*. [*Konzentration.*] The ratio of the amount of a component of interest to the total amount of the sample. In expressing the ratio, the units used for numerator and denominator should be named: $mg\,g^{-1}$, $\mu g\,l^{-1}$, $mol\,l^{-1}$; or the ratio may be expressed as a mole fraction. See also: PPM; APPARENT CONCENTRATION; CHARACTERISTIC CONCENTRATION; QUANTITY OF ANALYTE.

concentration ratio. [*Konzentrationsverhältnis.*] The ratio of the CONCENTRATIONS of two different elements in the same sample. Ordinarily the concentration in the numerator is that of the ANALYTE while the concentration in the denominator is that of the REFERENCE ELEMENT.

concomitant. [*Lösungspartner.*] Any component of the SAMPLE SOLUTION other than the SOLVENT and the ANALYTE. Concomitants can be classed as CONCOMITANTS IN THE SAMPLE and ADDITIVES.

concomitant in the sample; when not ambiguous, merely **concomitant**. [*Begleitstoff.*] A CONCOMITANT (see above) present in the original sample in unknown or varying concentration; a concomitant other than an ADDITIVE; a concomitant that may (though it does not necessarily) cause an INTERFERENCE. The totality of concomitants (in the sample) is called the MATRIX.

confidence interval. [*Vertrauensbereich.*] A range of values serving as an estimate of a statistical parameter such as a MEAN and expected to contain the true value of the parameter with a specified probability $1-P$. For example, let a given ANALYTICAL SAMPLE be subjected to n repeated analyses according to a specified procedure (in which SYSTEMATIC ERRORS can be assumed absent) in a laboratory participating in an INTERLABORATORY COMPARISON. The mean ANALYTICAL RESULT is $\bar{x}$ and the STANDARD DEVIATION is s. It can then be stated that the true analytical result μ lies between $\bar{x} - ts/\sqrt{n}$ and $\bar{x} + ts/\sqrt{n}$, that

is, in the confidence interval $\bar{x} \pm ts/\sqrt{n}$, with a specified probability or CON-
FIDENCE LEVEL $1 - P$, where t is a number, usually between 1 and 10, selected in
a table of Student's t distribution (in any textbook of statistics) so as to give the
desired value of the SIGNIFICANCE LEVEL P (often $0\cdot05$); t depends on n but as n
increases the t distribution approaches the NORMAL DISTRIBUTION. If n exceeds
about 30, t is nearly equal to the multiple of s that gives the desired value of P in
a normal distribution ($1\cdot96$ for $P = 0\cdot05$). Thus, if n is not too small, μ may be
expected to lie within the range $\bar{x} \pm 2s/\sqrt{n}$ with a probability $(1 - P)$ of about 95
per cent, and this range is called the '95 per cent confidence interval'. The
averaged analytical result is usually reported in the form $\bar{x} \pm 2s/\sqrt{n}$ or $\bar{x} \pm ts/\sqrt{n}$ (t
being chosen to give the 95 per cent confidence interval or confidence level); the
number of analyses n should also be stated. Compare: STANDARD ERROR (of the
mean).

confidence level. [*statistische Sicherheit*; *Aussagewahrscheinlichkeit.*] The probability $1 - P$ as defined under CONFIDENCE INTERVAL; the LEVEL OF SIGNIFI-
CANCE subtracted from unity.

confidence limits; fiducial limits. [*Vertrauensgrenzen.*] The end points of the CON-
FIDENCE INTERVAL; in the last example in that article, $\bar{x} - 2s/\sqrt{n}$ and $\bar{x} + 2s/\sqrt{n}$
are called the '95 per cent confidence limits'.

continuum; spectral continuum. [*Kontinuum*; *kontinuierliche Strahlung* or *Absorption.*] Radiation emitted or absorbed throughout an extensive contiguous
(continuous) wavelength range. Also called *continuous radiation*.

continuum source; spectral continuum source. [*Kontinuumstrahler.*] A LIGHT
SOURCE that emits a CONTINUUM. Examples are high-pressure xenon lamps,
hydrogen or deuterium lamps, and tungsten-filament lamps.

dark current. [*Dunkelstrom*; *Nullstrom.*] The output current of an unilluminated
PHOTODETECTOR, due to thermal effects, cosmic radiation, etc. Compare: PHO-
TOCURRENT.

depression. [*Depression.*] An INTERFERENCE in which the APPARENT CONCEN-
TRATION is lower than the true concentration of the ANALYTE. Compare: EN-
HANCEMENT.

desolvation. [*Verdampfung.*] Loss by a nebulised sample of its liquid components
by evaporation in the spray chamber of the NEBULISER or in the flame or
PLASMA; conversion of a MIST to a DRY AEROSOL.

desolvation interference. [*Verdampfungsbeeinflussung.*] An INTERFERENCE due to
the effect of dissolved CONCOMITANTS IN THE SAMPLE on the rate of DESOL-
VATION of a MIST. An example is the slowing of desolvation by an elevated salt
content.

detection limit; limit of detection. [*Nachweisgrenze.*] The CONCENTRATION or
QUANTITY OF ANALYTE corresponding in a given analytical procedure to the
smallest MEASURE x_L that can be accepted with reasonable confidence as a
genuine indication of the presence of the analyte and not merely an accidentally
high value of the BLANK MEASURE. The value of x_L is given by $x_L = \bar{x}_{bl} + k s_{bl}$,
where $\bar{x}_{bl}$ is the mean and s_{bl} the STANDARD DEVIATION of the blank measure

and k is a factor chosen according to the desired CONFIDENCE LEVEL. If the SENSITIVITY can be assumed to be constant at low concentrations, the (concentration) detection limit equals the ratio of the NET MEASURE $x_L - \bar{x}_{bl}$ to the sensitivity. For k the IUPAC recommends a value of 3, although 2 has often been used. At $k = 3$ the theoretical confidence level is over 99·8 per cent in a one-sided NORMAL DISTRIBUTION and for a single measurement of x_L. But if $\bar{x}_{bl}$ and s_{bl} are based on few measurements, and since the distribution may not be normal at low concentrations, the practical confidence level at $k = 3$ may be as low as 90 per cent. In reporting a detection limit the value of k must be stated as well as the TIME CONSTANT of the read-out system and any other instrumental conditions that affect the detection limit. Distinguish from: CHARACTERISTIC CONCENTRATION and SENSITIVITY.

diffusion flame. [*Diffusionsflamme.*] A flame that gets all its oxygen for combustion from the ambient air by diffusion. A PRIMARY COMBUSTION ZONE is absent.

digital. [*digital.*] An adjective describing a READING (or a technique of measurement or an instrument that provides a reading) that is presented in a discontinuous manner as a number of quantised units or light impulses; the reading is displayed on a panel, exhibited as a sequence of digits, or coded in some other way, for example as punched tape. Compare: ANALOGUE.

diluent. [*Verdünnungsmittel.*] An ADDITIVE that dilutes the sample in order to pass from the curved to the linear portion of the ANALYTICAL CURVE by lowering the concentration, or, if the sample is very small, to provide enough solution for the nebuliser, or to suppress INTERFERENCES by CONCOMITANTS IN THE SAMPLE.

direct-injection burner. [*Direkt-Zerstäuber-Brenner; Direktzerstäuber-Turbulenz-brenner.*] The combination of a burner with a nebuliser that injects the sample solution directly into the flame without an intermediate spray chamber. The oxidant and fuel emerge from separate (usually coaxial) ports and are mixed above them to produce a TURBULENT FLAME. The oxidant is most commonly used for aspirating and nebulising the sample. Formerly called *atomiser–burner* and *total-consumption burner*. When its function as a nebuliser is in mind, it may be called a *direct-injection nebuliser* [*Direkt-Zerstäuber*]. Distinguish from: PREMIX BURNER with CHAMBER-TYPE NEBULISER; REVERSED DIRECT-INJECTION BURNER.

dispersion. [*optische Dispersion, spektrale Dispersion.*] The capacity of an optical element such as a prism or grating in a spectral apparatus to act upon a beam of radiation comprising components of different wavelengths so as to separate them spatially in the order of their wavelengths, producing a spectrum of the radiation. The dispersion is measured in terms of the dimension characterising the spectrum. For example, the *angular dispersion* [*Winkeldispersion*] is $d\phi/d\lambda$, ϕ being the angle taken by the dispersed ray of wavelength λ; the *linear dispersion* [*Lineardispersion*] is $dx/d\lambda$, x being the distance along the wavelength axis of an image of the spectrum; the *reciprocal linear dispersion* is $d\lambda/dx$, commonly expressed in Å mm^{-1}. Distinguish from: RESOLVING POWER.

dissociation. [*Dissoziation.*] The splitting of a molecule into simpler fragments or atoms. In this book *dissociation* refers only to the splitting of molecules in the gas phase.

202

dissociation energy. [*Dissoziationsenergie.*] The minimal energy required to dissociate one molecule or radical (or one mole of these species) at zero kelvin in the perfect gas state.

doublet. See MULTIPLET.

drift. [*Drift.*] An unwanted slow change (usually of unknown origin) in the zero point, BASELINE, BLANK MEASURE, or SENSITIVITY, measured by the average change per unit time, for example by the slope of the baseline with respect to the time axis on a recording. Only slow changes are called drift, while quicker fluctuations are called NOISE. Drift may be classified as DARK-CURRENT drift, blank drift, etc.

dry aerosol. [*festes* (or *geschmolzenes*) *Aerosol.*] A gaseous suspension of condensed-phase particles of solute resulting from DESOLVATION of the droplets in a MIST.

EDTA. [*EDTA.*] Ethylenediaminetetraacetic acid [*Äthylendiamintetraessigsäure*], a widely used complexing agent for sample preparation.

efficiency of atomisation; overall efficiency of atomisation; symbol: ε_a. [*Gesamtwirkungsgrad der Atomdampferzeugung.*] The ratio of the rate at which ANALYTE passes through the cross section of the flame at the mean OBSERVATION HEIGHT in the form of free neutral or ionised atoms to the rate at which it is aspirated. Distinguish from: FRACTION ATOMISED, β_a. Compare: FRACTION DESOLVATED, β_s; FRACTION VOLATILISED, β_v; EFFICIENCY OF NEBULISATION, ε_n. We have: $\varepsilon_a = \varepsilon_n \beta_s \beta_v \beta_a$.

efficiency of fluorescence. See QUANTUM EFFICIENCY OF FLUORESCENCE.

efficiency of nebulisation; symbol: ε_n. [*Wirkungsgrad der Zerstäubung, Zerstäuberwirkungsgrad.*] The ratio of the rate at which ANALYTE in any form enters the bottom of the flame to the rate at which it is aspirated.

electrodeless-discharge lamp. [*elektrodenlose Gasentladungslampe.*] A type of gas discharge lamp containing a noble gas at low pressure and some volatile metal or salt but no sealed-in electrodes. The energy is supplied by means of a high-frequency electromagnetic field.

emission analysis. [*Emissionsanalyse.*] Analytical spectroscopy employing emission; analytical EMISSION SPECTROSCOPY.

emission spectroscopy. [*Emissionsspektroskopie*; shorter: *E-Methode.*] In an analytical context, analytical spectroscopy (spectrochemical analysis) employing characteristic emission of electromagnetic radiation by atoms, molecules or radicals. The ANALYTE (or an element whose concentration can be quantitatively related to it) is converted if necessary by operations such as DESOLVATION, VOLATILISATION, and DISSOCIATION to a vapour of atoms, molecules, or radicals, which are excited by collisions. In the emission process the EXCITATION ENERGY or part of it is converted into characteristic radiation, used for qualitative or quantitative analysis. Compare: ATOMIC EMISSION SPECTROSCOPY; FLAME EMISSION SPECTROSCOPY.

energy, radiant. See RADIANT ENERGY.

energy level; level. [*Energieniveau.*] A state of an atom or molecule corresponding to one of the various possible discrete values of its internal energy, and characterised by quantum numbers; a quantum state. The levels can be arranged in an *energy-level diagram* according to their quantum numbers and energy values. A distinction is made between the ground level (or GROUND STATE) and the excited levels (or EXCITED STATES). See further § 6.2.1. Compare: TERM.

enhancement. [*Erhöhung, Anhebung.*] An INTERFERENCE in which the APPARENT CONCENTRATION is higher than the true concentration of the ANALYTE. Compare: DEPRESSION.

error. See MISTAKE; RANDOM ERROR; SYSTEMATIC ERROR.

estimated standard deviation. See STANDARD DEVIATION.

evaluation. [*Auswertung.*] The mathematical conversion of the MEASURES or the READINGS into the desired ANALYTICAL RESULT (which may be a concentration) with the aid of the analytical function (see ANALYTICAL CURVE) and, if needed, with correction for INTERFERENCES.

evaluation curve. See ANALYTICAL CURVE.

excitation. [*Anregung.*] A process in which an atom or molecule, by gain of energy, is transferred to a higher ENERGY LEVEL or EXCITED STATE.

excitation energy. [*Anregungsenergie.*] The discrete amount of energy needed to raise an atom or molecule from the GROUND STATE to an EXCITED STATE.

excited state; excited level. [*angeregter Zustand.*] An ENERGY LEVEL corresponding to an internal energy above that of the GROUND STATE.

exposure. [*Bestrahlung.*] The product of IRRADIANCE and the duration of the irradiation. The unit is $J\,m^{-2}$.

FAAS. Abbreviation for flame atomic absorption spectroscopy; see FLAME AB-SORPTION SPECTROSCOPY.

FAFS. Abbreviation for flame atomic fluorescence spectroscopy; see FLAME FLU-ORESCENCE SPECTROSCOPY.

FAS. Abbreviation for FLAME ABSORPTION SPECTROSCOPY.

FES. Abbreviation for FLAME EMISSION SPECTROSCOPY.

FFS. Abbreviation for FLAME FLUORESCENCE SPECTROSCOPY.

fiducial limits. See CONFIDENCE LIMITS.

fill gas. See HOLLOW-CATHODE LAMP.

filter flame spectrometer. [*Filter-Flammenphotometer.*] A FLAME SPECTROMETER employing optical filters for spectral separation.

first resonance line. [*Grundlinie.*] The RESONANCE LINE corresponding to the transition between the GROUND STATE and the lowest EXCITED STATE for which an optical transition to the ground state is allowed.

flame. See DIFFUSION FLAME; FUEL-LEAN FLAME; FUEL-RICH FLAME; LAMINAR FLAME; PREMIXED FLAME; SPLIT FLAME; TURBULENT FLAME; UNPREMIXED FLAME. See chapter 3.

flame absorption spectroscopy; abbreviation: FAS. [*Flammenabsorptionsphoto-metrie.*] ATOMIC ABSORPTION SPECTROSCOPY employing a flame for ATOMISATION of the ANALYTE.

flame analysis. [*Flammenanalyse; Flammenmethode.*] A general (unofficial) term for ANALYTICAL FLAME SPECTROSCOPY or FLAME SPECTROMETRY.

flame background. [*Flammenuntergrund*; in FES: *Eigenemission* or *Eigenstrahlung der Flamme*; in FAS: *Eigenabsorption der Flamme*; in FFS: *Eigenstreuung der*

Flamme.] The RADIANT FLUX emitted, absorbed, or scattered by the flame itself when no liquid is nebulised into it. In FES it is due to the emission of a CONTINUUM and/or molecular BANDS, in FAS to the absorption of PRIMARY RADIATION by molecular bands, and in FFS to the scattering of primary radiation by molecules. Like BLANK BACKGROUND, the term *flame background* may refer to the entire spectrum of the flame or, in analogy to BLANK SIGNAL, it may refer to the corresponding SIGNAL observed within the SPECTRAL BANDWIDTH specified for the analysis, depending on whether WAVELENGTH MODULATION is employed.

flame emission spectroscopy; abbreviation: FES. [*Flammenemissionsphotometrie*; shorter: *E-Methode.*] Analytical EMISSION SPECTROSCOPY employing a flame as an excitation source. See FLAME SPECTROSCOPY.

flame fluorescence spectroscopy; abbreviation: FFS. [*Flammenfluoreszenzphotometrie.*] ATOMIC FLUORESCENCE SPECTROSCOPY employing a flame for ATOMISATION of the ANALYTE.

flame photometer. [*Flammenphotometer.*] An earlier term for FLAME SPECTROMETER. The IUPAC disapproves it.

flame photometry. [*Flammenphotometrie.*] An earlier term for analytical FLAME SPECTROSCOPY or FLAME SPECTROMETRY. The IUPAC disapproves it.

flame spectrograph. [*Flammenspektrograph.*] An apparatus for spectrochemical analysis employing a flame and a SPECTROGRAPH.

flame spectrometer. [*Flammenspektrometer.*] An apparatus for generally quantitative spectrochemical analysis employing a flame and a SPECTROMETER. Formerly called *flame photometer* or *flame spectrophotometer.*

flame spectrometry. [*Flammenphotometrie.*] Quantitative spectrochemical analysis by means of a FLAME SPECTROMETER; ANALYTICAL FLAME SPECTROSCOPY utilising a SPECTROMETER. Formerly called *flame photometry* or *flame spectrophotometry.*

flame spectrophotometer; flame spectrophotometry. [*Flammenspektrophotometer; Flammenspektrophotometrie.*] Earlier terms for FLAME SPECTROMETER and FLAME SPECTROMETRY. The IUPAC disapproves them.

flame spectroscopy; where the context is not unambiguous: **analytical flame spectroscopy.** [*Flammenphotometrie, analytische Flammenphotometrie*; informally, *Flammenanalyse.*] Qualitative or quantitative spectrochemical analysis employing a flame for one or more of the operations of DESOLVATION, VOLATILISATION, and DISSOCIATION of the ANALYTICAL SAMPLE. In FLAME EMISSION SPECTROSCOPY the flame serves to excite the atoms, radicals, or molecules produced by the analyte in the flame. In ATOMIC ABSORPTION SPECTROSCOPY and ATOMIC FLUORESCENCE SPECTROSCOPY, EXCITATION is not needed, provided that a RESONANCE LINE is used, but ATOMISATION is essential. Formerly generally called *flame photometry.* See also FLAME SPECTROMETRY; FLAME ANALYSIS.

flow. See LAMINAR FLOW; TURBULENT FLOW.

fluorescence. [*Fluoreszenz.*] Photon emission from atoms or molecules that have been raised to a higher excited level by the absorption of photons from a beam of PRIMARY RADIATION from a BACKGROUND SOURCE. (See EXCITATION; ENERGY LEVEL.) If the emission is from atoms it is *atomic fluorescence* [*Atomfluoreszenz*]. If the wavelengths of the absorbed and emitted radiation are identical, it is *resonance fluorescence* [*Resonanzfluoreszenz*]. Synonym: SECONDARY RADIATION. See also ATOMIC FLUORESCENCE SPECTROSCOPY; QUANTUM EFFICIENCY OF FLUORESCENCE.

fluorescence quenching. [*Fluoreszenzlöschung.*] The reduction of the number of photons emitted in atomic FLUORESCENCE below the number of photons absorbed from the PRIMARY RADIATION, due to conversion of the absorbed radiant energy into other forms of energy by collision of the photoexcited atoms with flame gas molecules. The collisions are said to *quench* the fluoresence. The quenching is measured by the QUANTUM EFFICIENCY OF FLUORESCENCE. See § 6.8.

fluorescence spectroscopy. [*Fluoreszenzphotometrie.*] See ATOMIC FLUORESCENCE SPECTROSCOPY.

flux. See RADIANT FLUX.

fraction atomised; symbol: β_a. [*Wirkungsgrad der Atomdampferzeugung.*] The ratio of the quantity of ANALYTE present as free neutral or ionised atoms in the observed volume of the flame to the quantity of analyte present there in the gaseous state (as atoms and molecules). Distinguish from: EFFICIENCY OF ATOMISATION.

fraction desolvated; symbol: β_s. [*Wirkungsgrad der Verdampfung.*] The ratio of the quantity of ANALYTE present in the observed volume of the flame in the form of DRY AEROSOL or vapour to the total quantity of analyte present there in all forms.

fraction volatilised; symbol: β_v. [*Wirkungsgrad der Verflüchtigung.*] The ratio of the quantity of ANALYTE present in the gaseous state (as free atoms or molecules) in the observed volume of the flame to the total quantity of analyte present there in the desolvated state (as vapour or DRY AEROSOL).

fuel-lean flame; oxidising flame. [*oxidierende Flamme.*] A flame to which more oxidant is added than corresponds to the stoichiometric ratio of oxidant to fuel. Also called simply *lean flame*.

fuel-rich flame; reducing flame. [*reduzierende Flamme.*] A flame containing fuel in excess of the stoichiometric ratio of fuel to oxidant, for the purpose of withholding oxygen from the atoms of an ANALYTE that tends to form a stable oxide in the flame, and thereby promoting its ATOMISATION. Also called simply *rich flame*.

Gaussian distribution. See NORMAL DISTRIBUTION.

grating flame spectrometer. [*Gitter-Flammenspektrometer.*] A FLAME SPECTROMETER employing a grating for spectral DISPERSION.

ground state; ground level. [*Grundzustand*; *Grundniveau.*] In an atom or molecule, the ENERGY LEVEL or quantum state of lowest internal energy.

half-intensity width. [*Halbwertsbreite, spektrale Halbwertsbreite*; *spektrale Bandbreite.*] The full width, for example in wavelength units, at half the peak height of any spectral profile such as a spectral line or the transmittance of a spectral selection device. The common but ambiguous synonym *halfwidth* is not recommended by the IUPAC. See also: LINE PROFILE; SPECTRAL BANDWIDTH.

height of observation. See OBSERVATION HEIGHT.

hollow-cathode lamp. [*Hohlkathodenlampe.*] A gas-discharge lamp in which accelerated electrons collide with atoms of the *carrier gas* or *fill gas* (usually a noble gas at low pressure), generating ions which, after further acceleration in the electric field, collide with the hollow cylindrical cathode, releasing (by cathodic sputtering) atoms of the cathode material that are then excited by the discharge and emit their characteristic spectrum. The lamp is commonly used as a BACKGROUND SOURCE for AAS and AFS. Compare: CATHODIC SPUTTERING CHAMBER.

I_0, I_t. The unattenuated INTENSITY of a LINE or CONTINUUM of the BACKGROUND SOURCE before entering the flame and its attenuated intensity after passing through the absorbing flame, respectively.

I_λ. SPECTRAL INTENSITY. One distinguishes between $(I_\lambda)_0$ and $(I_\lambda)_t$ in the same way as between I_0 and I_t. The spectral intensity at the centre λ_0 of a SPECTRAL LINE is denoted by I_{λ_0}.

impact bead; impact surface. [*Prallfläche.*] An object such as a glass ball supported directly in front of the SPRAYER NOZZLE to impart a finer drop size distribution to the MIST. Sometimes called *spray barrier.*

indirect nebuliser. [*Indirekt-Zerstäuber.*] An earlier name for CHAMBER-TYPE NEBULISER, the term now recommended by the IUPAC.

inner cone; inner zone. See PRIMARY COMBUSTION ZONE.

intensity; symbol: *I*. [*Intensität.*] A general term for any radiant quantity, lacking a specified unit or measured only in relative units; the relative strength of a SPECTRAL LINE, a background, a beam from a LIGHT SOURCE, etc. In comparing intensities, the same (relative) units must be used. Note, however, that RADIANT INTENSITY has a well defined meaning. See also SPECTRAL INTENSITY.

interconal zone. See INTERZONAL REGION.

interference. [*Beeinflussung.*] The effect of a CONCOMITANT IN THE SAMPLE on the MEASURE of the ANALYTE; a SYSTEMATIC ERROR due to an INTERFERENT. An interference exists when the measure for a given SAMPLE SOLUTION differs from that for the same concentration of analyte in the same chemical combination and solvent but without the interferent. If the interference is not corrected, the analysis will be in error. Usage requires the prepositions *of* and *on*: the interference *of* a concomitant *on* the analyte. See also: APPARENT CONCENTRATION; DEPRESSION; ENHANCEMENT; MATRIX EFFECT; DESOLVATION INTERFERENCE; IONISATION INTERFERENCE; SPECTRAL INTERFERENCE; VOLATILISATION INTERFERENCE.

interference curve. [*Beeinflussungskurve.*] A plot of the MEASURE or of the APPARENT CONCENTRATION of the ANALYTE (at fixed concentration) as a function of the concentration of an INTERFERENT.

interferent. [*Störelement.*] A CONCOMITANT IN THE SAMPLE that causes an INTERFERENCE in a given analysis.

interfering line. [*Störlinie.*] A SPECTRAL LINE causing SPECTRAL INTERFERENCE. A

207

special case is *line coincidence*, in which the true spectral LINE PROFILES of the interfering and analytical lines overlap at least partly.

interlaboratory comparison. [*Vergleich bei Rundversuchen.*] A trial in which samples of the same ANALYTICAL MATERIAL are analysed by several operators in different laboratories using different instruments and sometimes different methods; a round-robin trial. The STANDARD DEVIATION is generally greater than in an INTRALABORATORY COMPARISON. When several different methods are used, any that yield out-of-bounds values (having a statistically significant departure from the overall mean of the series of determinations by these methods) are to be suspected of SYSTEMATIC ERRORS. See REPRODUCIBILITY; PRECISION.

internal absorbance. See ABSORBANCE.

internal absorption factor; internal absorptance. See ABSORPTION FACTOR.

internal-reference element. See REFERENCE ELEMENT.

internal-reference-element technique. See REFERENCE-ELEMENT TECHNIQUE.

internal-reference line. See REFERENCE LINE.

internal standard. Earlier term for REFERENCE ELEMENT.

internal transmission factor; internal transmittance. See TRANSMISSION FACTOR.

interzonal region. [*Zwischenzone*; less frequent synonyms: *Zwischengas, mittlere Flammenzone*; earlier synonym: *reaktionsfreier Teil.*] The region of the flame between the PRIMARY and SECONDARY COMBUSTION ZONES; it is most commonly used for flame analysis. When the combustion zones are conical it may be called the *interconal zone.* It is best observed in a SEPARATED FLAME or a SHIELDED FLAME.

intralaboratory comparison. [*Wiederholbedingungen*, referring to the conditions of such a comparison.] A trial of PRECISION in which an operator repeatedly analyses a given SAMPLE SOLUTION with a single instrument and under the same operating conditions. He arrives at the mean ANALYTICAL RESULT and its STANDARD DEVIATION or CONFIDENCE LIMITS. SYSTEMATIC ERRORS are not detectable in an intralaboratory comparison. Distinguish from: INTERLABORATORY COMPARISON. Compare: REPRODUCIBILITY; REPEATABILITY.

ionic line; ion line. [*Ionenlinie.*] A SPECTRAL LINE emitted or absorbed by a free atomic ion, that is, a free atom that has lost one or more electrons. Distinguish from: ATOMIC LINE.

ionisation. [*Ionisation.*] The splitting from an atom or molecule of one or more free electrons, producing a (singly or multiply) charged positive atomic or molecular ion.

ionisation buffer. [*Normalisator, Normalisatorsubstanz*; *Puffer, Puffersubstanz.*] A SPECTROCHEMICAL BUFFER that readily ionises in the flame, added to suppress IONISATION INTERFERENCE by increasing the concentration of free electrons in the flame and thus repressing and stabilising the degree of ionisation of the ANALYTE and INTERFERENTS despite variation of their concentrations. Caesium salts are a good example. [Ionisation buffering is *Normalisierung.*]

ionisation energy. [*Ionisationsenergie.*] The least energy that must be supplied to an atom or molecule in the GROUND STATE in order to remove an electron from it.

208

ionisation interference. [*Ionisationsbeeinflussung.*] An INTERFERENCE of a CONCOMITANT IN THE SAMPLE on an ANALYTE both of which ionise markedly in the vapour phase in the flame or PLASMA. An ionising concomitant raises the electron concentration above what it would be in the absence of the concomitant, shifting the equilibrium among the analyte atom, analyte ion, and electron concentrations in the direction of reducing the degree of ionisation of the analyte. This in turn raises the INTENSITY of an ATOMIC LINE of the analyte, which depends on the concentration of analyte atoms.

irradiance. [*Bestrahlungsstärke.*] The ratio of the RADIANT FLUX striking a surface to the area irradiated. The unit is $W\,m^{-2}$ or (in practice) $W\,cm^{-2}$.

laminar flame. [*laminare Flamme.*] A flame characterised by LAMINAR FLOW of the gas, usually vertically upward. Compare: TURBULENT FLAME.

laminar flow. [*laminare Strömung.*] A mode of flow in which the lines of flow are more or less parallel and change only smoothly, if at all, in time and space. Eddy-free flow. Compare: TURBULENT FLOW.

lean flame. See FUEL-LEAN FLAME.

level. See ENERGY LEVEL.

level of significance; significance level. [*Signifikanzgrenze.*] The probability P as defined under NORMAL DISTRIBUTION and CONFIDENCE INTERVAL. Compare: CONFIDENCE LEVEL.

light modulation. See MODULATION.

light source. [*Strahlungsquelle.*] A device for generating certain kinds of radiation (including but not limited to visible light), distributed discretely or continuously over the spectrum; in AAS and AFS, the BACKGROUND SOURCE. See, in particular, HOLLOW-CATHODE LAMP, ELECTRODELESS-DISCHARGE LAMP, LINE SOURCE, CONTINUUM SOURCE.

limit of detection. See DETECTION LIMIT.

line. See SPECTRAL LINE.

line coincidence. See INTERFERING LINE.

line profile; spectral line profile. [*Linienprofil.*] The curve representing the distribution of the INTENSITY of a SPECTRAL LINE (in emission or absorption) over the narrow wavelength range within which the intensity is measurable; the SPECTRAL INTENSITY of a line plotted as a function of wavelength. See also SELF-REVERSAL.

line source; spectral line source. [*Linienstrahler.*] A LIGHT SOURCE that emits discrete SPECTRAL LINES; best used as a BACKGROUND SOURCE in AAS and AFS. Examples are METAL-VAPOUR LAMPS, ELECTRODELESS-DISCHARGE LAMPS, and HOLLOW-CATHODE LAMPS.

linewidth. [*Linienbreite.*] See SPECTRAL LINE.

mμ. Millimicron, an obsolete synonym of *nanometre* (see NM).

major constituent; main constituent. [*Hauptbestandteil.*] The constituent of a sample present at highest concentration, generally taken as an element. Distinguish: MATRIX.

material; starting material. [*Material; Ausgangsmaterial.*] Virtually synonymous with ANALYTICAL MATERIAL. The ANALYTICAL SAMPLE is prepared from it. If a representative sample is to be obtained from a markedly inhomogeneous starting material, several randomly selected equal portions are taken and mixed. See also SAMPLE SOLUTION.

matrix. [*Matrix*; occasionally: *Summe der Begleitstoffe.*] The aggregate of all CONCOMITANTS IN THE SAMPLE, often viewed in conjunction with their combined effect on the ANALYTE. See MATRIX EFFECT.

matrix effect. [*Matrix-Effekt.*] The combined effect or composite, unspecified INTERFERENCE of all CONCOMITANTS IN THE SAMPLE on the MEASURE of the ANALYTE under the prevailing analytical conditions.

mean. [*Mittelwert*; *arithmetisches Mittel.*] The arithmetic mean or average; the sum of a set of measurements divided by the number of measurements. See STANDARD DEVIATION; NORMAL DISTRIBUTION; STANDARD ERROR.

measure. [*Messwert.*] A quantitative value of the magnitude of the optical SIGNAL in analytical flame spectroscopy, namely the RADIANT FLUX emitted, absorbed, or fluoresced (within the SPECTRAL BANDWIDTH specified for the analysis) when an analytical solution is nebulised. Since the flux as such need not be determined absolutely, the measure can be and usually is expressed in other terms, for example, the corresponding PHOTOCURRENT (in FES and AFS), or the difference, ratio, or logarithmic ratio of the photocurrents corresponding to the un-attenuated and attenuated flux (in AAS). It is here assumed that the conditions of the optical train and photodetector are fixed. The measure is thus independent of the setting of the amplifier gain or the adjustment of the read-out device. The measure is derived, after correction for any such adjustments, from the READING, and the set of measures obtained during an analysis yields the ANALYTICAL RESULT through the process of EVALUATION. See also: NET MEASURE; BLANK MEASURE.

Méker burner. [*Méker-Brenner.*] A PREMIX BURNER consisting of an upright cylindrical tube topped by a usually loose burner cap with a horizontal plate (which may be up to 1 cm thick) having 10 to 100 narrow outlet ports for the gas mixture. Above each port there is usually a separate, tiny PRIMARY COMBUSTION ZONE, while the burnt gases from each of these zones merge into a single, thick flame which is surrounded by a single SECONDARY COMBUSTION ZONE (see § 3.4.1). Distinguish from: BUNSEN BURNER; SLOT BURNER.

metal-vapour lamp. A gas-discharge lamp filled with a noble gas and a metal vapour which is produced from the volatile element by the thermal effect of the gas discharge. When operated at a low pressure the lamp is a LINE SOURCE emitting the characteristic LINES of the metal atom.

mist; (wet) aerosol. [*flüssiges Aerosol.*] A mixture of gas and SAMPLE SOLUTION in the form of nebulised droplets. Compare: DRY AEROSOL.

mistake. [*grober Fehler, Ausreisser.*] An error resulting usually from the analyst's carelessness, avoidable in principle. Examples are confusion of the reference or sample solutions, accidental contamination, wrong weighing, etc. Mistakes should be distinguished from RANDOM and SYSTEMATIC ERRORS.

modulation; light (intensity) modulation; optical modulation. [*optische Modulation.*] The periodic interruption or attenuation of radiation, as by means of a rotating sector (*chopper*) placed in the appropriate beam or by electronic means, for the purpose of discriminating the (desired) SIGNAL thus modulated (e.g. the FLUORESCENCE induced by the modulated PRIMARY RADIATION in AFS) from another, (unwanted) unmodulated signal (e.g. the thermal emission of the flame) that may be simultaneously present. Compare: WAVELENGTH MODULATION.

monochromator. [*Monochromator.*] A device, usually part of a SPECTROMETER, that isolates a narrow wavelength region of the spectrum. In a common design, a beam of light entering through an *entrance slit* is focused, after DISPERSION by a prism or grating, upon an *exit slit*, and the wavelength of the light passing the exit slit is selected by turning the dispersing device. Compare: POLYCHROMATOR.

multiplet. [*Multiplett.*] A collection of closely spaced single SPECTRAL LINES belonging to the same transition between two spectroscopic TERMS. Also, a collection of closely spaced ENERGY LEVELS belonging to the same term. It is called a *singlet* [*Singulett*], *doublet* [*Dublett*], or *triplet* [*Triplett*] when the collection comprises 1, 2, or 3 members, respectively. See further § 6.2.1.

multiplier phototube. See PHOTOMULTIPLIER TUBE.

multislot burner. See SLOT BURNER.

nebulisation. [*Zerstäubung.*] The conversion of a solution into MIST by a NEBU-LISER.

nebulisation efficiency. See EFFICIENCY OF NEBULISATION.

nebuliser. [*Zerstäuber.*] See CHAMBER-TYPE NEBULISER; DIRECT-INJECTION BURNER; PNEUMATIC NEBULISER; SPRAYER NOZZLE.

net measure. [*Nutzmesswert.*] The MEASURE of the ANALYTE SIGNAL, usually equal to the total measure obtained during introduction of the ANALYTE minus the BLANK MEASURE.

net signal. See ANALYTE SIGNAL.

nm. Abbreviation for nanometre, 10^{-9} metre. Formerly called *millimicron*, mμ.

noise; scatter. [*Rausch, Rauschen; Streubereich, Streubreite.*] Random or statistical fluctuation or variation in any measurement or measured quantity (a VARIATE) such as a MEASURE, SIGNAL, or READING, revealed by repeating the measurement and/or by observing the signal continuously for a time. The numerical value of the scatter is the STANDARD DEVIATION of the measurement. In flame spectrometry, sources of scatter in the BLANK and NET MEASURES include: accidental external contamination of the solution; variation of impurities in the reagents; accidental weighing errors; random fluctuation of NEBULISATION, DESOLVA-TION, ATOMISATION, and EXCITATION; random fluctuation in the FLAME BACK-GROUND or in the LIGHT SOURCE; mechanical instability in the spectrometer; shot noise and low-frequency (flicker) noise in the PHOTOCURRENT and DARK CURRENT; amplifier noise; accidental reading and calibration errors, etc. The scatter caused by some of these sources depends on the RESPONSE TIME of the instrument, which must be specified when the performance of different instruments is to be compared. The total scatter equals the square root of the sum of the squares of the separate statistically independent contributions. The BLANK SCATTER alone determines the DETECTION LIMIT; the blank scatter and the scatter of the net measure determine the PRECISION. Distinguish from: DRIFT.

noise level. [*Störpegel.*] The magnitude of the NOISE represented by the rapid fluctuations observed during the course of a single READING. See also: SIGNAL-TO-NOISE RATIO.

normal distribution; Gaussian distribution. [*Normalverteilung*; *Gaussverteilung.*] A theoretical frequency distribution or probability-density function approximating the distribution of many random VARIATES, given by

$$f(x) = \frac{1}{\sqrt{2\pi}\sigma} \exp\left[-(x-\mu)^2/2\sigma^2\right],$$

where μ is the mean of the variate x and σ is its STANDARD DEVIATION. This represents a bell-shaped curve, the *normal curve*, symmetrical about μ. An example is the curve obtained by plotting against the MEASURE x the frequency of occurrence, $f(x)$, of individual measures observed on repeatedly aspirating one and the same sample solution in a flame spectrometer. The width of the curve, defined by σ, gauges the dispersion or SCATTER of the measures about their mean. The normal distribution has the property that among, for example, 1000 independent measures, 317 are expected to fall outside the range $\mu \pm \sigma$. Hence the probability P (called the LEVEL OF SIGNIFICANCE) of finding a measure more than σ distant from the mean is 31·7 per cent and the probability $1-P$ of finding it within $\pm\sigma$ of the mean is 68·3 per cent; similarly the probability of finding it within $\mu \pm 2\sigma$ is $1-P = 95·4$ per cent; for $\mu \pm 3\sigma$, $1-P = 99·73$ per cent, that is, in 99·73 per cent of all trials x is expected to lie within 3σ of the mean.

observation height (in the flame). [*Beobachtungshöhe in der Flamme.*] The difference in height between the top of the burner and the centre of the observed beam.

optical conductance. [*Lichtleitwert.*] A quantity embodying the geometrical factors that limit the RADIANT FLUX transmitted by an optical system such as a MONOCHROMATOR. For monochromatic radiation, this radiant flux equals the product of the optical conductance, the RADIANCE, and the TRANSMISSION FACTOR. Optical conductance is formally analogous to electric conductance.

optical modulation. See MODULATION.

outer cone; outer zone. See SECONDARY COMBUSTION ZONE.

oxidising flame. See FUEL-LEAN FLAME.

parts per million. See PPM.

photocurrent. [*Photostrom.*] That part of the output current of a PHOTODETECTOR that is induced by the incident radiation. Compare: DARK CURRENT.

photodetector. [*Strahlungsempfänger, Photodetektor.*] A detector or receiver of *light*, here understood as including the visible, ultraviolet, and infrared regions of the electromagnetic spectrum. Photoelectric detectors include: vacuum phototube, barrier-layer cell, photoconductor, PHOTOMULTIPLIER TUBE. See also PHOTOMETRY.

photometry. [*Photometrie.*] The measurement of light, in the narrower sense radiation perceived by the eye. In *subjective photometry* the eye itself serves as radiation detector; in *objective photometry* use is made of a photodetector and filter having in combination a relative spectral sensitivity resembling that of the standard eye. But the term *objective photometry* (or just *photometry*) has also been extended to the measurement of radiation outside the visible region of the spectrum; the IUPAC disapproves of this extension.

photomultiplier tube; photomultiplier; multiplier phototube. [*Photoelektronen-Vervielfacher*, abbreviated PEV; *Sekundärelektronen-Vervielfacher*, abbreviated SEV.] A PHOTODETECTOR employing the external photoelectric effect, in which quanta of light striking the photocathode release electrons that are amplified by secondary-electron emission at the successive dynodes in the photomultiplier tube and give rise to an anode current that exceeds the photocathode current by a large factor.

plasma. [*Plasma.*] Here used as an informal term for a high-temperature gas or vapour serving as the medium for a spectrochemical analysis.

pneumatic nebuliser. [*pneumatischer Zerstäuber.*] A nebuliser driven by compressed gas (usually air or oxygen), in which the liquid is aspirated through a capillary by the reduced pressure arising at the SPRAYER NOZZLE, where it is reduced to fine drops carried off by the gas stream.

polychromator. [*Polychromator*]. A device like a MONOCHROMATOR that permits simultaneous isolation of two or more narrow wavelength regions.

ppb. Parts per billion (by weight), a measure of CONCENTRATION in units of 10^{-9}; compare: PPM.

ppm. Parts per million (by weight), a measure of CONCENTRATION in units of 10^{-6}; the numerator and denominator of the ratio must have the same dimensions, e.g. $\mu g\, g^{-1}$ (micrograms of ANALYTE per gram of sample) or $mg\, kg^{-1}$. The term ppm should thus not be used for $mg\, l^{-1}$ or $\mu g\, ml^{-1}$.

precision. [*Präzision.*] The quality of an ANALYTICAL RESULT or procedure as characterised by its RANDOM ERRORS. We distinguish: (i) serial precision (REPEATABILITY of results obtained on the same SAMPLE SOLUTION with the same instrument on the same day by the same operator); (ii) day-to-day precision (precision of results obtained on the same sample solution with the same instrument and operator but on different days); (iii) interlaboratory precision (of results obtained in round-robin trials with different instruments and operators but with the same ANALYTICAL MATERIAL). (See INTERLABORATORY COMPARISON.) The (RELATIVE) STANDARD DEVIATION or CONFIDENCE INTERVAL serves as a numerical measure of the precision. Compare: ACCURACY.

premix burner. [*Brenner für vorgemischte Gase.*] A burner in which the fuel gas and oxidant are thoroughly mixed in the burner body before leaving the burner port(s). The burner usually produces an approximately LAMINAR FLAME and is commonly combined with a CHAMBER-TYPE NEBULISER. Distinguish from: DIRECT-INJECTION BURNER. See also: BUNSEN BURNER; MÉKER BURNER; SLOT BURNER.

premixed flame. [*vorgemischte Flamme.*] A flame in which the entering fuel gas and oxidant are thoroughly mixed in the burner body before reaching the PRIMARY COMBUSTION ZONE. See also PREMIX BURNER.

primary combustion zone; primary reaction zone; inner zone. [*primäre* or *innere Verbrennungszone*; *primäre* or *innere Reaktionszone.*] The region in which the first of the two oxidation steps takes place in the fuel gas–oxidant mixture entering a LAMINAR FLAME above a PREMIX BURNER. It is thin, conical (above a round port), usually quite luminous, and interior to the SECONDARY COMBUSTION ZONE. See § 3.4.1. Occasionally called the *inner cone* [*Innenkonus, Innenkegel*]. Above a DIRECT-INJECTION BURNER the primary combustion zone is less clearly defined because of the intense turbulence (see § 3.5.1). See also INTERZONAL REGION.

primary radiation. [*Primärstrahlung.*] The radiation from the BACKGROUND SOURCE entering the flame, usually at right angles to the direction of observation, that excites the FLUORESCENCE in AFS. Distinguish from: SECONDARY RADIATION.

prism flame spectrometer. [*Prismen-Flammenspektrometer.*] A FLAME SPECTROMETER utilising a prism for spectral DISPERSION.

quantity of analyte; symbol: q. [*Menge.*] The total amount of ANALYTE in the sample, measured in units of mass. Distinguish from: CONCENTRATION.

quantum efficiency of fluoresence. [*Fluoreszenzausbeute.*] The ratio of the number of photons emitted as FLUORESCENCE radiation in all directions to the number of photons absorbed from the PRIMARY RADIATION.

quenching. See FLUORESCENCE QUENCHING.

radiance. [*Strahldichte.*] The ratio of the RADIANT FLUX passing through a surface in a given direction to the product of the solid angle filled by the radiation and the area of the surface projected upon a plane normal to the direction considered. The unit is $W\,m^{-2}\,sr^{-1}$ or (in practice) $W\,cm^{-2}\,sr^{-1}$. See also: SPECTRAL INTENSITY.

radiant energy. [*Strahlungsenergie.*] Energy occurring in the form of radiation. The unit is W s or J. See also: SPECTRAL INTENSITY.

radiant flux. [*Strahlungsfluss.*] RADIANT ENERGY per unit time; radiant power. The unit is W. See also: SPECTRAL INTENSITY.

radiant intensity. [*Strahlstärke.*] The ratio of the RADIANT FLUX emitted by a source of radiation in a given direction to the solid angle filled by the radiation. The unit is $W\,sr^{-1}$. See also INTENSITY and SPECTRAL INTENSITY.

radiation, primary, secondary. See PRIMARY RADIATION, SECONDARY RADIATION.

random error. [*zufälliger Fehler.*] An error in a measurement resulting from uncontrollable, varying effects such as those listed under NOISE. The random error represents the SCATTER and is defined by the STANDARD DEVIATION of the measurements, usually assumed to have a NORMAL DISTRIBUTION. The PRECISION of an ANALYTICAL RESULT is characterised by the random error and is specified by stating the MEAN value along with the standard deviation. These two statistical quantities can be assessed more reliably as the number of repeated analyses increases. Compare: STANDARD ERROR (of the mean); CONFIDENCE INTERVAL. Distinguish from: SYSTEMATIC ERROR or BIAS.

range. See WORKING RANGE.

reading. [*Ablesung, Anzeige.*] An observed value of the magnitude of the output SIGNAL in the final stage of any measuring instrument, specifically here of a FLAME SPECTROMETER; this value may be expressed as the ANALOGUE or

DIGITAL indication [*Analoganzeige, Digitalanzeige*] of the *read-out device*, which may be a needle meter (whose scale is divided into arbitrary divisions, units of current, or units of concentration), an optically projected scale, a potentiometer balanced with the aid of a null meter, a strip-chart recorder, a digital display panel, a printer, a punched tape, etc. An analogue reading may have a reading error, namely the difference between the true and the observed positions of the indicator on the scale. The MEASURE is inferred from the reading; unlike the measure, the reading is not unaffected by the amplifier gain or certain other instrumental adjustments.

recording (of a spectrum). See SPECTROGRAM.

reducing flame. See FUEL-RICH FLAME.

reference element; in case of ambiguity: **internal-reference element.** [*Leitelement*; *Bezugselement.*] An element with whose concentration that of the ANALYTE is compared. It can be an essential constituent of the sample (e.g. iron in steel) or it can be an element added to the SAMPLE SOLUTION. Formerly called *internal standard* [*innerer Standard*]. [Reference-element concentration is *Leitkonzentration*; formerly *Bezugskonzentration.*]

reference-element technique; internal-reference-element technique. [*Leitlinienverfahren.*] The spectrochemical determination of an ANALYTE by comparing its MEASURE with that of a REFERENCE ELEMENT.

reference line; in case of ambiguity: **internal-reference line.** [*Leitlinie*; *Bezugslinie.*] A SPECTRAL LINE of the REFERENCE ELEMENT that is employed, by comparing its INTENSITY with that of a line of the ANALYTE, for determining the concentration of the analyte. The intensities of the two lines should respond in the same way to changes in the experimental conditions such as irregularities of NEBULISATION (compare Gerlach's *homologous lines*).

reference sample. [*Bezugsprobe.*] One of a series of homogeneous, synthetically prepared materials containing the ANALYTE in different, known concentrations and sometimes one or more CONCOMITANTS, submitted to the same procedure as the ANALYTICAL SAMPLE in a spectrochemical analysis for the purpose of establishing the ANALYTICAL CURVE and regulating the SENSITIVITY. Compare: REFERENCE SOLUTION; STANDARD SAMPLE.

reference solution. [*Bezugslösung*; *Referenz-Lösung.*] A REFERENCE SAMPLE in solution form, having the same SOLVENT as the SAMPLE SOLUTION. Besides their other functions, reference solutions help to guard against DRIFT by being nebulised at regular intervals during routine determinations. Formerly called *standard solution* [*Eichlösung, Standardlösung*]. [A reference solution used for checking sensitivity is *Kontroll-Lösung.*]

relative standard deviation; symbol: s_r. [*relative Standardabweichung.*] The ratio of the STANDARD DEVIATION of an ANALYTICAL RESULT (or more generally of any measured quantity) to its MEAN value. The number n of observations should always be reported. The same distinction is made as that between $s(S)$ and $s(D)$ (see STANDARD DEVIATION) or between serial and day-to-day PRECISION (REPEATABILITY and REPRODUCIBILITY, respectively). Formerly called *coefficient of variation* [*Variationskoeffizient*].

215

relative systematic error. See SYSTEMATIC ERROR.

repeatability. [*Wiederholbarkeit*; *Reproduzierbarkeit in der Serie.*] A measure of PRECISION defined by the (ESTIMATED) STANDARD DEVIATION $s(S)$ of the ANALYTICAL RESULT on a given SAMPLE SOLUTION obtained on the same day with the same instrument by the same operator.

reproducibility. [*Reproduzierbarkeit.*] PRECISION of INTERLABORATORY COMPARISON.

reproducibility of instrument. [*instrumentelle Reproduzierbarkeit.*] The PRECISION in which only the RANDOM ERROR of the instrument is taken into account. Random errors due to sampling, sample preparation, and the variation of errors of READING and EVALUATION among observers are not included. The instrument is assumed to be one that gives an objective read-out such as an ANALOGUE recording or a DIGITAL print-out. This reproducibility is measured by the (ESTIMATED) STANDARD DEVIATION $s(D)$ of the ANALYTICAL RESULT obtained with the same SAMPLE SOLUTION, instrument, and operator on different days. It will always be better than interlaboratory precision. Distinguish from: REPEATABILITY. See also NORMAL DISTRIBUTION; NOISE; INTERLABORATORY and INTRALABORATORY COMPARISON.

resolving power; (practical) resolution. [*Auflösungsvermögen*; *Auflösung.*] A measure of the difference between the wavelengths of two components of radiation that can barely be separated by a spectral instrument. It is given by $\lambda/\Delta\lambda$, where $\Delta\lambda$ is the HALF-INTENSITY WIDTH of a very narrow SPECTRAL LINE imaged by the instrument or the SPECTRAL BANDWIDTH of the instrument. The (practical) resolution of an analytical SPECTROMETER, governed by a compromise among conflicting desiderata (low NOISE, short RESPONSE TIME, rapid scanning, low sample consumption, low DETECTION LIMIT, etc), may be much poorer than the (theoretical) resolving power of the instrument under ideal conditions. The term *resolution* is applied also to other quantities such as dimensions on a photograph or time. Ambiguity is removed by the term *spectral resolution*, etc. Distinguish from: DISPERSION.

resonance fluorescence. See FLUORESCENCE.

resonance line. [*Resonanzlinie.*] An ATOMIC LINE corresponding to an optical transition between an EXCITED STATE and the GROUND STATE. See also FIRST RESONANCE LINE.

response time. [*Einstellzeit.*] The time needed by an instrument to reach a READING that is a specified fraction, for example 99 per cent, of the final value in the case of an abruptly applied SIGNAL, starting from zero. When a response time is reported, this fraction must be stated. If the fraction is $1-1/e=0.632$, the response time is called the TIME CONSTANT. It is helpful to mention which components of the instrument (such as the load resistor of a phototube) govern the response time.

result of analysis. See ANALYTICAL RESULT.

reversal. See SELF-REVERSAL.

reversed direct-injection burner. [*Direkt-Zerstäuber-Brenner mit innerer Brenngas-Zufuhr.*] A DIRECT-INJECTION BURNER in which the fuel is used for aspirating and nebulising the sample solution.

216

rich flame. See FUEL-FICH FLAME.

rise-velocity (of the flame gas). [*Steiggeschwindigkeit der Flammengase.*] The rate of ascent of the burned gas after passing the PRIMARY COMBUSTION ZONE. It may depend on the location within the flame. Distinguish from: BURNING VELOCITY.

sample. See ANALYTICAL SAMPLE.

sample solution. [*Analysenlösung; Bestimmungslösung.*] The solution to be analysed, prepared from the ANALYTICAL SAMPLE.

sampling-boat technique. [*Schiffchenmethode.*] The introduction of a solid or liquid sample directly into the flame in a small, usually elongate container with a handle (*sampling boat* [*Verdampfungsschiffchen*]), usually made from a thin sheet of high-melting, corrosion-resistant metal such as tantalum.

scatter. [*Streuung, Streubreite.*] See NOISE; BLANK SCATTER; STANDARD DEVIATION; NORMAL DISTRIBUTION; PRECISION.

scattered light. See STRAY LIGHT.

secondary combustion zone; secondary reaction zone; outer zone. [*sekundäre or äussere Verbrennungszone; sekundäre or äussere Reaktionszone.*] The region in which the hot gas of a LAMINAR FLAME comes into contact with the surrounding air and the second of the two oxidation steps takes place if the flame is approximately stoichiometric or FUEL-RICH. It surrounds the INTERZONAL REGION and bounds the flame externally. See § 3.4.3. Occasionally called (in a BUNSEN BURNER flame) the *outer cone* [*Aussenkonus*].

secondary radiation. [*Sekundärstrahlung.*] Synonym for FLUORESCENCE.

self-absorption. [*Selbstabsorption.*] The weakening of a RESONANCE LINE emitted by atoms in a PLASMA (such as a flame) due to absorption by GROUND-STATE atoms of the same kind that are in the plasma between the emitting atoms and the observer (see § 6.6.1). The absorbed radiant energy can be re-emitted in all directions, but in flames it is mostly converted into other forms of energy (heat) and is thus 'lost'. The degree of self-absorption depends, aside from the oscillator strength and the LINE PROFILE, on the thickness of the plasma and the (mean) concentration of atoms in the source. In the case of 'pure' self-absorption the plasma is homogeneous with respect to temperature or more generally with respect to the excitation conditions. Distinguish from: SELF-REVERSAL.

self-reversal. [*Selbstumkehr.*] A special case of SELF-ABSORPTION occurring when the radiating core of an emission source such as a flame or HOLLOW-CATHODE LAMP is surrounded by a mantle of ATOMIC VAPOUR (of the same element as the emitting atoms) in which comparatively little or no excitation is taking place (see § 6.6.2). An example is a flame whose outer zone is distinctly cooler than the centre. The reabsorption of the RESONANCE LINE radiation from the core by the mantle is then no longer fully balanced by emission in the mantle. Since the ABSORPTIVITY is greatest at the centre of a line, the unbalanced loss of radiation is greater there than in the lateral wings of the LINE PROFILE. This can create a minimum or *reversal dip* at the centre, and we speak of a *reversed line*. In an extreme case, when virtually only the wings remain, the resonance line may look like two separate, diffuse lines.

sensitivity. [*Empfindlichkeit.*] The first derivative of the ANALYTICAL CALIBRATION FUNCTION with respect to the concentration c, or the slope of the analytical calibration curve, $\mathrm{d}x/\mathrm{d}c$; or the reciprocal of the first derivative of the ANALY-TICAL EVALUATION FUNCTION $c = f(x)$, namely $[\mathrm{d}f(x)/\mathrm{d}x]^{-1}$. These two expressions, with specified units for numerator and denominator, represent a constant independent of concentration only if the analytical curve is a straight line. Otherwise the sensitivity varies with the concentration. Sensitivity data can be compared on an absolute basis only among AAS methods where x represents, for example, the ABSORBANCE. In FES and AFS the sensitivity includes the arbitrary amplifier gain, scale division size, etc, as factors when we take the READING x as a MEASURE for the signal (that is, at given amplifier gain). Distinguish from DETECTION LIMIT, which is not a ratio but a concentration or a mass. Distinguish also: CHARACTERISTIC CONCENTRATION. The term 'sensitivity' is often used loosely to designate factors such as amplifier gain that affect the READING; these uses are to be discouraged.

separated flame. [*geteilte Flamme.*] A flame in which the PRIMARY and SECONDARY COMBUSTION ZONES are separated by means of a tube above the burner and surrounding the flame; the zones are at the bottom and top of the tube, respectively, and the tube itself is occupied by the INTERZONAL REGION. Compare: SHIELDED FLAME. Both the separated flame and the shielded flame may be called a SPLIT FLAME; they also provide protection against contamination from the ambient air.

shielded flame. [*abgeschirmte Flamme*; *Flamme mit Schutzgasatmosphäre.*] A flame like the SEPARATED FLAME in which the combustion zones are separated by surrounding the flame with a flowing sheath of inert gas that issues from ports at the rim of the burner top.

signal. [*Signal.*] Any of a sequence of inter-related physical quantities serving to carry information of interest, for example in a FLAME SPECTROMETER during introduction of a SAMPLE, BLANK, or REFERENCE SOLUTION. As examples, in FES the ANALYTE causes an emission of light whose wavelength is characteristic for the kind of analyte and whose INTENSITY or, more precisely, RADIANT FLUX is related to its concentration. The radiation passing the optical train (lenses, stops, spectral selector) and striking the PHOTODETECTOR is called the *optical* or *primary signal*. The PHOTOCURRENT generated by this beam may undergo many transformations in an amplifier and much processing (as by an analogue–digital converter, background compensator, ratio-ing, logarithmic amplifier, integrator) before it appears as an output current, voltage or pulse train driving a read-out device. Each of these consecutive quantities—fluxes, currents, etc—is a signal. See also ANALYTE SIGNAL; BLANK SIGNAL; MEASURE; READING.

signal-to-noise ratio. [*Signal/Rausch Verhältnis.*] The ratio of the mean magnitude of a SIGNAL to the NOISE occurring in its measurement. The noise is usually taken to be the STANDARD DEVIATION in the determination of the signal and is expressed in the same units. If the signal is to be estimated from a fluctuating recorder trace or by watching a fluctuating needle, the standard deviation can be estimated by taking one-fourth of the range within which the signal remains 95 per cent of the time. The DETECTION LIMIT and the PRECISION depend on the signal-to-noise ratio. The signal-to-noise ratio is a reciprocal RELATIVE STANDARD DEVIATION.

218

significance level. See LEVEL OF SIGNIFICANCE.

singlet. See MULTIPLET.

slot burner. [*Schlitz-Brenner.*] A PREMIX BURNER whose cap is a rectangular metal block with a longitudinal machined slot facing upward and serving as outlet port for the gas mixture. The flame is therefore elongated and narrow. It is usually placed along the optical axis so that a long absorption path length is obtained in AAS. To diminish the effect of external contaminants such as dust on the observed region of the flame or to improve the laminarity, the burner is often provided with three slots (a *three-slot burner*); the optical beam then passes along the flame gas from the middle slot, while the flame gases from the two outer slots function to some extent as a protective gas shield. The general term for this type is *multislot burner* [*Mehrschlitz-Brenner*].

solute-volatilisation interference. See VOLATILISATION INTERFERENCE.

solvent. [*Lösungsmittel.*] A liquid used for dissolving the desired constituent of the ANALYTICAL SAMPLE, for diluting the SAMPLE SOLUTION, or for improving the transport or EXCITATION of the ANALYTE. Compare: DILUENT.

solvent blank. See BLANK SOLUTION.

source. See LIGHT SOURCE.

spectral band; band. [*(spektrale) Bande.*] A crowded group of adjacent SPECTRAL LINES due to radiative transitions in a free molecule or polyatomic radical. See also BAND HEAD; BAND LINE.

spectral bandwidth; bandwidth. [*spektrale Bandbreite; Halbwertsbreite.*] The HALF-INTENSITY WIDTH of the light transmitted by a spectral selection device such as a MONOCHROMATOR or filter. If the TRANSMITTANCE of monochromatic radiation is plotted against the wavelength, the curve shows a maximum τ_{max} at a wavelength λ_{max}. The ordinate $\tau = \tau_{max}/2$ intersects the curve at two points of wavelength λ_1 and λ_2. The bandwidth is $\Delta\lambda = |\lambda_1 - \lambda_2|$.

spectral continuum. See CONTINUUM.

spectral intensity, etc; symbol: I_λ, etc. [*spektrale Intensität.*] INTENSITY (or RADIANT INTENSITY, RADIANCE, RADIANT FLUX, etc) per unit wavelength at the wavelength λ. If the intensity is continuously distributed over a range of wavelengths, the intensity dI embraced within an infinitesimal wavelength interval $d\lambda$ can be given by $dI = I_\lambda \, d\lambda$. The spectral intensity I_λ is in general a function of λ. Note that the units of I and I_λ are different, since the unit of wavelength enters into I_λ. Integration (over the wavelength range under consideration) gives $I = \int I_\lambda \, d\lambda$. See also I_λ.

spectral interference. [*spektrale Beeinflussung.*] An INTERFERENCE caused by incomplete spectral separation of the radiation emitted or absorbed by the ANALYTE from other radiation due to or affected by the INTERFERENT. It can be detected by comparing the MEASURES of the BLANK SOLUTION and the solvent blank, and it usually depends strongly on the SPECTRAL BANDWIDTH and STRAY LIGHT of the instrument.

spectral line; line. [*Spektrallinie; Linie.*] Originally, an image of the entrance slit in monochromatic light, emitted or absorbed, as seen through a spectroscope. More generally, a very narrow band of frequencies (or range of wavelengths) of electromagnetic radiation resulting from a transition between two ENERGY LEVELS

219

in an atom (or molecule). The actual range of wavelengths covered by the line—the *true linewidth* or LINE PROFILE in the source—is usually much narrower than the observed *apparent* linewidth, which in a MONOCHROMATOR is determined largely by the slit width(s), dispersion and/or diffraction. This apparent broadening of the line by instrumental factors should be well distinguished from: (i) line broadening due to physical processes in the flame, including Doppler and collision broadening and broadening due to SELF-ABSORPTION (see § 6.3 and 6.6.1); and (ii) the natural line broadening (see § 6.3). The natural broadening of (resonance) lines is due to the finite optical lifetime of the pertinent EXCITED STATE of the atom. It determines the least possible linewidth, depending only on the nature of the line itself, and observable only under ideal conditons, for example in a thin layer of atoms at low temperature and pressure. See also ATOMIC LINE; IONIC LINE; BAND LINE; RESONANCE LINE; FIRST RESONANCE LINE; MULTIPLET; SELF-REVERSAL.

spectral line source. See LINE SOURCE.

spectral resolution. See RESOLVING POWER.

spectrochemical buffer. [*Puffer, Puffersubstanz.*] An ADDITIVE mixed with both the SAMPLE SOLUTION and the REFERENCE SOLUTIONS, often pipetted in with the DILUENT, to mitigate an INTERFERENCE by making the MEASURE of the ANALYTE less sensitive to variations in the INTERFERENT concentration. An IONISATION BUFFER is an example.

spectrogram. [*Spektrogramm.*] The record of a SPECTRUM obtained either with a SPECTROGRAPH or with a scanning SPECTROMETER and a recorder such as a strip-chart recorder; in the latter case the spectrogram may be called a *recording* of the spectrum.

spectrograph. [*Spektrograph.*] An instrument for photographing a SPECTRUM on a film or plate.

spectrometer. [*Spektrometer.*] An instrument used in spectrochemical analysis, employing optical filters or a dispersing system (see DISPERSION) such as a MONOCHROMATOR for spectral isolation of the desired radiation, and a photoelectric detector with associated electronics and a meter for measuring INTENSITIES at one or more wavelengths. The older term *spectrophotometer* is no longer approved by the IUPAC. See also FLAME SPECTROMETER; PHOTODETECTOR.

spectroscopic term. See TERM.

split flame. [*gespaltene Flamme.*] See SEPARATED FLAME; SHIELDED FLAME.

spray barrier. See IMPACT BEAD.

spray chamber. [*Zerstäuberkammer.*] See CHAMBER-TYPE NEBULISER.

sprayer nozzle; sprayer. [*Zerstäuberdüse.*] The part of a NEBULISER where the actual nebulisation takes place, that is, where the aspirated liquid is disrupted into droplets.

sputtering chamber. See CATHODIC SPUTTERING CHAMBER.

sr. See STERADIAN.

standard deviation; symbol: *s.* [*Standardabweichung.*] A measure of the dispersion of the NORMAL DISTRIBUTION, given by

$$s = \sqrt{\sum_{i=1}^{n} (\bar{x} - x_i)^2 / (n-1)},$$

where x_i is the ith individual measurement or value of a VARIATE in a set of n such measurements and $\bar{x}$ is their MEAN. The standard deviation is useful for describing the random variation of measurements that do not give exactly the same value when made repeatedly under the same conditions. An ANALYTICAL RESULT is a good example of such a measurement. The standard deviation serves as a measure of many kinds of variation such as NOISE or scatter, SIGNAL-TO-NOISE RATIO, RANDOM ERROR, PRECISION, REPEATABILITY, and REPRODUCI-BILITY. The quantity s is the *estimated standard deviation* [*geschätzte Standardabweichung*]; as n increases, s approaches the *true standard deviation σ* and $\bar{x}$ approaches the true mean μ; that is, the estimated mean and standard deviation become more reliable. The number n of observations should always be reported. A distinction is made between $s(D)$, the estimated standard deviation of measurements obtained on different days, and $s(S)$, that obtained on a series during the same day. See also RELATIVE STANDARD DEVIATION; STANDARD ERROR.

standard error. [*mittlerer Fehler des Mittelwertes.*] A measure of the uncertainty of an estimated MEAN value, given by the estimated STANDARD DEVIATION of the measurements entering into the mean divided by the square root of the number of measurements: $s_\mu = s/\sqrt{n}$. Since 'standard error' is often loosely used as synonymous with standard deviation, s_μ should be called in full the *standard error of the mean* to avoid ambiguity. It is an estimate of the standard deviation of a set of mean values all obtained in the same way. See also CONFIDENCE INTERVAL.

standard sample. [*Eichprobe.*] A synthetically prepared REFERENCE SAMPLE available from official bureaus (e.g. the National Bureau of Standards, Washington, DC, and the Chemisch-Technische Bundesanstalt, Berlin), containing the component of interest at a concentration guaranteed within specified tolerance limits and used for purposes of calibration or standardization.

standard solution. See REFERENCE SOLUTION.

starting material. See MATERIAL.

steradian; abbreviation: sr. [*Steradiant.*] A unit of solid angle, equal to $1/4\pi$ of the total solid angle of the sphere.

stray light; scattered light. [*Streulicht.*] Radiation reaching the PHOTODETECTOR from other than the desired direction, source, or wavelength. It includes radiation reaching the detector in AFS through scattering or reflection of the PRIMARY RADIATION from solid particles or mist droplets in the flame.

systematic error. [*systematischer Fehler.*] The difference $\Delta\bar{x}$ between the mean value $\bar{x}$ of an ANALYTICAL RESULT found by many repetitions of a given procedure for sample preparation and analysis (to minimise the RANDOM ERROR) and the 'true' or 'correct' value μ; the latter is found by a method recognised as reliable, or known because the analysis was done on a synthetic sample of known composition such as a STANDARD SAMPLE. The number of repetitions in such a test should be great enough so that the STANDARD ERROR of the mean is small relatively to $\Delta \bar{x}$. The *relative systematic error* is $\Delta \bar{x}_r = (\bar{x} - \mu)/\mu$. The systematic error is a measure of the BIAS and together with the random error characterises the ACCURACY.

term; spectroscopic term. [*Term.*] A group of one or more ENERGY LEVELS belonging to the same MULTIPLET. The term or term value in a *term diagram* corresponds to the weighted average value of the wavenumbers (or energies) of the associated, closely spaced energy levels. See further §6.2.1.

thermal radiator. [*thermischer Strahler.*] A gaseous or solid radiator in which all the particles are in a state of thermal equilibrium (see § 6.7.4).

three-slot burner. See SLOT BURNER.

time constant; RC-time. [*Zeitkonstante*; *RC-Zeit.*] The RESPONSE TIME for attainment of $1 - 1/e$ or $0 \cdot 632$ of the final READING of a measuring instrument, under the assumption that the inertia of the instrument can be described by a single exponential function.

total-consumption burner. See DIRECT-INJECTION BURNER.

transmission factor; transmittance; symbol: τ. [*Transmissionsgrad*; *Durchlässigkeit.*] The ratio of the RADIANT FLUX transmitted through a medium to the incident radiant flux. If window losses, etc, are excluded, we speak of the *internal transmission factor* or *internal transmittance* τ_i [*Reintransmissionsgrad*], referring to the sample medium alone.

triplet. See MULTIPLET.

turbulent flame. [*turbulente Flamme.*] A flame characterised by TURBULENT FLOW of the gas. Compare: LAMINAR FLAME.

turbulent flow. [*turbulente Strömung.*] A mode of flow in which the lines of flow vary strongly and irregularly in time and space while retaining an average general direction. Eddying or vortical flow. Compare: LAMINAR FLOW.

twin-nebuliser experiment. [*Zwei-Zerstäuberversuch.*] An arrangement in which the ANALYTE and a CONCOMITANT are fed separately by two parallel nebulisers into the same flame in order to distinguish between INTERFERENCE mechanisms occurring in the condensed and in the gaseous phase; see § 9.4.

unpremixed flame. [*nicht vorgemischte Flamme.*] A flame in which the fuel gas and oxidant do not mix until they enter the flame.

variance. [*Varianz*]. The square of the STANDARD DEVIATION. If several factors contributing to RANDOM ERROR are statistically independent, the variances due to each can be added to find the total variance due to all the factors jointly. See also NOISE.

variate. [*Veränderliche.*] A random variable; any quantity whose value may vary statistically. Examples: the MEASURE of a concentration; a galvanometer READING.

volatilisation. [*Verflüchtigung.*] Conversion of the DRY AEROSOL, containing the ANALYTE, to the vapour phase. Distinguish from: DESOLVATION, which precedes volatilisation.

volatilisation interference; solute-volatilisation interference. [*Verflüchtigungs-*

beeinflussung.] An INTERFERENCE of a CONCOMITANT that affects the rate of VOLATILISATION of the DRY AEROSOL. It usually takes the form of a DEPRESSION. An example is the retardation of the volatilisation of alkaline-earth metals by phosphorus–oxygen compounds.

volatiliser. [*Verflüchtigungshilfe*; *Bläsersubstanz.*] An ADDITIVE that facilitates or hastens VOLATILISATION, raising the FRACTION VOLATILISED.

wavelength modulation. The periodic variation of the central wavelength transmitted by the spectral selection device, for the purpose of discriminating the (desired) SIGNAL from the (FLAME or BLANK) BACKGROUND. Compare: MODULATION.

working curve. See ANALYTICAL CURVE.

working range; dynamic range. [*brauchbarer* (or *empfohlener* or *optimaler*) *Konzentrationsbereich* (or *Bereich*); *dynamischer Bereich.*] The range of an ANALYTICAL CURVE whose lower end is far enough above the DETECTION LIMIT for practical work and whose upper end does not reach such high concentrations that any serious loss of PRECISION is suffered, for example through strong curvature of the analytical curve due to saturation; in AFS an upper end may exist owing to inversion of the analytical curve (see § 8.5). The approximately linear part of the analytical curve is generally most useful for analysis, having a slope near 45° in doubly logarithmic coordinates.

zero line. See BASELINE.

zero suppression. Displacement of the zero of the meter, usually by electronic means, so that a relatively small difference between two signals can be read more precisely by increasing the amplifier gain.

Appendix 1:   Spectrograms

1. General

Spectrograms 1–44 show line-, band-, and continuum-emission spectra recorded by a spectrometer (relative photocurrent as a function of wavelength in nm) when solutions containing a metallic element, an alkaline-earth element plus a halogen (as in the indirect analysis of halogens), phosphorus or sulphur (as in the direct analysis of P and S by their chemiluminescent bands) are nebulised into a flame. These spectra are superposed upon the flame background spectrum, which may be influenced by the solvent used. Background-emission spectra are shown separately for a few flames in spectrograms 45–48. In spectrogram 49, the atomic line spectrum of Fe emitted by a Fe hollow-cathode lamp is shown as an example of a background source in AAS (and AFS); it is mainly the resonance lines in these spectra that produce a useful signal.

2. Sources and Acknowledgments

Most spectrograms are taken from the German 2nd edition of *Flammen-photometrie*, 1960, by R Herrmann and C Th J Alkemade, and reproduced by permission of Springer-Verlag, Berlin and Heidelberg. Technical specifications are given in the captions. The sources of the new spectrograms (8, 10–12, 40–44 and 49) are included in the pertinent captions. Most of the spectrograms taken from the German 2nd edition were recorded by Mr Paul T Gilbert Jr at Beckman Instruments, Fullerton, California; the rest of these older spectrograms (3, 4, 6, 9, 13, 18, 20, 21, 24, 25, 26 and 37) were recorded by Mr Hideo Watanabe of the same firm and spectrogram 2 by Mr Bruce E Buell of Union Oil Company, Brea, California. Parts of these spectrograms were also published in 1955 *Applied Spectroscopy* **9** 132, 1958 *Analytical Chemistry* **30** 1514 and 1962 **34** 635, and by the American Society for Testing and Materials in 1960 *Spec. Tech. Publication* no. 269.

3. Equipment

Most of the older spectrograms were recorded with a Beckman DU spectrometer with a photomultiplier tube. Below 700 nm the RCA photomultiplier tube type 1P28 was used, while above 700 nm the Beckman red-sensitive phototube, or the RCA C7160, or the Farnsworth 16 PMI photomultiplier tube was used. The older spectrograms were all obtained with the use of a direct-injection burner.

4. Materials

The solutions were aqueous except as noted. In a few cases, acid present in the solution attacked the nebuliser capillary, introducing impurity lines of palladium, iron, nickel, manganese, chromium and silver. In other cases impurities in the chemicals used are revealed by their spectra. The heavy sodium contamination in spectrogram 32 arose from the surface of the metallic aluminium sample. The mercury lines at 404·7, 435·8 and 546·1 nm are caused by stray light from the fluorescent room lamps. The carbon line in spectrogram 45 originates from the flame itself.

5. Intensities

In all of the recordings the base line or wavelength scale represents zero intensity, except as noted. (The stated zero suppression is the distance of the true zero line below the base line.) The intensity (ordinate) scale measured from the zero line is adjusted arbitrarily in each spectrogram to give a convenient range. In a few cases recordings at different sensitivity adjustments (gain or amplification) or slit widths are included on the same spectrogram for comparison. In most of the older spectra the blank (in most cases water) is shown along with the spectrum of the element in the flame to reveal the net line or band intensities, the continuum due to the element, and the background itself. When there is no measurable element continuum, as in spectrogram 26, the blank is exactly superposed on the spectrum of the element except at the lines. In spectrogram 23 the intensity scale of the blank is magnified with respect to that of the cobalt spectrum. In spectrograms 3 and 4 the background is not entirely correct, since the recordings for the photomultiplier tube and the red-sensitive phototube were deliberately joined. In general, spectra obtained with the same kind of flame, instrument and solvent (e.g. oxyhydrogen and water) can be compared with each other by using the blank intensity as a normalising factor.

6. Index of Spectrograms (in alphabetical order of element or flame)

Element: Ag 26; Al 32; B 31; Ba 11, 12, 13, 41; Be 5; Bi 38, 39; Br 42; C 45; Ca 8, 9; Cd 29; Ce 15; Cl 40; Co 22, 23; Cr 18; Cs 4; Cu 25, 41; F 41; Fe 21, 49; Hg 30; In 33, 34, 40, 42; La 14; Li 1; Mg 6, 7, 41; Mn 20; Mo 19; Na 2; Ni 24; P 43; Pb 37; Rb 3; S 44; Sn 36; Sr 10; Tl 35; V 17; Yb 16; Zn 27, 28.

Flame (with or without blank): air–acetylene 11, 12; air–hydrogen 27, 38; nitrous oxide–acetylene 48; oxyacetylene 28, 30; oxycyanogen 45–47; oxyhydrogen 2, 5–7, 9, 13–21, 23, 35, 36.

226

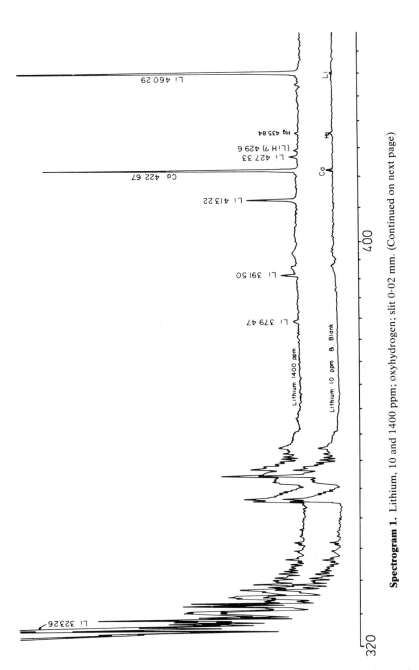

Spectrogram 1. Lithium, 10 and 1400 ppm; oxyhydrogen; slit 0·02 mm. (Continued on next page)

227

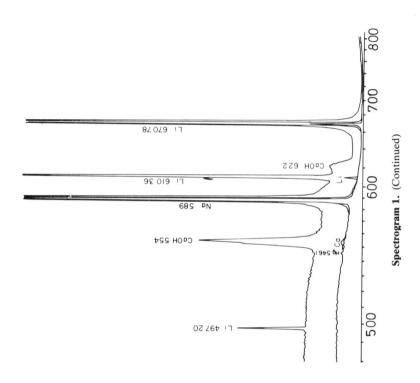

Spectrogram 1. (Continued)

228

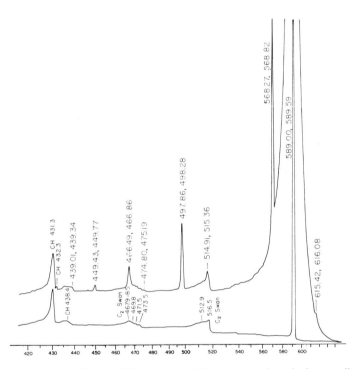

Spectrogram 2. Sodium, 800 ppm in naphtha–isopropanol; oxyhydrogen; slit 0·02 mm.

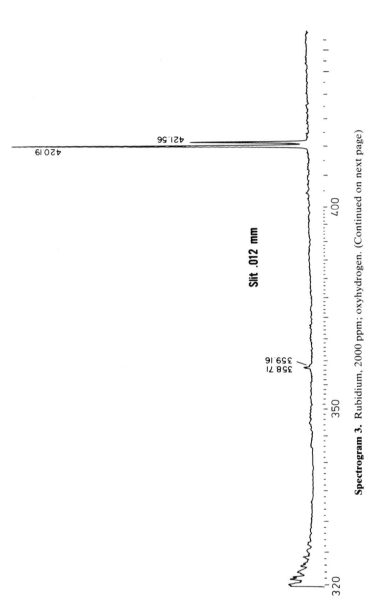

Spectrogram 3. Rubidium, 2000 ppm; oxyhydrogen. (Continued on next page)

230

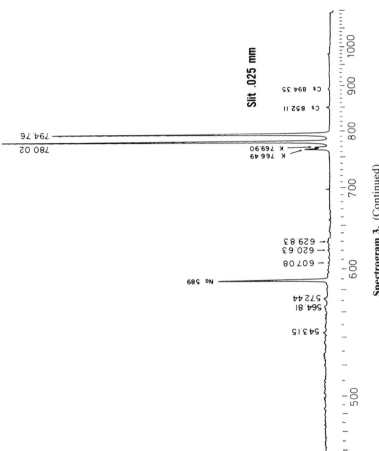

Spectrogram 3. (Continued)

Slit .025 mm

Cs 894.35
Cs 852.11
794.76
780.02
K 769.90
K 766.49
629.83
620.63
607.08
Na 589
572.44
564.81
543.15

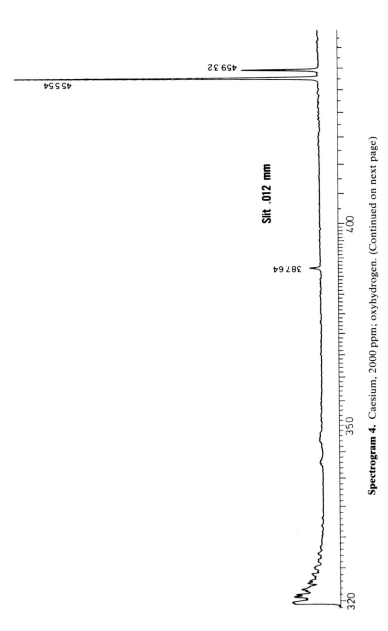

Spectrogram 4. Caesium, 2000 ppm; oxyhydrogen. (Continued on next page)

232

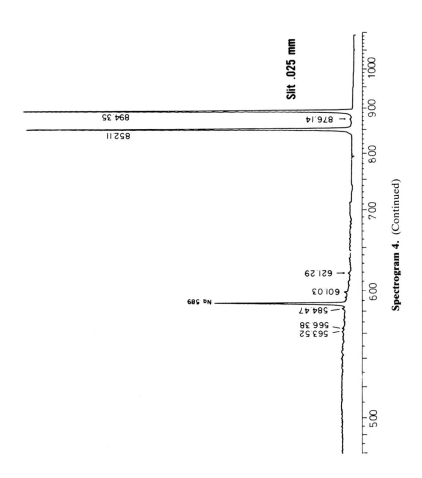

Slit .025 mm

894.35

876.14

852.11

621.29

601.03

Na 589

584.47

566.38

563.52

Spectrogram 4. (Continued)

233

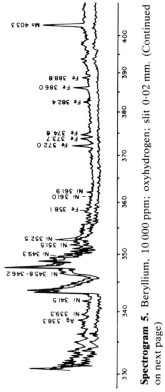

Spectrogram 5. Beryllium, 10 000 ppm; oxyhydrogen; slit 0·02 mm. (Continued on next page)

234

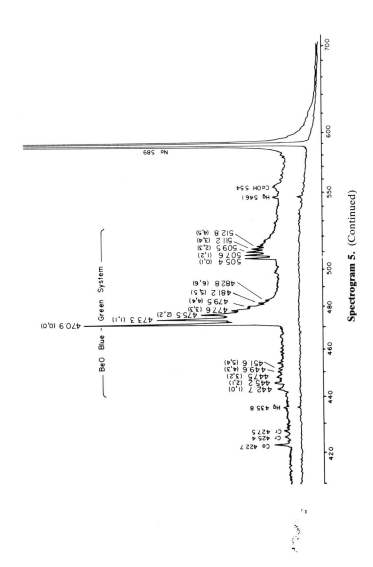

Spectrogram 5. (Continued)

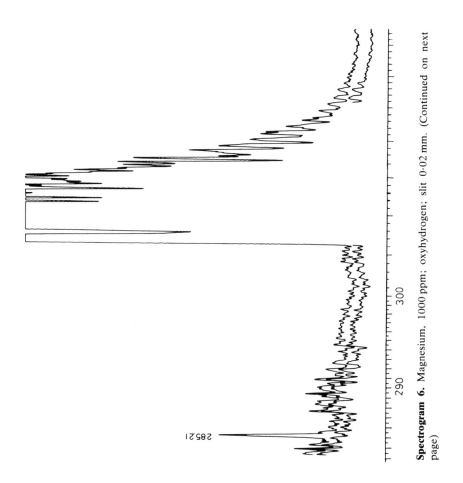

Spectrogram 6. Magnesium, 1000 ppm; oxyhydrogen; slit 0·02 mm. (Continued on next page)

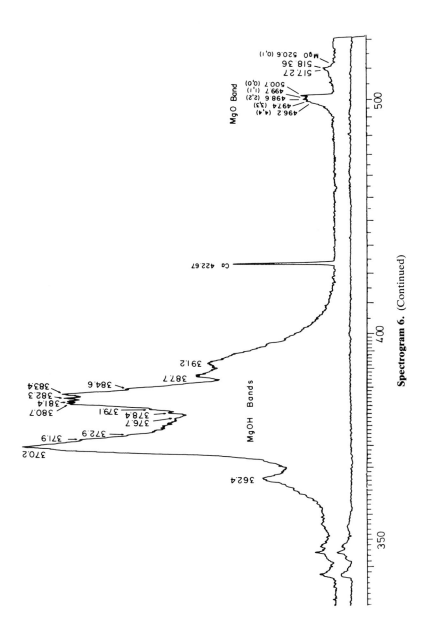

Spectrogram 6. (Continued)

237

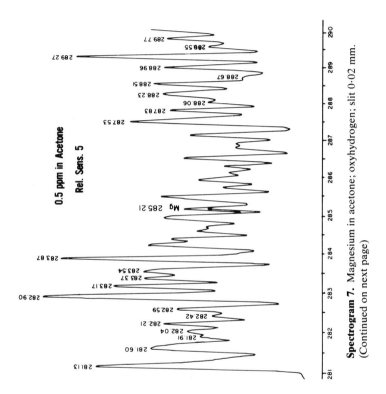

Spectrogram 7. Magnesium in acetone; oxyhydrogen; slit 0·02 mm.
(Continued on next page)

238

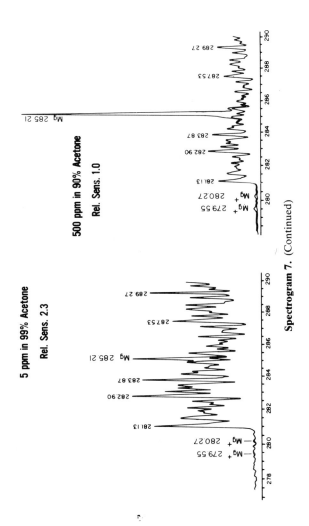

Spectrogram 7. (Continued)

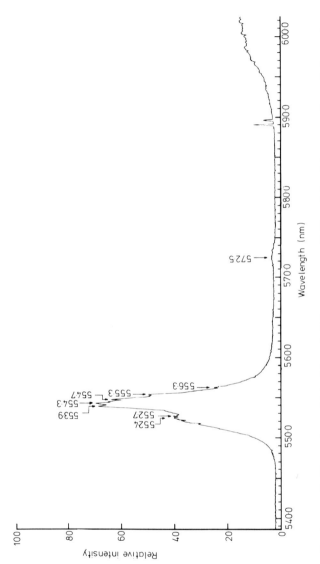

Spectrogram 8. Ca spectrum obtained by nebulising an aqueous $CaCl_2$ solution into an acetylene–air flame of about 2450 K, showing the CaOH bands in the visible region. The relative intensity was not corrected for the spectral response of the spectrometer. The monochromator bandwidth was about 0·1 nm. The flame background intensity amounted to about 2 scale units. (From the thesis of J van der Hurk, Utrecht, 1974, by permission of the author.) (Continued on next page)

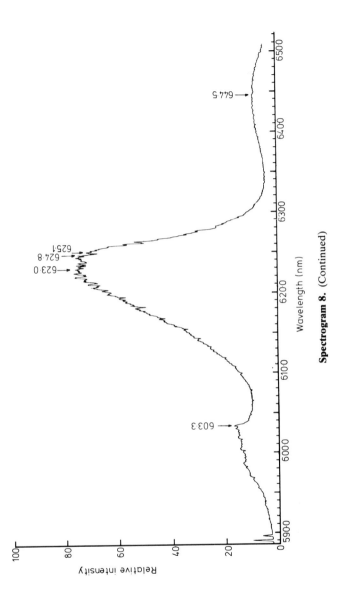

Spectrogram 8. (Continued)

241

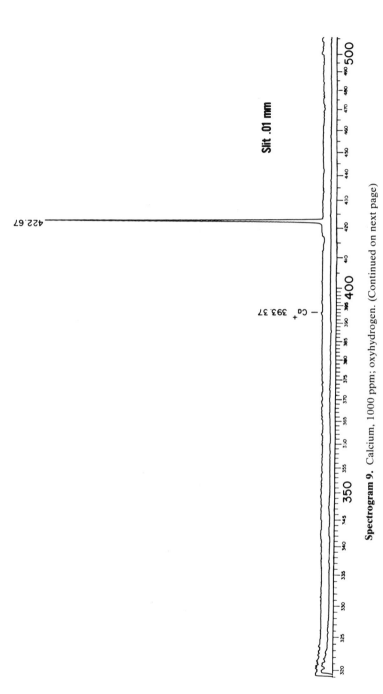

Spectrogram 9. Calcium, 1000 ppm; oxyhydrogen. (Continued on next page)

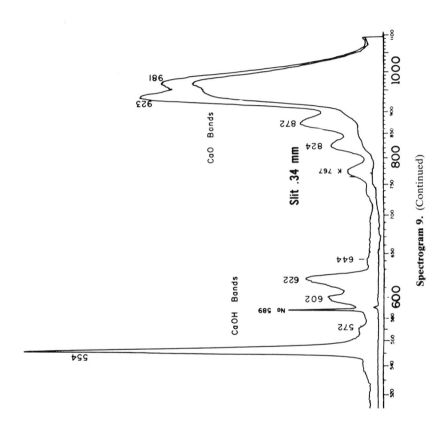

Spectrogram 9. (Continued)

243

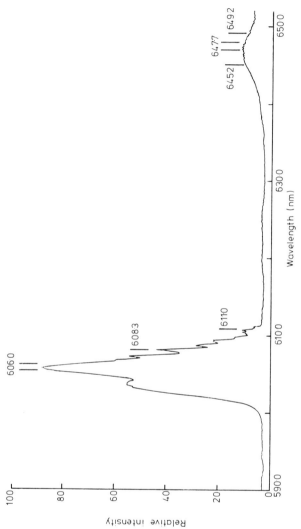

Spectrogram 10. Sr spectrum obtained by nebulising an aqueous $SrCl_2$ solution into an acetylene–air flame of about 2450 K, showing the SrOH bands in the visible region. The relative intensity was not corrected for the spectral response of the spectrometer. The monochromator bandwidth was about 0·1 nm. The flame background intensity amounted to about 2 scale units. The well resolved structure of the band peaks has been partially lost in the reproduction of the original recording. (From the thesis of J van der Hurk, Utrecht, 1974, by permission of the author.) (Continued on next page)

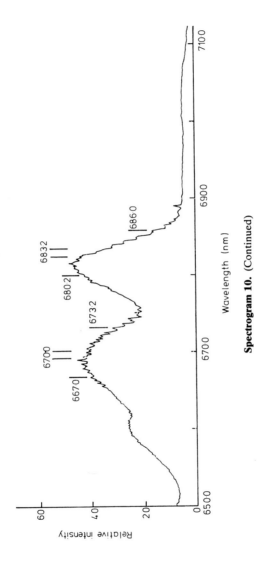

Spectrogram 10. (Continued)

245

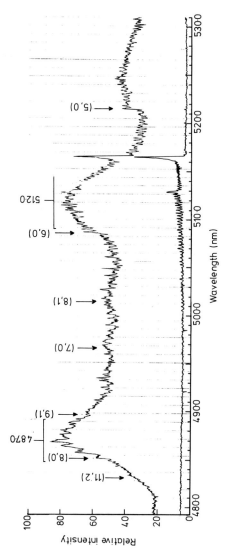

Spectrogram 11. Ba spectrum obtained by nebulising an aqueous $BaCl_2$ solution into a fuel-rich acetylene–air flame of 2446 K. The relative intensity was not corrected for the spectral response of the spectrometer. The monochromator bandwidth was about 0·05 nm. The lower curve shows the flame background spectrum with the (0, 0) and (1, 1) bands of C_2 at 516·5 and 512·9 nm respectively. The diffuse BaOH bands at 487·0 and 512·0 nm are indicated; the BaO band heads are indicated by their vibrational quantum numbers (v', v''). (From the thesis of J van der Hurk, Utrecht, 1974, by permission of the author.)

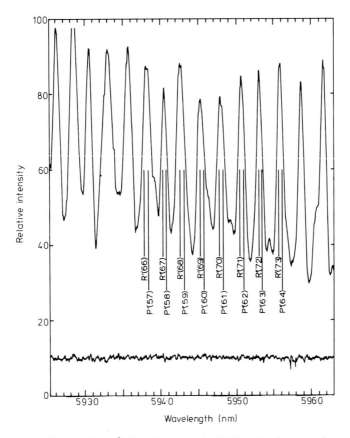

Spectrogram 12. Part of the (2, 1) emission band of BaO under the same observation conditions as in spectrogram 11 in a slightly more oxygen-rich flame. The lower curve shows the flame background spectrum. (From the thesis of J van der Hurk, Utrecht, 1974, by permission of the author.)

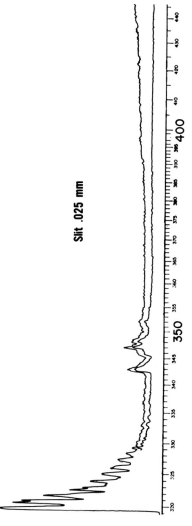

Spectrogram 13. Barium, 1000 ppm; oxyhydrogen. (Continued on next page)

248

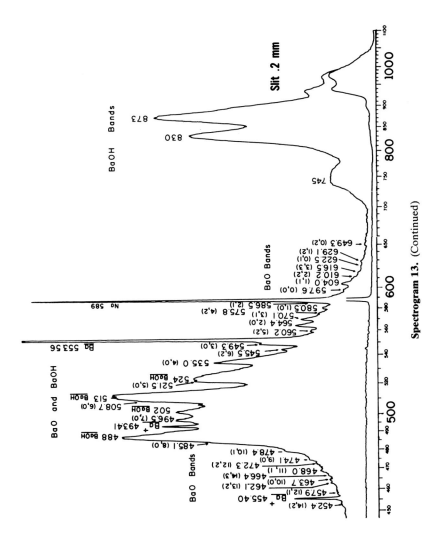

Spectrogram 13. (Continued)

249

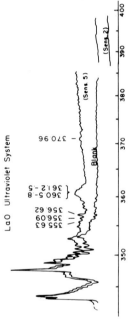

Spectrogram 14. Lanthanum, 1000 ppm; oxyhydrogen; slit 0·02 mm. (Continued on next page)

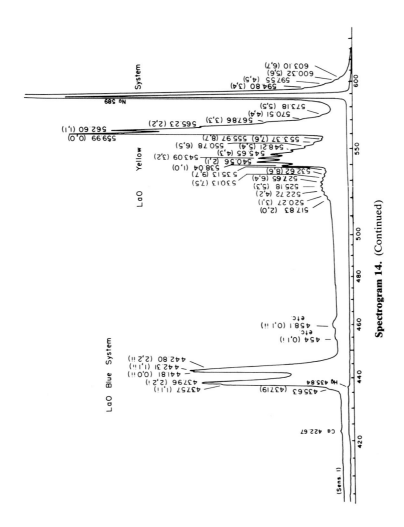

Spectrogram 14. (Continued)

251

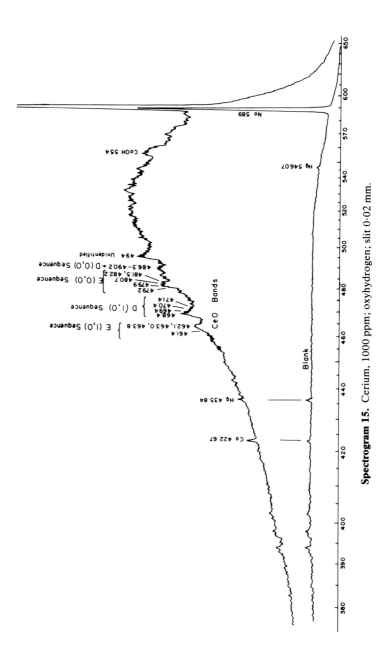

Spectrogram 15. Cerium, 1000 ppm; oxyhydrogen; slit 0·02 mm.

Na 589

CooH 554

Hg 546·07

494 Unidentified
486·3–490·2 + D (0,0) Sequence
481·5, 482·2
480·7
479·9
E (0,0) Sequence
479·2
471·4
470·4
470·4
469·4
D (1,0) Sequence
468·4
CeO Bands
461·4
462·1, 463·0, 463·8
E (1,0) Sequence

Hg 435·84

Ca 422·67

Blank

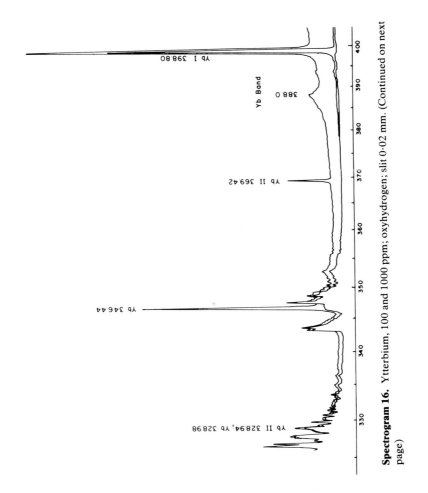

Spectrogram 16. Ytterbium, 100 and 1000 ppm; oxyhydrogen; slit 0·02 mm. (Continued on next page)

253

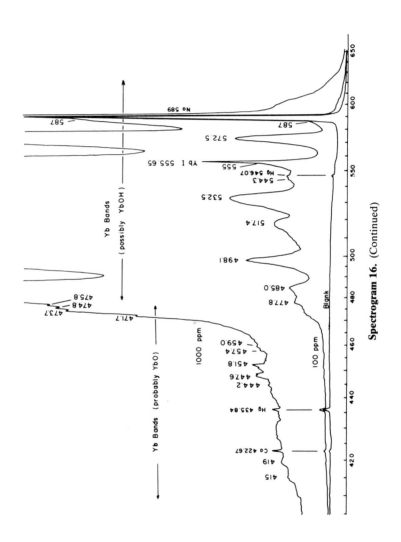

Spectrogram 16. (Continued)

254

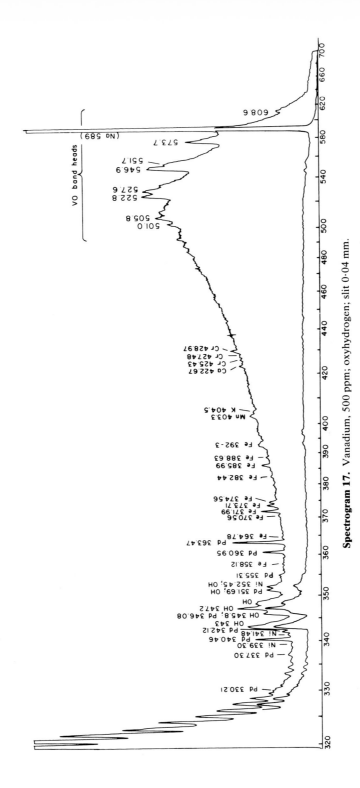

Spectrogram 17. Vanadium, 500 ppm; oxyhydrogen; slit 0·04 mm.

255

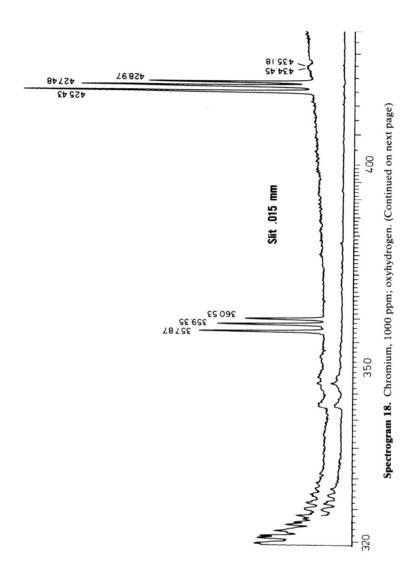

Spectrogram 18. Chromium, 1000 ppm; oxyhydrogen. (Continued on next page)

256

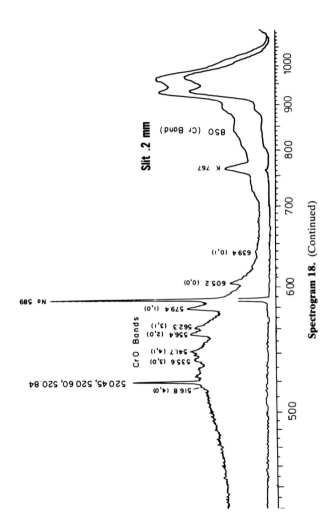

Spectrogram 18. (Continued)

Slit .2 mm

850 (Cr Band)

K 767

639.4 (0,1)

605.2 (0,0)

Na 589

579.4 (1,0)

CrO Bands

562.3 (3,1)

556.4 (2,0)

541.7 (4,1)

535.6 (3,0)

520.45, 520.60, 520.84

516.8 (4,0)

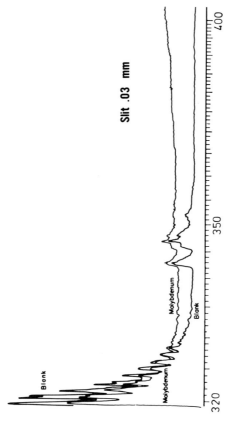

Spectrogram 19. Molybdenum, 5000 ppm; oxyhydrogen. (Continued on next page)

258

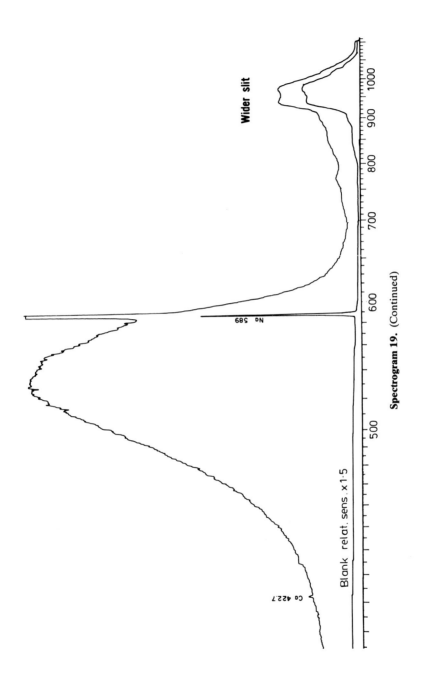

Wider slit

No. 589

Blank relat. sens. x 1.5

Ca 422.7

Spectrogram 19. (Continued)

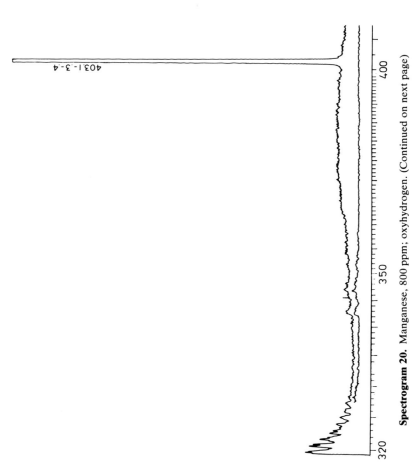

Spectrogram 20. Manganese, 800 ppm; oxyhydrogen. (Continued on next page)

403.1 - 3. - 4

320   350   400

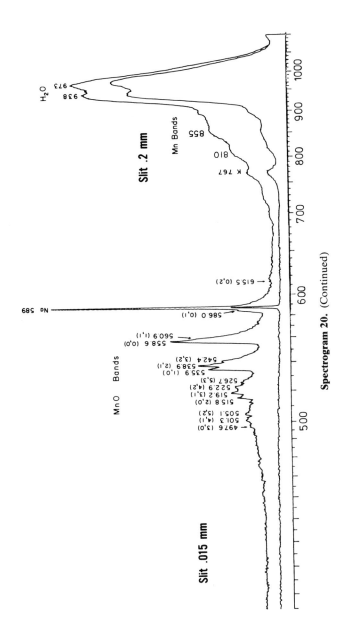

Spectrogram 20. (Continued)

261

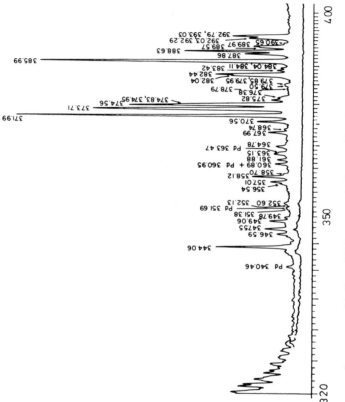

Spectrogram 21. Iron, 2500 ppm; oxyhydrogen. (Continued on next page)

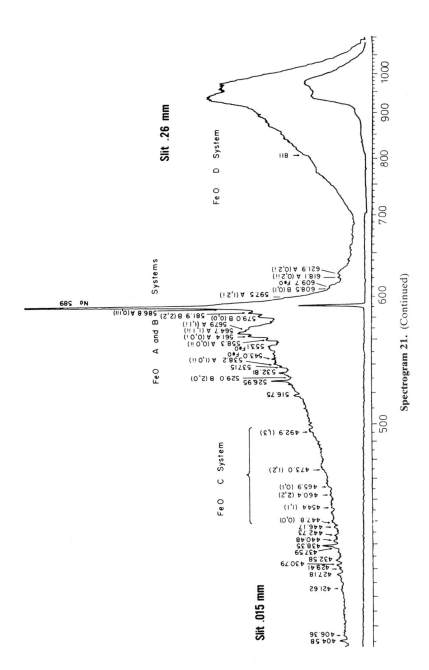

Spectrogram 21. (Continued)

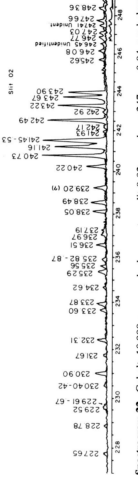

Spectrogram 22. Cobalt, 10 000 ppm; oxyhydrogen; true slit 0·02 mm above 247 nm, 0·04 mm below 250 nm; blank superposed. (Continued on next page)

264

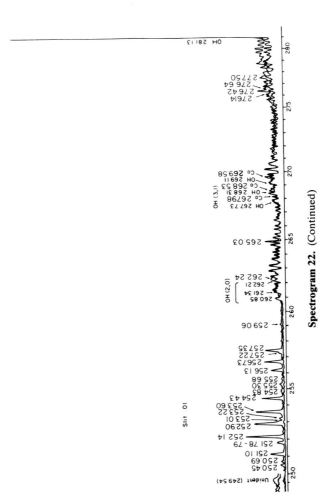

Spectrogram 22. (Continued)

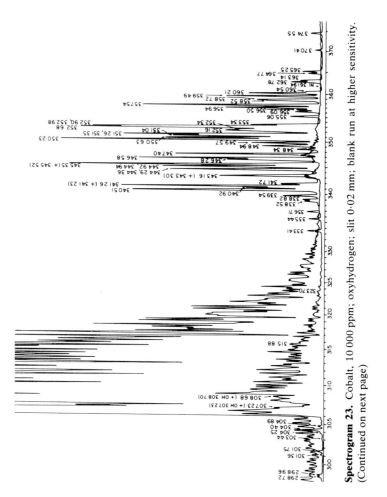

Spectrogram 23. Cobalt, 10 000 ppm; oxyhydrogen; slit 0·02 mm; blank run at higher sensitivity.
(Continued on next page)

266

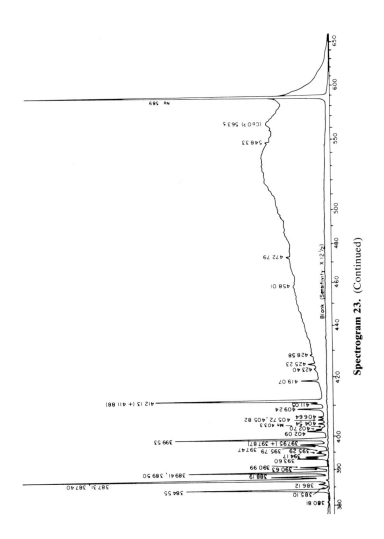

Spectrogram 23. (Continued)

267

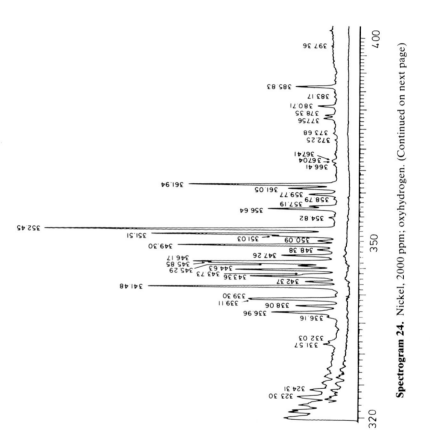

323 30
324 31
331.57
332 03
336.16
336.96
338 06
339 11
339 30
341 48
342 37
343 36 343 73
344 63
345 29
345 85
346 17
347 26
348 38
349.30
350 09
351 03
351.51
352 45
354 82
356 64
357.19
358 79 359 77 361 05
361.94
366 41
367 04
367 41
372 25
373 68
377 56
378 35
380 71
383 17
385 83
397 36

320   350   400

Spectrogram 24. Nickel, 2000 ppm; oxyhydrogen. (Continued on next page)

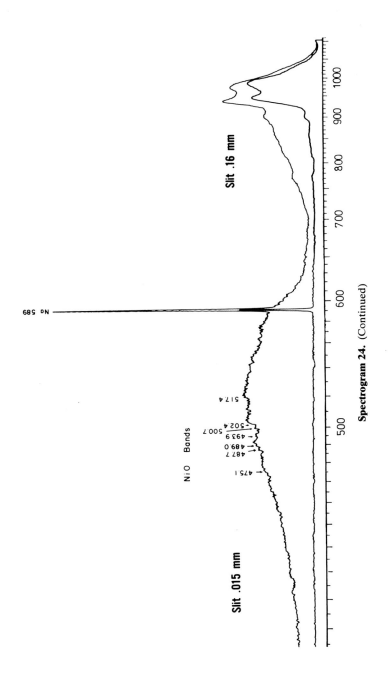

Spectrogram 24. (Continued)

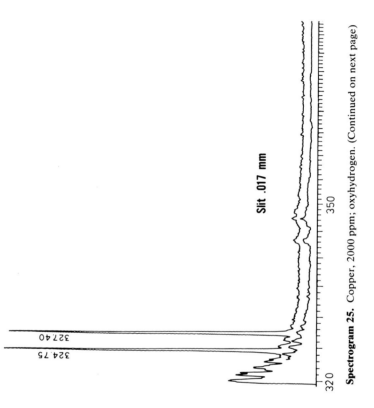

Spectrogram 25. Copper, 2000 ppm; oxyhydrogen. (Continued on next page)

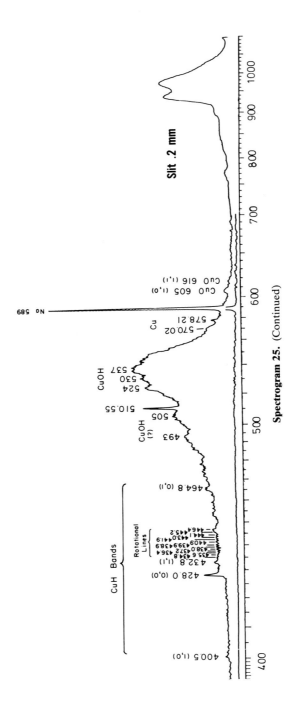

Spectrogram 25. (Continued)

271

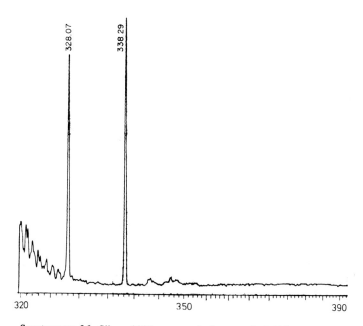

Spectrogram 26. Silver, 1000 ppm; oxyhydrogen; slit 0·015 mm.

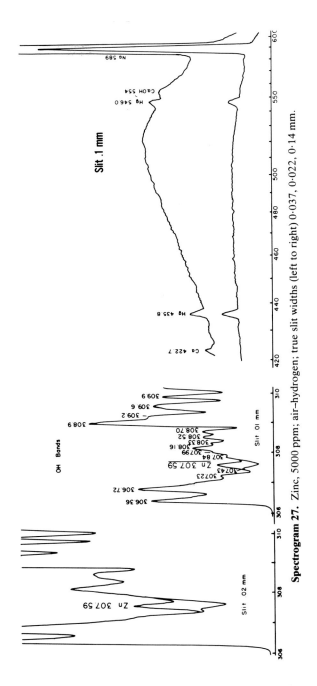

Spectrogram 27. Zinc, 5000 ppm; air–hydrogen; true slit widths (left to right) 0·037, 0·022, 0·14 mm.

273

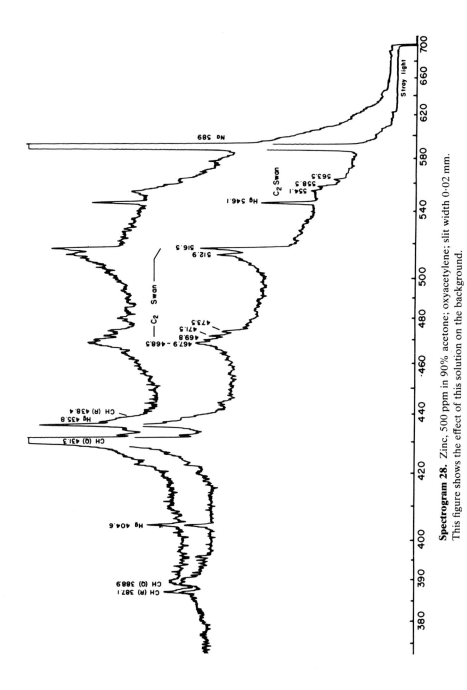

Spectrogram 28. Zinc, 500 ppm in 90% acetone; oxyacetylene; slit width 0·02 mm. This figure shows the effect of this solution on the background.

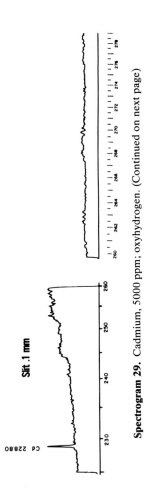

Spectrogram 29. Cadmium, 5000 ppm; oxyhydrogen. (Continued on next page)

275

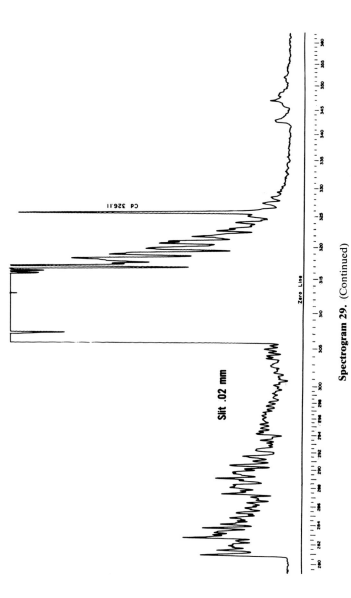

Spectrogram 29. (Continued)

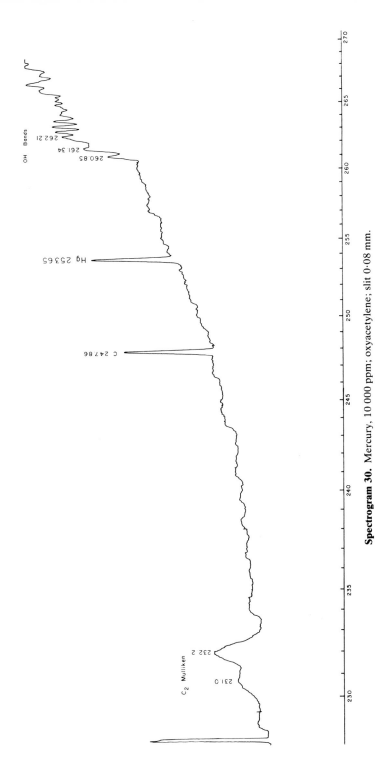

Spectrogram 30. Mercury, 10 000 ppm; oxyacetylene; slit 0·08 mm.

277

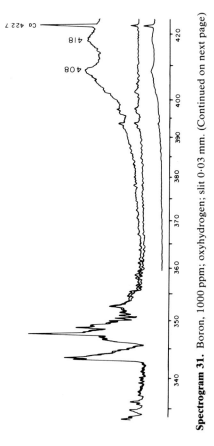

Spectrogram 31. Boron, 1000 ppm; oxyhydrogen; slit 0·03 mm. (Continued on next page)

278

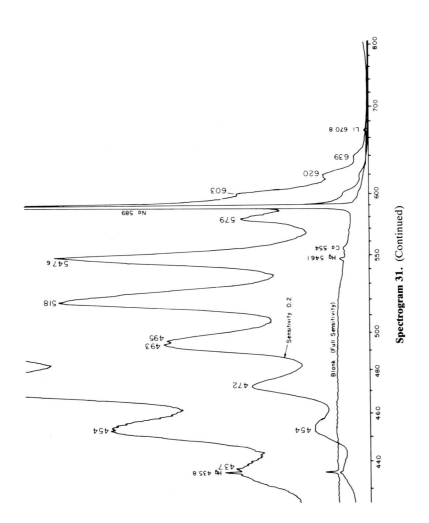

Spectrogram 31. (Continued)

279

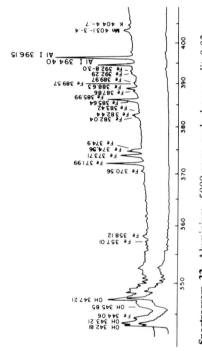

Spectrogram 32. Aluminium, 5000 ppm; oxyhydrogen; slit 0·02 mm.
(Continued on next page)

280

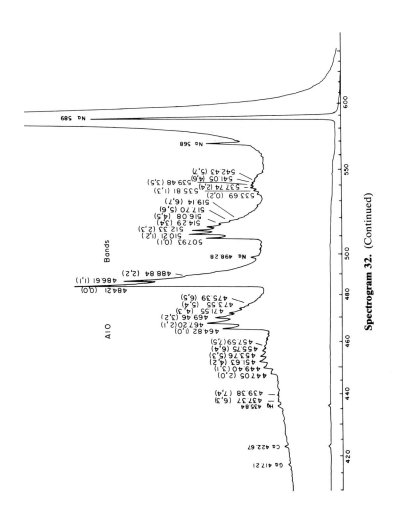

Spectrogram 32. (Continued)

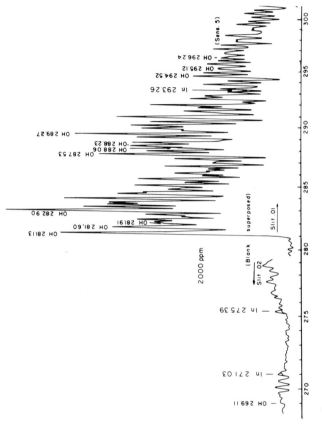

Spectrogram 33. Indium, 200 and 2000 ppm; oxyhydrogen; true slit width 0·04 mm below 280 nm, 0·02 mm above; different sections run at different instrument sensitivities as shown. (Continued on next page)

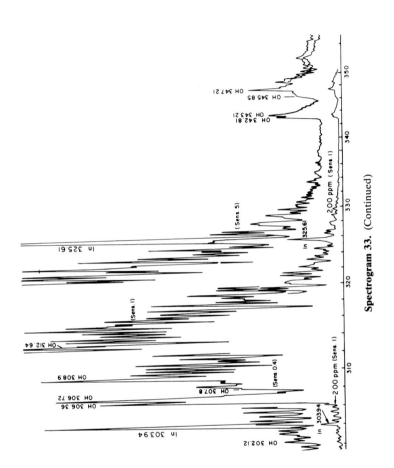

Spectrogram 33. (Continued)

283

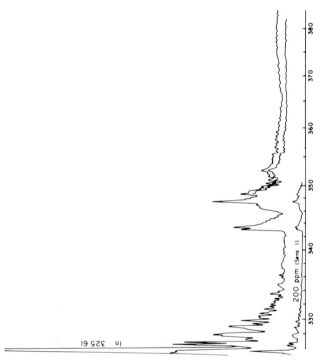

Spectrogram 34. Indium, 200 and 2000 ppm; oxyhydrogen; slit 0·02 mm; instrument sensitivities as shown. (Continued on next page)

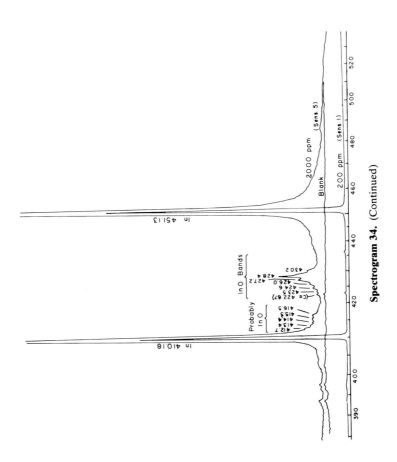

Spectrogram 34. (Continued)

285

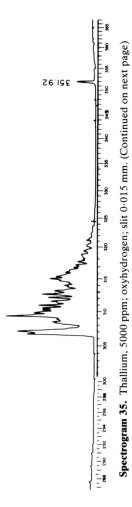

Spectrogram 35. Thallium, 5000 ppm; oxyhydrogen; slit 0·015 mm. (Continued on next page)

286

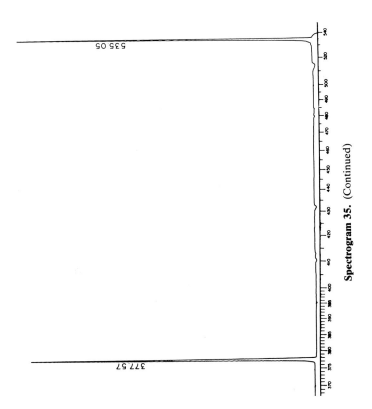

Spectrogram 35. (Continued)

535 05

377 57

287

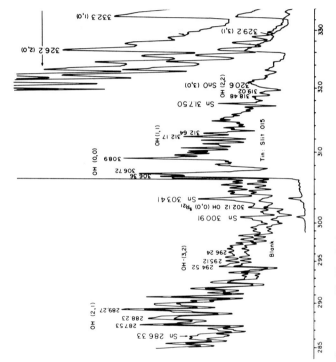

Spectrogram 36. Tin, 20 000 ppm; oxyhydrogen; the two runs (at different slit widths) represent different samples and flame adjustments. (Continued on next page)

288

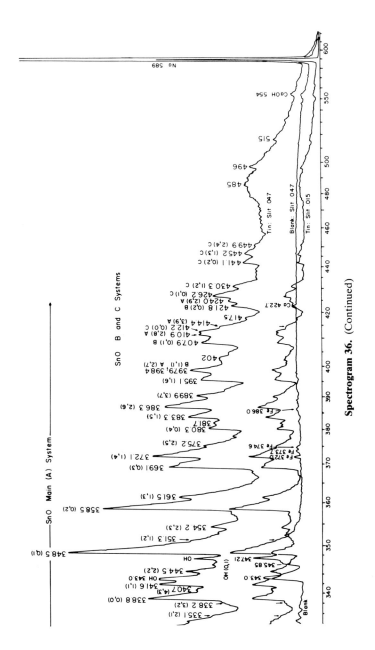

Spectrogram 36. (Continued)

289

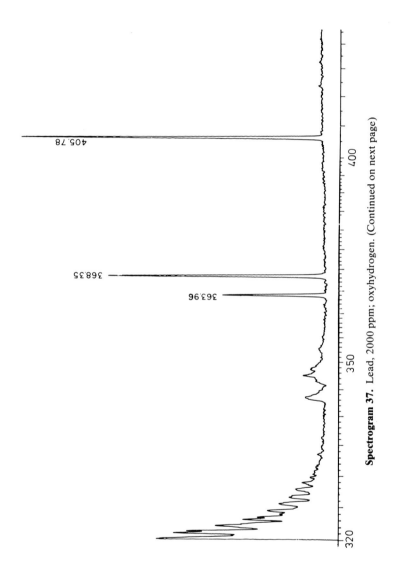

405.78

368.35
363.96

320   350   400

Spectrogram 37. Lead, 2000 ppm; oxyhydrogen. (Continued on next page)

290

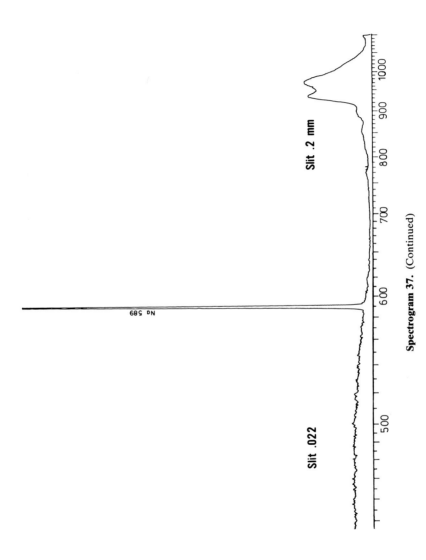

Slit .2 mm

Slit .022

Nd 589

1000 900 800 700 600 500

Spectrogram 37. (Continued)

291

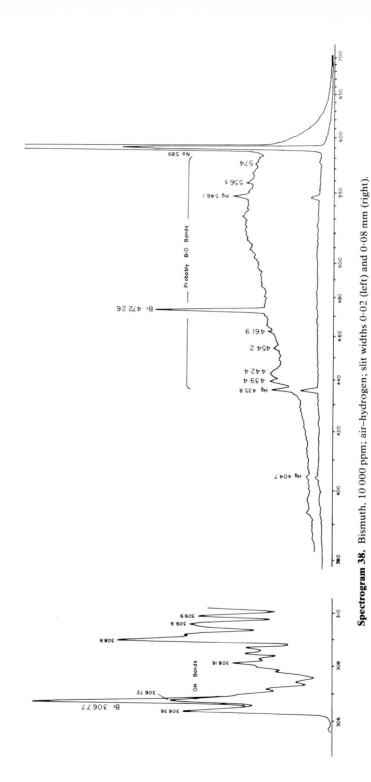

Spectrogram 38. Bismuth, 10 000 ppm; air–hydrogen; slit widths 0·02 (left) and 0·08 mm (right).

292

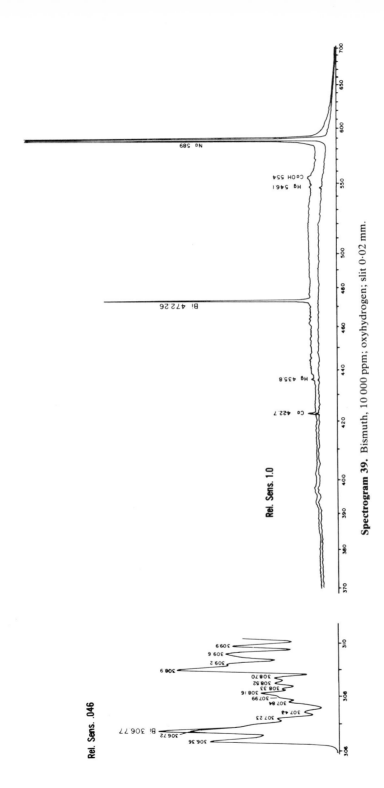

Spectrogram 39. Bismuth, 10 000 ppm; oxyhydrogen; slit 0·02 mm.

Rel. Sens. 1.0

Rel. Sens. .046

Bi 47226

Ca 4227

Hg 4358

Hg 5461

CaOH 554

Na 589

Bi 30677

306 36

306 72

307 23

307 44

307 84

307 99

308 16

308 33

308 52

308 70

308 9

309 2

309 6

309 9

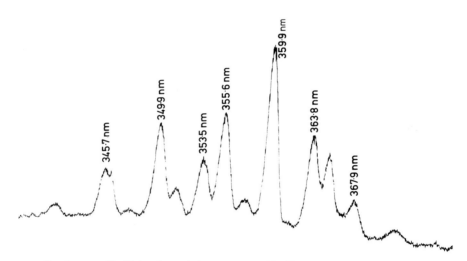

Spectrogram 40. Molecular emission spectrum of InCl bands in the upper zone of a split hydrogen–air flame between 345 and 370 nm. Monochromator: Hilger–Uvispek. Further details are given in 1968 *Z. Anal. Chem.* **241** 55. (Reproduced by permission of Springer-Verlag, Berlin and Heidelberg.)

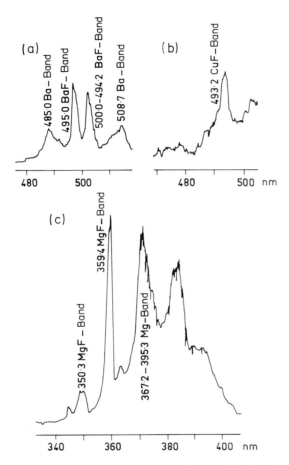

Spectrogram 41. Molecular emission spectrum of (a) BaF, (b) CuF, and (c) MgF bands in a hydrogen–oxygen flame. Direct-injection burner: Beckman 4040; monochromator: Bausch and Lomb 0·5 m Type 534 UB; bandwidth varying between 0·08 and 0·5 nm. Further details are given in 1972 *Z. Anal. Chem.* **258** 277. (Reproduced by permission of Springer-Verlag, Berlin and Heidelberg.)

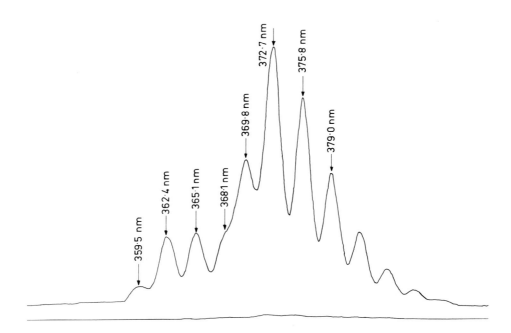

Spectrogram 42. Molecular emission spectrum of InBr bands in the upper zone of a split hydrogen–air flame between 358 and 391 nm. Monochromator: Hilger–Uvispek. The details are the same as in spectrogram 40. (Supplied by R Herrmann at Fachgruppentagung 'Anal. Chem.' of 'Gesellschaft Deutscher Chemiker', Freiburg, April 1969.)

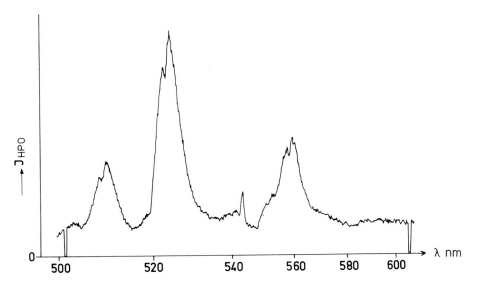

Spectrogram 43. Molecular chemiluminescence spectrum of HPO bands in a cool hydrogen–air and nitrogen flame between 500 and 600 nm. Water-cooled burner; monochromator: Hilger–Uvispek with glass prism. Further details are given in 1971 *Deutsche Lebensmittel-Rundschau* **67** 243. (Reproduced by permission of Wissenschaftl. Verlagsgesellschaft mbH, Stuttgart, Germany.)

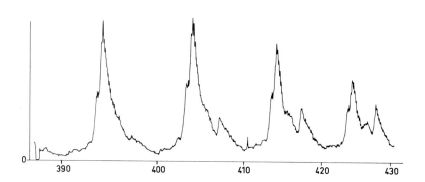

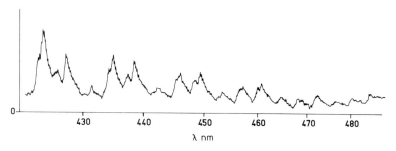

Spectrogram 44. Molecular chemiluminescence spectrum of S_2 bands between 390 and 480 nm in a cool hydrogen–air and nitrogen flame produced with a water-cooled glass burner. The details are the same as in spectrogram 43. (Reproduced by permission of Wissenschaftl. Verlagsgesellschaft mbH, Stuttgart, Germany.)

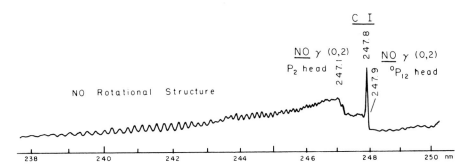

Spectrogram 45. Carbon line in the empty oxycyanogen flame; slit 0·02 mm.

299

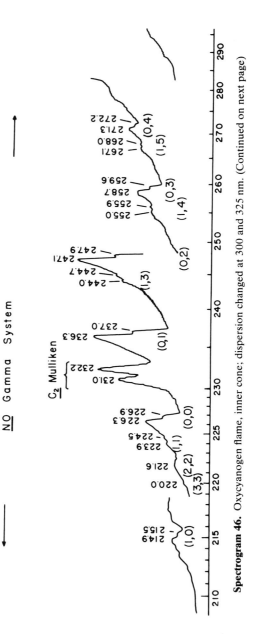

Spectrogram 46. Oxycyanogen flame, inner cone; dispersion changed at 300 and 325 nm. (Continued on next page)

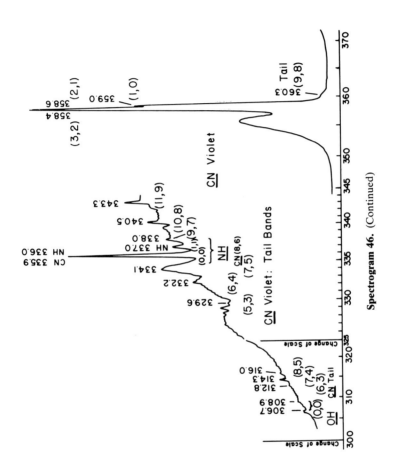

Spectrogram 46. (Continued)

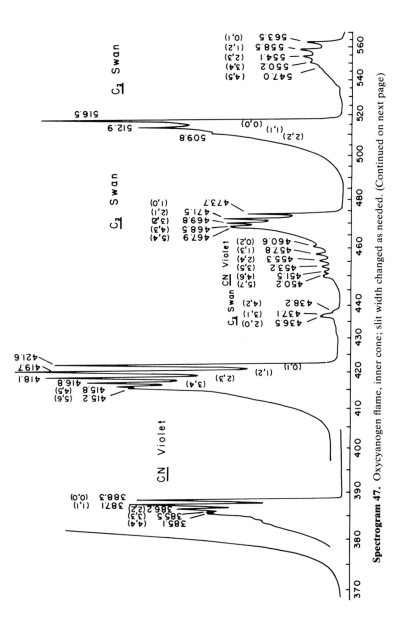

Spectrogram 47. Oxycyanogen flame, inner cone; slit width changed as needed. (Continued on next page)

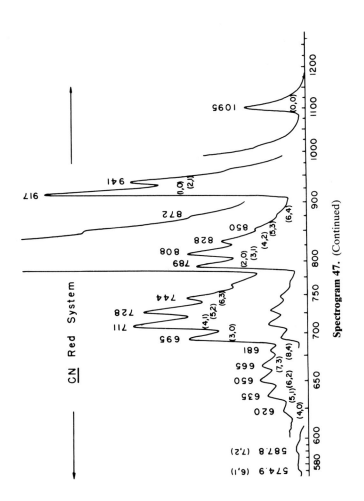

Spectrogram 47. (Continued)

CN Red System

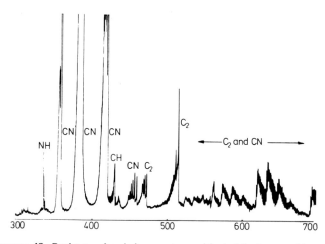

Spectrogram 48. Background emission spectrum of fuel-rich nitrous oxide–acetylene flame.

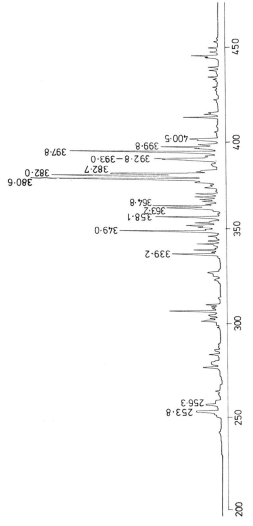

Spectrogram 49. Spectrum of a Fe hollow-cathode lamp between 200 and 700 nm operated at 25 mA. Same monochromator as in spectrogram 41; bandwidth 0·25 nm. (Recorded by B Gutsche, Giessen, Germany; unpublished.) Correction: for line Fe 380·6, read Fe 372·0 nm. The other wavelength data should be shifted in accordance with this. (Continued on next page)

305

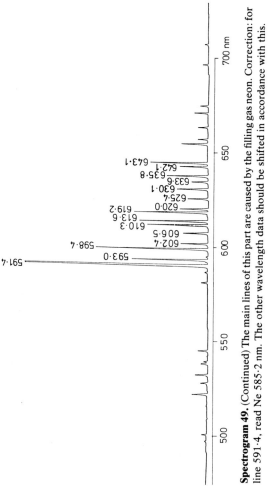

Spectrogram 49. (Continued) The main lines of this part are caused by the filling gas neon. Correction: for line 591·4, read Ne 585·2 nm. The other wavelength data should be shifted in accordance with this.

Appendix 2
Wavelength Tables

1. Introduction

Listed in the following table (table 6) are wavelengths given in nm of about 7000 atomic lines or molecular bands belonging to 70 atomic or 85 molecular species that can be analysed by atomic or molecular emission (E), atomic absorption (A), or atomic fluorescence (F) methods in flames. Bands that are emitted by the flame itself are, as a rule, not included. The data were collected at the Abteilung für Medizinische Physik of the University of Giessen.

2. Details

The wavelength data of lines (or bands) refer to air, that is, no corrections for the refractive index of the air have been made. Intensity estimates are not listed because differences in temperature, reducing conditions, observation height, etc, entail large differences in relative intensities. In addition, differences in the spectral bandwidth of the monochromators used affect the ratio of the recorded line and (quasicontinuous) band intensities. We have excluded lines of, for example, the noble gas atoms, which do not appear in flames but may appear in hollow-cathode lamps used as background sources in AAS and AFS.

Emission band heads of flat bands are not well defined; in such cases, the wavelength is rounded off to 1 nm. On the other hand, it may be possible that two lines have nearly the same position in the spectrum. To distinguish them, the wavelength is expressed often to more than two decimal places. The emitter of the molecular bands observed in a flame (e.g. CaO, CaOH) may be uncertain; in these cases, the uncertain component is written in parentheses, for example, CaO(H). If nothing is known about the composition of the molecule, only the element symbol (e.g. W) is used, while the symbol b (= band) is given after the wavelength. Errors are possible especially where bands and atomic lines coincide, such as in complex spectra.

Because of the complexity of the molecular spectra, especially in the case of polyatomic molecules, we have listed only the band heads or peaks and other general characteristics of the molecular bands, such as their degradation to shorter or longer wavelengths. For more detailed descriptions of some molecular bands we refer to the spectrograms in Appendix 1 and to the following books:

Gaydon A G 1968 *Dissociation Energies and Spectra of Diatomic Molecules* 3rd edn (London: Chapman and Hall)

Gaydon A G 1974 *The Spectroscopy of Flames* 2nd edn (London: Chapman and Hall)

Gatterer A, Junkes J and Salpeter E W 1957 *Molecular Spectra of Metallic Oxides* (Cittá del Vaticano)

Pearse R W B and Gaydon A G 1976 *The Identification of Molecular Spectra* 4th edn (London: Chapman and Hall)

Rosen B (ed) 1970 *Spectroscopic Data relative to Diatomic Molecules* (Oxford: Pergamon)

3. Abbreviations Used

Method

A = atomic line observed in absorption in flames and/or useful in AAS analysis

E = atomic line or molecular band observed in emission in flames

F = atomic line observed in fluorescence in flames and/or useful in AFS analysis

Characters

a = line of the neutral atom, not p, f or u

b = peak or head of a molecular band (added if molecular compound is not specified)

d = double line

f = intersystem combination (intercombination) line

m = multiple (more than three components) line

p = resonance line or line with lower level close to ground level (if line is listed as A or F, the symbol p often has been dropped)

r = head of a band degraded to longer wavelengths

t = triple line

u = unclassified (as to neutral atom or ion)

v = head of a band degraded to shorter wavelengths

w = peak of a wide or diffuse band

x = sharp peak of or in a headless band formed of bunched rotational lines

y = rotational line or group of a few rotational lines

Flames

AA = air–acetylene flame

AG = air–gas flame

AH = air–hydrogen flame

AOA = air–oxygen–hydrogen flame

AOHA = air–oxygen–hydrogen–acetylene flame
   FH = fluorine–hydrogen flame
      i = inner cone or reaction zone of flame
   OA = oxyacetylene flame
   OAz = fuel-rich (incandescent) oxyacetylene flame
   OC = oxycyanogen flame
   OG = oxygen–gas flame
   OH = oxyhydrogen flame
   PH = perchloryl fluoride–hydrogen flame

Solvents

   AA = acetylacetone
   Ac = acetone
   Al = alcohol
   Bu = butanol (4% in water)
   C = chloroform
   Cu = copper solution
   D = dimethylformamide
   G = gasoline
   H = hexone (methyl isobutyl ketone)
   Hp = heptane
   IO_3 = iodate solution
   M = methanol
   n = non-aqueous solvent
   N = naphtha
   P = isopropanol
   q = background emission

4. Acknowledgments

Grateful acknowledgments are made to Mrs M Brinkmann, Mr W Frenzel, Mrs M Höhne and Mrs I Matuszewski of the Abteilung für Medizinische Physik of the University of Giessen for collecting and revising the material, and to Dr J van der Hurk of the Fysisch Laboratorium of the University of Utrecht for checking and extending the data on the Ca, Sr, and Ba bands.

5. Sources; References

The flame spectra were taken from the following sources:

Angino E E and Billings G K 1967 *Atomic Absorption Spectrometry in Geology* (Amsterdam: Elsevier)

Fassel V A, Curry R H and Kniseley R N 1962 Flame spectra of the rare-earth elements, *Spectrochim. Acta* **18** 1127–53

Gilbert P T Jr (formerly of Beckman Instruments Inc, Fullerton, California) private communications

Herrmann R and Alkemade C Th J 1963 *Chemical Analysis by Flame Photometry* transl Paul T Gilbert Jr (New York and London: Wiley)

Mavrodineanu R and Boiteux H 1965 *Flame Spectroscopy* (New York and London: Wiley)

Mossotti V G and Fassel V A 1964 The atomic spectra of the lanthanide elements, *Spectrochim. Acta* **20** 1117–27

Parsons M L and McElfresh P M 1971 *Flame Spectroscopy: Atlas of Spectral Lines* (New York: Plenum)

In doubtful cases, the following more general wavelength tables were consulted:

Harrison G R 1963 *Wavelength Tables* (Cambridge, Mass: MIT Press)

Kayser H and Ritschl R 1939 *Tabellen der Hauptlinien der Linienspektren aller Elemente* 2nd edn (Berlin and Heidelberg: Springer)

Kroonen J and Vader D 1963 *Line Interference in Emission Spectrographic Analysis* (Amsterdam: Elsevier)

Zaidel A N, Prokof'ev V K and Raiskii S M 1961 *Tables of Spectrum Lines* (Oxford and New York: Pergamon)

Table 6 Index of flame spectra by wavelength

λ		Species		Class
177·495		P		A
184·9	p	Hg		E
189·0		As		A, F, E
193·696		As		A, E
195·38		Bi		A
195·94		Bi		A
196·026		Se		A, F, E
197·2		As		A, F, E
202·12		Bi		A
202·4		Cu		A, E
202·5		Mg		E
203·985		Se		A, F
204·169		Ge		A
204·376		Ge		A
204·937		Pt		A
206·17		Bi	AHn	A, F
206·297		Se		A, F
206·520		Ge		A
206·75		Pt		A
206·833		Sb	AHn	A, F, E
206·865		Ge		A
207·336		Al		A
207·479		Se		A, F
208·882		Ir		A
208·99	r	CO	OHinq	E
209·263		Ir		A
209·423		Ge		A
211·03		Bi	AHn	A, E
211·31	r	CO	OHinq	E
211·39		Sn	AHn	E
212·739		Sb		A
212·75		Sb	AHn	A, E
212·83		CO	OHinq	E
213·856		Zn		A, F, E
213·98	a	Sb	AHn	A, E
214·275		Te		A, F, E
214·423		Pt		A
214·44		Cd II	OHn	E
214·50	a	Sb	AHn	A, E
214·77		Pt		E
214·87		Sn	AHn	E
214·91	v	NO	AOHA	E
215·02	r	CO	OHinq	E
215·14	a	Sn	AHn	A, E
215·25		Sn		E
215·49	v	NO	AOHA	E
216·4		Se		A
216·51		Cu		A, E
216·517		Pt		A
216·68		Fe	AHn	A, E
217·00		Pb	OHin	A, E
217·30	r	CO	OHinq	E
217·467		Pt		A
217·56	a	Pb		E
217·589		Sb	AHn	A, F, E
217·81		Fe	AHn	E
217·89		Cu		A, E
217·93	a	Sb	AHn	E
218·17		Cu		A, E
218·65		Fe	AHn	E
219·18	p	Fe	AHn	E
219·45	a	Sn	AHn	E
219·68	r	CO	OHinq	E
219·87		Ge		A
219·93	f	Sn	AHn	E
220·04	p	Fe	AHn	E
220·35		Pb	OHin	E
220·797		Si		A, E
220·85	a	Sb	AHn	A, E
220·97	a	Sn	AHn	A, E
221·088		Si		A
221·54		Si(O)		E
221·58	r	CO	OHinq	E
221·667		Si		A
222·08	a	Sb	AHn	A, E
222·15	r	CO	OHinq	E
222·49	a	Sb	AHn	A, E
222·57		Cu		A, E
222·83		Bi	AHn	A, F, E
223·06	f	Bi	AHn	A, F, E
223·17	a	Sn	AHn	A, E
223·63	r	Si(O)	OAzn	E
223·74	a	Pb	OHin	A, E
223·83	r	CO	OHinq	E
223·94	v	NO	AOHA	E
224·43		Cu		A, E
224·54	v	NO	AOHA	E
224·61		Sn	AHn	A, F, E
224·69	a	Pb	OHin	E
224·72	r	CO	OHinq	E
225·9		Te		A, F
226·17	r	CO	OHinq	E
226·28	v	NO	AOHA	E
226·502		Cd		E
226·89	a	Sn	AHn	A, E
226·94	v	NO	AOHA	E
227·39	r	CO	OHinq	A, E
227·65	a	Co	OH	A, E
227·66		Bi	AHn	A, F, E
227·72	r	Si(O)	OAzn	E
228·61	r	CO	OHinq	E
228·67	f	Sn	AHn	E
228·70		Ni		E
228·78	a	Co	OH	A, E
228·802		Cd	OC	A, F, E
228·81	a	As	AHn	F, E
228·82	dv	P(O)	AHn	E
228·99		Ni		A, F
229·25	p	Fe	AHn	E
229·49	dv	P(O)	AHn	E
229·52	f	Co	OH	A, E
229·60	f	Co	OH	E
229·67	a	Co	OH	E
229·78	p	Fe	AHn	E
229·89	r	Si(O)	OAzn	E
229·92	p	Fe	AHn	E
230·08	p	Ni	AHn	E
230·17	dv	P(O)	AHn	E
230·40	a	Co	OH	E
230·42	p	Co	OH	E
230·424		Ba II		E
230·69	dv	P(O)	AHn	E
230·90		Co	OH	A, E
231·0	x	C_2	OAq	E
231·096		Ni		A, F, E
231·147		Sb	AHn	A, F, E
231·15	r	CO	OHinq	E
231·23		Ni	AHn	F, E
231·37	dv	P(O)	AHn	E
231·40		Ni	AHn	A, F, E
231·598		Tl		A
231·68	p	Co	OH	E
231·72	p	Ni	AHn	E
231·72	f	Sn	AHn	E
231·976		Ni II		A
232·003		Ni	AHn	A, F, E
232·06	dv	P(O)	AHn	E
232·14	p	Ni	AHn	E
232·2	x	C_2	OAq	E
232·31	p	Co	OH	E
232·58		Ni	AHn	A, F, E
233·00	p	Ni	AHn	E
233·24	a	Pb	OHin	A, E
233·48	p	Sn	AHn	E
233·527		Ba		E
233·60	p	Co	OH	E
233·75		Ni	AHn	A, E
233·79	r	CO	OHinq	E
233·87	p	Co	OH	E
234·24	r	Si(O)	OAzn	E
234·43	r	Si(O)	OAzn	E
234·55		Ni	AHn	A, F, E
234·62	f	Co	OH	E
234·75		Ni		A
234·861		Be	OAzn	A, F, E
234·98	a	As	AHn	F, E

λ		Element		
235·29	a	Co	OH	E
235·48		Sn	AHn	A, F, E
235·56	p	Co	OH	E
235·7		Hg		E
235·71		Pt		E
235·82	p	Co	OH	E
235·87	p	Co	OH	E
236·06	p	Ni	AHn	E
236·21	p	Ni	AHn	E
236·33		NO	AOHA	E
236·379		Co		E
236·45	r	Si(O)	OAzn	E
236·47	p	Cr		E
236·51	p	Co	OH	E
236·57		Si(O)		E
236·59	p	Cr	AHn	E
236·68	p	Cr	AHn	E
236·70		Al		A, E
236·73	dv	P(O)	AHn	E
236·97	a	Co	OH	E
237·02	v	NO	AOHA	E
237·186		Co	OH	E
237·277		Ir		A
237·313		Al		A
237·33		Al		A, E
237·36	p	Fe	AHn	E
237·52	dv	P(O)	AHn	E
237·97		Tl		A, E
237·99	dv	P	AHn	E
238·05	p	Co	OH	E
238·07	p	Sn	AHn	E
238·12	f	As	AHn	F, E
238·20		Fe II	GCN	E
238·3		PO		E
238·32	a	Te	AHn	A, F, E
238·35	dv	P(O)	AHn	E
238·49	p	Co	OH	E
238·579		Te		A, F, E
238·79	dv	P(O)	AHn	E
238·79	r	Si(O)	OAzn	E
238·88	a	Pb	OHin	E
239·20	p	Co	OH	E
239·38	a	Pb	OHin	E
239·56		Fe II		E
239·63	dv	P(O)	AHn	E
239·709		W		E
239·86		Ca		A, E
239·96	a	Pb	OHin	E
240·09	a	Bi	AHn	E
240·18	p	Ni	AHn	E
240·19	f	Pb	OHin	E
240·22	f	Co	OH	E
240·725		Co		A, F, E
240·82	a	Sn	AHn	E

λ		Element		
241·102		Co		A
241·16		Co	OH	A, F, E
241·17	a	Pb	OHin	E
241·38	b	Si(O)	OAzn	E
241·45	p	Co	OH	A, F
241·53		Co	OH	A, F, E
241·91	a	Co	OH	A, E
241·93	p	Ni	AHn	E
242·17	a	Co	OH	E
242·17	a	Sn	AHn	E
242·49		Co	OH	A, F, E
242·80		Au	OH	A, E
242·81		Sr		A
242·92	p	Co	OH	E
242·95	a	Sn	AHn	A, F, E
243·22		Co	OH	A, F, E
243·27		Ta II	OCn	E
243·583		Co		A
243·63	f	Si		E
243·67	p	Co	OH	E
243·67		Pt		E
243·72	f	As	AHn	A, F, E
243·73	v	AsO	AHn	F, E
243·85	v	AsO	AHn	E
243·90		Co	OH	A, E
244·00	v	NO	AOHA	E
244·01	a	Pt	OC	A, E
244·16		Cu		A, E
244·38	a	Pb	OHin	E
244·55	a	Sb	AHin	E
244·62	a	Pb	OHin	E
244·70	v	NO	AOHA	E
244·79		Pd	OA	A, E
245·007		Ga		A
245·097		Pt		A
245·2		W		A
245·46	dv	P(O)	AHn	E
245·62	a	Co	OH	E
245·65	f	As	AHn	F, E
245·76	a	Fe	AHn	E
245·90	r	SiO	OAzn	E
246·08	p	Co	OH	E
246·26	p	Fe	AHn	E
246·42	dv	P(O)	AHn	E
246·420	p	Co	OH	E
246·44		Yb		A, E
246·5		Yb		A, E
246·51	a	Fe	AHn	E
246·74		Pt		A, E
246·77	p	Co	OH	E
246·89		Fe	AHn	E
247·03	a	Co	OH	E
247·11	v	NO	AOHA	E
247·29	p	Fe	AHn	E

λ		Element		
247·39	p	Co	OH	E
247·44		Pd		E
247·48	a	Fe	AHn	E
247·51		Ir		A, E
247·64	a	Pb	OHin	E
247·64	f	Pd	OA	A, E
247·66	a	Co	OH	E
247·687		Ni		A
247·79	dv	P(O)	AHn	E
247·86	a	C	OAq	E
247·87	v	NO	AOHA	E
247·98	p	Fe	AHn	E
248·19	r	SiO	OAzn	E
248·33		Fe	AHn	A, F, E
248·34	a	Sn	AHn	E
248·36	a	Co	OH	E
248·68		SiO	OAzn	E
248·717	a	Pt	OC	A, E
248·81	p	Fe	AHn	E
248·98		Fe		E
249		SiO		E
249·01		Pt		E
249·0644	p	Fe	AHn	E
249·1155		Fe		E
249·21		Cu	AHn	A, E
249·29	f	As	AHn	F, E
249·47	a	Be	OAzn	E
249·54		Co		E
249·55	a	Fe	AHn	E
249·56	a	Co	OH	E
249·57	f	Sn	AHn	E
249·58		Pt		E
249·678		B		A, E
249·773		B		E
249·796	f	Ge	AHn	A, E
249·85		Pt		E
250·070		Ga		A
250·11		Fe	AHn	A, E
250·26		Ir		E
250·298		Ir		A, E
250·36	dv	AsO	AHn	E
250·45	a	Co	OH	E
250·47	v	AsO		E
250·69	a	Co	OH	E
250·69		Si	OAzn	A, F, E
250·73		CS	AOAn	E
250·778		V		A
250·85		Pt		E
250·99	r	SiO	OAzn	E
251·05		Sb		F
251·08	p	Fe	AHn	E
251·10	a	Co	OH	E
251·12	r	CS	AOAn	E
251·165		V		A

251·18		Co		E	255·58	r CS	AOAn	E	
251·195		V		A	255·68	a Co	OH	E	
251·43		Si	OAzn	A, F, E	255·68	Ho		E	
251·56		Pt		E	255·90	v NO	AOHA	E	
251·61		Si	OAzn	A, F, E	255·94	Ta		A	
251·714		V		A	256·02	In		A, E	
251·79	a	Co	OH	E	256·13	a Co	OH	E	
251·81	p	Fe	AHn	E	256·38	r SiO	OAzn	E	
251·87	dv	P(O)	AHn	E	256·5	mb Sb(O)	AH	E	
251·87		Ho		A, E	256·73	p Co	OH	E	
251·92	p	Si	OAzn	A, F, E	256·80	p Al	OHn	E	
251·96		V		A	256·947	Sr		A	
252·1363		Co	OH	A, F, E	256·97	v AsO	AHn	E	
252·28		Fe	AHn	A, F, E	257·09	v AsO	AHn	E	
252·32	r	CS	AOAn	E	257·14	Zr II	OCn	E	
252·41		Si	OAzn	A, F, E	257·16	f Sn	AHn	E	
252·43	p	Fe	AHn	E	257·22	a Co	OH	E	
252·45	a	Bi	AHn	E	257·27	r CS	AOAn	E	
252·62		V		A	257·35	a Co	OH	E	
252·74	p	Fe	AHn	E	257·4	mr Sb(O)	AH	E	
252·85	a	Sb	AHn	E	257·4	V		A	
252·85		Si	OAzn	A, F, E	257·51	Al		A, E	
252·90		Co	OH	A, F, E	257·541	Al		A	
252·91	p	Fe	AHn	E	257·56	r CS	AOAn	E	
252·94	dv	P(O)	AHn	E	257·61	Mn II	OC	E	
253·00	r	CS	AOAn	E	257·73	f Pb	OHin	E	
253·01	a	Co	OH	E	258·014	Tl		A	
253·018		V		A	258·45	a Fe	AHn	E	
253·07		Te		A, F	258·65	Ho		E	
253·22	a	Co	OH	E	258·71	r SiO	OAzn	E	
253·38		Ho		E	258·75	v NO	AOAzn	E	
253·40		P		E	258·96	r CS	AOAn	E	
253·56		P		E	259·06	a Co	OH	E	
253·56	p	Fe	AHn	E	259·25	Ge	AHn	A, F, E	
253·60	p	Co	OH	E	259·29	Ho		E	
253·649		Co		A	259·37	Mn II	OC	E	
253·652	f	Hg	AHn	A, F, E	259·44	a Sn	AHn	E	
253·87	r	CS	AOAn	E	259·57	v NO	AOHA	E	
253·92		Pt		E	259·57	dv P(O)	AHn	E	
254·04	dv	P(O)	AHn	E	259·805	Sb		F, E	
254·10	p	Fe	AHn	E	259·8062	f Sb	AHn	F, E	
254·3971		Ir		A, E	259·94	Fe		E	
254·43	p	Co	OH	E	259·96	a Fe	AAi	E	
254·49		Co		E	260·515	Ti		A	
254·60	p	Fe	AHn	E	260·57	Mn II	OC	E	
254·64		Sn		A	260·59	r CS	AOAn	E	
254·66	f	Sn	AHn	A, F, E	260·639	W		A	
254·83	a	Co	OH	E	260·68	a Fe	AAi	E	
254·95	r	CS	AOAn	E	260·69	W		A	
255·00	v	NO	AOHA	E	260·78	Ta		A	
255·135		W		A	260·82	Ta		A	
255·30	a	Co	OH	E	260·86	a Zn	OHn	E	
255·33		P		E	260·863	Ta		A	
255·493		P	AHn	E	260·899	Tl		A	

261·0	OH		E
261·13	Ir		E
261·148	Ti		A
261·21	Ru		E
261·23	Sb		E
261·37	a Pb	OHin	E
261·42	a Pb	OHin	A, E
261·54	Lu		A
262·16	r CS	AOAn	E
262·21	OH		E
262·24	a Co	OH	E
262·35	a Fe	AAi	E
262·37	Dy		E
262·79	a Bi	AHn	F, E
262·80	Pt		A, E
263·0	C		E
263·13	Ru		E
263·58	a Fe	AAi	E
263·7	Os		A
263·93	Pt		E
263·942	Ir		A
263·97	Ir		A, E
264·11	Ti		A
264·14	Hf		E
264·15	Au		E
264·21	Dy		E
264·30	Ru		E
264·4	Os		A
264·426	Ti		A
264·48	r SiO	OAzn	E
264·664	Ti		A
264·69	Pt		A, E
264·747	Ta		A
264·795	Au		A
265·03	a Co	OH	E
265·047	Be		E
265·055	Be	OAzn	E
265·09	Pt		E
265·12	Ge	AHn	A, F, F
265·13	Ru		E
265·16	Ge	AHn	A, F, F
265·18	Ru		E
265·26	Sb		E
265·654	W		A
265·804	W		E
265·90	Rh		E
265·95	Pt	OC	A, E
265·99	Ga		E
266·12	f Sn	AHn	E
266·189	Ta		A
266·1983	Ir		A, E
266·21	Ta		A
266·26	r CS	AOAn	E
266·32	a Pb	OHin	E

266·48		Ir		A, E		272·36	p	Fe	AHn	E		278·50	f	Sn	AAi	E
266·557		Tl		A		272·435		W		A		278·57		Ho		E
266·79	f	Fe	AAi	E		272·72		Sb		E		278·81	a	Fe	AHn	E
266·90	r	SiO	OAzn	E		272·80	a	Fe	AHn	E		279·33		Pt		E
267·06		Sb		F, E		272·99		Pt		E		279·48	p	Mn	OA	A, F, E
267·14	v	NO	AOHA	E		273·36	a	Fe	AHn	E		279·55		Mg II	OC	A, E
267·19		Yb		E		273·40		Pt		A, E		279·73		Ir		E
267·265		Yb		A		273·57		Ru		E		279·77		Ir		E
267·60		Au	OH	F, E		273·73	p	Fe	AHn	E		279·83		Mn	OA	A, F, E
267·70	r	CS	AOAn	E		274·13	p	Li	OH	E		280·11		Mn	OA	A, F, E
267·71		Pt		A, E		274·24	p	Fe	AHn	E		280·20		Pb	OHin	E
267·716		Cr		E		274·41	p	Fe	AAi	E		280·27		Mg II	OC	A, E
267·91	a	Fe	AAi	E		274·50		As		F, E		280·32		Pt		E
267·98	a	Co	OH	E		274·83	a	Au	OH	A, E		280·69		Os		A
268·0	v	NO	AOHA	E		275·01	p	Fe	AHn	E		281·05		Zn		E
268·04	dp	Na	OHn	E		275·37	a	Fe		E		281·13	r	OH		E
268·14		W		A		275·39	a	In	OHn	A, E		281·33	a	Fe	AAi	E
268·28		Sb		E		275·39		Pt		E		281·36	f	Sn	AAi	E
268·42	a	Zn	OHn	E		275·40	a	Fe	AAi	E		281·60	r	OH	OHq	E
268·517		Ta		E		275·44	a	Fe	AAi	E		281·615		Mo		E
268·53	a	Co	OH	E		275·46		Ge	AHn	A, F, E		282·022		Hf		E
268·80		Dy		E		275·49		Pt		E		282·32	a	Pb	OHin	E
269·01	f	Fe	AAi	E		275·50	r	SiO	OAzn	E		282·33	a	Fe	AAi	E
269·13		Ge	AHn	A, F, E		275·63	p	Fe	AHn	E		282·44		Ir		E
269·32		CS	AOAn	E		275·83		Ta		A, E		282·56	a	Fe	AAi	E
269·37	r	SiO	OAzn	E		276·14	a	Co	OH	E		282·616		Tl		A
269·42		Ir		E		276·18	f	Sn	AAi	E		282·90	r	OH	OHq	E
269·58	a	Co	OH	E		276·18	a	Fe	AHn	E		283·03		Pt		A, E
269·68		Bi		F, E		276·20	a	Fe	AHn	E		283·138		W		A
269·706		Nb		A, E		276·23		W		A		283·17	x	OH	OHq	E
269·84		Pt		E		276·31	f	Pd	OA	A, E		283·24	a	Fe	AAi	E
270·24		Pt		A, E		276·34		Ru		E		283·31		Pb	OHin	A, F, E
270·59		Pt		A, E		276·42	a	Co	OH	E		283·563		Cr		E
270·65		Sn	AHn	F, E		276·44		Ho		E		283·87	x	OH	OHq	E
270·66	a	Fe	AAi	E		276·64	a	Co	OH	E		284·00	a	Sn	AHn	A, F, E
270·9		Tl		A		276·75	a	Fe	AHn	E		284·325		Cr		E
270·92		Ru		E		276·79		Tl	OHn	A, F, E		284·40	a	Fe	AAi	E
270·96		Ge	AHn	A, F, E		276·9		Te		A		284·97		Ir		A, E
271·03		In	OHn	A, E		276·99	f	Sb	AHn	F, E		285·06	f	Sn	AHn	E
271·067		Tl		A		277·17		Pt		E		285·18	a	Fe	AAi	E
271·25	a	Zn	OHn	E		277·21	p	Fe	AAi	E		285·213		Mg	OC	A, F, E
271·32	v	NO	AOHA	E		277·50	a	Co	OH	E		285·28	p	Na	OHn	E
271·39		In		E		277·54		In		E		285·3		Pt		A
271·46		Os		A		277·588		Ta		A		285·30	p	Na	OHn	E
271·467		Ta		A, E		277·67	a	Mg	AAi	E		286·04		As		F, E
271·75		Rh		E		277·82	a	Fe	AAi	E		286·33		Sn	AHn	A, F, E
271·85		Rh		E		277·83	a	Mg	AAi	E		286·637		Hf		A
271·89		Sb		E		277·98	a	Mg	AAi	E		286·66		Ru		E
271·89		W		A		277·98	f	Sn	AHn	E		287·33	a	Pb	OHin	E
271·90		Pt		A, E		278·02	a	As	AHn	F, E		287·424		Ga	OHn	A, E
271·90		Fe	AHn	A, F, E		278·05	r	SiO	OAzn	E		287·50		Ru		E
271·97		Ga		A, E		278·05		Bi		F, E		287·53	r	OH	OHq	E
272·09	p	Fe	AHn	E		278·14	a	Mg	AAi	E		287·79	a	Sb		F, E
272·22	v	NO	AOHA	E		278·30		Mg	AAi	E		287·939		W		A

314

288·116		Ir		A	296·58		Sc		E	301·20	a	Ni	OHn	E
288·16	p	Si	OAzn	E	296·69		Fe	AAi	A, E	301·36	p	Co	OH	E
288·65		Ru		E	297·01	p	Fe	AAi	E	301·48	da	Cr	OC	E
288·96		U		A	297·05		Yb		E	301·52	a	Cr	OC	E
289·27	r	OH	OHq	E	297·11	a	Cr	AAi	E	301·53		Sc		E
289·39		Pt		E	297·32		Tm		A, E	301·72		Ru		E
289·645		W		A	297·32	p	Fe	AAi	E	301·75	p	Co	OH	E
289·79		Pt		E	297·339		Tm		A	301·76	a	Cr	OC	E
289·80	a	Bi	AHn	F, E	297·40		Sc		E	301·80		Os		A, E
289·87		As		F	297·45		Y		E	301·85	a	Cr	OC	E
290	x	OH	OHq	E	297·55	a	Cr	AAi	E	301·88	a	Cr	OC	E
290·441		Hf		A	297·69		Ru		E	301·91	p	Ni	OHn	E
290·475		Hf		A	298·06		Cd	OHn	E	301·93		Sc		E
290·89		Ru		E	298·07		Sc		E	302·049		Fe		A, F
290·91		Os		A, E	298·08	a	Cr	AAi	E	302·053		Hf		A
291·1		W		A	298·14	f	Fe	AAi	E	302·06	d	Fe	OHn	A, F, E
291·139		Lu		E	298·16	p	Ni	OHn	E	302·07	a	Cr	OC	E
291·35	f	Sn	OHn	E	298·36		Fe	AAi	A, E	302·09		Ru		E
291·83	a	Tl	OHn	A, E	298·41	p	Ni	OHn	E	302·11	p	Fe	OHn	E
291·9	x	OH	OHq	E	298·42		Y		E	302·12		OH		E
292·15		Tl	OHn	A, E	298·539		Zr		A	302·16	a	Cr	OC	E
292·36		V		A, E	298·55		Er		A, E	302·17		Y		E
292·48		Ir		A, E	298·60		Cr		E	302·2		Cr		E
292·781		Nb		A, E	298·65		Cr		E	302·228		Y		E
292·98		Pt		A, E	298·72		Co	OH	A, E	302·44	a	Cr	OC	E
293·18		Sr		A	298·86	a	Cr	OC	E	302·46	f	Bi	AHn	F, E
293·26		In	OHn	A, E	298·89		Ru		E	302·58	p	Fe	AAi	E
293·69		Fe	AAi	A, E	298·90	f	Bi	AHn	F, E	302·79		Pd		E
293·83	a	Bi	AHn	F, E	298·92		Lu		A, E	302·92		Au		E
294·077		Hf		A	298·96	f	Co	OH	A, E	302·95		Zr		A
294·09		Ho		E	299·13		Dy		E	302·98		Sb		E
294·13	p	Fe	OC	E	299·22	p	K	OHn	E	303·02	a	Cr	AAi	E
294·199		Ti		A	299·26	p	Ni	OHn	E	303·07		Os		A
294·32		Ir		E	299·33	a	Bi	AHn	F, E	303·07		Sc		E
294·36		Ga	OHn	A, E	299·44	p	Fe	AAi	E	303·14		Cr		E
294·39	p	Ni	OHn	E	299·45	f	Ni	OHn	E	303·16	a	Fe	AAi	E
294·42		Ga	OHn	A, E	299·50		Ru		E	303·19	p	Ni	OHn	E
294·44		W		A, E	299·645		W		A	303·28		Sn		E
294·52	r	OH	OHq	E	299·66		Cr		E	303·285		As		F
294·698		W		A	299·67		Dy		E	303·35		Ru		F
294·738		W		A	299·80		Pt		A, E	303·382		V		E
294·79	p	Fe	AAi	E	300·05	f	Co	OH	E	303·41		Sn	AHn	F, E
294·826		Ti		A	300·09		Cr.		E	303·42		Cr		E
294·84		Y		E	300·10	p	Fe	AAi	E	303·44	f	Co	OH	E
295·068		Hf		A, E	300·10		Dy		E	303·48	p	K	OHn	E
295·088		Nb		A, E	300·23		Pt		E	303·70		Cr		E
295·39	p	Fe	AAi	E	300·249		Ni		A, F, E	303·74	p	Fe	OHn	E
295·61		Ti		A, E	300·36		Ni	OHn	F, E	303·79	p	Ni	OHn	E
295·74	p	Fe	AAi	E	300·51		Cr		E	303·9		In		E
296·46		Dy		E	300·66		Ru		E	303·906	a	Ge	AHn	E
296·488		Hf		A	300·7		Fe		E	303·936		In		A, E
296·49		Y		E	300·81	p	Fe	OHn	E	304·03		Ru		E
296·52		Ru		E	300·91		Sn	AHn	F, E	304·08		Cr		E
296·53	p	Fe	AAi	E	301·175		Zr		A	304·25		Ru		E

Wavelength		Element		
304·25	p	Co	OH	E
304·26		Pt		A, E
304·27	a	Fe	AAi	E
304·28		Ru		E
304·31	p	V	OC	E
304·36	p	V	OC	E
304·40		Co	OH	A, E
304·49		V		E
304·50		Ni		E
304·53		Y		E
304·68		Tm		E
304·76	p	Fe	OHn	E
304·85		Ru		E
304·88		Ru		E
304·89	p	Co	OH	E
305·08		Ni	OHn	A, F, E
305·08		V		E
305·22	p	V	OC	E
305·36		V	OC	A, E
305·39	a	Cr	AAi	E
305·4		V		E
305·43	p	Ni	OHn	E
305·49		Ru		E
305·63		V	OC	A, E
305·74	a	Fe	OC	E
305·76	p	Ni	OHn	E
305·87		Os		A, E
305·89		Eu		E
305·91	p	Fe	AAi	E
306·05		V	OC	A, E
306·18		Co		E
306·37	r	OH	OHq	E
306·46	p	Ni	OHn	E
306·47		Pt	OC	A, E
306·48		Ru		E
306·64		V	OC	A, E
306·77		Bi	AHn	A, F, E
306·78	r	OH	OHq	E
307·2		Zn		E
307·23	p	Co	OH	E
307·288		Hf		A, E
307·38	p	V	OC	E
307·59	f	Zn	AH	A, E
308·08	p	Ni	OHn	E
308·11		Tm		E
308·14		Lu		A, E
308·17	x	OH	OHq	E
308·22		Al	OH	A, E
308·26	p	Co	OH	E
308·68	p	Co	OH	E
308·96	p	Co	OH	E
308·99		OH	OHq	E
309·271		Al		A, F, E
309·284		Al		A, F, E
309·31		V II	OC	E
309·418		Nb		A, E
309·6	x	OH	OHq	E
309·95	x	OH	OHq	E
310·16		Ni	OHn	F, E
310·19		Ni		F, E
310·2		V II		E
310·20	p	K	OHn	E
310·21	x	OH	OHq	E
310·22	p	K	OHn	E
310·23		V	OC	E
310·56	r	AsO	OHq	E
310·60	x	OH	OHq	E
310·61		Eu		E
310·68		AsO	OHq	E
311·07		V	OC	E
311·14		Eu		A, E
311·21		Mo		A, E
311·23	x	OH	OHq	E
311·41	f	Ni	OHn	E
311·78	x	OH	OHq	E
311·84		V II	OC	E
311·84		Lu		A, E
312		OH		E
312·11		V II	OC	E
312·14	p	Co	OH	E
312·26	x	OH	OHq	E
312·28		Au		A, E
312·53		V II	OC	E
312·83	x	OH	OHq	E
313		Oh		E
313·03		V II	OC	E
313·042		Be		E
313·079		Nb		A, E
313·107		Be		E
313·126		Tm		E
313·26		Mo	OAn	A, E
313·3		Mo		E
313·3		V II		E
313·41		Ni	OHn	F, E
313·73	p	Co	OH	E
313·99		Co	OH	E
314·71	p	Co	OH	E
314·75	x	OH	OHq	E
315·38		Ho		E
315·42		Ho		E
315·82		Mo	OAn	A, E
315·88	p	Co	OH	E
315·93		Ho		E
316·34		Nb		A, E
316·96	x	OH	OHq	E
317·03		Mo	OAn	A, E
317·06	r	AsO		E
317·13		Lu		E
317·24	r	AsO		E
317·26		Tm		E
317·28		Tm		E
317·5	a	Sn	AHn	F, E
317·5		Sb		E
317·9		AgH		E
317·93		Ca II		E
318·05		Tm		E
318·34		V	OC	A, E
318·40		V	OC	A, E
318·54		V	OC	A, E
318·63		Ho		E
318·64		Ti		A, E
319·07		V II	OC	E
319·157		W		A
319·19		Ti		A, E
319·39	a	V	OC	E
319·40		Mo	OAn	A, E
319·80	p	V	OC	E
319·99		Ti		A, E
320·10		Ho		E
320·24	p	V	OC	E
320·40		Pt		E
320·58		SnO	AH	E
320·617		Ho		A, E
320·74	p	V	OC	E
320·78	r	AsO		E
320·88		Mo	OAn	A, E
320·91	r	AsO		E
321·05		Eu		A, E
321·24		V		E
321·28		Eu		A, E
321·37		Eu		E
321·70	p	K	OHn	E
321·75	p	K	OHn	E
321·92	f	Co	OH	E
322·0		AgH		E
322·00		Ho		E
322·078		Ir		E
322·80		Rb	OHn	E
322·91	p	Rb	OHn	E
322·975		Tl	OHn	A, E
322·98	r	Sn	AH	E
323·0		SnO		E
323·206		Os		A
323·25		Sb		E
323·26		Li	OH	A, E
323·30		Ni	OHn	A, E
323·33		Ho		E
323·37		Tm		E
323·387		Ho		A
323·452		Ti		E
323·46		Ni	OHn	E
323·51		Eu		E

316

323·70	f	Co	OH	E
324·14		Eu		E
324·15		Tm		E
324·228		Y		E
324·27	a	Pd	OHn	A, E
324·31	f	Ni	OHn	E
324·60		Eu		E
324·75		Eu		E
324·75		Cu	OHn	A, F, E
324·85	p	Ni	OHn	E
325·16		Pd		E
325·57		Sc		A, E
325·61		In	OHn	A, E
325·86		In	OHn	A, E
325·88	f	Pd	OHn	E
326·11		Cd	AH	A, F, E
326·2		SnO		E
326·229		Os		A
326·23	a	Sn	AHn	F, E
326·24	r	Sn	AH	E
326·31		Rh		E
326·32	a	V	OC	E
326·479		Er		A
326·673		Gd		A
326·75		Sb		E
326·77		V II	OC	E
326·79		Os		A
326·95		Ge	AHn	E
326·99		Sc		A, E
327·11	f	Ni	OHn	E
327·112		V		E
327·16		V II	OC	E
327·16	a	Rh	OH	E
327·305		Zr		E
327·36		Sc		A, E
327·40		Cu	OHn	A, E
327·5		AgH		E
327·61		V II	OC	E
327·8		AsO		E
327·89		Lu		A, E
327·91	r	AsO		E
328·06	a	Rh	OH	E
328·07		Ag	AH	A, F, E
328·17		Lu		A, E
328·23	a	Zn	OHn	E
328·27	p	Ni	OHn	E
328·33		V		E
328·36	f	Rh	OH	E
328·69	p	Ni	OHn	E
328·72	a	Pd	OHn	E
328·86	p	Cs	OHn	E
328·93		Yb		E
328·94		Yb II	OHn	E
328·98		Yb		E

329		SnO		E
329·059		Th		E
329·10		Tm		A, E
329·18	r	Sn	AH	E
329·203		Th		E
329·81		V		E
329·90		Tm		E
330·03		AuH		E
330·16		Os	OC	A, E
330·16		Ru		E
330·19		Pt		E
330·21	a	Pd	OHn	E
330·23		Na	OHn	A, E
330·28	da	Zn	OHn	E
330·30	p	Na	OHn	E
331·21		Lu		A, E
331·242		Er		A
331·32		Cs	OH	E
331·44		Eu		E
331·50		Pt		A, E
331·52		Ru		E
331·57	f	Ni	OHn	E
331·88		Ta		E
332·03	f	Ni	OHn	E
332·13	a	Be	OAzn	E
332·22		Eu		E
332·23	f	Ni	OHn	E
332·3		SnO		E
332·31	f	Rh	OH	E
332·35	r	Sn	AH	E
332·38		Pt		E
332·99	a	Mg	AAi	E
333·0		AgH		E
333·06	f	Sn	AHn	F, E
333·15		Er		E
333·22	a	Mg	AAi	E
333·3		Co		E
333·41	a	Co	OH	E
333·43		Eu		A, E
333·67	a	Mg	AAi	E
333·72	f	Co	OH	E
333·749		La		E
333·85	a	Rh	OH	E
333·96		Ru		E
334·18		Ti		A, E
334·197		Nb		E
334·31		Er		E
334·371		Nb		E
334·396		Nb		E
334·42	a	Rh	OH	E
334·43		Er		E
334·47	r	SnO	AH	E
334·50	r	Te	AH	E
334·50	a	Zn	OHn	E

334·56	a	Zn	OHn	E
334·59		Zn		E
334·6		TeO		E
334·65		Nd		E
334·74	p	Cs	OH	E
334·87	p	Cs	OH	E
334·87	p	Rb	OH	E
334·904		Ti		E
334·906		Nb		E
334·952		Nb		E
334·99		Tm		E
335·04		Eu		E
335·09	p	Rb	OHn	E
335·14	r	Sn	AH	E
335·37		Eu		E
335·44	a	Co	OH	E
335·46		Ti		A, E
335·474		Nb		E
335·7		AgH		E
335·842		Nb		E
335·862		Gd		A
335·87		Nd		E
335·9		CN		E
335·95		Lu		A, E
335·99		Rh	OH	E
336·0		Lu		E
336·0	v	MnF	FH	E
336·0	x	NH	OCq	E
336·16	p	Ni	OHn	E
336·22		Gd		A
336·22	a	Rh	OH	E
336·26		Tm		E
336·58	f	Ni	OHn	E
336·62	f	Ni	OHn	E
336·67		Er		E
336·71		Co	OH	E
336·80		Er		E
336·84	a	Rh	OH	E
336·85		Ir		E
336·85		Ru		E
336·96		Ni	OHn	A, E
337·0	x	NH	OCq	E
337·03	f	Co	OH	E
337·04		Ti		E
337·14		Ti		A, E
337·20	p	Ni	OHn	E
337·23	a	Rh	OH	E
337·27		Er		A, E
337·30	a	Pd	OHn	E
337·42		Ni	OHn	E
337·47	r	Sn	AH	E
337·65		Lu		A, E
337·74		Ti		A, E
337·758		Ti		A

338·06	a	Ni	OHn	E
338·07	a	Pd	OHn	E
338·09	p	Ni	OHn	E
338·17	r	Sn	AH	E
338·20		Er		E
338·27	r	Te	AH	E
338·29		Ag	AH	A, F, E
338·3		Te(O)		E
338·37		Er		E
338·46		Mo		E
338·50		Tm		E
338·50		Dy		E
338·52	a	Co	OH	E
338·55		Lu		E
338·59		Ti		E
338·82	a	Co	OH	E
338·83	r	SnO	AH	E
338·95		Ru		E
339·10		Ni	OHn	A, E
339·19		Zr		E
339·25		Ru		E
339·30	p	Ni	OHn	E
339·31		Tm		E
339·35		Dy		E
339·54	f	Co	OH	E
339·6		AgH		E
339·68		Lu		A, E
339·68		Rh	OH	A, E
339·72		Bi		F, E
339·75		Tm		E
339·78		Tm		E
339·81	p	Cs	OH	E
339·89		Ho		A, E
339·96		Er		E
339·97	a	Rh	OH	E
340·0	p	Cs	OH	E
340·0		MnOH		E
340·17		Ru		E
340·20		Dy		E
340·36		Cd		E
340·43		Mo		E
340·46	a	Pd	OHn	A, F, E
340·51	a	Co	OH	A, E
340·65		Rh		E
340·69		Ta		E
340·69	r	Sn	AH	E
340·77		Dy		E
340·81		Pt		E
340·92	a	Co	OH	E
340·93		Ru		E
340·96	p	Ni	OHn	E
341		SnO		E
341·00		Tm		A, E
341·23	u	Rh	OH	E
341·23		Co	OH	A, E
341·25		Co		E
341·25		Tm		E
341·26	p	Co	OH	E
341·26		Co		A
341·26		Tm		E
341·35	p	Ni	OHn	E
341·39	f	Ni	OHn	E
341·48		Ni	OHn	A, F, E
341·49		Ho		E
341·58	r	Sn	AH	E
341·64		Ho		A, E
341·66		Tm		A, E
341·72	a	Co	OH	E
341·74		Ru		E
341·8		InCl	AH	E
341·84		Ho		E
342·02		Rh		E
342·12	a	Pd	OHn	E
342·2		Te(O)		E
342·24	r	Te	AH	E
342·247		Gd		E
342·28		Er		E
342·37	p	Ni	OHn	E
342·39		Gd		A
342·41		Ho		E
342·44		Er		E
342·79		Pt		E
342·81	r	OH	OHq	E
342·83	p	Ru	OHn	E
342·86		Ru		E
342·92		Er		E
342·93		Tm		E
342·95		Ru		E
343		CN		E
343·08		Ru		E
343·16		Co	OH	A, E
343·21	r	OH	OHq	E
343·25		Eu		E
343·27		Ru		E
343·30	a	Co	OH	E
343·34	a	Pd	OHn	E
343·36	p	Ni	OHn	E
343·43		Dy		E
343·50		Rh	OH	A, E
343·67	p	Ru	OHn	E
343·70		Ho		E
343·70		Ir		E
343·73	f	Ni	OHn	A, E
343·84		Ru		E
343·9	v	MnF	FH	E
344·05		Rh		E
344·06		Fe	OHn	A, E
344·10	p	Fe	OHn	E
344·14	a	Pd	OHn	E
344·15		Tm		E
344·26		Er		E
344·29		Co	OH	A, E
344·30		Tm		E
344·36	a	Co	OH	E
344·39	p	Fe	OC	E
344·46	r	Sn	AH	E
344·55		Dy		E
344·63	p	Ni	OHn	E
344·6722	p	K	OHn	E
344·68		Er		E
344·71		Mo		E
344·73		Tm		E
344·7701	p	K	OHn	E
344·90		Ir		E
344·92	a	Co	OH	E
344·93		Ho		E
344·94	a	Co	OH	E
345·09		Er		E
345·1		AgH		E
345·12		Ho		E
345·18		Re	OAzn	A, E
345·29	f	Ni	OHn	E
345·31		Ho		E
345·350		Co		A, E
345·36		Er		E
345·36		Tm		E
345·38		Ho		E
345·4		Co		E
345·43		Dy		E
345·52	p	Co	OH	E
345·52	a	Rh	OH	E
345·60		Ho		A, E
345·64		Mo		E
345·7		InCl	AH	E
345·70		Eu		E
345·71		Rh		E
345·74		AuH		E
345·79		Rh		E
345·85	p	Ni	OHn	E
346·0		Ni		E
346·047		Re	OAzn	A, E
346·08		Pd	OHn	E
346·09		Dy		E
346·17	f	Ni	OHn	A, F, E
346·20	a	Rh	OH	E
346·22		Tm		E
346·25		Er		E
346·28	a	Co	OH	E
346·38	r	Te	AH	E
346·44	a	Yb	OHn	E
346·45		Sr II	OHn	E
346·45		Er		E

λ		Element			Class
346·47		Re	OAzn		A, E
346·51		Er			E
346·58	p	Co	OH		E
346·59	p	Fe	OHn		E
346·62		Cd			E
346·63		U			A
346·68		Mo			E
346·75	f	Ni	OHn		E
346·75		Tm			E
346·77		Cd			E
346·78		Eu			E
346·94		Er			E
346·95	f	Ni	OHn		E
347·07	a	Rh	OH		E
347·19		Dy			E
347·21	r	OH	OHq		E
347·25	p	Ni	OHn		E
347·40		Co	OH		A, E
347·42		Ho			E
347·48	a	Rh	OH		E
347·55	p	Fe	OHn		E
347·63		Er			E
347·65		Fe			E
347·66		Tm			E
347·67	p	Fe	OHn		E
347·69	p	Cs	OH		E
347·82		Dy			E
347·89	a	Rh	OH		E
348·01	p	Cs	OH		E
348·07		Er			E
348·09		Tm			E
348·12	a	Pd	OHn		E
348·34	a	Co	OH		E
348·34		Pt			E
348·38	p	Ni	OHn		E
348·4	x	OH	OHq		E
348·45	r	Sn	AH		E
348·48		Ho			A, E
348·53		Pt			E
348·59	f	Ni	OHn		E
348·73		Tm			E
348·93		Er			E
348·937		U			A, E
348·94	a	Co	OH		E
348·98		Pd			E
349·06	p	Fe	OHn		E
349·13	p	Co	OH		E
349·30	p	Ni	OHn		E
349·449		Dy			E
349·49		Tb			E
349·57	a	Co	OH		E
349·78	p	Fe	OHn		E
349·87	f	Rh	OH		E
349·894		Ru	OHn		A. E
349·9		InCl	AH		E
349·91		Er			A
349·99		Tm			E
350		HCO			E
350·00		Tm			E
350·007		U			A
350·09	p	Ni	OHn		E
350·11	p	Ba	OH		E
350·23		Co	OH		A, E
350·25		Rh			A, E
350·26	f	Ni	OHn		E
350·26		Co			A, E
350·27		Er			E
350·3	dv	MgF	PH		E
350·38		SrO			E
350·56		Er			A, E
350·63	a	Co	OH		A, E
350·73	a	Rh	OH		A, E
350·734		U			A
350·739		Lu			A
350·77	f	Ni	OHn		E
350·78	r	Te	AH		E
350·79		Ho			E
350·84		Lu			E
350·9		Zr			A
350·917		Tb			E
350·98		Co			E
351·0		SnO			E
351·03	p	Ni	OHn		E
351·04	p	Co	OH		E
351·07		Ho			E
351·09		Bi			F, E
351·10		Dy			E
351·26	a	Co	OH		E
351·29	r	Sn	AH		E
351·35		Co	OH		A, E
351·36		Ir			A, E
351·36		Gd			A, E
351·38	f	Fe	OHn		E
351·39	f	Ni	OHn		E
351·40		Tm			E
351·45		Ru			E
351·462		U			A, E
351·49		Er			E
351·51		Ni	OHn		F, E
351·55		Ho			E
351·56		Dy			E
351·6		AgH			E
351·66		Tb			E
351·67	r	TeO	AH		E
351·69	a	Pd	OHn		E
351·76		Tm			E
351·8	v	MnF	FH		E
351·83	a	Co	OH		E
351·89		Tb			E
351·92	a	Tl	OHn		E
351·96		Ru			E
351·96		Zr			A, E
351·98	p	Ni	OHn		E
352·00		Er			E
352·01	p	Co	OH		E
352·11		Dy			E
352·13	f	Fe	OHn		E
352·16	f	Co	OH		E
352·20		Ir			E
352·25		Er			E
352·34	f	Co	OHn		E
352·34	f	Ni	OHn		E
352·45		Ni	OHn		A, F, E
352·54		SrO			E
352·60	p	Fe	OHn		E
352·62		Fe			E
352·68		Er			E
352·685		Co	OH		A, E
352·80	p	Ni	OHn		E
352·80	p	Rh	OH		E
352·86		Os			E
352·9033		Co	OH		A, E
352·94	a	Tl	OHn		E
352·981	a	Co	OH		E
353·00	r	SnO	AH		E
353·17		Dy			A, E
353·18	a	Mn	OA		E
353·20	a	Mn	OA		E
353·21	a	Mn	OA		E
353·27		Ho			E
353·34	p	Co	OH		E
353·40		Dy			E
353·5	b	Ba			E
353·5		InCl	AH		E
353·53		Nb			E
353·55		Tm			E
353·60		Dy			E
353·79		Tm			E
353·81	a	Rh	OH		E
353·85		Dy			E
353·875		Th			E
353·94		Ru			E
353·95		Er			A, E
354		SnO			E
354·14		Ho			E
354·19		Rh			E
354·24	r	Sn	AH		E
354·33	a	Co	OH		E
354·39	a	Rh	OH		E
354·5		LaO			E
354·58		Er			E
354·6		AgH			E

354·66		SrO		E
354·76		Zr		A, E
354·78	a	Mn	OA	E
354·78		Dy		E
354·80	a	Mn	OA	E
354·82	a	Mn	OA	E
354·82		Ni	OHn	E
354·95	f	Rh	OH	E
355·06	p	Co	OH	E
355·08		U		A
355·15	p	Ni	OHn	E
355·26		Y		E
355·27	p	Co	OH	E
355·3		LaO		E
355·31	a	Pd	OHn	E
355·6		InCl	AH	E
355·63		Er		E
355·64		La	OH	E
355·70		Er		E
355·80		Er		A, E
355·85	f	Fe	OHn	E
355·87		Er		E
355·88	f	Co	OH	E
355·9		LaO		E
355·90		Ir		E
355·98		Os		E
356·00		CN	OHin	E
356·05	r	Te	AH	E
356·08		Ce II	OC	E
356·09	a	Co	OH	E
356·09		Os		E
356·09		LaO		E
356·1	b	Te(O)		E
356·10	v	La	OH	E
356·18	f	Ni	OHn	E
356·34		Er		E
356·36		Er		E
356·38		Tm		E
356·41		Rh		E
356·50	f	Co	OH	A, E
356·51		Er		E
356·54	f	Fe	OHn	E
356·61	v	LaO	OH	E
356·64	a	Ni	OHn	E
356·66		U		A, E
356·71		SrO		E
356·73		Tm		E
356·78		Lu		A, E
356·92		Er		E
356·94	a	Co	OH	E
356·95	a	Mn	OA	E
356·98	a	Mn	OA	E
356·98		Tm		E
356·99		Er		E
357·00		Mn	OA	E
357·01	f	Fe	OHn	E
357·02	a	Rh	OH	E
357·04		Ho		E
357·06		Ru		E
357·07		Er		E
357·10		Dy		E
357·12	a	Pd	OHn	E
357·13		Dy		E
357·19	p	Ni	OHn	E
357·27	a	Pb	OHin	E
357·37		Ir		E
357·443		La		A, E
357·45	r	Sn	AH	E
357·5		SnO		E
357·50	a	Co	OH	A, E
357·54		Co	OH	A, E
357·58		Nb		E
357·79		Mn		E
357·83		Er		E
357·87		Cr	OAn	A, F, E
357·91		Ho		E
357·94		Er		E
358·02		Nb		E
358·04		Er		E
358·12	a	Fe	OHn	E
358·18		Ho		E
358·3		AgH		E
358·30		CuH		E
358·31		Rh	OH	E
358·39	v	CN	OHin	E
358·45		Y II		E
358·48		Co	OH	E
358·48		U		A, E
358·5		SnO		E
358·50		Dy		E
358·52	a	Co	OH	E
358·54	r	Sn	AH	E
358·57		Er		E
358·59	v	CN	OHin	E
358·60		Tm		E
358·63		Sm		E
358·66		Er		E
358·69		SrO		E
358·70	a	Fe	OHn	E
358·71	p	Rb	OHn	E
358·72	a	Co	OH	E
358·79	f	Ni	OHn	E
358·83		Er		E
358·85		Eu		E
358·92		Eu		E
358·92	a	Ru	OHn	E
358·97		Ho		E
359·04	v	CN	OHin	E
359·07		Er		E
359·16	p	Rb	OHn	E
359·26		Sm		E
359·29		Y		E
359·30	a	Ru	OHn	E
359·30		Ho		E
359·349		Cr	OAn	A, F, E
359·4	dv	MgF	PH, OH	E
359·44		Ir		E
359·45		Dy		E
359·49	p	Co	OH	E
359·5	v	MnF	FH	E
359·50		Dy		E
359·58		Er		E
359·6		Ru		E
359·61		Bi		E
359·618	p	Ru	OHn	E
359·619	a	Rh	OH	E
359·715	f	Rh	OH	E
359·77	p	Ni	OHn	E
359·86		Tm		E
359·89		Tb		E
359·9		InCl	AH	E
359·94		Ho		E
359·98		Ru		E
360·00		Ho		E
360·05		Cd		E
360·07		Y		E
360·09		Ho		E
360·11		Zr		A, E
360·21	p	Co	OH	E
360·23	f	Ni	OHn	E
360·42		Sm		E
360·44		LaO	OH	E
360·48		Gd		E
360·52		Gd		E
360·533		Cr	OAn	A, E
360·536	f	Co	OH	E
360·57		Ho		E
360·59		Er		E
360·59		Rh		E
360·68		Te	AH	E
360·7	r	TeO		E
360·71		SrO		E
360·74		Er		E
360·74		Ta		E
360·76		Sm		E
360·81	v	La	OH	E
360·87		Tm		E
360·89	a	Fe	OHn	E
360·93	f	Ni	OHn	E
360·94		Er		E
360·94		Sm		E
360·95	a	Pd	OHn	A, E

360·98	İr		E
361·05	p Ni	OHn	E
361·07	Dy		E
361·1	LaO		E
361·10	Y II		E
361·11	Dy		E
361·14	La	OH	E
361·15	p Cs	OH	E
361·24	Er		E
361·25	Er		E
361·25	a Rh	OH	E
361·27	p Ni	OHn	E
361·29	Cd		E
361·308	La		A
361·38	Sc		E
361·45	Cd		E
361·47	v La	OH	E
361·47	r SnO	AH	E
361·5	LaO		E
361·658	Er		A
361·74	p Cs	OH	E
361·75	W		A, E
361·88	a Fe	OHn	E
361·9	LaO		E
361·94	a Ni	OHn	E
362	SnO		E
362·01	Er		E
362·01	v La	OH	E
362·09	Y		E
362·26	v La	OH	E
362·3	LaO		E
362·35	Ho		E
362·386	Zr		A
362·4	InBr	OH	E
362·4	MgOH	AH	E
362·42	Tm		E
362·47	f Ni	OHn	E
362·50	a Co	OH	E
362·66	a Rh	OH	E
362·7	MgO(H)		E
362·70	SrO		E
362·78	f Co	OH	E
362·78	Er		E
362·80	Er		E
362·81	Pt		E
362·87	Ir		E
362·92	Sm		E
362·93	Er		E
363·0	MnOH		E
363·07	Sc		E
363·09	Ho		E
363·14	p Co	OH	E
363·15	Fe	OHn	E
363·32	Er		E
363·429	Sm		E
363·46	Er		E
363·47	Pd	OHn	A, E
363·49	p Ru	OHn	E
363·52	Ti		A
363·55	Ti	OHn	A, E
363·6	Te		E
363·62	Lu		E
363·63	Er		E
363·66	a Cr	OC	E
363·7	AgH		E
363·8	InCl	AH	E
363·82	U		A
363·84	Tm		E
363·86	Er		E
363·88	Pt		E
363·89	Tb		E
363·90	Er		E
363·95	a Rh	OH	E
363·96	a Pb	OHin	E
364·0	b Mn	AH	E
364·12	Ho		E
364·268	Ti		E
364·35	Dy		E
364·53	Tb		E
364·54	Dy		A, E
364·60	SrO		E
364·60	MgO(H)		E
364·67	Tm		E
364·68	Dy		E
364·77	Lu		E
364·77	p Co	OH	E
364·78	a Fe	OHn	E
364·953	La		A
365·1	InBr	OH	E
365·15	Sm		E
365·19	AuH		E
365·25	p Co	OH	E
365·35	Ti		A, E
365·462	Gd		A
365·49	Rh		E
365·70	a Co	OH	E
365·73	Sm		E
365·80	Rh	OH	A, E
365·81	Ti		E
365·916	U		A
365·95	Er		E
366·13	Sm		E
366·14	f Ru	OHn	E
366·15	r Te	AH	E
366·20	f Te	OHn	E
366·22	Ho		E
366·28	Er		E
366·29	Ho		E
366·34	Ru		E
366·41	p Ni	OHn	E
366·44	Er		E
366·49	Dy		E
366·62	Sm		E
366·62	a Rh	OH	E
366·66	Ho		E
366·68	Dy		E
366·79	Ho		E
366·89	Ti		E
366·90	Ho		E
366·92	f Ni	OHn	E
366·95	Ru		E
366·95	Ho		E
366·96	Tb		E
367·007	U		A, E
367·04	p Ni	OHn	E
367·08	Sm		E
367·1	MnF	FH	E
367·15	SrO		E
367·16	Ti		E
367·2	MgF	OH	E
367·40	Gd		E
367·41	f Ni	OHn	E
367·44	Dy		E
367·57	V		E
367·8	SnO		E
367·83	Ru		E
367·84	Dy		E
367·89	Er		E
367·9	InBr	OH	E
367·91	Ho		E
367·921	Gd		A
367·97	Ho		E
367·99	Fe	OHn	A, E
368·10	Rh		E
368·22	Hf		A, E
368·26	Er		E
368·30	a Co	OH	E
368·31	p Fe	OHn	E
368·347	a Pb	OHin	E
368·4	Gd		E
368·4	MgF	PH	E
368·40	Er		E
368·412	Gd		A, E
368·42	Er		E
368·51	Ho		E
368·57	Dy		E
368·71	Tb		E
368·75	a Fe	OHn	E
368·78	Sm		E
368·79	Yb		E
368·84	p Ni	OHn	E
368·84	Eu		E

369·00		Sm		E	371·0		AgH		E	373·41		Tm II		E
369·02		SrO		E	371·02		Y		E	373·47		Yb		E
369·03	f	Pd	OHn	E	371·08	r	Te	AH	E	373·49	a	Fe	OHn	F, E
369·06		Ho		E	371·1	mv	MgCl	OA, C	E	373·49		Ho		E
369·07	p	Rh	OH	E	371·17		Tb		E	373·53	a	Rh	OH	E
369·14	r	SnO	AH	E	371·28		Ho		E	373·63		Ho		E
369·19		Ho		E	371·30		Nb		E	373·68	f	Ni	OHn	E
369·236		Rh		A, E	371·30	a	Rh	OH	E	373·71	p	Fe	OHn	E
369·26		Er		A, E	371·35		Gd		A, E	373·71		Sm		E
369·31	f	Co	OH	E	371·382		Nb		E	373·73		Rh		E
369·35		Tb		E	371·4		TeO		E	373·85		Ir		E
369·39	f	Ni	OHn	E	371·59		Er		E	373·92	f	Ni	OHn	E
369·41		Yb		E	371·74		Gd		A, E	373·92		Sm		E
369·46		Ho		E	371·79		Tm		A, E	373·93		Dy		E
369·48		Dy		E	371·8	r	Te	AH	E	373·97		Gd		E
369·48		Er		E	371·86		Ho		E	373·98		Nb		A, E
369·5		Mn	AH	E	371·89	a	Pd	OHn	E	373·99	f	Pb	OHin	E
369·55		Rh		E	371·9		MgOH	AH	E	374·0	b	Mn	AH	E
369·6		MgO(H)		E	371·93	u	Ru	OHn	E	374·0		AgH		E
369·72		Er		E	371·93		Er		E	374·00		Dy		E
369·739		Nb		E	371·99		Fe		A, E	374·10		Ti		E
369·76		Er		E	372·07		Ho		E	374·106		Ti		A
369·78		Nb		E	372·10		Sm		E	374·11		Dy		E
369·8		InBr	OH	E	372·12	r	Sn	AH	E	374·164		Ti		A
369·99		Pt		E	372·13		Nd		E	374·178		Nb		A, E
370·00		Ho		E	372·20		Sm		E	374·23		Nb		A, E
370·01		Tb		E	372·25	p	Ni	OHn	E	374·23	a	Ru	OHn	E
370·026		Tm		E	372·26	p	Fe	OHn	E	374·24		Ho		E
370·09		Rh	OH	A, E	372·49		Eu II		A, E	374·34		Fe		E
370·11		Tb		E	372·62		Nb		A, E	374·36		Tb		E
370·12		Ho		E	372·69	p	Ru	OHn	E	374·40		Tm		A, E
370·13		Tb		E	372·7		InBr	OH	E	374·42		Rh		E
370·13		Tm		E	372·723		Nb		A, E	374·48		Gd		E
370·2		MgOH	AH	E	372·76	a	Fe	OHn	E	374·50		Tb		E
370·3		MgO(H)		E	372·79		Dy		E	374·54		Sm		E
370·30		Tb		E	372·8		CaCl	OA, C	E	374·55		Co	OH	E
370·36		V		A, E	372·80		Ru	OHn	A, E	374·56	p	Fe	OHn	E
370·41	a	Co	OH	E	372·81		Sm		E	374·56		Ru		E
370·47		V		A, E	372·9		MgOH	AH	E	374·59		Fe		E
370·50		V		E	372·98		Ti		A, E	374·6	v	MnF	FH	E
370·50		Tb		E	373		HCO		E	374·60		Er		E
370·56	p	Fe	OHn	E	373·04	p	Ru	OHn	E	374·7		Fe		E
370·57		Er		E	373·07		Sm		E	374·72		Ir		E
370·65		Er		E	373·08		Cr		E	374·75		Er		E
370·69		Sm		E	373·1		MgO(H)		E	374·81		Ho		E
370·7		MgO(H)		E	373·12		Sm		E	374·82	a	Rh	OH	E
370·75		Dy		E	373·14		Ho		E	374·83	p	Fe	OHn	E
370·78	p	Fe	OHn	E	373·15		Tb		E	374·85		Sm		E
370·84		Tb		E	373·20		Cr		E	374·89		Er		E
370·85		SrO		E	373·20		Ho		E	374·90	a	Cr	OC	E
370·88	a	Co	OH	E	373·23		Gd		E	374·90	f	Ni	OHn	E
370·92	a	Fe	OHn	E	373·24	a	Co	OH	E	374·95	a	Fe	OHn	E
370·97	v	La	OH	E	373·26		Gd		E	374·99	a	Co	OH	E
370·97		Ho		E	373·33	p	Fe	OHn	E	375·1		MgO(H)		E

322

λ		Element		Class
375·18		Tm		E
375·23		SnO	AH	E
375·24		SrO		E
375·25		Os		E
375·28		Ti		A, E
375·32		CaO		E
375·36		Ti		E
375·56		Rh		E
375·60		Er		E
375·64		Sm		E
375·70		Dy		E
375·79		Gd		E
375·8		InBr	OH	E
375·82	a	Fe	OHn	E
375·93		Tb		E
375·955		Nb		A, E
375·98		Ru		E
376·00		Sm		E
376·00	a	Ru	OHn	E
376·01		Tb		E
376·04		Rh		E
376·064		Nb		A, E
376·11		Tb		E
376·13		Tm II		E
376·19		Tm II		A, E
376·19		Er		E
376·22		Gd		E
376·38	a	Fe	OHn	E
376·51	f	Rh	OH	E
376·51		Tb		E
376·7		MgOH	AH	E
376·7	r	Te(O)	AH	E
376·72	a	Fe	OHn	E
376·74		Ru		E
376·76		Dy		E
376·839		Gd		A
376·845		W		A
376·87		Er		E
376·90		Ho		E
376·94		Tb		E
377·00		Rh		E
377·01		Y		E
377·07		Ir		E
377·10		Dy		E
377·16		Ti		E
377·26		Dy		E
377·3		TeO		E
377·30		Dy		E
377·32		CaO		E
377·33		Sm		E
377·41		Ho		E
377·43		Y		E
377·47		Dy		E
377·47		Ho		E
377·48		Er		E
377·48		Ho		E
377·5	mr	CaCl	OA, C	E
377·52		Er		E
377·53		SrO		E
377·56	f	Ni	OHn	E
377·57		Tl	OHn	A, F, E
377·6	mv	MgCl	OA, C	E
377·61		Ho		E
377·63		Cu		E
377·7	r	Te	AH	E
377·76		Hf		A, E
377·76		Ru		E
377·81	f	Ni	OHn	E
377·81		Rh		E
377·86		V		E
377·86		Er		E
377·93	r	SnO	AH	E
378·1		AgH		E
378·11		Tm		E
378·14		Dy		E
378·22		Os		E
378·26		Sm		E
378·30		Gd		A, E
378·35	f	Ni	OHn	E
378·35		Tb		E
378·4		MgOH	AH	E
378·61	a	Ru	OHn	E
378·62	a	Dy		E
378·70		Nb		A, E
378·748		Nb		A, E
378·75		Nd		E
378·79	a	Fe	OHn	E
378·84		Dy		E
378·85	a	Rh	OH	E
378·87		Y		E
378·99		Tb		A, E
379·0		InBr	OH	E
379·01		Nb		A, E
379·03		V		E
379·05	p	Ru	OHn	E
379·1		MgOH	AH	E
379·11		Er		E
379·12		Nb		A, E
379·15		Ho		E
379·15		Er		E
379·22		Rh		E
379·29		Er		E
379·29		Ho		E
379·32	a	Rh	OH	E
379·36		V		E
379·36	p	Ni	OHn	E
379·36		Nd		E
379·37		CaO		E
379·47	a	Li	OH	E
379·49		V		E
379·50	a	Fe	OHn	E
379·52		Tm		E
379·576		Tm		E
379·62		Tb		E
379·637		Gd		A
379·67		Ho		E
379·72		Ho		E
379·77		Dy		E
379·81		Nb		A, E
379·81	a	Ru	OHn	E
379·82		Er		E
379·83		Mo	OAn	A, E
379·85	a	Fe	OHn	E
379·85		Tm		E
379·86		Er		E
379·89	p	Ru	OHn	E
379·92	a	Pd	OHn	E
379·93	a	Rh	OH	E
379·93	p	Ru	OHn	E
379·95	a	Fe	OHn	E
379·99		V		E
380·0	b	Mn	AH	E
380·01		Ir		A, E
380·1		MgO(H)		E
380·10	f	Sn	AHn	F, E
380·27	v	Sn	AH	E
380·29		Nb		A, E
380·3		SnO		E
380·34		V		E
380·39		Sm		E
380·41		Ho		E
380·59		Rh		E
380·67	a	Mn	OA	E
380·68	a	Rh	OH	E
380·7		MgO	AH	E
380·71	f	Ni	OHn	E
380·75		V		E
380·77		Tm		E
380·81	a	Co	OH	E
380·85		V		E
380·96		V		E
380·98		Dy		E
380·98		Sm		E
381·0		MgO(H)		E
381·03		Er		A, E
381·07		Ho		A, E
381·18		Ho		E
381·22		Dy		E
381·27		Ru		E
381·30	f	Fe	OHn	E
381·34		V		E
381·38		Sm		E

381·4		MgO(H)	AH	E
381·44		Ra II		E
381·47		CaO		E
381·5		MgO(H)		E
381·54		Er		E
381·58	a	Fe	OHn	E
381·62		Dy		E
381·63	a	Co	OH	E
381·65		Rh		E
381·7	r	SnO	AH	E
381·74		Dy		E
381·78		V		E
381·82	a	Rh	OH	E
381·82		V		E
381·83		Sm		E
381·87		Pt		E
381·90		Ru		E
381·94		Dy		E
381·96		Eu		E
381·97	r	Te	AH	E
381·99		V		E
382·00	b	Te(O)		E
382·04	a	Fe	OHn	E
382·14		Dy		E
382·14		V		E
382·17		Ho		E
382·18		Dy		E
382·2		MgO(H)		E
382·20		VO		E
382·21		Ru		E
382·23	a	Rh	OH	E
382·24		SrO		E
382·28		V		E
382·29		Sm		E
382·3		MgO(H)	AH	E
382·32		V		E
382·35	a	Mn	OA	E
382·39		Mn		E
382·44		Fe	OHn	A, E
382·488		Nb		E
382·49		Ru		E
382·56		Ho		E
382·56		Dy		E
382·588	a	Fe	OHn	E
382·63		Tm		E
382·68		Er		E
382·7		TeO		E
382·78		Fe		E
382·8	mr	CaCl	OA, C	E
382·85	a	Rh	OH	E
382·85		V		A, E
382·92		Ho		E
382·94	a	Mg	AAi	E
383·00		Er		E
383·02		Tb		E
383·09		Nd		E
383·17	f	Ni	OHn	E
383·18		Ru		E
383·23	a	Pd	OHn	E
383·23	a	Mg	AAi	E
383·28		Sm		E
383·3		AgH		E
383·32		SnO	AH	E
383·34		Tb		E
383·39	a	Rh	OH	E
383·4	b	Mg	AH	E
383·40		Tb		E
383·42	a	Fe	OHn	E
383·44	a	Mn	OA	E
383·44		Sm		E
383·5	r	Te	AH	E
383·59		Zr		E
383·62		CaO		E
383·83	a	Mg	AAi	E
383·97		Ru		E
384·00	b	Mn	AH	E
384·04		Dy		E
384·04	a	Fe	OHn	E
384·04		V		A
384·07		V		A, E
384·08		Tm		E
384·09		Dy		E
384·11	a	Fe	OHn	E
384·11		Lu		E
384·20	a	Co	OH	E
384·32		Gd		E
384·43		Dy		E
384·44		V		E
384·5	mv	MgCl	OA, C	E
384·55	a	Co	OH	E
384·56		SrO		E
384·6		MgOH	AH	E
384·62		Sm		E
384·67		Sm		E
384·69		Dy		E
384·73		V		E
384·8		MgO(H)		E
384·80		Tm II		E
384·87		Sm		E
384·98		Ho		E
384·99		Er		E
385·00	a	Fe	OHn	E
385·08		Fe		E
385·09	v	CN	OHin	E
385·10	a	Co	OH	E
385·24		Ho		E
385·3		CN		E
385·32		Sm		E
385·41		CaO		E
385·45		Sm		E
385·47	v	CN	OHin	E
385·5		MgO(H)		E
385·53		V		A, E
385·58		V		A
385·59		Er		E
385·64	p	Fe	OHn	E
385·65	a	Rh	OH	E
385·76		Ru		E
385·76		Ho		E
385·83		CaO		E
385·83	f	Ni	OHn	E
385·84		Dy		E
385·86		Sm		E
385·9		MgO(H)		E
385·93		Ho		E
385·99		Fe	OHn	A, E
386·01		Sm		E
386·12	a	Co	OH	E
386·19	v	CN	OHin	E
386·26		Ho		E
386·3		SnO		E
386·30		Ho		E
386·308		Er		A, E
386·38		Zr		A, E
386·41		Mo	OAn	A, E
386·48		V		E
386·49	r	SnO	AH	E
386·55	a	Fe	OHn	E
386·69		Gd		E
386·76		V		E
386·78		Ru		E
386·79		W		E
386·88		Dy		E
387·1		CH		E
387·14	v	CN	OHin	E
387·2	r	CH	OHnq	E
387·21		Dy		E
387·25	a	Fe	OHn	E
387·31	a	Co	OH	E
387·35		Co		E
387·40		Dy		E
387·40	a	Co	OH	E
387·47		Er		E
387·48		Ho		E
387·50		V		E
387·639		Cs		E
387·68	a	Co	OH	E
387·7		MgO(H)	AH	E
387·73	a	Rh	OH	E
387·74		Sm		E
387·76	r	SnO	AH	E

λ		species	notation	class
387·8	r	Pb	OH	E
387·80		Fe		E
387·85		Pu		A
387·863	p	Fe	OHn	E
387·9	r	Te	AH	E
388·0		YbO	OH	E
388·17		Sm		E
388·19	a	Co	OH	E
388·2	mr	CaCl	OA, C	E
388·2		MgO(H)		E
388·3		CN		E
388·31		Tm		A, E
388·34	v	CN	OHin	E
388·34		Tm II		A, E
388·4		TeO		E
388·42	r	SnO	AH	E
388·51		Tb		E
388·52		Cr		E
388·54		Zr		E
388·628	p	Fe	OHn	E
388·68	a	Cr	OAn	E
388·73		Tm		E
388·745		Tm		A
388·78		Nd		E
388·82		Tb		E
388·86	p	Cs	OH	E
388·9	r	CH	OHnq	E
388·93	u	Ba	OH	E
389·0	b	Mn	AH	E
389·01		V		E
389·03		Zr		A, E
389·036		U		A
389·04		Ho		E
389·07	r	Te	AH	E
389·10		Ho		A, E
389·13		Zr		E
389·19		Dy		E
389·26		Er		A, E
389·26		Ho		E
389·28		V		E
389·28		Dy		E
389·40	r	SrCl	OG	E
389·40		Cr		E
389·41	a	Co	OH	E
389·42	f	Pd	OHn	E
389·47		CaO		E
389·50	a	Co	OH	E
389·57	p	Fe	OHn	E
389·62		Er		E
389·66		Tm		E
389·71		SrO		E
389·85		Dy		E
389·86		La		A
389·91		Dy		E
389·93	r	SnO	AH	E
389·971	p	Fe	OHn	E
390·13		Tb		A, E
390·22		V		A, E
390·25		Ir		E
390·27		Ho		E
390·27		Er		E
390·29	a	Cr	OC	E
390·29		Fe		E
390·30		Mo	OAn	A, E
390·39		Er		E
390·39		Ho		E
390·44		Ho		E
390·5		AgH		E
390·54		Er		A, E
390·54		Ho		E
390·59		Dy		E
390·63		Er		E
390·63	a	Co	OH	E
390·65	p	Fe	OHn	E
390·71		Eu II		E
390·75		Sc		E
390·79		Tb		E
390·84		Pr		E
390·88	a	Cr	OC	E
390·95		Ho		E
390·98		V		E
390·99	f	Co	OH	E
391·0	r	Pb	OH	E
391·18		Sc		A, E
391·18		Ho		E
391·2		MgO(H)	AH	E
391·28	a	Rh	OH	E
391·30	f	Ni	OHn	E
391·35		Rh		E
391·36		Dy		E
391·4		MgO(H)		E
391·50	a	Li	OH	E
391·54		Tb		E
391·54		Ir		E
391·57		CaO		E
391·64		Tm		A, E
391·72	a	Fe	OA	E
391·73		Dy		E
391·83	r	SrCl	OG	E
391·92	a	Cr	OC	E
391·94		Ho		E
391·95	r	Sn	AH	E
391·96		SrO		E
392·0		SnO		E
392·03	p	Fe	OHn	E
392·22	a	Rh	OH	E
392·28	f	Co	OH	E
392·29	p	Fe	OHn	E
392·45		Ti		E
392·52		Sm		E
392·59	f	Ru	OHn	E
392·75		La		A, E
392·79	p	Fe	OHn	E
392·86		Cr		E
392·98		Ti		E
393·0	b	Mn	AH	E
393·01		Dy		E
393·030	p	Fe	OH	E
393·04		Eu		E
393·18		Ru		E
393·23		Tb		E
393·33		Sc		A, E
393·36		Ru		E
393·367		Ca		A, E
393·4	b	Te(O)		E
393·42	f	Rh	OH	E
393·47		Gd		E
393·48		Ir		E
393·59		Nd		E
393·60		Co	OH	E
393·67		Dy		E
393·70		Er		A, E
393·71		Dy		E
393·71	r	SrCl	OG	E
393·76		Tb		E
393·80		Dy		E
393·82		Dy		E
393·88		Ho		E
394·0		S_2	OH	E
394·06	r	Te	AH	E
394·09		Co		E
394·17	a	Co	OH	E
394·2		MgO(H)		E
394·25		Ho		E
394·27	a	Rh	OH	E
394·275		Ce		E
394·29		Tb		E
394·36		Tb		E
394·38		U		A, E
394·40		Al	OAn	A, E
394·44		Er		A, E
394·47		Dy		A, E
394·53	a	Co	OH	E
394·55		Gd		A, E
394·76	r	Sn	AH	E
394·77		Ti		A, E
394·80		Er		E
394·86		Ti		A, E
394·91		La		E
394·92		Tm		A, E
395·01		Tb		E
395·02		Ru		E

395·04		Ru		E	397·47	a	Co	OH	E	400·58		Dy		E
395·04		Tb		E	397·5	b	Mn	AH	E	400·60		Dy		E
395·05		Ho		E	397·53		Rh		E	400·75		Te		E
395·10	r	SnO	AH	E	397·58		Ho		E	400·76		La		E
395·18		Sm		E	397·67		Er		E	400·797		Er		A, E
395·18		Tb		E	397·69		Ho		E	400·875		W		A, E
395·2	r	Te(O)	AH	E	397·70		Er		E	400·89		Ti		E
395·23		Co		E	397·82		Sm		E	400·94		SrCl	OG	E
395·29	a	Co		E	397·87		Co		E	401·0	r	TeO	AH	E
395·3		MgF	OH	E	397·87	r	SnO	AH	E	401·00		Tb		E
395·35		Ti		A	397·94		Ru		E	401·08		Tb		E
395·4		TeO		E	397·95	f	Co	OH	E	401·10	f	Co	OH	E
395·5		Mn	AH	E	398·13		Dy		E	401·225		Nd		E
395·5	r	Pb	OH	E	398·17		Ti		A, E	401·239		Co		E
395·57		Ho		A, E	398·23		Er		E	401·25		Er		E
395·62		Dy		E	398·24		Ti		A, E	401·38		Dy		E
395·63		Ti		A, E	398·34	r	SrCl	OG	E	401·53		La		E
395·64		Er		E	398·39	a	Cr	OC	E	401·81	a	Mn	OA	E
395·73		Tb		E	398·39	r	Sn	AH	E	401·9	r	SnO	AH	E
395·79	a	Co	OH	E	398·44		Rh		E	401·93		Co		E
395·81		Tm II		E	398·49		Ru		E	402·0	b	Mn	AH	E
395·82		Ti		A, E	398·71	a	Co	OH	E	402·04		Sc		A, E
395·86	f	Pd	OHn	E	398·76		Er		E	402·05		Er		E
395·89	a	Rh	OH	E	398·80	r	Pb	OH	E	402·09	a	Co	OH	E
395·94		CaO		E	398·80		Yb	OHn	A, E	402·15		Er		E
395·96		Ho		E	398·97		Ti		A, E	402·19		Er		E
396·153		Al		A, F, E	399·0		AgH		E	402·22		Ru		E
396·16	r	SrCl	OG	E	399·00		Sm		E	402·33		Gd		E
396·21		Sm		E	399·10		Sm		E	402·36		Sc		E
396·26		Dy		E	399·17	a	Co	OH	E	402·37		Dy		E
396·26		Tb		E	399·3	r	InJ	OH	E	402·37		Tb		E
396·28		Ti		E	399·35		Dy		E	402·45		Ti		E
396·42		Ti		E	399·419		Yb		E	402·47		Tb		E
396·45		Er		E	399·45		Dy		E	402·49		Dy		E
396·63		Er		E	399·53	f	Co	OH	E	402·6	r	SnO	AH	E
396·64		Pt		E	399·56		Rh		E	402·70	f	Co	OH	E
396·75		Dy		E	399·61		Rh		E	402·72		Ho		E
396·83		Dy		E	399·66		Tm		A	402·84		Dy		E
396·84		Dy		A, E	399·66		Sc		A, E	403·08		Mn	OA	A, E
396·84		Lu		E	399·67		Dy		E	403·17		Ho		E
396·84		Y		E	399·79	f	Co	OH	E	403·2		Mn		E
396·847		Ca II		A, E	399·82		Ho		E	403·23		Tb		E
396·90		Gd		E	399·83		Sm		E	403·30		Ga	OHn	A, F, E
396·9		MgO(H)		E	399·86		Ti		A, E	403·31	p	Mn	OA	E
396·93	a	Fe	OHn	E	399·89		Dy		E	403·45	p	Mn	OA	E
397·19		Eu		E	399·95		Ho		E	403·61		Er		E
397·22	f	Ni	OHn	E	400·00		Dy		E	403·72		La		A, E
397·26		Er		A	400·048		Dy		A	403·76		Er		E
397·28		AuH		E	400·3	r	Te	AH	E	403·76		Ho		E
397·30		Er		A, E	400·33		Ho		E	403·8	r	Pb	OH	E
397·36	f	Ni	OHn	E	400·40		Er		E	403·88		Dy		E
397·36		Er		E	400·5	r	CuH	OH	E	403·88		Tb		E
397·38		Dy		E	400·52	a	Fe	OHn	E	403·9		AgH		E
397·46		Sm		E	400·54		CuH		E	403·94		Tb		E

λ		Element		Code
403·98		Y		E
404·08		Ho		A, E
404·14	a	Mn	OA	E
404·27		U		A, E
404·414		K		A, E
404·44		Tm		E
404·5	b	Mn	AH	E
404·50		Gd		A, E
404·54	f	Co	OH	E
404·54		Ho		E
404·56		W		E
404·58	a	Fe	OHn	E
404·59		Dy		A, E
404·69		Er		E
404·71		Tb		E
404·721	p	K	OHn	E
404·76		Y		E
404·78		Sc		E
404·88		Mn	OA	E
404·89		Dy		E
404·93		Dy		E
405·005		U		A
405·19		$^{11}BO_2$		E
405·36		Gd		A, E
405·39		Ho		A, E
405·41		Tb		E
405·44		Lu		E
405·45		Sc		A, E
405·45		Sm		E
405·47		Gd		E
405·50		Dy		E
405·55	a	Mn	OA	E
405·72	f	Co	OH	E
405·72		$^{11}BO_2$		E
405·78	a	Pb	OHin	A, F, E
405·8		Co		E
405·82		Gd		A, E
405·89	a	Mn	OA	E
405·894		Nb		A, E
405·95		Er		E
406·0	r	Te(O)	AH	E
406·03		Tb		E
406·03		La		E
406·15		Tb		A, E
406·19		CuH		E
406·23		Sm		E
406·255		U		A
406·35	a	Mn	OA	E
406·36	a	Fe	OHn	E
406·41		Ru		E
406·45		Ru		E
406·47		La		E
406·55		$^{11}BO_2$		E
406·64	f	Co	OH	E
406·80		Ho		E
406·82		Au		E
407·0	b	Mn	AH	E
407·01		Tb		E
407·12		Tb		E
407·17	a	Fe	OHn	E
407·18		Ho		E
407·43		W		A, E
407·5	r	Te(O)	AH	E
407·61	a	Co	OH	E
407·67		Ru		E
407·737		Y		A, E
407·77		Sr II	OHn	A, E
407·79		Dy		A, E
407·79		Er		E
407·87		Gd		A, E
407·9		SnO	AH	E
407·91		La		A, E
407·92	a	Mn	OA	E
407·92		Dy		E
407·9241		Mn		E
407·9422	a	Mn	OA	E
407·973		Nb		A, E
407·98		Sm		E
408·0	b	B	OHn	E
408·06	f	Ru	OHn	E
408·12		Tb		E
408·15		$^{11}BO_2$		E
408·24		Sc		A, E
408·26	a	Co	OH	E
408·28	f	Rh	OH	E
408·2944	a	Mn	OA	E
408·3628	a	Mn	OA	E
408·37		Y		E
408·43		Rh		E
408·51		Dy		E
408·63	a	Co	OH	E
408·66		Tb		E
408·672		La		A
408·73		Pd		E
408·73		Ho		E
408·75		Ho		E
408·76		Er		A, E
408·96		La		E
409·04		Gd		E
409·05		V		E
409·06		$^{11}BO_2$		E
409·22		Tb		E
409·24	a	Co	OH	E
409·26		V		E
409·27		Gd		E
409·29		Er		E
409·4	r	Lu	OHn	E
409·40		Tb		E
409·418	a	Tm	OAn	A, E
409·5		LuO		E
409·54		V		E
409·6	r	Lu	OHn	E
409·78		Ru		E
409·81		Er		E
409·9		InJ	OH	E
409·98		Dy		E
409·98		V		E
409·99		Sm		E
410·02		Gd		E
410·04		Nb		A, E
410·075		Pr		E
410·09		Nb		A, E
410·109		Ho		A, E
410·17		Ru		E
410·18	p	In	OHn	E
410·21		V		E
410·23		Y		A, E
410·34		Tb		E
410·38		Ho		A, E
410·38		Dy		E
410·476		In		A, F, E
410·48		La		E
410·50		Dy		E
410·51		V		E
410·53		Tb		A, E
410·58	a	Tm	OAn	A, E
410·59		Dy		E
410·65		Ho		E
410·7	r	Lu	OHn	E
410·73		Ho		E
410·8		AgH		E
410·86		Ho		A, E
410·89	r	SnO	AH	E
410·98		V		E
411·0		Lu		E
411·05	a	Co	OH	E
411·18		V		A, E
411·19		Ho		E
411·22		SnO		E
411·25		Tb		E
411·27	a	Ru	OHn	E
411·27		Th		E
411·28		Tb		E
411·51		V		A, E
411·61		U		A
411·64		V		E
411·87		Pt		E
411·88	a	Co	OH	E
411·99		Tb		E
412·0	r	CuCl	AH, Cu	E
412·02		Ho		E
412·13	a	Co	OH	E

412·15		Bi		F, E
412·16		SnO		E
412·17	a	Rh	OH	E
412·2	r	SnO	AH	E
412·32		Er		E
412·35		V		E
412·38		Nb		A, E
412·47		Lu		E
412·48		Er		E
412·5	b	Te	AH	E
412·52		Sm		E
412·56		Ho		E
412·61		Dy		E
412·64	r	InO	OH	E
412·7		LuO		E
412·716		Ho		A, E
412·80		V		E
412·83		Y		A, E
412·89	a	Rh	OH	E
412·91		Dy		E
412·96		Eu	OHn	A, E
413·01		Tb		E
413·04		Dy		E
413·066		Ba		E
413·1		TeO		E
413·15		Er		E
413·20		V		E
413·21	a	Fe	OHn	E
413·22	f	Co	OH	E
413·22	a	Li	OH	E
413·24		Tb		E
413·30		Sc		E
413·38		Dy		E
413·41		Gd		E
413·41		Dy		E
413·44		V		E
413·45		Ho		E
413·47		Dy		E
413·48	r	InO	OH	E
413·53	f	Rh	OH	E
413·53		Tb		E
413·55		Sm		E
413·58		Os		E
413·6	r	Te	AH	E
413·62		Ho		A, E
413·70		La		E
413·71		Nb		A, E
413·759		Nb		A, E
413·83		Tm		E
413·85		Dy		E
413·90		Tb		E
413·944		Nb		E
413·95		Dy		E
413·97		Nb		E

413·97		Tb		E
414·03		Sc		E
414·15		Dy		E
414·15		Tb		E
414·25	r	InO(H)	OH	E
414·28		Y		A, E
414·33		Tb		E
414·35		Tb		E
414·39	a	Fe	OHn	E
414·4	r	SnO	AH	E
414·42		Ru		E
414·5		LuO		E
414·52		Sm		E
414·6	r	Pb	OH	E
414·60		Dy		E
414·66		Dy		E
414·69		Tb		E
415·0		YbO(H)	OH	E
415·06		Gd		E
415·11		Er		A, E
415·12		Sm		E
415·2	v	CN	OHin	E
415·23		Sc		E
415·25		Nb		A, E
415·39		U		A, E
415·40		Lu		E
415·49	r	InO(H)	OH	E
415·77		Gd		E
415·78		Dy		E
415·8	v	CN	OHin	E
415·82		Tb		E
415·85		Tb		E
415·86		Tm		E
415·93		Dy		E
416·0	b	Pr	OHn	E
416·02		La		E
416·29		$^{11}BO_2$		E
416·30		Pr		E
416·303		Ho		A, E
416·36		Nb		A, E
416·40		Ho		E
416·46		Pt		E
416·46		Nb		A, E
416·51		Sc		E
416·53	r	InO(H)	OH	E
416·7	r	Sr	OG	E
416·72		Gd		E
416·75		Ru		E
416·75		Y		E
416·79		Dy		A, E
416·8	v	CN	OHin	E
416·81		Nb		A, E
416·89		$^{11}BO_2$		E
416·91		Tb		E

416·93		Tb		E
416·98		Pd		E
417·19		Dy		E
417·21		Ga	OHn	A, F, E
417·26		Tb		E
417·28		Tb		E
417·34		Ho		E
417·4	r	SnO	AH	E
417·41		Y		E
417·55		Gd		E
417·94		V		E
418·0	b	B	OHn	E
418·1	v	CN	OHin	E
418·25		V		E
418·33		Sm		E
418·36		Dy		E
418·5	r	Te(O)	AH	E
418·57		Er		E
418·66		Ce		E
418·68	a	Dy	OH	A, E
418·71		Tb		E
418·73		La		A, E
418·76	a	Tm	OAn	A, E
418·8	r	CuCl	AH	E
418·80		Tb		E
418·98		V		E
419·0		AgH		E
419·0	w	YbO	OH	E
419·07		Er		E
419·07		Gd		A, E
419·07	f	Co	OH	E
419·07		Er		A, E
419·09		Dy		E
419·16		Gd		E
419·16		Dy		E
419·43		Ho		E
419·48		Dy		A, E
419·65	f	Rh	OH	E
419·67		Tb		E
419·7	v	CN	OHin	E
419·76	a	Ru	OHn	E
419·79		Dy		E
419·84	f	Co	OH	E
419·89	f	Ru	OHn	E
419·99	a	Ru	OHn	E
420·13		Dy		E
420·19		Rb	OHn	A, E
420·20	a	Fe	OHn	E
420·22		Dy		E
420·25		Dy		E
420·37	a	Tm	OAn	A, E
420·37		Tb		E
420·5		Te		E
420·5	r	Te(O)	AH	E

420·50		Dy		E
420·505		Eu		A, E
420·51		CaO		E
420·57		Sm		E
420·60	a	Ru	OHn	E
420·64		Pb		E
420·64		Tb		E
420·647		Pu		A
420·76		Dy		E
420·76	a	Co	OH	E
420·8		Pu		A
420·98		V		E
421·1	r	CuCl	AH, Cu	E
421·11	f	Rh	OH	E
421·172	a	Dy	OH	A, E
421·21		Ru		E
421·30	f	Pd	OHn	E
421·31		Dy		E
421·35		Tb		E
421·43		Dy		E
421·51		Tb		E
421·51		Dy		E
421·55	s	Sr II	OHn	A, E
421·56	p	Rb	OHn	E
421·6	v	CN	OHin	E
421·62	f	Fe	OHn	E
421·75		Tb		E
421·77		SnO		E
421·80		Dy		E
421·84		Er		E
421·86		Sm		E
421·91		Tb		E
421·93		Sm		E
422·11		Dy		E
422·19		CaO		E
422·22		Ho		E
422·22		Dy		E
422·26		Tm		E
422·34		Ho		E
422·41	r	InO	OH	E
422·42		Tb		E
422·51		Dy		E
422·58		Gd		E
422·61		Sm		E
422·67		Cd		F
422·673		Ca		A, E
422·70		Ho		A, E
422·9	r	Pb	OH	E
423·03		Ru		E
423·07		Sm		E
423·18		Tb		E
423·20		Dy		E
423·28		Tb		E
423·32	r	InO	OH	E
423·40	f	Co	OH	E
423·51	a	Mn	OA	E
423·53	a	Mn	OA	E
423·53		Tb		E
423·7	y	CH	OHnq	E
423·59		Y		E
423·9	r	Lu	OHn	E
423·98		Dy		E
424·0	r	SnO	AH	E
424·04		Sm		E
424·08		CaO		E
424·2	r	Lu	OHn	E
424·22		Tm II		E
424·3	y	CH	OHnq	E
424·38	r	InO	OH	E
424·42		Sm		E
424·59		Dy		E
424·9	y	CH	OHnq	E
425·0	r	Lu	OHn	E
425·08	a	Fe	OC	E
425·2	f	Co	OH	E
425·3	r	Lu	OHn	E
425·43		Cr	OHn	A, E
425·44		Ho		A, E
425·52		Tb		E
425·6	y	CN	OHnq	E
425·64		Nd		E
425·76	r	InO	OH	E
425·81		Dy		E
425·82		Tb		E
425·85		Dy		E
425·9	r	CuCl	AH, Cu	E
426·01		Gd		E
426·05	a	Fe	OC	E
426·085		Os		A
426·18		CaO		E
426·2	y	CH	OHnq	E
426·2	r	SnO	AH	E
426·36		Tb		E
426·40		Ho		E
426·60		Gd		E
426·60		Ho		E
426·63		Tb		E
426·63		Sm		E
426·70		Gd		E
426·8	y	CH	OHnq	E
426·8	r	Te(O)	AH	E
426·80	f	Co	OH	E
426·83		Dy		E
426·939		W		A
427·03		InO	OH	E
427·17		Tm		E
427·18	a	Fe	OHn	E
427·18		Sm		E
427·3		AgH		E
427·33	a	Li	OH	E
427·4	y	CH	OHnq	E
427·41		Gd		E
427·48		Cr	OAn	A, E
427·67		Dy		E
428·0	y	CH	OHnq	E
428·0	r	CuH	OH	E
428·02		La		E
428·1	r	CuCl	AH	E
428·1	r	Sr	OG	E
428·22		Sm		E
428·26	dr	In	OH	E
428·28		Sm		E
428·3		InO		E
428·30		Ca	OH	E
428·35		Sm		E
428·46		CaO		E
428·58	f	Co	OH	E
428·6	y	CH	OHnq	E
428·65		Er		E
428·87	a	Rh	OH	E
428·94		Ca		E
428·97		Cr	OAn	A, E
429·19		Dy		E
429·2	y	CH	OHnq	E
429·21	f	Fe	OHn	E
429·41		Fe		E
429·46		W		A, E
429·58		Cr	OC	E
429·67		Sm		A, E
429·7	y	CH	OHnq	E
429·70	a	Cr	OC	E
429·77		Cr		E
429·77		Ru		E
429·8		CrO		E
429·8		Dy		E
429·8		Sm		E
429·83		Tb		E
429·89		Er		E
429·90		Ca		E
429·91		Sm		E
430		Dy		E
430·05		Ti		E
430·10		Ti		E
430·12		Sm		E
430·15	r	InO	OH	E
430·21		W		A, E
430·25	a	Ca	OH	E
430·26		$^{11}BO_2$		E
430·3	r	SnO	AH	E
430·3	r	Sr	OG	E
430·32	f	Co	OH	E
430·51		$^{11}BO_2$		E

λ		Species		
430·54		Sr II	OHn	E
430·59		Ti		E
430·62		V		E
430·63		Gd		E
430·72		Tb		E
430·75	u	Cr	OC	E
430·77		Ca		E
430·79	a	Fe	OHn	E
430·95		CaO		E
430·98		V		E
431·0		Dy		E
431·0		Er		E
431·04		Tb		E
431·28		Sm		E
431·38		Gd		E
431·38		Sm		E
431·5	v	CH	OHnq	E
431·7	r	Pb	OH	E
431·8		Tb		E
431·8	r	Te(O)	AH	E
431·87	a	Ca	OH	E
431·88		Tb		A, E
431·95		Sm		E
432·0		Dy		E
432·02		Tb		E
432·06	a	Cr	OC	E
432·1		CrO		E
432·11		Gd		E
432·12		Cr		E
432·16		Cr		E
432·4	x	CH	OHnq	E
432·44		Sm		E
432·51		Sm		E
432·557		Gd		E
432·58	a	Fe	OHn	E
432·61		Sm		E
432·648		Tb		A, E
432·71		Gd		E
432·8	r	CuH	OH	E
432·8		AgH		E
433·0	p	V	OC	E
433·00		Sm		E
433·13		Er		E
433·14		Sm		E
433·21		Tb		E
433·28	p	V	OC	E
433·3	r	CuCl	AH	E
433·4	y	CH	OHnq	E
433·41		Sm		E
433·50		$^{11}BO_2$		E
433·54		$^{11}BO_2$		E
433·61		Sm		E
433·65		Tb		E
433·76	a	Cr	OAn	E
433·84		Tb		A, E
433·9	y	CH	OHnq	E
433·93		Sm		E
433·94	a	Cr	OAn	E
433·94		AuH		E
433·97		Cr		E
433·99		Sm		E
434·00		$^{11}BO_2$		E
434·05		$^{11}BO_2$		E
434·06		Tb		E
434·10	p	V	OC	E
434·169		U		A
434·25		Tb		E
434·3	y	CH	OHnq	E
434·3	r	SnO	AH	E
434·3	r	Te(O)	AH	E
434·35		Nd		E
434·45	a	Cr	OAn	E
434·6	r	Sr	OG	E
434·646		Gd		A, E
434·66		NbO		E
434·663		Gd		A, E
434·772		Dy		E
434·8	y	CH	OHnq	E
434·80		ScO		E
434·81		$^{11}BO_2$		E
434·83		Er		E
434·90		$^{11}BO_2$		E
435·07		Ho		E
435·08		Sm		E
435·11		Cr		E
435·18	a	Cr	OAn	E
435·2	y	CH	OHnq	E
435·26		AlO		E
435·29	p	V	OC	E
435·4	r	CuCl	AH	E
435·5		Ho		E
435·58		Sm		E
435·59		V		E
435·6	y	CuH	OH	E
435·6	r	IO	OH, IO$_3$	E
435·6		LaO	OHn	E
435·68		Tb		E
435·78		Sm		E
435·84	a	Hg	OH	E
435·96	a	Cr	OC	E
435·99		Tm	OAn	A, E
436·0		Dy		E
436·12		Ru		E
436·29		Sm		E
436·4	y	CuH	OH	E
436·41		Nd		E
436·5	v	C$_2$	OHnq	E
436·5	r	SnO	AH	E
436·59		Sm		E
436·67		Dy		E
436·80		V		E
437·0	b	B	OHn	E
437·1	v	C$_2$	OHnq	E
437·13	a	Cr	OC	E
437·2	y	CuH	OH	E
437·2	r	LaO	OHn	E
437·20		Gd		E
437·22		Ru		E
437·38		Gd		E
437·4	r	AlO	OAn	E
437·48		NbO		E
437·48	f	Rh	OH	E
437·59	f	Fe	OHn	E
437·6	r	La(O)	OHn	E
437·9	r	Sn(O)	AH	E
437·92	a	V	OC	A, E
438·0		CuH	OH	E
438·0		Dy		E
438·0		La(O)		E
438·04		Sm		E
438·13		ScO		E
438·2		La(O)	OHn	E
438·2	v	C$_2$	OHnq	E
438·21		Er		E
438·3	r	Te(O)	AH	E
438·35	f	Fe	OHn	E
438·4	r	La	OHn	E
438·47		V	OC	A, E
438·50		Cr		E
438·54		Ru		E
438·56		Ru		E
438·62		Sm		E
438·64		Er		E
438·64		Tm	OAn	A, E
438·76		LaO		E
438·82		Gd		E
438·9	y	CuH	OH	E
439·00	a	V	OC	A, E
439·01	a	Na	OHn	E
439·04		Ru		E
439·09		Dy		E
439·10		Ru		E
439·2	r	La(O)	OHn	E
439·33		Sm		E
439·34	a	Na	OHn	E
439·38	r	AlO	OAn	E
439·4		BiO	AH	E
439·44		Tm		E
439·52	a	V	OC	E
439·6	r	La(O)	OHn	E
439·7		AgH		E
439·73		Sm		E

439·9	y	CuH	OH	E	442·43		Sm		E	445·2	y	CuH	OH	E
440·0	r	La(O)	OHn	E	442·4		BiO	AH	E	445·2		SnO	AH	E
440·0	r	Sr	OG	E	442·45		Er		E	445·2		Cu		E
440·06	a	V	OC	E	442·54	a	Ca	OH	A, E	445·29		Sm		E
440·11		Sm		E	442·58		CaO		E	445·3	r	LaO	OHn	E
440·18		Gd		E	442·60		V		E	445·4	r	Pb	OH	E
440·31		Gd		E	442·67		Er		A, E	445·48		Ca	OH	A, E
440·31		Sm		E	442·7	r	BeO	OH	E	445·5	r	Lu	OHn	E
440·32		Ho		E	442·73	f	Fe	OHn	E	445·59	a	Ca	OH	E
440·39		CaO		E	442·8	r	LaO	OHn	E	445·61		Nd		E
440·48	f	Fe	OHn	E	442·85		V		E	445·66		Ca		E
440·5	r	La(O)	OHn	E	442·9	r	SnO	AH	E	445·67		Sm		E
440·5		RaO	OHn	E	442·96		Sm		E	445·74		V		E
440·5		Ho		E	442·98		V		E	445·8	r	La	OHn	E
440·66	a	V	OC	E	443·0	y	CuH	OH	E	445·92		Sm		E
440·76		V		E	443·01		Gd		E	445·97		V		E
440·8	r	Te(O)	AH	E	443·06		Gd		E	446·0	r	Te(O)	AH	E
440·82		V		E	443·29		NbO		E	446·0		Dy		E
440·85	a	V	OC	E	443·3	r	LaO	OHn	E	446·00		Ru		E
440·88		Pr		E	443·30		Sm		E	446·03	a	V	OC	E
440·9	y	CuH	OH	E	443·38		Sm		E	446·17	f	Fe	OHn	E
440·93		Er		A, E	443·4	r	CuCl	AH, Cu	E	446·20	a	Mn	OA	E
441·0	r	La(O)	OHn	E	443·40		Gd		E	446·3	r	Gd(O)	OHn	E
441·0	r	Pb	OH	E	443·43		Sm		E	446·38		Sm		E
441·00		Ru		E	443·50	a	Ca	OH	A, E	446·4	y	CuH	OH	E
441·1	r	SnO	AH	E	443·553		Eu II		E	446·4	r	La	OHn	E
441·15		Sm		E	443·57	a	Ca	OH	E	446·73		Sm		E
441·16		Gd		E	443·61		V		E	447·0	v	C_2	OHnq	E
441·2	r	CuCl	AH, Cu	E	443·66		AuH		E	447·0		Ho		E
441·2		Dy		E	443·78		V		E	447·08		Sm		E
441·4		NdO	OHn	E	443·8	r	LaO	OHn	E	447·1	r	AlO	OAn	E
441·46		LaO		E	443·98		Ru		E	447·1		TbO	OH	E
441·47		Gd		E	444·0		Dy		E	447·13		LaO		E
441·51	f	Fe	OHn	E	444·1	y	CuH	OH	E	447·3	r	Tb	OHn	E
441·53		ScO		E	444·17	a	V	OC	E	447·36		Pd		E
441·64		V		E	444·18	a	Sm	AA	E	447·5	r	BeO	OH	E
441·79		Gd		E	444·2		YbO	OH	E	447·6		B		E
441·8	r	LaO	OHn	E	444·2271		Sm	AA	E	447·6		YbO	OH	E
441·87		Er		E	444·23		Fe	OHn	E	447·61		Gd		E
441·9	y	CuH	OH	E	444·26		Pt		E	447·74		Sm		E
441·93	a	Sm	AA	E	444·3	r	LaO	OHn	E	447·78		Nd		E
441·99		V		E	444·32		Sm		E	447·8	r	FeO	OHn	E
442·0		Ho		E	444·42	a	V	OC	E	448·0		Nd(O)		E
442·0		DyO	OHn	E	444·46		Ho		E	448·03		Sm		E
442·02		Gd		E	444·49		Nd		E	448·09		Nd		E
442·04		Os		A, E	444·51		Sm		E	448·1		Gd(O)	OHn	E
442·05		Sm		E	444·55		Pt		E	448·18		Nd		E
442·1	r	Sr	OG	E	444·8	r	LaO	OHn	E	448·22	f	Fe	OHn	E
442·11		Sm		E	444·93		Ru		E	448·56		ScO		E
442·15		V		E	444·99		CaO		E	448·6		TbO		E
442·24		Gd		E	445·0		Ho		E	448·69		Gd		E
442·3		LaO	OHn	E	445·01		ScO		E	448·7	r	Te(O)	AH	E
442·31		Tb		E	445·16	a	Mn	OA	E	448·8	r	IO	OH, IO_3	E
442·33		Sm		E	445·2	r	BeO	OH	E	448·9	r	Lu	OHn	E

331

449·00	Sm		E
449·10	Gd		E
449·30	Tb		E
449·4	r AlO	OAn	E
449·4	r CuCl	AH	E
449·42	a Na	OHn	E
449·5	Ho		E
449·6	r BeO	OH	E
449·6	r YO	OHn	E
449·63	Er		E
449·68	$^{11}BO_2$		E
449·69	a Cr	OAn	E
449·7	B		E
449·73	Gd		E
449·73	Nd		E
449·77	a Na	OHn	E
449·77	Ho		E
449·9	r Gd(O)	OHn	E
449·9	r Sn(O)	AH	E
449·91	Sm		E
450·0	NbO		E
450·0	r Lu	OHn	E
450·0	Dy		E
450·05	Y		E
450·2	r Sc	OHn	E
450·25	$^{11}BO_2$		E
450·3	ScO		E
450·30	$^{11}BO_2$		E
450·33	Sm		E
450·61	$^{11}BO_2$		E
450·63	Gd		E
450·9	r LuO	OHn	E
450·9	r Pb	OH	E
451·04	NbO		E
451·1	NdO	OHn	E
451·13	In	OHn	A, F, E
451·13	Sm		E
451·2	r Lu	OHn	E
451·5	YbOH		E
451·6	r AlO	OAn	E
451·6	BeO		E
451·6	r CuCl	AH	E
451·7	r Be	OH	E
451·8	b Yb	OH	E
451·85	Lu		A, E
451·86	Er		E
451·96	Gd		E
451·963	Pr		E
452·0	PrO	OHn	E
452·1	r Lu	OHn	E
452·19	ScO		E
452·2	r SnO	AH	E
452·25	Sm		E
452·25	Eu		E
452·27	Er		E
452·3	r Sr	OG	E
452·31	Sm		E
452·40	r BaO	AH	E
452·47	a Sn	AH	E
452·5	YO		E
452·6	r Lu	OHn	E
452·6	r Y	OHn	E
452·72	Nd		E
452·72	Y		E
452·74	Sm		E
452·77	Y		E
452·87	f Rh	OH	E
452·98	Nd		E
453·0	$B(_2O_3)$	OHn	E
453·11	Er		E
453·15	Dy		E
453·22	LaO		E
453·23	Pr		E
453·24	Sm		E
453·3	r LuO	OHn	E
453·32	Ti		E
453·38	Sm		E
453·45	Ho		E
453·6	BeO		E
453·6	r Sc	OHn	E
453·615	Dy		E
453·62	LaO		E
453·653	Dy		E
453·69	DyO		E
453·7	ScO		E
453·75	Sm		E
453·78	Gd		E
453·78	r BaO	AH	E
453·8	r AlO	OAn	E
454	b B		E
454·0	r Lu	OHn	E
454·02	NbO		E
454·02	LaO		E
454·1	r SnO	AH	E
454·12	Pr		E
454·17	a Na	OHn	E
454·2	BiO	AH	E
454·20	Nd		E
454·20	Gd		E
454·4	r FeO	OHn	E
454·4	r Sr	OG	E
454·47	LaO		E
454·52	a Na	OHn	E
454·6	SmOH	OHn	E
454·60	a Cr	OC	E
454·62	LuO		E
454·82	Nd		E
454·91	Nd		E
454·91	LaO		E
454·95	La		E
455·00	Sm		E
455·09	Dy		E
455·16	Dy		E
455·22	Pr		E
455·3	r BeO	OH	E
455·36	LaO		E
455·4	r Pb	OH	E
455·403	Ba II		A, E
455·45	a Rn	OHn	E
455·5	Dy(O)		E
455·54	Cs		E
455·6	r Te(O)	AH	E
455·66	Sm		E
455·75	AlO	OAn	E
455·83	LaO		E
455·91	ScO		E
455·96	Nd		E
456·04	Nd		E
456·1	r LuO	OHn	E
456·18	Nd		E
456·2	r YO	OHn	E
456·25	Ho		E
456·30	LaO		E
456·5	r Sr	OG	E
456·51	Dy		E
456·55	Dy		E
456·67	Sm		E
456·73	Nd		E
456·79	La		E
456·95	NbO		E
456·95	Sm		E
456·96	Dy		E
457·0	dr Dy	OHn	E
457·0	r Lu	OHn	E
457·0	r SnO	AH	E
457·00	La		E
457·1	r TbO	OHn	E
457·1	r ScO	OHn	E
457·11	f Mg	OH	E
457·4	YbO	OH	E
457·42	dr Dy	OHn	E
457·43	Ta		E
457·5	r LuO	OHn	E
457·6	r AlO	OAn	E
457·71	V		E
457·78	Dy	OH	E
457·9	Dy(O)		E
457·94	r BaO	AH	E
458·01	f Co	OH	E
458·01	Cr		E
458·04	V		E
458·1	r LaO	OHn	E

λ		Species	Notes	Ref
458·13		Gd		E
458·158		Sm		A, E
458·174		Sm	AA	E
458·44	f	Ru	OHn	E
458·5	r	LaO	OHn	E
458·59	a	Ca	OH	E
458·63		V		E
458·66		Nd		E
458·7	r	IO	OH, IO$_3$	E
458·7		TbO		E
458·87		Ca		E
458·9	r	LaO	OHn	E
458·93		Dy		E
459·0	b	Tb	OHn	E
459·0		YbO	OH	E
459·1	r	Lu	OHn	E
459·1	r	SnO	AH	E
459·19		Dy		E
459·2		RaO	AH	E
459·32	p	Cs	OH	E
459·38		AlO		E
459·4	r	LaO	OHn	E
459·40		Eu	OHn	A, E
459·41	p	V	OC	E
459·46		Nd		E
459·67		Sm		E
459·8	b	Dy(O)	OHn	E
459·8	r	LaO	OHn	E
459·83		Sm		E
459·90		Tm		E
460·0		PrO	OHn	E
460·08		Cr		E
460·27		LaO		E
460·29	a	Li	OH	E
460·3	r	La	OHn	E
460·37		YO		E
460·38		Nd		E
460·4	r	FeO	OHn	E
460·4	r	YO	OHn	E
460·54	a	Mn	OA	E
460·66		Er		A, E
460·7	r	ScO	OHn	E
460·7	r	TbO	OHn	E
460·73		LaO		E
460·733		Sr	OHn	A, E
460·98		Nd		E
461		TbO		E
461·12		Sm		E
461·2		CeO	OHn	E
461·2	r	SnO	AH	E
461·20		LaO		E
461·22		Dy		E
461·34		Cr		E
461·5		TbO		E
461·5		Sm(O)		E
461·6	mr	Gd(O)	OHn	E
461·6	r	Te	AH	E
461·61		Cr		E
461·9		BiO	AH	E
461·9	b	Tb	OHn	E
462·0		NdO		E
462·1	r	CeO	OHn	E
462·1	r	BaO	AH	E
462·19		Nd		E
462·35		Dy(O)		E
462·4	r	Ce	OHn	E
462·42		Nd		E
462·65		Nd		E
462·7	b	Ce	OHn	E
462·72		Eu	OHn	A, E
462·79		Nd		E
462·9		NdO		E
462·94		Sm		E
463		CeO		E
463·12		Nd		E
463·15		Dy(O)		E
463·4		GdO	OHn	E
463·4	r	SnO	AH	E
463·42		Nd		A, E
463·56		La		E
463·56		Pr		E
463·68	r	BaO		E
463·72		Nd		E
463·85	a	K	OHn	E
463·9	r	Ba(O)	AH	E
463·91		Nd		E
463·98		Pr		E
464		CeO		E
464·0	r	Te(O)	AH	E
464·11		Nd		E
464·158		K	OHn	E
464·217		K	OHn	E
464·37		Y		E
464·4	v	Lu	OHn	E
464·44		ScO		E
464·5		NdO	OHn	E
464·54		Sm		E
464·62	a	Cr	OAn	E
464·64		Nd		E
464·69		Pr		E
464·80		Sm		E
464·82	r	AlO	OAn	E
464·84	r	CuH	OH	E
464·94		Sm		E
464·96		Nd		E
465·02	dr	YO	OHn	E
465·10		Nd		E
465·128	a	Cr	OAn	E
465·216	a	Cr	OAn	E
465·23		Nd		E
465·3	r	Gd	OHn	E
465·4	r	Lu	OHn	E
465·47		Nd		E
465·8	r	Pb	OH	E
465·9	r	FeO	OHn	E
466·0	r	Lu	OHn	E
466·09		Pr		E
466·1	r	SnO	AH	E
466·11	a	Ta	OCn	E
466·17		LuO		E
466·187		Eu		E
466·188		Eu	OHn	E
466·2		Lu(O)		E
466·25		La		A
466·35		Sm		E
466·4	r	BaO	AH	E
466·485		Na	OHn	E
466·859	a	Na	OHn	E
467·0		PrO	OHn	E
467·0		GdO	OHn	E
467·0770		Sm		E
467·083	a	Sm	AA	E
467·1	r	Gd	OHn	E
467·10		Nd		E
467·2	r	AlO	OAn	A, E
467·2	dr	LuO	OHn	E
467·21		Sm		E
467·3	r	ScO	OHn	E
467·31		Er		E
467·39		Nd		E
467·48		Y		E
467·48		Pr		E
467·50	a	Rh	OH	E
467·55		Nd		E
467·63		YO	OHn	E
467·82		Cd		E
467·9	v	C$_2$	OHnq	E
468·01	a	Zn	OHn	E
468·03	r	BaO		E
468·05	f	W	OC	E
468·15		Sm		E
468·19		Tm		E
468·2		Yb	OH	E
468·23		RaII	AH	E
468·34		Nd		E
468·4	r	CeO	OHn	E
468·4	r	LuO	OHn	E
468·40		Nd		E
468·5	v	C$_2$	OHnq	E
468·58		$^{11}BO_2$		E
468·78		Pr		E
468·78	a	Zr	OCn	E

333

468·85		NbO		E		472·283		Bi		F		475·07		Sm		E
468·85		Nd		E		472·38		Dy		E		475·1	r	NiO	OHn	E
468·87	a	Sm	AA	E		472·42		Tm		E		475·18	a	Na	OHn	E
468·90		CuH		E		472·5		YbOH		E		475·2	r	Tb	OHn	E
469·03		Nd		E		472·6		Gd(O)		E		475·31		Dy		E
469·3	r	Sr	OG	E		472·6	r	Lu	OHn	E		475·31		Sc		E
469·4	r	IO	OH, IO₃	E		472·65		Nd		E		475·39	a	K	OHn	E
469·48		¹¹BO₂		E		472·7		TbO		E		475·4	r	AlO	OAn	E
469·5	r	AlO	OAn	E		472·73		¹¹BO₂		E		475·40	a	Mn	OA	E
469·5		CeO	OHn	E		472·79	f	Co	OH	E		475·45		K		E
469·57		Pr		E		472·84	a	SmOH	AA	A, E		475·5	r	BeO	OH	E
469·6	r	LuO	OHn	E		472·877		Sc		E		475·5		Tb		E
469·64		Nd		E		472·923		Sc		E		475·58		Nd		E
469·8	v	C₂	OHnq	E		473·0		PrO	OHn	E		475·6	r	Gd(O)	OHn	E
469·83		Nd		E		473·0	r	FeO	OHn	E		475·6		YbOH		E
469·9	r	Te	AH	E		473·06		Pr		E		475·62		Dy		E
470·23		Dy		E		473·17		Nd		E		475·74	a	K	OHn	E
470·3		Mg		A		473·3	r	BeO	OH	E		475·74		Dy		E
470·34		Dy		E		473·3		YbOH		E		475·85		Nd		E
470·4		CeO	OHn	E		473·33	u	Tm	OAn	A, E		475·93		Nd		E
470·6	r	Pb	OH	E		473·40		Sc		E		476·0	r	YbO	OH	E
470·69		Nd		E		473·41		CuH		E		476·026		Sm		A, E
470·7	dr	YO	OHn	E		473·46		Dy		E		476·04		Nd		E
470·7		Yb		E		473·49		Nd		E		476·09		Y		E
470·7	r	ScO	OHn	E		473·5		Lu	OHn	E		476·15	a	Mn	OA	E
470·8	r	LuO	OHn	E		473·5	r	Gd	OHn	E		476·24		Mn	OA	E
470·9	r	BeO	OH	E		473·5		Pu		A		476·26		Er		E
470·95		Ru		E		473·5		TbO		E		476·4	r	LuO	OHn	E
471·0		Sm		E		473·6	r	Al	OAn	E		476·4		SmOH		E
471·1		Te(O)	AH	E		473·6		Gd(O)		E		476·5	r	YbOH	OH	E
471·29		Dy		E		473·66		Pr		A, E		476·586	a	Mn	OA	E
471·31		Pr		E		473·7	v	C₂	OHnq	E		476·643	a	Mn	OA	E
471·35		Dy		E		473·76		Sc		E		476·68		La		A, E
471·5	b	B(₂O₃)	OHn	E		473·8	r	YbO	OH	E		476·8		YbOH		E
471·5	v	C₂	OHnq	E		474·0	w	EuO	OHn	E		476·9		NdO	OHn	E
471·6	r	AlO	OAn	E		474·09	a	K	OHn	E		477·019		Sm		E
471·6	r	YbOH	OH	E		474·10	r	BaO		E		477·019		Nd		E
471·61	a	Sm	AA	E		474·10		Sc		E		477·029		Sm		E
471·669		Sm		E		474·10	a	Sr	OHn	E		477·22		Nd		E
471·7	r	Gd(O)	OHn	E		474·2	r	ScO	OHn	E		477·4	r	Te(O)	AH	E
471·709	a	Sm	AA	E		474·2		YbOH		E		477·4		YbOH		E
471·8		RaO		E		474·38		Sc		E		477·480		Dy	OHn	E
471·867		Sm		E		474·41		Pr		E		477·5		Tm		E
471·90		Nd		A, E		474·43	a	K	OHn	E		477·53	b	Dy		E
471·95		¹¹BO₂		E		474·5	dr	YO	OHn	E		477·6	r	BeO	OH	E
472	b	B		E		474·5		Tm		E		477·8	mr	YbOH	OH	E
472·013		SmO		E		474·51	a	Rh	OH	E		477·84		Nd		E
472·1	r	Lu	OHn	E		474·79	a	Na	OHn	E		477·85	b	Sc	OHn	E
472·22		Bi		F, E		474·9	r	Pb	OH	E		477·935		Sc		E
472·22	a	Zn	OHn	E		474·9	r	YbOH	OH	E		477·94		Nd		E
472·23	a	Sr	OHn	E		474·9	r	LuO	OHn	E		478		Dy(O)		E
472·255	a	Bi	AH	F, E		474·97		Nd		E		478·0	r	LuO	OHn	E
472·27		Er		E		475·0		RaO	OH	E		478·0	r	Tb	OHn	E
472·27	r	BaO	AH	E		475·0		Tm		E		478·31	a	Sm	AA	A, E

478·34		Mn	OA	E	483·56		Nd		E	488·17		Nd		E
478·4		TbO		E	483·66		Nd		E	488·2	r	CuCl	AH	E
478·41	r	BaO		E	484·0	r	YbOH	OH	E	488·2	r	Ce	OHn	E
478·587		Sm		E	484·02		Tm		E	488·21		$^{11}BO_2$		E
478·59	a	SmOH	AA	E	484·17	a	Sm	AA	E	488·24		LuO		E
478·689	a	K	OHn	E	484·20	dr	Y	OHn	E	488·378		Sm		A, E
478·74		Nd		E	484·23	r	AlO	OAn	E	488·38		Nd		E
478·99		Sm		E	484·5	r	IO	OH, IO_3	E	488·398		Sm		A, E
479·109	a	K	OHn	E	484·6	b	Ce	OHn	E	488·50		Nd		E
479·130		Dy		E	484·8		Ti		E	488·8	r	AlO	OAn	E
479·15		Sc		E	484·83		Sm		E	488·92		Re		E
479·150		Dy		E	484·9	r	TiO	OH	E	489	r	NiO	OHn	E
479·2		CeO	OHn	E	484·988	a	K	OHn	E	489·1		TbO		E
479·2	r	Tb	OHn	E	485		IO		E	489·1	b	Ce	OHn	E
479·5	r	BeO	OH	E	485·0		Sn(O)	AH	E	489·10		Nd		E
479·5		TbO		E	485·0		YbOH	AH	E	489·16		$^{11}BO_2$		E
479·8	r	Gd(O)	OHn	E	485·0		BaF	OH	E	489·18		$^{11}BO_2$		E
479·9		CeO	OHn	E	485·0	b	Tm		E	489·2	r	Tm	OHn	E
479·99		Cd	OH	E	485·06	r	BaO	AH	E	489·3	mr	GdO	OHn	E
480·00		MnOH		E	485·08		La		E	489·3		TmO		E
480·016		K		E	485·1	r	Pb	OH	E	489·32		Nd		E
480·2	r	Te(O)	AH	E	485·148	p	V	OC	E	489·4	r	ScO	OHn	E
480·5		Ti		E	485·33		Nd		E	489·50		LuO		E
480·519		K		E	485·53		Nd		E	489·61		Pr		E
480·6	dr	TiO	OH	E	485·57	r	Gd(O)	OHn	E	489·65		BaO		E
480·66		Nd		E	485·74		Er		E	489·69		Nd		A, E
480·768		Ce	OHn	E	485·8	r	VO	OHn	E	489·8	db	Tm	OHn	E
480·81		Pr		E	485·8		YbOH		E	490		CeO		E
481		CeO		E	485·81		ScO		E	490·00		$^{11}BO_2$		E
481·05	a	Zn	OHn	E	485·95		Nd		E	490·15		Nd		E
481·19	a	Sr	OHn	E	486·01		LuO		E	490·18		Nd		E
481·2	r	BeO	OH	E	486·03		Ho		E	490·3	db	Tm	OHn	E
481·275		Ta		E	486·3	r	CeO	OHn	E	490·4	r	V	OHn	E
481·3	r	VO	OHn	F	486·3	r	Te(O)	AH	E	490·49		Sm		E
481·5		TmO	OHn	E	486·361	a	K	OHn	E	490·69		Pr		E
481·5	r	CeO	OHn	E	486·47		V	OC	E	490·69		$^{11}BO_2$		E
481·7	r	GdO	OHn	E	486·64	r	AlO	OAn	E	490·71		Eu		E
481·7	r	Pb	OH	E	486·67		Nd		E	490·72		Nd		E
481·82	dr	YO	OHn	E	486·83		YO		E	490·72		LuO		E
482·0		PrO	OHn	E	486·92		Nd		E	490·77		Nd		E
482·07		Er		E	486·98	a	K	OHn	E	490·94		Tm(O)		E
482·17		Gd		E	487·0	r	YO	OHn	E	490·95		Gd		E
482·35	a	Mn	OA	E	487·0		BaOH		E	491·0	mr	GdO	OHn	E
482·4	r	Yb	OH	E	487·08		LuO		E	491·00		Nd		E
482·59		Ra	OH	E	487·3	r	Ce	OHn	E	491·04		Sm		E
482·745		V	OC	E	487·548		V		E	491·14		Eu		E
482·8	r	BeO	OH	E	487·63	a	Sr	OHn	E	491·34		Nd		E
483·0	w	NdO	OHn	E	487·7	r	NiO	OHn	E	491·40		Pr		A, E
483·0		TmO		E	487·83	r	BaO		E	491·46		NdO		E
483·00	r	BaO		E	487·979		Nd		E	491·47		Dy		E
483·16		V	OC	E	488		NdO		E	491·70		$^{11}BO_2$		E
483·21	a	Sr	OHn	E	488·01		Dy		E	491·89		Sn		E
483·24		V	OC	E	488·10		$^{11}BO_2$		E	492·0		PrO	OHn	E
483·50	r	Gd(O)	OHn	E	488·16		V	OC	E	492·17		LuO		E

335

492·40	Sm		E
492·45	Pr		E
492·45	Nd		E
492·5	r Ce	OHn	E
492·8	mr GdO	OHn	E
492·9	r FeO	OHn	E
492·93	¹¹BO₂		E
493·0	b B	OHn	E
493·0	w CuOH	AH	E
493·1	r Ce	OHn	E
493·12	ErO		E
493·2	r CuF	PH	E
493·25	¹¹BO₂		E
493·4	r Ce	OHn	E
493·41	BaII	OAn	A, E
493·48	Ho		E
493·55	LuO		E
493·60	Pr		E
493·75	TmO		E
493·9	r NiO	OHn	E
493·90	Ho		E
493·97	Pr		E
494	BO₂		E
494	CeO		E
494·03	Pr		E
494·1	r CeO	OHn	E
494·1	r Te(O)	AH	E
494·13	¹¹BO₂		E
494·17	r BaO		E
494·2	BaF	OH	E
494·20	a K	OHn	E
494·48	Nd		E
494·6	b Tm	OHn	E
494·63	Sm		E
494·67	Nd(O)		E
494·7	r Ce	OHn	E
494·9	r Gd	OHn	E
494·97	La		A, E
494·98	LuO		E
495·0	BaF	OH	E
495·0	B(O)	OHn	E
495·0	B(O)	OHn	E
495·02	Nd		E
495·07	Nd		E
495·08	a K	OHn	E
495·1	r BaF	PH	E
495·1	r Ce	OHn	E
495·1	r VO	OHn	E
495·136	Pr		A, E
495·25	Nd		E
495·47	Nd		E
495·59	¹¹BO₂		E
495·6	dr TiO	OH	E
495·6	YbOH		E
495·60	a K	OHn	E
495·69	Er		E
495·9	NdO	OHn	E
496	Sn(O)	AH	E
496	B(O)		E
496·02	Pr		E
496·1	YbOH		E
496·23	v Tm	OHn	E
496·23	a Sr	OHn	E
496·3	v MgO	OHn	E
496·33	Nd		E
496·4	r IO	OH,IO₃	E
496·50	a K	OHn	E
496·54	¹¹BO₂		E
496·54	r BaO		E
496·6	YbOH		E
496·7	a Rb	OHn	E
496·97	Nd		E
497·0	Ba		E
497·1	YbOH		E
497·13	Er		E
497·2	a Li	OH	E
497·36	¹¹BO₂		E
497·39	Sm		E
497·49	Pr		E
497·5	v MgO	OHn	E
497·5	YbOH		E
497·5	r Te	AH	E
497·54	Nd		E
497·57	Pr		E
497·59	Sm		E
497·6	r MnO	OHn	E
497·6	NdO		E
497·64	Pr		E
497·85	a Na	OHn	E
497·95	Er		E
497·99	Ho		E
498·1	YbOH	AH	E
498·17	a Ti	OCn	E
498·28	a Na	OHn	E
498·3	a Rb	OHn	E
498·4	r Pb	OH	E
498·5	YbOH		E
498·6	v Mg	OHn	E
498·77	b Er	OHn	E
498·86	Nd		E
499·11	a Ti	OCn	E
499·2	b Lu	OHn	E
499·60	Er		E
499·7	v MgO	OHn	E
499·75	¹¹BO₂		E
499·95	a Ti	OCn	E
500	Er(O)		E
500	(Pd)		E
500	MgO		E
500·0	BaF	OH	E
500·0	r Ti	OH	E
500·1	r BaF	PH	E
500·11	Lu		E
500·24	r Er	OHn	E
500·3	TiO		E
500·4	r Lu	OHn	E
500·6	r Te	AH	E
500·7	v MgO	OHn	E
500·7	r NiO	OHn	E
500·7238	Er		E
500·90	b Er		E
501·1	r VO	OHn	E
501·2	YbOH		E
501·24	r BaO	AH	E
501·3	r MnO	OHn	E
501·45	Nd		E
501·50	Gd		E
501·7	a Rb	OHn	E
501·85	Pr		E
501·9	r Lu	OHn	E
501·97	Pr		E
502·0	BaOH	AH	E
502·0	PrO	OHn	E
502·21	Dy		E
502·3	a Rb	OHn	E
502·4	r NiO	OHn	E
502·5	YO	OHn	E
502·69	Pr		E
502·71	Nd		E
502·739	U		E
502·94	Nd		E
503	SmOH		E
503·33	Pr		E
503·42	v Er	OHn	E
503·5	r Lu	OHn	E
503·5	r Te(O)	AH	E
503·7	a Rb	OHn	E
503·99	Ti		E
504	ErO		E
504·01	Nd		E
504·1	Lu	OHn	E
504·13	v ErO	OHn	E
504·19	b Dy		E
504·262	Dy		E
504·38	Pr		E
504·42	Sm		E
504·55	Pr		A, E
504·95	Sm		E
504·96	b Er	OHn	E
505·0	w CuOH	AH	E
505·0	YO	OHn	E
505·1	r MnO	OHn	E

505·34		Dy		E	510·0	r	Er	OHn	E	513·90	b	Dy(O)	E
505·34		Pr		E	510·0		LuO		E	513·98		Pr	E
505·44	r	BeO	OH	E	510·0	mr	Gd	OHn	E	514·0	dv	Dy	OHn	E
505·68		Nd		E	510·2	r	AlO	OAn	E	514·1		TbO	E
505·7	r	VO	OHn	E	510·31		Nd		E	514·1	r	BeO	OH	E
505·8	b	Lu	OHn	E	510·34		Gd		E	514·21	dv	Ho(O)	OHn	E
505·8		V		E	510·5	r	HoO	OHn	E	514·27		Ho(O)	E
506·08		Tm		E	510·5	r	VO	OHn	E	514·3	r	AlO	OAn	E
506·09		Sm		E	510·53		Nd		E	514·3		DyO	E
506·1	r	CuF	PH	E	510·55	a	Cu	OHn	E	514·4		YbOH	E
506·46		Ti		E	510·91		Dy(O)		E	514·44		Dy(O)	E
506·6	r	BaCl	OHn	E	511·04	f	Fe	OHn	E	514·46		$^{11}BO_2$	E
506·69	b	Dy(O)		E	511·2	r	BeO	OH	E	514·54		La	E
506·7	r	Er	OHn	E	511·2	r	Y	OHn	E	514·55		Ho(O)	E
506·76	b	Ho(O)		E	511·22	a	K	OHn	E	514·59		$^{11}BO_2$	E
506·98	db	Ho	OHn	E	511·39		Tm		E	514·6	dr	CaF	PH	E
507·02		Sc		E	511·43		Eu		E	514·75	r	Ho(O)	OHn	E
507·119		Sm		E	511·7	b	Gd(O)	OHn	E	514·909		Na	A
507·18		Nd		E	511·71		Sm		A, E	514·95		Nd	E
507·3	b	Er	OHn	E	511·8	r	Te(O)	AH	E	514·96		Dy(O)	E
507·58	a	Rb	OHn	E	511·83		Ho(O)		E	515		Eu(O)	E
507·6	r	BeO	OH	E	511·87		$^{11}BO_2$		E	515·07	a	Rb	OHn	E
507·76		Dy		E	511·88	r	HoO	OHn	E	515·1	v	Er	OHn	E
507·8	r	Er	OHn	E	512·0	r	Lu	OHn	E	515·27	a	Cs	OH	E
507·8	dr	YO	OHn	E	512·0		BaOH		E	515·3	r	Y	OHn	E
507·9	r	A1O	OAn	E	512·16		Dy(O)		E	515·364		Na	A
508·1	v	Er	OHn	E	512·21		Sm		E	515·4		YO	E
508·15		Sc		E	512·3	r	AlO	OAn	E	515·63		$^{11}BO_2$	E
508·26		Dy		E	512·45		Er		E	515·7	r	PrO	OHn	E
508·31		Dy		E	512·5		BaF	PH	E	515·70	r	Ho(O)	OHn	E
508·37		Sc		E	512·6	r	Ho	OHn	E	515·72		Sm	E
508·42	a	K	OHn	E	512·77		$^{11}BO_2$		E	515·77		$^{11}BO_2$	E
508·55		Sc		E	512·8	r	BeO	OH	E	515·8		DyO	E
508·58	a	Cd	OH	E	512·9	v	C_2	OHn	E	515·80		MnO	E
508·67	r	BaO	AH	E	512·90		Eu		E	515·86		La	A, E
508·69		Sc		E	513		PrO		E	515·9	r	MnO	OHn	E
508·71		Pr		E	513·0		BaOH	AH	E	516·0		Dy	E
508·83		Sm		E	513·1	r	IO	OH,IO$_3$	E	516·0		Sn(O)	AH	E
508·9	db	Er	OHn	E	513·2	r	Ho	OHn	E	516·00		Eu	E
508·90	a	Rb	OHn	E	513·2	a	Rb	OHn	E	516·1	r	AlO	OAn	E
508·93	b	Dy(O)		E	513·3		Pr		A, E	516·1	r	Lu	OHn	E
508·97	b	Er(O)		E	513·34		Eu		E	516·2	r	Pb	OH	E
509·2	b	Ho	OHn	E	513·34		Pr		E	516·23	r	Ho(O)	OHn	E
509·5	r	BeO	OH	E	513·37	r	ScO	OHn	E	516·49	a	Rb	OHn	E
509·7	r	ScO	OHn	E	513·4		Dy(O)		E	516·5	v	C_2	OHnq	E
509·7	b	Ho	OHn	E	513·4		Gd(O)		E	516·5		SnOH	E
509·71	a	K	OHn	E	513·5	v	Er	OHn	E	516·61		$^{11}BO_2$	E
509·8	v	C_2	OHnq	E	513·6	r	Dy	OHn	E	516·63		Ho(O)	E
509·8	r	Ni	OHn	E	513·6	r	Gd	OHn	E	516·67		Eu	E
509·86	b	Dy(O)		E	513·7	r	Pr	OHn	E	516·7	r	BaCl	OG	E
509·9		Gd(O)		E	513·8	r, v	BaCl	OG	E	516·7		TiO	E
509·92		Sc		E	513·8		NdO	OHn	E	516·73	a	Mg	OH	E
509·92	a	K	OHn	E	513·8	r	Pb	OH	E	516·74	v	Er(O)	OHn	E
510		BeO		E	513·9	r	Te(O)	AH	E	516·75	a	Fe	OHn	E

λ		Species		Class
516·8	r	Cr(O)	OHn	E
516·8	dr	Ti	OH	E
516·83	r	Ho	OHn	E
516·88		$^{11}BO_2$		E
516·93		Dy(O)		E
517		(YbOH)		E
517·03	r	LuO	OHn	E
517·10	r	ScO	OHn	E
517·19		Dy(O)		E
517·27	a	Mg	OH	A, E
517·27		Sm		E
517·27		Er		E
517·37		Ti		E
517·4		YbOH	AH	E
517·4	r	NiO		E
517·52	db	ErO	OHn	E
517·54		Sm		E
517·6	db	AlO	OHn	E
517·67		ErO		E
517·73		La		E
517·89		Er(O)		E
517·9	dr	LaO	OHn	E
518·0	b	B	OHn	E
518·0		YbOH		E
518·0		BO_2		E
518·07		$^{11}BO_2$		E
518·09	db	Er(O)	OHn	E
518·25		$^{11}BO_2$		E
518·3		YbOH		E
518·35		$^{11}BO_2$		E
518·36	a	Mg	OH	A, E
518·36	db	ErO	OHn	E
518·47		Er		E
518·5		LuO		E
518·57	b	Ho	OHn	E
518·6	b	Lu	OHn	E
518·70		Sm		E
518·885		Ca		A
518·98		Ho(O)		E
519·1	r	AlO	OAn	E
519·25		MnO	OHn	E
519·28	v	Dy(O)	OHn	E
519·29		Ti		E
519·44		Pr		E
519·53		DyO		E
519·53	a	Rb	OHn	E
519·53		Pr		E
519·56	r	Ho(O)	OHn	E
519·61		$^{11}BO_2$		E
519·67	a	Cs	OH	E
519·80		Nd		E
519·97		Nd		E
520	w	EuO	OHn	E
520		NiO		E
520		Zn		E
520·0	w	RaOH		E
520·0		LuO		E
520·0		PO		E
520·059		Sm		A, E
520·2	b	Lu	OHn	E
520·27		La		E
520·4	dr	La	OHn	E
520·43		Nd		E
520·44	b	Dy(O)	OHn	E
520·45	a	Cr		E
520·6	r	Lu	OHn	E
520·6	v	MgO	OHn	E
520·60	a	Cr	OAn	E
520·65		Er		E
520·72		$^{11}BO_2$		E
520·84	a	Cr	OAn	E
520·9	r	IO	OH, IO_3	E
520·9	r	Sc	OHn	E
520·95	da	Cs	OH	E
521·03		Ti		E
521·07		Sm		E
521·13	r	Ho(O)	OHn	E
521·18		La		E
521·32		Nd		E
521·41		Dy(O)		E
521·47	r	BaO	AH	E
521·5		LuO		E
521·50		Eu		E
521·77	r	LuO	OHn	E
522·34		Eu		E
522·6	b	Tm	OHn	E
522·8	dr	LaO	OHn	E
522·80		Pr		E
522·82	dr	VO	OHn	E
522·83		TmO		E
522·84		MnO		E
522·9	r	Cr(O)	OHn	E
522·90		$^{11}BO_2$		E
522·98		Dy(O)		E
523·0		Tm		E
523·06		Dy(O)		E
523·3	r	Lu	OHn	E
523·40	a	Rb	OHn	E
523·42		La		E
523·5		LuO		E
523·62		Dy		E
523·86	u	Sr	OHn	E
524		HPO	OH	E
524·0		CuOH	AH	E
524·0	r	BaCl	OG	E
524·0		BaOH	AH	E
524·7		YbOH		E
524·7	v	Gd	OHn	E
524·71	mb	Dy	OHn	E
524·76	a	Cr	OC	E
524·85		Dy		E
524·95		Dy(O)	OHn	E
525·0		NdO		E
525·06		Dy(O)'		E
525·14		Ho		E
525·18		LaO		E
525·189		Sm		E
525·3	dr	La	OHn	E
525·4		Gd(O)		E
525·4		YbOH		E
525·6	r	Lu	OHn	E
525·61		Tm(O)		E
525·66	a	Cs	OH	E
525·69	u	Sr	OHn	E
525·7	r	Tm	OHn	E
525·8	b	Ho	OHn	E
525·8	r	Pb	OH	E
525·9	v	Gd	OHn	E
526·00	a	Rb	OHn	E
526·08		Dy		E
526·3		DyO		E
526·32		Tb(O)		E
526·34		Ho(O)		E
526·42	a	Cr	OAn	E
526·56		Sm		E
526·64		Eu		E
526·7	r	MnO	OHn	E
526·74		Tm(O)		E
526·95	a	Fe	OHn	E
526·99		Tb(O)		E
527·0	v	HoO	OHn	E
527·12		La		E
527·14	a	Sm	AA	A, E
527·19		Eu		E
527·27		Tb(O)		E
527·3	r	Dy(O)	OHn	E
527·42		Ho(O)		E
527·5		Gd(O)		E
527·5	dr	YO	OHn	E
527·55		Re		E
527·6	r	VO	OHn	E
527·7		YbOH		E
527·7	v	Gd	OHn	E
527·8	dr	LaO	OHn	E
528·0	r	Te(O)	AH	E
528·2	v	Tm	OHn	E
528·28		Eu		E
528·29		Sm		A, E
528·34		Nd		E
528·5	r	Dy(O)	OHn	E
528·56		Pr		E
528·69		Ho		E

338

528·81		Nd		E	532·68	r	Gd(O)	OHn	E	535·76		Eu		E
529·0	r	FeO	OHn	E	532·7	b	Nd	OHn	E	535·8	r	AlO	OAn	E
529·11		Tm(O)		E	532·72	r	Tb	OHn	E	535·8		ScO		E
529·16		Nd		E	532·79		TmO		E	535·88	r	Ho	OHn	E
529·2		Gd(O)		E	532·8	dr	La	OHn	E	535·94	r	MnO	OHn	E
529·2	dr	CaF	PH	E	532·805	a	Fe	OHn	E	535·95	a	K	OHn	E
529·3	r	Cr(O)	OHn	E	532·92	dr	TmO	OHn	E	535·99		Ho		E
529·4	b	Tm	OHn	E	533·01		Ho		E	536		TbO		E
529·6	r	Dy	OHn	E	533·1	r	Pb	OH	E	536·01		Ho(O)		E
529·6		YbOH		E	533·25	db	Dy(O)	OHn	E	536·1		Tm(O)	OHn	E
529·669	a	Cr	OAn	E	533·32	r	Ho	OHn	E	536·16		Dy(O)		E
529·798		Cr	OAn	E	533·34		Dy(O)		E	536·26	a	Rb	OHn	E
529·8	dr	CaF	PH	E	533·433		Nd		E	536·26	r	Tb(O)	OHn	E
529·88		Nd		E	533·61	r	Gd(O)	OHn	E	536·32		Dy		E
529·9		Dy(O)		E	533·69		Dy(O)		E	536·45		Dy		E
529·9	b	Ho	OHn	E	533·7		AlO	OAn	E	536·5	b	Tm	OHn	E
530		PrO		E	533·83		Dy		E	536·61	r	Gd(O)	OHn	E
530		LaO		E	533·9	v	Tm	OHn	E	536·66	r	BaO	AH	E
530·0		CuOH	AH	E	533·97	a	K	OHn	E	536·7	dr	PrO	OHn	E
530·02		TmO		E	534·0		YbOH		E	536·83		Sm		E
530·14	a	Cs	OH	E	534·0	dr	TbO	OHn	E	536·86	r	TbO	OHn	E
530·14		Ho		E	534·09	a	Cs	OH	E	536·9		TmO		E
530·15		Dy		E	534·1		Gd(O)		E	536·95		TmO		E
530·19		Sc		E	534·12		Sm		A, E	537·0	w	CuOH	AH	E
530·27	dv	TmO	OHn	E	534·29		Sc		E	537·02		Tm		E
530·3	v	Gd	OHn	E	534·30	a	K	OHn	E	537·15	a	Fe	OHn	E
530·3	dr	La	OHn	E	534·58	a	Cr	OAn	E	537·16	v	Er	OHn	E
530·4		Ho	OHn	E	534·59	r	Ho(O)	OHn	E	537·26	r	TbO	OHn	E
530·4	dr	Y	OHn	E	534·59	r	Gd(O)	OHn	E	537·33		VO		E
530·59	b	TmO		E	534·7		TmO	OHn	E	537·44	r	Tm	OHn	E
530·71		Tm		A, E	534·80		Sm		E	537·69		GdO	OHn	E
530·8	r	IO	OH,IO₃	E	534·83	a	Cr	OAn	E	537·7	r	AlO	OAn	E
530·81		MnO		E	534·9	dr	Dy(O)	OHn	E	537·77		Nd		E
530·93		Dy		E	534·91		Sm		E	537·8	r	Pr	OHn	E
531		IO		E	534·95		Nd		E	537·85	r	TbO	OHn	E
531·1		Gd(O)		E	534·96		Tb(O)		E	538		TmO		E
531·2		Dy(O)		E	534·97	r	BaO	AH	E	538·0		PrO		E
531·24		TmO		E	534·97		Sc		E	538·0	tb	HoO	OHn	E
531·3	b	Nd	OHn	E	535·04	a	Cs	OH	E	538·1	dr	LaO	OHn	E
531·44	db	Dy	OHn	E	535·05	a	Tl	OHn	A, F, E	538·14		HoO		E
531·47		HoO	OHn	E	535·05	dr	Tb(O)	OHn	E	538·2	r	FeO	OHn	E
531·61		DyO		E	535·06		Sm		E	538·2		PrO		E
531·96		HoO	OHn	E	535·2	r	PrO	OHn	E	538·25		LaO		E
532·06		Sm		A, E	535·2		Gd(O)		E	538·31	r	GdO	OHn	E
532·1	v	BaCl	OG	E	535·2		TmO	OHn	E	538·41		DyO		E
532·24	a	Rb	OHn	E	535·24		Tb(O)		E	538·49		Ho		E
532·3	r	Ni	OHn	E	535·4	dr	LaO	OHn	E	538·5		CaOH		E
532·31		DyO		E	535·4	r	Pb	OH	E	538·68	r	TbO	OHn	E
532·32	a	K	OHn	E	535·42		Tm(O)		E	538·95	r	MnO		E
532·45		VO	OHn	E	535·58	r	Gd(O)	OHn	E	539·06		Rb	OHn	E
532·47		TmO		E	535·6	r	CrO	OHn	E	539·2		TmO		E
532·5	r	YbOH	AH	E	535·610		Sc		E	539·39		DyO		E
532·6	r	Ho	OHn	E	535·69	r	TbO	OHn	E	539·4	r	AlO	OAn	E
532·6		LaO		E	535·7	r	ScO	OHn	E	539·40		ErO		E

339

				E
539·47	f	Mn	OA	E
539·5	r	ScO	OHn	E
539·55		Dy		E
539·6		TmO	OHn	E
539·62	r	TbO	OHn	E
539·8		PrO		E
539·82		DyO		E
539·9		Gd(O)		E
539·99		Tb		E
540		SmOH		E
540·0		Dy(O)		E
540·05	r	TmO	OHn	E
540·09	mb	ErO	OHn	E
540·23		TmO		E
540·25		Lu		E
540·27		Eu		E
540·27	r	BaO		E
540·3	r	Lu	OHn	E
540·36		DyO		E
540·36		Sm		E
540·37		Er(O)		E
540·38		$^{11}BO_2$		E
540·42	mr	Dy	OHn	E
540·5		LuO		E
540·5	r	Gd	OHn	E
540·5		TbO		E
540·5	r	Pb	OH	E
540·52	a	Sm		E
540·56		LaO		E
540·58	a	Fe	OHn	E
540·67	a	Cs	OH	E
540·72		$^{11}BO_2$		E
540·8	r	Ni	OHn	E
540·87		Dy(O)		E
540·98	a	Cr	OAn	E
541·0		Gd(O)		E
541·0	r	AlO	OAn	E
541·1		YbOH		E
541·21		Er(O)		E
541·38		Tm(O)		E
541·39	a	Cs	OH	E
541·4		PrO		E
541·5	r	Tb	OHn	E
541·5		TmO	OHn	E
541·6	r	Gd	OHn	E
541·63		Sm		E
541·7		CrO	OHn	E
541·98	r	BaO	AH	E
542		TbO		E
542·0		Dy(O)		E
542·07		$^{11}BO_2$		E
542·1		Gd(O)		E
542·1	db	Er	OHn	E
542·15		Sm		E
542·2		YbOH		E
542·28		Er(O)		E
542·3	r	Al	OAn	E
542·33		Dy		E
542·4	r	MnO	OHn	E
542·5	b	Rh(O)	OH	E
542·5	b	Tm	OHn	E
542·5		CaOH		E
542·51		VO		E
542·7	r	Gd	OHn	E
542·8	r	Tb	OHn	E
542·96		Er(O)		E
542·97	a	Fe	OHn	E
543	r	FeO	OHn	E
543·1		LaO		E
543·15	a	Rb	OHn	E
543·25	f	Mn	OA	E
543·27		Tb(O)		E
543·31		LaO		E
543·4	x	Ca	OH	E
543·4	b	Tm	OHn	E
543·5		GdO		E
543·5	r	ScO	OHn	E
543·6		TbO		E
543·60		$^{11}BO_2$		E
543·75		$^{11}BO_2$		E
543·8	db	Er	OHn	E
543·8	dr	Yb	OH	E
543·9	r	Gd	OHn	E
543·91		Er(O)		E
544·0	b	Ca	OH	E
544·0		Tb		E
544·0		YbOH		E
544·0	w	Nd	OHn	E
544·33	r	TbO	OHn	E
544·47		Er(O)		E
544·6		Gd(O)		E
544·6	db	Er	OHn	E
544·71		$^{11}BO_2$		E
544·9	r	Lu	OHn	E
544·9	r	Mn	OHn	E
544·9		TiO		E
545·0	dr	Gd	OHn	E
545·10		Dy		E
545·13		Er(O)		E
545·15		Eu		E
545·29		Eu		E
545·3	db	Er	OHn	E
545·30		Sm		E
545·4		GdO		E
545·43		Er(O)		E
545·46	r	BaO	AH	E
545·5	db	Dy(O)	OHn	E
545·51		La		E
545·68		$^{11}BO_2$		E
545·7	v	Tb	OHn	E
545·7		LuO		E
545·8	dr	LaO	OHn	E
545·9	b	Er	OHn	E
545·9	r	Pb	OH	E
546·0		YbOH		E
546·01		$^{11}BO_2$		E
546·02		Pr		E
546·07	a	Hg	OH	E
546·13		MnO		E
546·15		Gd(O)		E
546·2		TbO		E
546·2	x	Ca	OH	E
546·4	r	Lu	OHn	E
546·5		Gd(O)		E
546·59	a	Cs	OH	E
546·64		Y		E
546·67		Sm		E
546·7	r	Tb	OHn	E
546·82		$^{11}BO_2$		E
546·9		YbOH		E
546·9	r	Gd	OHn	E
546·93		VO		E
547	b	B		E
547·0	b	Er	OHn	E
547·09		$^{11}BO_2$		E
547·1		Tb(O)	OHn	E
547·17		Tb(O)		E
547·3	r	Tm	OHn	E
547·51		Gd(O)		E
547·6		BO_2		E
547·6	dr	Gd(O)	OHn	E
547·8	b	Nd	OHn	E
547·93	dr	Er(O)	OHn	E
548·06	r	Gd	OHn	E
548·09	a	Sr	OHn	E
548·11		Er(O)		E
548·14		Tb(O)		E
548·19		Sc		E
548·33	a	Co	OH	E
548·43		LaO		E
548·46		Sc		E
548·54		Sm		E
548·7		TmO		E
548·7	r	Lu	OHn	E
548·75	r	Gd(O)	OHn	E
548·8		Tm(O)	OHn	E
548·86		Eu		E
549·25		DyO	OHn	E
549·27	r	BaO	AH	E
549·37		Sm		E
549·5	r	IO	OH, IO_3	E
549·73	db	Er(O)	OHn	E

340

549·82	Sm		E
549·90	Er(O)		E
550	Mo		E
550	NbO		E
550	b Ce		E
550	NdO		E
550	UO$_2$		E
550·04	r Tm(O)	OHn	E
550·12	Tb(O)		E
550·13	La		A, E
550·2	v C$_2$	OHnq	E
550·29	a Cs	OH	E
550·4	r Te(O)	AH	E
550·42	a Sr	OHn	E
550·64	Mo		E
550·84	Dy(O)		E
550·9	dr LaO	OHn	E
550·97	r BaO	AH	E
551·05	Eu		E
551·11	Sm		E
551·21	Sm		E
551·3	mb Er(O)	OHn	E
551·42	Sc		E
551·61	Sm	AA	E
551·73	r VO	OHn	E
551·911	Ba		A
552	ErO		E
552·0	SmOH		E
552·03	Tm(O)	OHn	E
552·05	Sc		E
552·1	x Ca(OH)OH		E
552·18	a Sr	OHn	E
552·30	a Ce	OGn	E
552·36	TmO	OHn	E
552·39	Nd		E
552·57	Nd		E
552·6	x CaOH	OH	E
552·75	Y		E
552·79	FeO		E
552·8	b Er(O)	OHn	E
552·846	Mg		A
552·90	Nd		E
553·0	Tm		E
553·0	EuOH	OHn	E
553·1	TmO	OHn	E
553·14	r FeO	OHn	E
553·3	La		E
553·3	r Ca(OH)OH		E
553·3	r IO	OH, IO$_3$	E
553·30	Mo		E
553·35	r Er(O)	OHn	E
553·38	Tb		E
553·48	a Sr	OHn	E
553·5	dr La	OHn	E
553·548	Ba		E
553·556	p Ba	OH	E
553·63	LaO		E
553·76	Tb(O)		E
553·9	r CaOH	OH	E
554·0	CaOH	OH	E
554·1	v C$_2$	OHnq	E
554·1	x CaOH	OH	E
554·2	YbOH		E
554·2	x CaOH	OH	E
554·32	FeO		E
554·4	x CaOH	OH	E
554·44	db Dy(O)	OHn	E
554·48	db Er(O)	OHn	E
554·52	Na		E
554·58	Er		E
554·6	x CaOH	OH	E
554·7	x CaOH	OH	E
554·72	Dy		E
554·74	Eu		E
554·78	BaO		E
554·8	x CaOH	OH	E
554·89	Sm		E
555·0	r Tb(O)	OHn	E
555·0	CaOH		E
555·0	YbOH	AH	E
555·04	Sm		E
555·1	CaOH		E
555·17	ZrO		E
555·17	b Er(O)	OHn	E
555·2	BaOH		A, E
555·4	v CaOH	OH	E
555·65	Yb	OHn	E
555·7	x CaOH	OH	E
555·70	b Er(O)	OHn	E
555·77	r Tm(O)	OHn	E
555·83	r Tb	OHn	E
555·9	x CaOH	OH	E
556·1	dr LaO	OHn	E
556·11	Nd		E
556·13	Sm		E
556·2	x CaOH	OH	E
556·26	b Dy(O)	OHn	E
556·29	r Er(O)	OHn	E
556·34	Sm		E
556·37	Ho(O)		E
556·4	r BiO	AH	E
556·41	r CrO	OHn	E
556·57	dr Ho(O)	OHn	E
556·6	tb TmO	OHn	E
556·63	r Tb(O)	OHn	E
556·73	Tm(O)		E
556·73	dv Er(O)	OHn	E
556·77	VO		E
556·84	a Cs	OH	E
556·85	Er		E
557·03	Eu		E
557·04	Mo		E
557·16	r TmO	OHn	E
557·37	a Cs	OH	E
557·5	r Tb	OHn	E
557·71	Eu		E
557·78	r TmO	OHn	E
557·88	a Rb	OHn	E
557·98	Ho(O)		E
558·00	Eu		E
558·0	db Er	OHn	E
558·1	TmO		E
558·17	Er		E
558·18	Y		E
558·23	Ho		E
558·28	r FeO	OHn	E
558·44	r Ho	OHn	E
558·5	v C$_2$	OHnq	E
558·65	r MnO	OHn	E
558·8	dr LaO	OHn	E
558·86	r TmO	OHn	E
559·2	db HoO	OHn	E
559·34	r Tb	OHn	E
559·6	db Er	OHn	E
559·7	PrO	OHn	E
559·8	r Ti	OH	E
560	ErO		E
560·0	LaO		E
560·01	dr LaO	OHn	E
560·06	r Tb	OHn	E
560·24	r BaO	AH	E
560·24	LaO		E
560·3	TbO		E
560·72	Ho(O)		E
560·99	db Ho	OHn	E
561·0	PrO		E
561·0	r MnO	OHn	E
561·11	Nd		E
561·2	r PrO	PHn	E
561·27	v Er(O)	OHn	E
561·40	r FeO	OHn	E
561·7	SmOH		E
561·79	Gd		E
561·8	r Pb	OH	E
561·80	VO		E
562	TiO		E
562·05	Nd		E
562·13	FeO		E
562·18	SmO		E
562·33	r CrO	OHn	E
562·41	FeO		E
562·60	a Sm	AA	E

562·63	r	Ho(O)	OHn	E	566·2		PrO		E	569·923		Ce		E
562·7	dr	LaO	OHn	E	566·26		Tb(O)		E	570·02	a	Cu	OHn	E
562·76		V		E	566·3	r	Zr	OH.	E	570·02		Sc		E
562·8	r	Er	OHn	E	566·37	a	Cs	OH	E	570·1	r	BaO(H)	AH	E
562·8		PrO	OHn	E	566·4	v	Er	OHn	E	570·2	r	Tb	OHn	E
562·86		LaO		E	566·4	r	Gd	OHn	E	570·36	a	V	OC	E
562·9	r	Zr	OH	E	566·4	dr	Tb	OHn	E	570·36		Dy(O)		E
563·0		SnO	AH	E	566·59		Sm(O)		E	570·48		Er		E
563·04		Y		E	566·8	r	Te(O)	AH	E	570·5	r	PrO	OHn	E
563·14		Tm		A, E	566·97		Nd		E	570·5	tb	Dy(O)	OHn	E
563·2		Gd(O)		E	566·99		VO		E	570·5	b	Ho	OHn	E
563·3	r	SrF		E	567·06		SrF	OH	E	570·6	db	Er	OHn	E
563·4	v	Er	OHn	E	567·18		Sc		E	570·62		Sm		E
563·4	r	Zr	OH	E	567·2	v	Ho	OHn	E	570·67		Sm		E
563·5	v	Ho	OHn	E	567·28		MnO		E	570·7	dr	LaO	OHn	E
563·5	w	Co	OH	E	567·29	r	BaO	AH	E	570·70	a	V		E
563·5	v	C_2	OHnq	E	567·4	b	Er	OHn	E	570·81		Er(O)		E
563·5	r	Tb	OHn	E	567·5	r	PrO	OHn	E	571·0		EuOH	OHn	E
563·52		Cs	OH	E	567·58		Tm		A, E	571·00	r	BaO		E
563·7	r	SrF	FH	E	567·59		Nd		E	571·09		MnO		E
563·9	r	TbO	OHn	E	567·7		Ho	OHn	E	571·1	v	Ho	OHn	E
563·9	r	BiO	AH	E	567·8	r	Pb	OH	E	571·33	r	HoO	OHn	E
563·95		Dy		E	567·9	r	FeO	OHn	E	571·37	r	BaO		E
564		ZrO		E	568		TbO		E	571·38	r	YO	OHn	E
564·0	r	Tm	OHn	E	568·0	dr	LaO	OHn	E	571·57		ErO	OHn	E
564·1	r	SrF	FH	E	568·1	dr	GdO	OHn	E	571·59	r	GdO	OHn	E
564·3	tr	Ho	OHn	E	568·1	b	Nd	OHn	E	571·60		SmO		E
564·32		Gd		E	568·1	r	Sm	OHn	E	571·81	r	Zr	OH	E
564·41	r	BaO	AH	E	568·1	r	Tb	OHn	E	571·85		DyO		E
564·43		Ho(O)		E	568·11		LaO		E	571·9	x	Ca	OH	E
564·5	r	Tb	OHn	E	568·26	a	Na	OHn	A, E	572·0		CaOH	OH	E
564·5	v	Er	OHn	E	568·3		SmOH		E	572·0	r	PrO	OHn	E
564·5	r	PrO	OHn	E	568·4	b	Tm	OHn	E	572·33	v	ErO	OHn	E
564·58	a	Eu	OHn	E	568·4	b	Er	OHn	E	572·38		VO		E
564·7	r	FeO	OHn	E	568·51		CrO		E	572·4	r	ScO	OHn	E
564·81	a	Rb	OHn	E	568·6		CuF	PH	E	572·40	r	Zr	OH	E
565		ErO		E	568·68		Sc		E	572·44	a	Rb	OHn	E
565·13		Gd(O)		E	568·8	v	Ho	OHn	E	572·5		YbOH	AH	E
565·2	x	Ca	OH	E	568·82	a	Na	OHn	A, E	572·55		Sm		E
565·20		Dy		E	569·1	r	PrO	OHn	E	572·6		NdO	OHn	E
565·23		LaO		E	569·26		Dy(O)		E	572·70		V		E
565·3	dr	Gd(O)	OHn	E	569·3	b	Er	OHn	E	572·70		Er		E
565·37	a	Rb	OHn	E	569·4	tb	Dy(O)	OHn	E	572·73	r	Tb	OHn	E
565·4	dr	La	OHn	E	569·6	v	Ho(O)	OHn	E	572·9	dr	Ho(O)	OHn	E
565·5		EuOH		E	569·62		Gd		E	572·9		DyO	OHn	E
565·58		Gd(O)		E	569·7	x	Ca	OH	E	572·92		Nd		E
565·6	v	Ho	OHn	E	569·70		Ce		E	572·93	r	SmO	OHn	E
565·73		Dy(O)		E	569·7994		Gd		E	573		TbO		E
565·8	r	Zr	OH	E	569·8	r	Sm	OHn	E	573·0	r	Er	OHn	E
565·9		HoO		E	569·8	r	YO	OHn	E	573·0	r	IO	OH, IO_3	E
565·90	r	BaO	AH	E	569·85	a	V		E	573·02		HoO		E
565·98		Sm		E	569·9	dr	GdO	OHn	E	573·1	r	PrO	OHn	E
566		ErO		E	569·9	r	Tb	OHn	E	573·1	r	YO	OHn	E
566·0		HoO	OHn	E	569·90	a	Ru	OHn	E	573·2	dr	LaO	OHn	E

342

λ		species		E		λ		species		E		λ		species		E
573·29		Sm		E		577·01	r	BaO		E		580·2	r	Sc	OHn	E
573·3		YbOH		E		577·1	r	Gd(O)	OHn	E		580·2	r	Tb	OHn	E
573·31		Dy		E		577·13		Sm(O)		E		580·20	a	K	OHn	E
573·4	dr	Gd(O)	OHn	E		577·2	r	SrF	OH	E		580·3		Cs		E
573·49		Dy(O)		E		577·21		Nd		E		580·4	v	Ho	OHn	E
573·7	db	Ho(O)	OHn	E		577·27		ScO		E		580·51	r	BaO	AH	E
573·7	dr	VO	OHn	E		577·3	r	ScO	OHn	E		580·58		$^{11}BO_2$		E
573·7	r	ScO	OHn	E		577·37		Sm		E		580·7		GdO	OHn	E
573·7		ErO	OHn	E		577·5	r	SrF		E		580·73		Dy(O)		E
573·76		HoO		E		577·6	r	Sc	OHn	E		580·8	r	Fe	OHn	E
573·86		Tb		E		577·61		Nd		E		580·8	v	Sm	OHn	E
574		SmOH		E		577·69		Dy		E		580·92		ZrO		E
574		ZrO		E		577·7	r	PrO	OHn	E		581·0	r	CaCl	OA, C	E
574·0		BiO	AH	E		577·77	a	Ba	OH	E		581·1	dr	ScO	OHn	E
574·0	mb	Dy	OHn	E		577·8	r	SrF	OH	E		581·25	a	K	OHn	E
574·06		La		E		577·83		Sm		E		581·3	x	Ca	OH	E
574·57	a	Cs	Oh	E		577·85		ZrO		E		581·32		$^{11}BO_2$		E
574·6	r	Gd(O)	OHn	E		577·9	r	SrF	OH	E		581·5	r	Tb	OHn	E
574·64	r	Sm(O)	OHn	E		577·92		Sm		E		581·76	r	BaO		E
574·7	r	Y(O)	OHn	E		577·92		Er		E		581·9	r	FeO	OHn	E
574·72		Dy(O)		E		578·0		CaOH		E		581·9	v	Gd(O)	OHn	E
574·81	r	Zr	OH	E		578·2		SrF	OH	E		581·9	r	YO	OHn	E
574·84		Dy(O)		E		578·21		Cu	OHn	E		581·94		Sm		E
574·87		CrO		E		578·23		$^{11}BO_2$		E		582		SrF	OH	E
574·9	r	PrO	OHn	E		578·25		Tm(O)		E		582·0	v	Ho	OHn	E
574·9		TmO	OHn	E		578·26	a	K	OHn	E		582·0	mb	SmO	OHn	E
574·91		Nd		E		578·3	r	YO	OHn	E		582·1	r	Tb	OHn	E
574·96		Nd		E		578·37		Eu		E		582·5		CaOH		E
575·07		MnO		E		578·47		Dy(O)		E		582·67		Er		E
575·1	b	Er(O)	OHn	E		578·49		Nd		E		582·67		Nd		E
575·13		Gd(O)		E		578·5	r	SrF	OH	E		582·92	r	BaO		E
575·2	v	Ho(O)	OHn	E		578·64		VO		E		583·0	r	CaF	FH	E
575·4	r	Zr	OH	E		578·82		Nd		E		583·1	v	Ho	OHn	E
575·51	v	Tb	OHn	E		578·83		Sm		E		583·10		Eu		E
575·54		Sm		E		578·84	r	Gd(O)	OHn	E		583·19		$^{11}BO_2$		E
575·6	r	Gd(O)	OHn	E		578·92		La		E		583·2	v	Gd	OHn	E
575·8		YbOH		E		579		BO_2		E		583·20		Dy		E
575·84	r	BaO	AH	E		579·0	r	FeO	OHn	E		583·21	a	K	OHn	E
575·9		TiO		E		579·0	b	B	OHn	E		583·4	tb	DyO	OHn	E
576·055		Th		E		579·07		$^{11}BO_2$		E		583·5	v	Ho	OHn	E
576·1	r	Sc	OHn	E		579·09		Sm		E		583·7	r	Sc	OHn	E
576·1	dr	Ti	OH	E		579·1	r	PrO	OHn	E		583·8	r	VO	OHn	E
576·18		La		E		579·13		La		E		583·8	r	Y	OHn	E
576·27		Er		E		579·4		CrO		E		583·88	a	Cs	OH	E
576·3	r	PrO	OHn	E		579·40		Dy		E		583·9	r	Tb	OHn	E
576·30		Tm(O)		E		579·52		MnO		E		583·91		Nd		E
576·42		Y(O)	OHn	E		579·8	r	Sc	OHn	E		583·92		Sm(O)		E
576·42		Tm		E		580·0		CuOH		E		584·0	r	Sc	OHn	E
576·5	r	ScO	OHn	E		580·0	r	YO	OHn	E		584·0	db	Sm	OHn	E
576·52	a	Eu	OHn	E		580·00		Tm(O)		E		584·09		Sm		E
576·65		$^{11}BO_2$		E		580·00		Nd		E		584·2	r	YO	OHn	E
576·79		Er		E		580·00		Sm		E		584·4		Gd	OHn	E
576·8	v	Tb(O)	OHn	E		580·02		Eu		E		584·44		Dy(O)		E
576·93		La		E		580·02		Sm		E		584·51	a	Cs	OH	E

343

λ		Species		Ref	λ		Species		Ref	λ		Species		Ref
584·7		Gd(O)		E	589·5	dr	La	OHn	E	594·25		VO		E
584·7		SmOH		E	589·51		Sm		E	594·27		Gd(O)		E
584·76	a	Cs	OH	E	589·59		Na	OHn	A, F, E	594·3	mb	SmO	OHn	E
584·77		ScO		E	589·67		LuO		E	594·3	b	Sr	AH	E
584·9	v	Ho	OHn	E	589·89		Sm		E	594·4	r	Mn	OHn	E
584·91	dr	ScO	OHn	E	590·2	r	Tb(O)	OHn	E	594·5		SrO(H)		E
585·00		Er		E	590·26		Sm		E	594·58		DyO		E
585·2	r	CrO	OHn	E	590·30	r	Fe	OHn	E	594·7	mb	SmO	OHn	E
585·2		SrF	OH	E	590·5	b	Gd(O)	OHn	E	594·80		Ho		E
585·4	v	Ho	OHn	E	590·60		Sm		E	595·0	dr	LaO	OHn	E
585·52		Er		E	590·9	r	Tb(O)	OHn	E	595·0		Gd(O)		E
585·56	dr	Dy(O)	OHn	E	590·90		Sm		E	595·0		SmO		E
585·72	r	Sm(O)	OHn	E	591·0	r	Mn	OHn	E	595·0		PrO		E
585·73		Dy(O)		E	591·0		Ho		E	595·52		SmO		E
585·73	r	TbO	OHn	E	591·1		GdO	OHn	E	595·6	tr	GdO	OHn	E
585·81		Gd		E	591·1		FeO		E	595·6	r	Y	OHn	E
585·9	r	YO	OHn	E	591·1	r	Pb	OH	E	595·73		GdO		E
586·0	r	MnO	OHn	E	591·14	a	Gd	OA	E	595·85		Sm		E
586·0		Y		E	591·22	r	YO	OHn	E	595·9	r	Sc	OHn	E
586·02		Ho		E	591·31		CrO		E	596·0	db	TbO	OHn	E
586·2	b	Gd	OHn	E	591·54		U		E	596·14		Dy(O)		E
586·45	r	BaO	AH	E	591·63		Sm		E	596·15		Tb(O)		E
586·47	v	Ho	OHn	E	591·8	r	Sc	OHn	E	596·21	r	Sm(O)	OHn	E
586·5	r	Tb	OHn	E	591·81	r	Gd(O)	OHn	E	596·3	r	Pr	OHn	E
586·6		Gd(O)		E	591·9		FeO		E	596·37		Eu		E
586·6		LaO		E	592·1		LaO		E	596·4		Gd(O)		E
586·7		YbOH		E	592·1	r	TbO	OHn	E	596·43		Tm		A
586·77		Sm		E	592·1	r	Dy	OHn	E	596·5	r	Ca	OH	E
586·8	r	FeO	OHn	E	592·12		Nd		E	596·6	tb	Tb(O)	OHn	E
586·8	b	Dy(O)	OHn	E	592·17		Ho		E	596·71		Eu		E
586·8	dr	La	OHn	E	592·2	dr	La	OHn	E	596·9	dr	ScO	OHn	E
586·86		SmO	AA	E	592·4	mb	Sm	OHn	E	597·0	tb	SrO(H)	AH	E
587·0		YbOH	AH	E	592·5		PrO		E	597·1	x	Ca	OH	E
587·10		Sm		E	592·5	r	Tb	OHn	E	597·1	r	LuO	OHn	E
587·41		Sm		E	592·51	r	BaO		E	597·1	r	Nd(O)	OHn	E
587·53	r	BaO		E	592·7	r	GdO	OHn	E	597·12		Tm		E
587·59		Sm		E	592·70		Sm(O)		E	597·17	a	Ba	OH	E
587·6	r	Tb	OHn	E	592·81	r	ScO	OHn	E	597·2	dr	Gd(O)	OHn	E
587·6	r	Y(O)	OHn	E	593·06		La		E	597·2	r	Y(O)	OHn	E
587·8		CN		E	593·1	r	YO	OHn	E	597·2	r	Pr	OHn	E
587·8	r	Sc	OHn	E	593·18		Tm(O)		E	597·21	r	Sm(O)	OHn	E
588·0	b	Ho	OHn	E	593·3	v	Tb(O)	OHn	E	597·27		Eu		E
588·1	r	MnO	OHn	E	593·4	r	CaCl	OA, C	E	597·27		Ho		E
588·29		Ho		E	593·48		FeO		E	597·28	u	Ba		E
588·4	db	Dy	OHn	E	593·5	x	Ca	OH	E	597·35		Ho		E
588·69	r	BaO	AH	E	593·9	r	YO	OHn	E	597·4	r	Nd(O)	OHn	E
588·8		ScO	OHn	E	594		Eu(O)		E	597·45		Dy		E
588·83		Mo		E	594·0	r	TbO	OHn	E	597·5	r	FeO	OHn	E
588·9	r	V	OHn	E	594·0	dr	Dy	OHn	E	597·59		CrO		E
588·97		Nd		E	594·04		TbO		E	597·60		Gd(O)		E
588·995		Na		A	594·05		Dy		E	597·63	r	BaO	AH	E
589		LaO		E	594·09		Sm(O)		E	597·7	dr	LaO	OHn	E
589·15		Tm(O)		E	594·2	dr	Gd(O)	OHn	E	597·77		ZrO		E
589·4	r	Y	OHn	E	594·2	r	Pr	OHn	E	597·9	dr	Tb(O)	OHn	E

597·9	dr	Dy(O)	OHn	E	601·8195		Eu		E	604·7	r	SmO	OHn	E
597·93		Sm		E	601·9	r	Pr(O)	OHn	E	604·9	x	Ca	OH	E
598		TbO		E	601·9	v	Gd(O)	OHn	E	604·97		Dy		E
598·0		EuOH	OHn	E	601·9	r	Sm	OHn	E	605		CuO		E
598·1	r	SmOH	OHn	E	601·95	a	Ba	OH	E	605·0	r	Dy	OHn	E
598·15		Tb(O)		E	602·0		CaOH	OH	E	605·0	v	TbO	OHn	E
598·29		Ho		A, E	602·0	w	(RaOH)		E	605·2	r	CrO	OHn	E
598·3	r	Pr	OHn	E	602·0	w	YbOH	AH	E	605·3		NdO	OHn	E
598·49	r	BaO	AH	E	602·0	r	Y(O)	OHn	E	605·4	r	YO	OHn	E
598·6	v	Ca(O)	AH	E	602·0		PrO		E	605·4		Gd(O)		E
598·7		GdO		E	602·13		ZrO		E	605·5	r	Pr	OHn	E
598·8	r	Y(O)	OHn	E	602·2	v	Ca(OH)OH		E	605·5	r	Sc		E
598·8	dr	Sm(O)	OHn	E	602·2	b	Tb(O)	OHn	E	605·50		Lu		E
598·8		GdO	OHn	E	602·6	r	Pr	OHn	E	605·6	r	TbO	OHn	E
598·85		Dy		E	602·6	r	Tb	OHn	E	605·63		Dy		E
598·89		Sm		E	602·7	r	Nd	OHn	E	605·73		Eu		E
599		YO		E	602·7	db	Dy(O)	OHn	E	605·8	db	DyO	OHn	E
599·0	b	Nd(O)	OHn	E	602·72		Sm(O)		E	605·95		Dy(O)		E
599·1	r	Pr	OHn	E	602·77		Dy		E	606·0	dr	Cu	OH	E
599·28		Eu		E	602·9	x	Ca(OH)OH		E	606·0		SrOH		E
599·3	r	Sm(O)	OHn	E	602·90		Eu		E	606·09		Pr(O)		E
599·3	r	LuO	OHn	E	603		PrO		E	606·1	dr	La	OHn	E
599·51		Sm		E	603·0	b	B	OHn	E	606·29		Gd(O)		E
599·76		VO		E	603·0		Dy(O)		E	606·31	a	Ba	OH	E
599·8	db	Nd(O)	OHn	E	603·0		SmO		E	606·4	v	CaF	FH	E
599·9	b	Tb	OHn	E	603·0662		Mo		E	606·4	db	Gd	OHn	E
600		Ni		E	603·3	dr	LaO	OHn	E	606·4	r	ScO	OHn	E
600·0		LuO		E	603·41	a	Cs	OH	E	606·5	b	Sm	OHn	E
600·00		Nd		E	603·45		LaO		E	606·58		Gd(O)		E
600·1	db	Gd	OHn	E	603·45		Tb(O)		E	606·8	v	Ca(O)	OH	E
600·18	b	Gd(O)		E	603·5	v	Ca(OH)OH		E	606·8	r	Tb	OHn	E
600·2	v	NdO	OHn	E	603·5	v	Gd(O)	OHn	E	607·0074		Sm		E
600·2	r	Sc	OHn	E	603·5		SmO	OHn	E	607·08	a	Rb	OHn	E
600·3	r	Ca(O)	OH	E	603·6	r	Pr(O)	OHn	E	607·1	r	Sm(O)	OHn	E
600·4	r	Y(O)	OHn	E	603·6	tr	Tb(O)	OHn	E	607·1	dr	TbO	OHn	E
600·42		Sm		E	603·62	r	ScO	OHn	E	607·17		Pr(O)		E
600·45		Lu		E	603·7	r	BiO	AH	E	607·3	r	ScO	OHn	E
600·5	dr	La(O)	OHn	E	603·7	r	YO	OHn	E	607·3	r	Y	OHn	E
600·5	dr	Pr	OHn	E	603·8		CaOH		E	607·6	v	Ca(O)	OH	E
600·57	tv	Dy(O)	OHn	E	603·8		PrO		E	607·66		Tb(O)		E
600·6	v	Ca(O)	OH	E	603·96	r	BaO		E	607·7	dr	SrOH	OG	E
600·67		LaO		E	603·97		V		E	607·8	dr	Dy(O)	OHn	E
600·7	b	Gd(O)	OHn	E	603·99		Tb(O)		E	607·8	mr	Tb(O)	OHn	E
600·70	b	Dy(O)		E	604		SrOH		E	607·89		Dy(O)		E
600·76		Nd		E	604·20	r	Pr	OHn	E	607·9		Tb(O)		E
601		ErO		E	604·28		CrO		E	607·9	r	Sc(O)	OHn	E
601		GdO		E	604·28		GdO		E	607·9	mb	Sm	OHn	E
601·0		NdO		E	604·3	db	Dy	OHn	E	608·0		TbO		E
601·05	a	Cs	OH	E	604·37		Dy		E	608·1	tb	Gd	OHn	E
601·08		Dy		E	604·4	tb	GdO	OHn	E	608·14		Sm(O)		E
601·4	r	LuO	OHn	E	604·46		Eu		E	608·14		V		E
601·4	v	Nd	OHn	E	604·49		Sm		E	608·17		Ho		E
601·4	b	Tb(O)	OHn	E	604·5	r	TbO	OHn	E	608·19		Gd(O)		E
601·7	r	Sc(O)	OHn	E	604·51	r	Cu	OH	E	608·38		Eu		E

608·4	db Dy	OHn		E
608·41	Sm			E
608·47	FeO			E
608·5	mv SrOH	OG		E
608·64	VO			E
608·82	Dy			E
608·9	r Sm	OHn		E
608·9	r Y	OHn		E
609	FeO			E
609	SmO			E
609	Gd(O)			E
609·0	v SrOH	OG		E
609·02	V			V
609·06	Pr(O)			E
609·1	DyO			E
609·1	db Tb(O)	OHn		E
609·2	v Ca	OH		E
609·2	r Sc	OHn		E
609·5	b Fe	OHn		E
609·5	r SnH	AH		E
609·5	tv SrOH	OG		E
609·66	Tb			E
609·7	FeO			E
609·7	r YO	OHn		E
609·76	Gd(O)			E
609·8	v Ca(O)	OH		E
609·8	r Pr(O)	OHn		E
609·8	dr Tb	OHn		E
609·9	dr Gd	OHn		E
609·93	Eu			E
609·99	b Sm(O)			E
609·9918	Sm			E
610·1	dv SrOH	OG		E
610·2	r Sc	OHn		E
610·23	r BaO			E
610·3	b Tb(O)	OHn		E
610·36	a Li	OH		E
610·48	Gd(O)			E
610·8	r Y	OHn		E
610·83	r Gd(O)	OHn		E
610·84	Sm			E
610·99	b Fe(O)	OHn		E
610·99	ScO			E
611·0	dv SrOH	OG		E
611·0	r ScO	OHn		E
611·08	a Ba	OH		E
611·16	V			E
611·2	b Tb(O)	OHn		E
611·3	r Sm(O)	OHn		E
611·4	v SrOH			E
611·40	Gd			E
611·5	x Ca	OH		E
611·5	r Gd	OHn		E
611·5	r YO	OHn		E
611·6	r Sc	OHn		E
611·62	r BaO			E
611·76	Tb(O)			E
611·77	Sm			E
611·95	V			E
612	GdO			E
612·1	db Dy(O)	OHn		E
612·2	v Tb	OHn		E
612·222	Ca			A
612·7	r Y	OHn		E
613	NdO			E
613·0	r Pr	OHn		E
613·2	r Sm	OHn		E
613·2	r Y(O)	OHn		E
613·2	r Gd	OHn		E
613·3	v Nd	OHn		E
613·4	r Tb	OHn		E
613·5	r PrO	OHn		E
613·6	r Gd	OHn		E
613·8	r Sm	OHn		E
613·91	VO			E
614	Gd(O)			E
614·0	r Tb	OHn		E
614·0	r Sc	OHn		E
614·17	Ba II	OAn		E
614·64	Pr(O)			E
614·7	r Cu	OH		E
614·8	Y(O)	OHn		E
614·9	r Sc	OHn		E
614·9	db Sm	OHn		E
615	YO			E
615·0	CuOH	OH		E
615·2	r Gd	OHn		E
615·4	r Sc(O)			E
615·42	a Na	OHn		A, E
615·47	MnO			E
615·6	tr Gd	OHn		E
615·62	Sm(O)			E
615·73	Gd(O)			E
615·95	Sm			E
615·96	a Rb	OHn		E
616·07	a Na	OHn		A, E
616·1	r Pb	OH		E
616·129	Ca			A
616·2	mr Cu	OH		E
616·3	r Pr	OHn		E
616·5	r Sm	OHn		E
616·51	r YO	OHn		E
616·51	r BaO			E
616·73	Dv CrO			E
616·84	Dy			E
616·85	r SmO	OHn		E
616·96	r BaO			E
617	SmOH			E
617·16	$^{11}BO_2$			E
617·2	r GdO	OHn		E
617·49	Sm			E
617·6	r MnO	OHn		E
617·6	r GdO	OHn		E
618	PrO			E
618·0	NdO	OHn		E
618·1	r FeO	OHn		E
618·1	r Sc	OHn		E
618·2	r YO	OHn		E
618·20	Pr			E
618·3	r Gd(O)	OHn		E
618·5	v CaCl	OA, C		E
618·8	dr La	OHn		E
618·81	r Sc(O)	OHn		E
618·81	Eu			E
619·16	VO			E
619·3	r ScO	OHn		E
619·4407	Sm			E
619·5	r SmO	OHn		E
619·50	DyO			E
619·50	Eu			E
619·56	PrO			E
619·91	V			E
620·0	SrOH			E
620·0	b B	OHn		E
620·0	r SmO	OHn		E
620·0	r YO	OHn		E
620·1	dr GdO	OHn		E
620·22	$^{11}BO_2$			E
620·30	GdO			E
620·4	r Mn(O)	OHn		E
620·63	a Rb	OHn		E
620·80	PrO			E
621·0	b Ra	OH		E
621·06	Sc			E
621·2	v CaCl	OA, C		E
621·2	r Gd	OHn		E
621·31	a Cs	OH		E
621·38	V			E
621·5	r Ti	OH		E
621·6	dr La	OHn		E
621·63	V			E
621·76	a Cs	OH		E
621·8	r YO	OHn		E
621·9	r FeO	OHn		E
621·9	NdO	OHn		E
622	CaOH			E
622·0	w YbOH	AH		E
622·0	r Sc	OHn		E
622·1	dr GdO	OHn		E
622·101	Er			E
622·2	r Ti			E
622·26	Y			E

622·45		V		E
622·47	r	BaO	AH	E
622·9	r	SmO	OHn	E
622·91		CrO		E
622·94	r	ScO	OHn	E
622·94		ZrO		E
623·0	w	EuOH	OHn	E
623·0		ScO		E
623·0		CaOH		E
623·07		V		E
623·2	r	GdO	OHn	E
623·31	r	ScO	OHn	E
623·32		V		E
623·7	r	YO	OHn	E
623·72	r	Mn	OHn	E
623·8	db	SmO	OHn	E
623·94		Sc		E
623·97		Sc		E
624·0	w	Sr	AH	E
624·2	r	Gd(O)	OHn	E
624·28		V		E
624·3	db	SmO	OHn	E
624·31		V		E
624·5	dr	La	OHn	E
624·5	r	TbO	OHn	E
624·53	r	GdO	OHn	E
624·7		NdO	OHn	E
624·7		RaOH		E
624·72		VO		E
624·99		La		E
625·0		CuOH		E
625·1	r	Pb	OH	E
625·1	r	Y	OHn	E
625·18		V		E
625·21	r	GdO	OHn	E
625·58		YO		E
625·6	tb	SmO	OHn	E
625·9	r	PrO	OHn	E
625·90		Dy		E
625·99		Sc		E
626·0		LuO		E
626·0		SrOH		E
626·09		ZrO		E
626·2	r	Sc	OHn	E
626·22		Eu		E
626·3	r	GdO	OHn	E
626·31	r	Pr	OHn	E
626·69		Eu		E
626·7	r	Gd(O)	OHn	E
626·7		PrO		E
626·9	r	NdO	OHn	E
626·9		RaOH		E
627·1	r	PrO	OHn	E
627·2	r	ScO	OHn	E
627·3	r	Gd	OHn	E
627·33		SmO	OHn	E
627·4	dr	La	OHn	E
627·46		V		E
627·5	r	ScO	OHn	E
627·5	r	YO	OHn	E
627·51		MnO		E
627·63		Sc		E
627·818		Au		A
627·89		FeO		E
627·9	r	Pr	OHn	E
628		NdO		E
628·0	b	Eu	OHn	E
628·1	v	Nd	OHn	E
628·3	r	SrF	OH	E
628·4	r	GdO	OHn	E
628·5	v	CaF	OH	E
628·5		RaOH		E
628·51		V		E
628·7	r	NdO	OHn	E
629·0	r	GdO	OHn	E
629·10	r	BaO	AH	E
629·13		Eu		E
629·2	r	TbO	OHn	E
629·28		V		E
629·28		ZrO		E
629·41		CrO		E
629·42	r	Gd	OHn	E
629·55	r	YO	OHn	E
629·59		FeO		E
629·64		V		E
629·7	r	NdO	OHn	E
629·80	r	PrO	OHn	E
629·83	a	Rb	OHn	E
629·97		Eu		E
630·2	dr	La	OHn	E
630·33		VO		E
630·41		Nd		E
630·5		CoO		E
630·52		NdO		E
630·53		Ho		E
630·56		Sc		E
630·6	r	GdO	OHn	E
630·6	mv	SrF		E
630·63	a	Rb	OHn	E
630·87		Er		E
631·04		Nd		E
631·2		GdO		E
631·3		NdO	OHn	E
631·47		MnO		E
631·51		NdO		E
631·60	r	GdO	OHn	E
631·6	r	Y	OHn	E
631·61		ScO		E
631·8	r	Pr	OHn	E
631·8	r	ScO	OHn	E
631·83		Pr(O)		E
631·84		YO		E
632·25		Nd(O)		E
632·59		La		E
632·61		Er		E
632·8		Sm(O)		E
632·9	r	Gd	OHn	E
632·9		RaOH		E
633·1	dr	La	OHn	E
633·2	b	Nd	OHn	E
633·5	v	SmOH	OHn	E
633·55		YO		E
633·57		Eu		E
633·59		Gd(O)		E
633·6	r	Pr	OHn	E
633·63		Nd(O)		E
633·8		Gd	OHn	E
633·8	r	Y	OHn	E
633·9	r	Sm	OHn	E
634·0		CoO		E
634·1	r	Tb	OHn	E
634·14		Pr(O)		E
634·2	r	Pb	OH	E
634·45		Sm(O)		E
634·48		Sc		E
634·49		ZrO		E
634·52		Nd(O)		E
634·8	r	Sm	OHn	E
634·9	r	Tb	OHn	E
634·9	r	Nd	OHn	E
634·9		RaOH		E
634·90		Dy(O)		E
635		SmO		E
635·0		Tb		E
635·00		Eu		E
635·2		NdO		E
635·2		TbO		E
635·2	r	Gd	OHn	E
635·2	r	Y	OHn	E
635·2	r	Pr	OHn	E
635·3	r	Sm	OHn	E
635·4	b	Nd	OHn	E
635·46	a	Co	OH	E
635·58		Eu		E
635·7	r	Pr	OHn	E
635·80		BaO		E
635·84		Nd(O)		E
635·9	v	SrCl	PH	E
636·0		La	OHn	E
636·04		VO		E
636·1	b	Nd	OHn	E
636·1	v	Sm	OHn	E

636·2	r	Sc	OHn	E
636·2	v	SrCl	PH	E
636·23	a	Zn	OHn	E
636·3	r	PrO	OHn	E
636·37		CrO		E
636·52		Nd(O)		E
636·74		Sm		E
636·92		Eu		E
637·0	db	NdO	OHn	E
637·0	r	Y	OHn	E
637·4	r	Sm	OHn	E
637·66		$^{11}BO_2$		E
637·7	r	Pr	OHn	E
637·83		ZrO		E
637·88		Sc		E
637·9	db	SmO	OHn	E
638·2	r	Nd	OHn	E
638·27		Eu		E
638·38		Eu		E
638·5196		Nd		E
638·6		NdO		E
638·68		Dy		E
638·7	r	Y	OHn	E
638·8	b	Sm	OHn	E
638·9	b	Nd	OHn	E
639·0	b	B	OHn	E
639·1	r	Sm	OHn	E
639·22		Dy(O)		E
639·4		SrF	OH	E
639·4	r	Cr(O)	OHn	E
639·42		La		E
639·5		Pr(O)		E
639·54	r	TbO	OHn	E
639·6		SmO	OHn	E
639·60		$^{11}BO_2$		E
639·76		Nd		E
640·00		Fe		E
640·09		Eu		E
640·20		Y		E
640·4	b	Nd(O)	OHn	E
640·6	db	Sm(O)	OHn	E
640·6	r	Y	OHn	E
640·61		Eu		E
640·84	r	ScO	OHn	E
641·0	r	Nd	OHn	E
641·01		Eu		E
641·09		La		E
641·13		Eu		E
641·23		ZrO		E
641·33		Sc		E
641·85		VO		E
641·9	v	SrF	OH	E
642·0	r	Sc	OHn	E
642·1		SmO	OHn	E
642·19		Dy		E
642·31	r	BaO	AH	E
642·4	r	Sc	OHn	E
642·4	r	Y	OHn	E
642·5	v	Sm(O)	OHn	E
642·5	r	NdO	OHn	E
642·5909		Sm		E
642·8	r	Pb	OH	E
643·0		NdO		E
643·03		FeO		E
643·20	a	Cs	OH	E
643·50		Y		E
643·50		Dy(O)		E
643·61	r	Tb(O)	OHn	E
643·62		Dy(O)		E
643·7	r	Sc	OHn	E
643·9	r	Sm(O)	OHn	E
644·0		CaOH	OH	E
644·0		Tm(O)		E
644·21	r	Nd(O)	OHn	E
644·52		FeO		E
644·54		Nd(O)		E
644·62	r	ScO	OHn	E
644·8	v	Nd(O)	OHn	E
644·81		Sc		E
645		SrOH		E
645·0	v	Tb(O)	OHn	E
645·0	b	Sm(O)	OHn	E
645·09	a	Ba	OH	E
645·2	r	Cr	OHn	E
645·3	v	Nd(O)	OHn	E
645·3		CaOH		E
645·42		Nd		E
645·45		La		E
645·48		Tb(O)		E
645·57		Nd(O)		E
645·59		La		E
645·6	dr	Sm(O)	OHn	E
645·68		Sm(O)		E
645·79		Eu		E
645·8	r	Pr	OHn	E
645·8	r	Sc	OHn	E
646·0	r	Tb(O)	OHn	E
646·0		SrOH		E
646·02		Tm		E
646·2	dr	Sc	OHn	E
646·21		Sm(O)		E
646·4	r	Nd(O)	OHn	E
646·5	w	Sr	AH	E
646·82	r	Y	OHn	E
646·9	r	Sm(O)	OHn	E
647		Sc(O)		E
647·0	w	Eu	OHn	E
647·14		Tb(O)		E
647·26	a	Cs	OH	E
647·36		Sm(O)		E
647·37		ZrO		E
647·4	r	PrO	OHn	E
647·43	db	Sm(O)	OHn	E
647·6	r	Sc	OHn	E
647·8	dr	V	OHn	E
647·82	r	Sm(O)	OHn	E
647·9		Pr(O)		E
648		EuOH		E
648·29	u	Ba	OH	E
648·5	dr	SmOH	OHn	E
648·5	v	Tb(O)	OHn	E
648·5	r	Y	OHn	E
648·53		Sm(O)		E
648·54		ScO		E
648·56		Nd		E
648·91		Yb		E
649·3	b	Nd(O)	OHn	E
649·31	r	BaO	AH	E
649·4	r	PrO	OHn	E
649·6	r	Sc	OHn	E
649·60	r	Tb(O)	OHn	E
649·69		BaI	OAn	E
649·81		MnO		E
649·88	a	Ba	OH	E
649·88		Sm(O)		E
649·89		Nd(O)		E
650		CN		E
650		TbO		E
650·0	dv	NdO	OHn	E
650·0		Sc(O)		E
650·0		Lu(O)		E
650·12	r	Y	OHn	E
650·15		Nd		E
650·3		Sc	OHn	E
650·81		ZrO		E
651·0	trv	SmO	OHn	E
651·18		CrO		E
651·2	mvb	SrF	OH	E
651·24		Sm(O)		E
651·4	r	Pr(O)	OHn	E
651·58		Nd(O)		E
651·8	r	Sc	OHn	E
651·8	db	Nd(O)	OHn	E
651·83		YO		E
652		SmO		E
652·38		MnO		E
652·41		FeO		E
652·5	r	SmO	OHn	E
652·6	r	Sc	OHn	E
652·73	a	Ba	OH	E
652·8		SrF	OH	E
652·80		Sm		E

652·99		Sm(O)		E
653·1	r	V	OHn	E
653·2	mr	Sm	OHn	E
653·2		CoO		E
653·28		VO		E
653·30		Sm(O)		E
653·53	r	Sc	OHn	E
653·6	r	Y	OHn	E
654·0	db	Sm(O)	OHn	E
654·0		YbOH		E
654·16	v	Nd	OHn	E
654·30		ZrO		E
654·31		La		E
655		PrO		E
655·0		EuO	OHn	E
655·18		Sm		E
655·36		Sm		E
655·4	r	Y	OHn	E
655·7	r	Sm(O)	OHn	E
655·73		Y		E
655·77		MnO		E
655·8	r	Sc	OHn	E
656·25		LaO		E
656·32	r	BaO	AH	E
656·352		Sm		E
656·4	b	Nd(O)	OHn	E
656·53	r	Sm(O)	OHn	E
656·67		FeO		E
657	r	ScO	OHn	E
657·3	r	Y	OHn	E
657·32		LaO		E
657·34		Pr(O)		E
657·5		SmO		E
657·51		CrO		E
657·6	r	Sc	OHn	E
657·85		La		E
657·93		Dy		E
658·0	r	Sm(O)	OHn	E
658·0	v	NdO	OHn	E
658·053		Sm		E
658·2	r	Nd	OHn	E
658·34		Er		E
658·38		LaO		E
658·5		Sm	OHn	E
658·6	db	Nd(O)	OHn	E
658·65	a	Cs	OH	E
658·89		VO		E
658·89		Sm		E
658·9	r	Pr(O)	OHn	E
659·1	r	Sc	OHn	E
659·44		LaO		E
659·47		MnO		E
659·53	a	Ba	OH	E
659·84	r	NdO	OHn	E
660		NdO		E
660		SrOH		E
660·0		SmOH	OHn	E
660·01	r	Nd	OHn	E
660·1	r	Sc	OHn	E
660·49		Er		E
660·49		Ho		E
660·51		LaO		E
660·74		Ho		E
660·8	db	Nd	OHn	E
660·9	r	Pr(O)	OHn	E
661·0	r	Sc	OHn	E
661·1	r	Y	OHn	E
661·2	r	Nd(O)	OHn	E
661·4		SrCl	PH	E
661·65		La		E
661·8	r	Sc	OHn	E
661·90	r	Sm(O)	OHn	E
662·0	w	Eu	OHn	E
662·1	b	Nd	OHn	E
662·4	b	Nd	OHn	E
662·5	r	Ca	OA	E
662·87	a	Cs	OH	E
662·89		Ho		E
663·0	r	Nd(O)	OHn	E
663·1	r	Pr(O)	OHn	E
663·1	r	Y	OHn	E
663·3	mv	SrF	OH	E
663·4	r	Nd(O)	OHn	E
663·4	r	Sm(O)	OHn	E
663·45	r	BaO	AH	E
663·55		MnO		E
663·80	b	Nd(O)	OHn	E
663·9	r	Ca		E
663·98		La		E
664·0		PrO		E
664·34		Dy		E
664·36		CrO		E
664·44		La		E
664·5	r	Sc	OHn	E
664·59		Sm(O)		E
664·64		VO		E
664·76	dr	Sm(O)	OHn	E
664·8	r	Nd(O)	OHn	E
665		RaOH		E
665·08		La		E
665·1	v	Sm(O)	OHn	E
665·1		CoO		E
665·15		FeO		E
665·2	r	NdO	OHn	E
665·3		CaOH		E
665·3		RaOH		E
665·44	r	Sc	OHn	E
665·5	b	Nd(O)	OHn	E
665·6	r	Sm(O)	OHn	E
665·6		SrF	OH	E
666·10	r	Sc	OHn	E
666·14		La		E
666·43	r	Nd(O)	OHn	E
666·92	v	Sm(O)	OHn	E
667·0		CoO		E
667·14		Sm		E
667·5		Nd(O)		E
668·0	mb	SmO	OHn	E
668·0	r	Nd	OHn	E
668·1	r	Ti	OH	E
668·3	r	Nd	OHn	E
668·760		Y		E
668·8	r	LuO	OHn	E
669		SrOH		E
669·1	r	Sc	OHn	E
669·39		Eu		E
669·4		Nd(O)	OHn	E
669·90		Nd(O)		E
670		$B(_2O_3)$		E
670		FeO		E
670·02	v	Sm	OHn	E
670·1	r	Sc	OHn	E
670·19		Nd		E
670·45		BaO		E
670·56		Nd(O)		E
670·6	r	Sc	OHn	E
670·60		VO		E
670·75		Nd(O)		E
670·78		Li	OH	A, F, E
671·20		Nd(O)		E
671·4	r	Ti	OH	E
671·51		CrO		E
671·64		Nd		E
671·65	b	Sm	OHn	E
671·9	r	Y	OHn	E
672·0	w	Sr	AH	E
672·2		PrO	OHn	E
672·33	a	Cs	OH	E
672·53		Nd(O)		E
672·59		Sm		E
672·8	r	Sm(O)	OHn	E
673		TiO		E
673·07		Gd		E
673·38		Nd(O)		E
673·5		Tb(O)		E
673·67		Nd(O)		E
673·8	r	Sc	OHn	E
674·43	v	Sm(O)	OHn	E
674·5	v	SrCl	PH	E
674·50		Ho		E
674·7	r	Ti	OH	E
674·8	r	Nd(O)	OH	E

674·82	r	Sc	OHn	E
674·9	r	LuO	OHn	E
675·1	r	Sm	OHn	E
675·2	r	Sc	OHn	E
675·61		Sm(O)		E
676·11		Sm(O)		E
676·62		VO		E
676·92		Nd(O)		E
677·13		LaO		E
677·2	r	Cr	OHn	E
677·4	r	Nd(O)	OHn	E
677·53		Sm		E
677·91		Sm		E
678·1	r	Ti	OH	E
678·28	r	BaO	AH	E
678·39		LaO		E
678·7	r	Sc	OHn	E
679		TbO		E
679·37		Y		E
679·6	r	Sm(O)	OHn	E
679·64		LaO		E
679·96		Yb		E
679·85		Ca		E
680·0		SrF	OH	E
680·0	dr	Sc	OHn	E
680·09		Nd(O)		E
680·27		Eu		E
680·29		Sm		E
680·8	r	Sm(O)	OHn	E
680·89		LaO		E
681·0	r	LuO	OHn	E
681·54		Sm		E
681·61		Eu		E
681·8		PrO		E
681·9	v	Nd(O)	OHn	E
682		SrOH		E
682·0	b	B(2O3)	OHn	E
682·0	w	Sr	AH	E
682·0	r	Sm(O)		E
682·15		LaO		E
682·2	r	Pr	OHn	E
682·4	r	Sc	OHn	E
682·47	a	Cs	OH	E
682·78		Sm		E
682·82		Gd		E
683		SmO		E
683·0	r	Cr	OHn	E
683·0		CaOH		E
683·0		FeO		E
683·14	v	Nd	OHn	E
683·54		Dy		E
683·9	b	Nd	OHn	E
684·0	w	EuOH	OHn	E
684·23		Nd		E

684·63	b	Nd	OHn	E
685		CrO		E
685·1	r	Pr	OHn	E
685·39		Sm		E
685·7	r	Nd	OHn	E
685·72	r	BaO	AH	E
686·10		Sm		E
686·4	r	Sc	OHn	E
686·45		Eu		E
686·46		Eu	OHn	E
686·47		Nd(O)		E
687·05	a	Cs	OH	E
687·2	r	Sm(O)	OHn	E
688·0	v	Nd(O)	OHn	E
688·0		MnO		E
688·22	r	Pr(O)	OHn	E
688·7	db	Nd(O)	OHn	E
688·7		PrO		E
689·15		CrO		E
689·259		Sr		A
689·3	r	Pr	OHn	E
689·37		Nd(O)		E
689·40		VO		E
689·6	r	Cr	OHn	E
689·71	r	Nd(O)	OHn	E
690·12		Sm(O)		E
690·15		Pr(O)		E
690·42		Nd(O)		E
690·6	r	Sc	OHn	E
690·64		Pr(O)		E
690·8	r	Nd(O)	OHn	E
691·0		CoO		E
691·13	a	K	OHn	E
691·3		MnO		E
691·8	r	Nd(O)	OHn	E
691·82		Sm(O)		E
691·9	r	H2O	OHq	E
692·2	r	Sc	OHn	E
692·4	r	Pr	OHn	E
692·42		Nd(O)		E
693·0	r	PrO	OHn	E
693·1	tb	Nd(O)	OHn	E
693·15	r	BaO	AH	E
693·7	r	Sc	OHn	E
693·90	a	K	OHn	E
694·0	db	Nd	OHn	E
694·26		NdO		E
694·5	db	Nd(O)	OHn	E
695·0	tr	CN	OCq	E
695·0	r	PrO	OHn	E
695·2	r	V	OHn	E
695·3		MnO		E
695·62	r	CaO	OA	E
695·72		CrO		E

695·72	r	Nd(O)	OHn	E
695·81		Dy		E
696·3	r	Sc	OHn	E
696·418		K		E
696·469		K		E
696·84	r	CaO		E
697·02	b	Nd	OHn	E
697·33	a	Cs	OH	E
697·62		Pr		E
697·9	b	V	OHn	E
698·0		CaOH		E
698·29	r	CaO	OA	E
698·35	a	Cs	OH	E
698·72	r	PrO	OHn	E
699·1	v	Pr	OHn	E
699·1	r	Sc	OHn	E
699·2	v	Nd(O)	OHn	E
699·5	r	La(O)	OHn	E
699·6		MnO		E
699·7	r	NdO	OHn	E
700·44		Nd		E
700·5	r	Sc	OHn	E
700·55		Pr(O)		E
700·71		BaO		E
701·1	r	Nd(O)	OHn	E
701·1	r	V	OHn	E
701·12	r	LaO	OHn	E
701·81	r	Pr(O)	OHn	E
701·91		Nd(O)		E
702·0		EuOH	OHn	E
702·1	db	Nd	OHn	E
702·2		FeO		E
702·31		Nd(O)		E
702·5	r	LaO	OHn	E
702·6	r	Sc	OHn	E
702·7	b	Nd	OHn	E
702·75		CrO		E
703·31	r	NdO	OHn	E
703·6	r	Sc	OHn	E
703·8	b	V	OHn	E
703·9	r	Pr(O)	OHn	E
704	w	SrOH	AH	E
704·1	r	LaO	OHn	E
704·4		MnO		E
705·5	r	LaO	OHn	E
705·6	r	Nd(O)	OHn	E
705·7	dr	Ti	OH	E
706·0		LuO		E
706·00	a	Ba	OH	E
706·4	b	Nd(O)	OHn	E
707·0	r	V	OHn	E
707·0		SrOH		E
707·02		VO		E
707·08	r	LaO	OHn	E

λ					λ					λ				
708·2	r	Sc	OHn	E	725·43		BaO		E	743		LaO		E
708·5	r	LaO	OHn	E	725·5	r	Nd(O)	OHn	E	743·1	dr	Pr(O)	OHn	E
709·0	dr	Ti	OH	E	725·72	r	LaO	OHn	E	743·4	dr	V	OHn	E
709·2	db	Nd(O)	OHn	E	726·5		FeO		E	743·43		LaO		E
709·4	r	Sc	OHn	E	726·8		MnO		E	743·53		VO		E
709·5		PrO	OHn	E	726·8	r	Nd(O)	OHn	E	743·65		Pr(O)		E
709·74	r	BaO		E	727·55		CeO		E	743·95		CrO		E
710·0	r	NdO	OHn	E	727·6	r	Sc	OHn	E	744·0	tr	CN	OCq	E
710·0		V		E	727·9	r	Pr(O)	OHn	E	744·0		CoO		E
710·1	r	LaO	OHn	E	728·0	tr	CN	OCq	E	744·0	tb	Nd(O)	OHn	E
710·65		Eu		E	728·5	r	Nd(O)	OHn	E	744·04	w	BaO	AH	E
710·8		VO	OHn	E	728·92		LaO		E	744·40		VO		E
711·0	tr	CN	OCq	E	729·68		Pr(O)		E	745		PrO		E
711·11		Nd(O)		E	729·72		CeO		E	745·4		V	OHn	E
711·5	db	Nd	OHn	E	729·9	r	H₂O	OHq	E	746·5	r	LaO	OHn	E
711·6	r	LaO	OHn	E	730·1		MnO		E	746·60		VO		E
712		NdO		E	730·83		CaO		E	746·65		CeO		E
712·03	r	Pr	OHn	E	731	w	SmOH	OHn	E	746·7		Nd(O)	OHn	E
712·60	r	NdO	OHn	E	731·2	r	Cr	OHn	E	747·0		V		E
712·7	dr	Ti	OH	E	731·41		CeO		E	747·2	r	VO	OHn	E
712·9	r	Pr	OHn	E	731·8	v	Nd(O)	OHn	E	748·5	r	Pr	OHn	E
713·16	r	LaO	OHn	E	731·85	r	CaO	OA	E	749·0	r	V	OHn	E
713·17		VO		E	732·02		PrO	OHn	E	749·0	r	Eu	OH	E
713·3	r	Nd(O)	OHn	E	732·48		BeO		E	749·2	r	V		E
713·8	r	Pr(O)	OHn	E	732·7	r	Pr	OHn	E	749·7	r	LaO	OHn	E
714·7	r	La	OHn	E	732·76	r	CaO		E	749·8	r	Nd(O)	OHn	E
715		TiO		E	733·0		EuOH		E	750·0		CoO		E
715·5	b	Sm(O)	OHn	E	733·5		Tm(O)		E	750·06	r	SrO	AA	E
715·8	r	Nd(O)	OHn	E	733·69		BaO		E	750·7	r	La	OHn	E
715·90	r	TiO	OH	E	734·0		MnO		E	750·95		CeO		E
716·3	r	LaO	OHn	E	734·3	r	Pr(O)	OHn	E	751·5		Sm(O)		E
716·5	r	H₂O	OHq	E	734·3	v	Nd(O)	OHn	E	751·7	b	V	OHn	E
717·2	tr	Nd(O)	OHn	E	734·69		CeO		E	752·28	r	SrO	AA	E
717·66		BaO		E	734·7		SrH	AA	E	752·35		BaO		E
717·9	r	LaO	OHn	E	735·04	r	Pr(O)	OHn	E	752·7		FeO		E
718		SmOH		E	735·07		CeO		E	752·82	r	LaO	OHn	E
718·6	r	Nd(O)	OHn	E	737·4	dr	V	OHn	E	752·9	r	Nd(O)	OHn	E
719·0	r	Cr	OHn	E	737·5		Cr	OHn	E	753·4	r	V	OHn	E
719·4	r	LaO	OHn	E	737·63	r	PrO	OHn	E	753·9	r	Pr(O)	OHn	E
720·0		Tm(O)		E	738·0	r	LaO	OHn	E	753·9	r	La	OHn	E
721·0	b	B(₂O₃)	OHn	E	738·00		CeO		E	754·16	r	SrO	AA	E
721·1	r	La	OHn	E	738·4		Mn		E	754·69		BeO		E
721·3		Nd(O)		E	739·4	r	V	OHn	E	755·6	r	Nd(O)	OHn	E
722·0		CoO		E	739·78		CeO		E	755·8	r	Sr	AA	E
722·0		SrOH		E	740·1		Pr(O)	OHn	E	756·0	r	LaO	OHn	E
722·52	r	La	OHn	E	740·4	r	LaO	OHn	E	758·0		EuOH		E
722·7		Nd(O)	OHn	E	740·5	r	V	OHn	E	759·0	w	SmOH	OHn	E
722·9		SmOH	OHn	E	740·5	r	Nd(O)	OHn	E	759·2	r	Nd(O)	OHn	E
723·58		CeO		E	741·1	r	Nd(O)	OHn	E	759·23	r	LaO	OHn	E
723·6	db	Nd	OHn	E	741·32	r	Pr	OHn	E	759·41		Pr(O)		E
724·1		MnO		E	741·80	r	V	OHn	E	761·01		BaO		E
724·5	r	Nd(O)	OHn	E	742·0	r	Nd(O)	OHn	E	761·89	a	Rb	OHn	E
724·55		Pr(O)		E	742·48		CeO		E	762·2	r	Nd(O)	OHn	E
724·91	r	CrO	OHn	E	742·8		FeO		E	762·5	r	LaO	OHn	E

763		NdO		E
763·0	w	Fe	OHn	E
765·6		MnO		E
765·78		LaO		E
766·3	r	PrO	OHn	E
766·49		K	OHn	A, F, E
766·5	r	La	OHn	E
766·82	r	Pr(O)	OHn	E
767·3		MnO		E
768·5		CoO		E
769·0		FeO		E
769·11		LaO		E
769·7		MnO		E
769·8979		K		F, E
770·2	r	La	OHn	E
770·8	r	Pr(O)	OHn	E
771·22		CaO		E
771·4	r	Cr	OHn	E
771·4	r	Pr	OHn	E
771·55	r	CaO	OA	E
771·57		CeO		E
772·11	r	CaO	OA	E
772·50		LaO		E
772·9		MnO		E
773·8	r	La	OHn	E
775·0	b	$B(_2O_3)$	OHn	E
775·77	a	Rb	OHn	E
775·88		LaO		E
776·5		EuOH		E
776·7		MnO		E
777·5		FeO		E
777·7	r	Cr	OHn	E
779·0	w	Fe	OHn	E
779·2	r	Nd	OHn	E
780·023		Rb	OHn	A, E
781·21		CrO		E
781·52		BaO		E
782·7	r	Pr	OHn	E
783·18		CeO		E
784·3	r	Cr	OHn	E
785		Nd(O)		E
785·1	r	V	OHn	E
785·28	r	SrO		E
786·5	r	V	OHn	E
786·6	r	Pr(O)	OHn	E
787·3	r	Pr(O)	OHn	E
787·3	r	Sr	AA	E
787·7	r	LaO	OHn	E
787·94		CeO		E
788·23	r	SrO	AA	E
788·49		CrO		E
788·5		PrO		E
789·0	tr	CN	OCq	E
789·2	tr	CN	OCq	E
789·7	r	V	OHn	E
790·0		NiO		E
790·20	r	SrO	AA	E
790·51		BaO		E
791·0	r	Cr	OHn	E
791·1	r	LaO	OHn	E
791·5		PrO		E
791·9	r	V	OHn	E
792		LaO		E
792·0	w	Fe	OHn	E
792·2	r	Pr(O)	OHn	E
792·70		CeO		E
792·85		VO	OHn	E
793·5		CoO		E
793·97		VO		E
794·5	r	LaO	OHn	E
794·76		Rb	OHn	E
794·77		VO		E
795·33		BeO		E
796·1	r	V	OHn	E
797·34	r	VO	OHn	E
797·46		CeO		E
797·93		Pr(O)		E
798·0	r	LaO	OHn	E
798·21		VO		E
798·3		NiO		E
798·64	r	Pr(O)	OHn	E
799·0		Pr		E
799·6	v	Nd(O)	OHn	E
800		PrO		E
800·0		VO	OHn	E
800·4	v	Nd	OHn	E
801·0		MnOH		E
801·5	r	LaO	OHn	E
801·57	a	Cs	OH	E
802·11		CeO		E
804·8	r	Pr(O)	OHn	E
805·02	r	LaO	OHn	E
805·5		PrO		E
807·6	r	Pr(O)	OHn	E
807·90	a	Cs	OH	E
808·0	tr	CN	OCq	E
808·61	r	LaO		E
809·4		NiO		E
809·7	r	H_2O	OHq	E
810·0	w	Mn(O)	OH	E
811	mb	FeO	OHn	E
812·2	r	LaO	OHn	E
812·65	a	Li	OH	E
813·7		FeO		E
814·0		EuOH		E
814·0		Tm(O)		E
815·31	r	CaO	OA	E
815·9	r	LaO	OHn	E
816·47		CaO		E
817·3	r	Pr(O)	OHn	E
818·33	a	Na	OHn	E
818·42	r	Pr	OHn	E
818·7		NiO		E
818·74		V		E
819		TiO		E
819·0		Sm(O)		E
819·48	a	Na	OHn	E
819·6	r	La	OHn	E
820·4		CoO		E
820·60		BeO		E
821·6	v	Nd	OHn	E
822·0	b	B	OHn	E
823·0		FeO		E
823·31		LaO		E
823·4	dr	PrO	OHn	E
823·53		Pr(O)		E
824		CaO		E
824·7		CoO		E
825·78	r	SrO	AA	E
827·07		LaO		E
827·22	r	SrO	AA	E
828·0	tr	CN	OCq	E
828·04		Pr(O)		E
830		Nd(O)		E
830·0		BaOH	AH	E
830·0		NiO		E
830·2		FeO		E
833·9	r	Cr	OHn	E
834·0		Fe	OHn	E
836·0	r	Cr	OHn	E
836·5		MnOH		E
838·0	b	Nd	OHn	E
838·6	r	Pr(O)	OHn	E
838·6	r	Cr	OHn	E
839·2		NiO		E
839·21	r	Pr(O)	OHn	E
839·61		CeO		E
840·6	r	La	OHn	E
841·7	r	Cr	OHn	E
842·7	r	Pr	OHn	E
843·2	r	Pr	OHn	E
844·9	r	Cr	OHn	E
845·07		CeO		E
845·4	r	LaO	OHn	E
846·0		Tm(O)		E
846·85		BeO		E
847·67	r	Pr(O)	OHn	E
848·88	r	Pr(O)	OHn	E
849·0	r	LaO	OHn	E
849·4		CoO		E
849·4	r	Pr(O)	OHn	E
849·52		CeO		E

850·0	tr	CN	OCq	E
850·0		Cr(O)		E
850·2		NiO		E
850·35	a	K	OHn	E
850·52	a	K	OHn	E
851		PrO		E
852·09		VO		E
852·111		Cs	OH	A, F, E
852·7	r	LaO	OHn	E
853·8	r	V	OHn	E
855·0	w	Mn(O)	OH	E
855·5	r	Pr(O)	OHn	E
856·4	r	LaO	OHn	E
857·4	dr	V	OHn	E
858·0	w	Fe	OHn	E
858·0		MnOH		E
859·7	r	V	OHn	E
859·86		CeO		E
860·0		NiO		E
860·1		LaO	OHn	E
860·5	r	V	OHn	E
861·0		Tm(O)		E
862·2	r	Pr(O)	OHn	E
862·5	r	V	OHn	E
863·0		EuOH		E
863·8	r	LaO	OHn	E
864·2	r	V	OHn	E
865·22	r	CaO	OA	E
865·79		VO		E
866·1	db	Nd	OHn	E
866·46		CaO		E
866·8	r	V	OHn	E
866·96		CaO		E
867·0		BaOH		E
867·7	r	La	OHn	E
870·00	r	SrO	AA	E
870·2	r	V	OHn	E
870·5		CoO		E
871·0		MnOH		E
871·34		BeO		E
871·5	r	La	OHn	E
872		CaO		E
872·25	r	SrO		E
873		VO		E
873·0		BaOH	AH	E
873·9		NiO		E
875·4	r	La	OHn	E
876·14	a	Cs	OH	E
879·28		LaO		E
882		$B(_2O_3)$		E
883·23		LaO		E
884·3		NiO		E
890·22	a	K	OHn	E
890·40	a	K	OHn	E
891·6	r	H_2O	OHq	E
892		VO		E
894·35		Cs	OH	E
895·0		NiO		E
897·4	r	H_2O	OHq	E
898·5		CoO		E
900·27		BeO		E
907·0		NiO		E
912·5		CoO		E
912·9	r	H_2O	OHq	E
917·0	tr	CN	OCq	E
918·3	r	H_2O	OHq	E
919·5		NiO		E
922·9	r	CaO	OA	E
927·7	r	H_2O	OHq	E
930		PrO		E
930		TiO		E
933·3	r	H_2O	OHq	E
938		H_2O	OHq	E
941·0		CN	OCq	E
944·0	r	H_2O	OHq	E
948·5	r	H_2O	OHq	E
955·9	r	H_2O	OHq	E
958·0	tr	CN	OCq	E
959·56	a	K	OHn	E
959·78	a	K	OHn	E
961·0	r	H_2O	OHq	E
966·9	r	H_2O	OHq	E
973		H_2O	OHq	E
980·7	v	CaO	OA	E
983·5	v	CaO	OA	E
1046		VO		E
1095·0	tr	CN	OCq	E
1102·1	da	K	OHn	E
1169·02	a	K	OHn	E
1176·9	a	K	OHn	E
1177·30	a	K	OHn	E
1243·2	a	K	OHn	E
1252·2	a	K	OHn	E

The following table (table 7) of flame lines is arranged in alphabetical order of element symbols. The sources and references for the preceding table are also valid here. To save space, we have deleted all additional notes to method, characters, flames and solvents. Additional information, if required, can be found by simply referring to the corresponding wavelength in table 6. If the lines for elements such as Cl in InCl bands, for example, are sought, then these can be found by referring to In.

We would like to thank Mr W Frenzel and Mrs I Matuszewski of the Abteilung für Medizinische Physik of the University of Giessen for the new reorganisation of the material.

Table 7 Index of flame spectra arranged in alphabetical order of element symbols

Ag	257·541	541·0	264·15	437·0	515·63
	308·22	542·3	264·795	447·6	515·77
317·9	309·271		267·60	449·68	516·61
322·0	309·284	**As**	274·83	449·7	516·88
327·5	350		302·92	450·25	518·0
328·07	394·40	189·0	312·28	450·30	518·07
333·0	396·1527	193·696	330·03	450·61	518·25
335·7	435·26	197·2	345·74	453·0	518·35
338·29	437·4	228·81	365·19	454	519·61
339·6	439·38	234·98	397·28	468·58	520·72
345·1	447·1	238·12	406·82	469·48	522·90
351·6	449·4	243·72	433·94	471·5	540·38
354·6	451·6	243·73	443·66	471·95	540·72
358·3	453·8	243·85	627·818	472	542·07
363·7	455·75	245·65		472·73	543·60
371·0	457·6	249·29	**B**	488·10	543·75
374·0	459·38	250·36		488·21	544·71
378·1	464·82	250·47	249·67	489·16	545·68
383·3	467·2	256·97	249·773	489·18	546·01
390·5	469·5	257·09	345	490·00	546·82
399·0	471·6	274·50	405·19	490·69	547
403·9	473·6	278·02	405·72	491·70	547·09
410·8	475·4	286·04	406·55	492·93	547·6
419·0	484·23	289·87	408	493·0	576·65
427·3	486·64	303·285	408·15	493·25	578·23
432·8	488·8	310·56	409·06	494	579
439·7	507·9	310·68	416·29	494·13	579·07
	510·2	317·06	416·89	495·0	579·2
	512·3	317·24	418·0	495·59	580·58
Al	514·3	320·78	430·26	496	581·32
	516·1	320·91	430·51	496·54	583·19
207·336	517·6	327·8	433·50	497·36	603·0
236·70	519·1	327·91	433·54	499·75	617·16
237·313	533·7		434·00	511·87	620·0
237·33	535·8	**Au**	434·05	512·77	620·22
256·80	537·7		434·81	514·46	637·66
257·51	539·4	242·795	434·90	514·59	639·0

639·60	532·1	656·32	732·48	221·58	426·18
670	534·97	659·53	754·69	222·15	428·30
682·0	536·66	663·45	795·33	223·83	428·46
721·0	540·27	670·45	820·60	224·72	428·94
775·0	541·98	678·28	846·85	226·17	429·90
822·0	545·46	683·0	871·34	227·39	430·25
882	549·27	685·72	900·27	231·0	430·77
	550·97	693·15		231·15	430·95
Ba	551·9115	700·71	**Bi**	232·2	431·87
	553·548	706·00		233·79	440·39
230·424	553·555	709·74		247·86	442·54
233·527	554·78	717·66	195·38	249·29	442·58
350	560·24	725·43	195·94	263·0	443·50
350·11	564·1	733·69	202·12	436·5	443·57
353·5	565·90	744·04	206·17	437·1	444·99
388·93	567·29	752·35	211·03	438·2	445·48
413·066	570·1	761·01	222·83	447·0	445·66
452·40	571	781·52	223·06	467·9	458·59
453·78	571·37	790·51	227·66	468·5	458·87
455·403	572·0	830·0	240·09	469·8	514·6
457·94	575·84	867·0	252·45	471·5	518·885
462·1	577·01	873·0	262·79	473·7	529·2
463·68	577·77		269·68	509·8	529·8
463·9	580·51	**Be**	278·05	512·9	538·5
466·4	581·76		289·80	516·5	542·5
468·03	582·92	234·861	293·83	550·2	543·4
472·27	586·45	249·47	298·90	554·1	544·0
474·10	587·53	265·0470	299·33	558·5	546·2
478·41	588·69	265·0550	302·46	563·5	550·0
483	592·51	313·042	306·77		552·1
485·0	597·17	313·107	339·72	**Ca**	552·6
485·06	597·28	332·13	350		553·3
487·0	597·63	442·7	351·09	239·86	553·9
487·83	598·49	445·2	359·61	317·93	554·0
489·65	601·95	447·5	412·15	372·8	554·1
493·41	603·96	449·6	439·4	375·32	554·2
494·17	606·31	451·6	442·4	377·32	554·4
494·2	610·23	451·7	454·2	377·5	554·6
495·0	611·08	453·6	461·9	379·37	554·7
495·1	611·62	455·3	472·22	381·47	554·8
496·54	614·17	470·9	472·2552	382·8	555·4
497·0	616·5	473·3	472·2831	383·62	555·7
500·0	616·51	475·5	556·4	385·41	555·9
500·1	616·96	477·6	563·9	385·83	556·2
501·24	622·47	479·5	574·0	388·2	565·2
502·0	629·10	481·2	603·7	389·47	569·7
506·6	635·80	482·8		391·57	571·9
508·67	642·31	505·44	**C**	393·3666	578·0
512·0	645·09	507·6		395·94	581·0
512·5	648·29	509·5	208·99	396·8468	581·3
513·0	649·31	510	211·31	420·51	582·5
513·8	649·69	511·2	212·83	422·19	583·0
516·7	649·88	512·8	215·02	422·67	593·4
521·4	652·73	514·1	217·30	424·08	593·5
524·0			219·68		

596·5	228·80	729·72	385·09	238·49	276·64
597·1	298·06	731·41	385·3	239·20	277·50
598·6	326·11	734·69	385·47	240·22	298·72
600·3	340·36	735·07	386·19	240·725	298·96
600·6	346·62	738	387·14	241·102	300·05
602·0	346·77	739·78	388·3	241·16	301·36
602·2	360·05	742·48	388·34	241·45	301·75
602·9	361·29	746·65	415·2	241·53	303·44
603·5	361·45	750·95	415·8	241·91	304·25
603·8	422·67	771·57	416·8	242·17	304·40
604·9	467·82	783·18	418·1	242·49	304·89
606·4	479·99	787·94	419·7	242·92	306·18
606·8	508·58	792·70	421·6	243·22	307·23
607·6		797·46	425·6	243·583	308·26
609·2		802·11	587·8	243·67	308·68
609·8	**Ce**	839·61	650	243·90	308·96
611·5		845·07	695·0	245·62	312·14
612·222	356·08	849·52	711·0	246·08	313·73
616·129	394·275	859·86	728·0	246·420	313·99
618·5	401·239		744·0	246·77	314·71
621·2	418·66		789·0	247·03	315·88
622·0	461·2		808·0	247·39	321·92
623·0	462·1	**CH**	828·0	247·66	323·70
628·5	462·4		850·0	248·36	333·3
644·0	462·7	387·1	917·0	249·54	333·41
645·3	463	387·2	941·0	249·56	333·72
662·5	464	388·9	958·0	250·45	335·44
663·9	468·4	423·7	1095·0	250·69	336·71
665·3	469·5	424·3		251·10	337·03
695·62	470·4	424·9	**Co**	251·18	338·52
696·84	479·2	426·2		251·79	338·82
698·0	479·9	426·8	227·65	252·1363	339·54
698·29	480·7677	427·4	228·61	252·90	340·51
730·83	481	428·0	228·78	253·01	340·92
731·85	481·5	428·6	229·52	253·22	341·23
732·76	484·6	429·2	229·60	253·60	341·25
771·22	486·3	429·7	229·67	253·649	341·26
771·55	487·3	431·5	230·40	254·43	341·72
772·11	488·2	432·4	230·42	254·49	343·16
815·31	489·1	433·4	230·90	254·83	343·30
816·47	490	433·9	231·68	255·30	344·29
824	492·5	434·3	232·31	255·68	344·36
865·22	493·1	434·8	233·60	256·13	344·92
866·46	493·4	435·2	233·87	256·73	344·94
866·96	494		234·62	257·22	345·350
872	494·1		235·29	257·35	345·4
922·9	494·7	**CN**	235·56	259·06	345·52
980·7	495·1		235·82	262·249	346·28
983·5	550	335·9	235·87	265·03	346·58
	552·30	343	236·379	267·98	347·40
Cd	569·70	356	236·51	268·53	348·34
	569·923	358·39	236·97	269·58	348·94
214·44	723·58	358·59	237·19	276·14	349·13
226·502	727·55	359·04	238·05	276·42	349·57

350·23	389·41	722·0	390·29	579·4	269·32
350·26	389·50	744·0	390·88	585·2	
350·63	390·63	750·0	391·92	591·31	**Cs**
350·98	390·99	768·5	392·86	597·59	
351·04	392·28	793·5	398·39	604·28	328·86
351·26	393·60	820·4	425·43	605·2	331·32
351·35	394·09	824·7	427·48	616·73	334·74
351·83	394·17	849·4	428·97	622·91	334·87
352·01	394·53	870·5	429·58	629·41	339·81
352·16	395·23	898·5	429·70	636·37	340·0
352·34	395·29	912·5	429·77	639·4	347·69
352·685	395·79		429·8	645·2	348·01
352·9033	397·47	**Cr**	430·75	651·18	361·15
352·981	397·87		432·06	657·51	361·74
353·34	397·95	236·47	432·1	664·36	387·639
354·33	398·71	236·59	432·12	671·51	388·86
355·06	399·17	236·68	432·16	677·2	455·54
355·27	399·53	267·716	433·76	685	459·32
355·88	399·79	283·563	433·94	689·15	515·27
356·09	401·10	284·325	433·97	689·6	519·67
356·50	401·93	297·11	434·45	695·72	520·95
356·94	402·09	297·55	435·11	702·75	525·66
357·50	402·70	298·08	435·18	719·0	530·14
357·54	404·54	298·60	435·96	724·91	534·09
358·48	405·72	298·65	437·13	731·2	535·04
358·52	405·8	298·86	438·50	737·5	540·67
358·72	406·64	299·66	449·69	743·95	541·39
359·49	407·61	300·09	454·60	771·4	546·59
360·21	408·26	300·51	458·01	777·7	550·29
360·5356	408·63	301·48	460·08	781·21	556·84
362·50	409·24	301·52	461·34	784·3	557·37
362·78	411·05	301·76	461·61	788·49	563·52
363·14	411·88	301·85	464·62	791·0	566·37
364·77	412·13	301·88	465·1285	833·9	574·57
365·25	413·22	302·07	465·2158	836·0	580·3
365·70	419·07	302·16	516·8	838·6	583·88
368·30	419·84	302·2	520·45	841·7	584·51
369·31	420·76	302·44	520·60	844·9	584·76
370·41	423·40	303·02	520·84	850·0	601·05
370·88	425·2	303·14	522·9		603·41
373·24	426·80	303·42	524·76	**CS**	621·31
374·55	428·58	303·70	526·42		621·76
374·99	430·32	304·08	529·3	250·73	643·20
380·81	458·01	305·399	529·6686	251·12	647·26
381·63	472·79	357·87	529·7976	252·32	658·65
384·20	548·33	359·349	534·58	253·00	662·87
384·55	563·5	360·533	534·83	253·87	672·33
385·10	630·5	363·66	535·6	254·95	682·47
386·12	634·0	373·08	540·98	255·58	687·05
387·31	635·46	373·20	541·7	257·27	697·33
387·35	653·2	374·90	556·41	257·56	698·35
387·40	665·1	388·52	562·33	258·96	801·57
387·68	667·0	388·68	568·51	260·59	807·90
388·19	691·0	389·40	574·87	262·16	852·111

876·14	510·55	366·49	396·84	421·43	475·31
894·35	524·0	366·68	397·38	421·51	475·62
	530·0	367·44	398·13	421·80	475·74
Cu	537·0	367·84	399·35	422·11	477·4804
	568·6	368·57	399·45	422·22	477·53
202·4	570·02	369·48	399·67	422·51	478
216·51	578·21	370·75	399·89	423·20	479·1299
217·89	580·0	372·79	400·00	423·98	479·1500
218·17	604·51	373·93	400·048	424·59	488·01
222·57	605	374·00	400·58	425·81	491·47
224·43	606·0	374·11	400·60	425·85	502·21
244·16	614·7	375·70	401·38	426·83	504·19
249·21	615·0	376·76	402·37	427·67	504·262
324·75	616·2	377·10	402·49	429·19	505·34
327·40	625·0	377·26	402·84	429·8	506·69
358·30		377·30	403·88	430	507·76
377·63	**Dy**	377·47	404·59	431·0	508·26
400·5		378·14	404·89	432·0	508·31
400·54	262·37	378·62	404·93	434·772	508·93
406·19	264·21	378·84	405·50	436·0	509·86
412·0	268·80	379·77	407·79	436·67	510·91
418·8	296·46	380·98	407·92	438·0	512·16
421·1	299·13	381·22	408·51	439·09	513·4
425·9	299·67	381·62	409·98	441·2	513·6
428·0	300·10	381·74	410·38	442·0	513·90
428·1	338·50	381·94	410·50	444·0	514·0
432·8	339·35	382·14	410·59	446·0	514·3
433·3	340·20	382·18	412·61	450·0	514·44
434·8	340·77	382·56	412·91	453·15	514·96
435·4	343·43	384·04	413·04	453·615	515·8
435·6	344·55	384·09	413·38	453·653	516·0
436·4	345·43	384·43	413·41	453·69	516·93
437·2	346·09	384·69	413·47	455·09	517·19
438·0	347·19	385·84	413·85	455·16	519·28
438·9	347·82	386·88	413·95	455·5	519·53
439·9	349·449	387·21	414·15	456·51	520·44
440·9	351·10	387·40	414·60	456·55	521·41
441·2	351·56	389·19	415·78	456·96	522·98
441·9	352·11	389·28	415·93	457·0	523·06
443·0	353·17	389·85	416·79	457·42	523·62
443·4	353·40	389·91	417·19	457·78	524·71
444·1	353·60	390·59	418·36	457·9	524·85
445·2	353·85	391·36	418·68	458·93	524·95
446·4	354·78	391·73	419·09	459·19	525·06
449·4	357·10	393·01	419·16	459·8	526·08
451·6	357·13	393·67	419·48	461·22	526·3
464·84	358·50	393·71	419·79	462·35	527·3
468·90	359·45	393·80	420·13	463·15	528·5
473·41	359·50	393·82	420·22	470·23	529·6
488·2	361·07	394·47	420·25	470·34	529·9
493·0	361·11	395·62	420·50	471·29	530·15
493·2	364·35	396·26	420·76	471·35	530·93
505·0	364·54	396·75	421·172	472·38	531·2
506·1	364·68	396·83	421·31	473·46	531·44

358

531·61	594·05	342·92	365·95	398·76	504
532·31	594·58	344·26	366·28	400·40	504·13
533·25	597·45	344·68	366·44	400·797	504·96
533·34	597·9	345·09	367·89	401·25	506·7
533·69	598·85	345·36	368·40	402·05	507·3
533·83	600·57	346·25	368·42	402·15	507·8
534·9	600·70	346·45	369·26	402·19	508·1
536·16	601·08	346·51	369·48	403·61	508·9
536·32	602·7	346·94	369·72	403·76	508·97
536·43	602·77	347·63	369·76	404·69	510·0
538·41	603·0	348·07	370·57	405·95	512·45
539·39	604·3	348·93	370·65	407·79	513·5
539·55	604·37	349·91	371·59	408·76	515·1
539·82	604·97	350·27	371·93	409·29	516·74
540·0	605·0	350·56	374·60	409·81	517·27
540·36	605·63	351·49	374·75	412·32	517·52
540·42	605·8	352·00	374·89	412·48	517·67
540·87	605·95	352·25	375·60	413·15	517·89
542·0	607·8	352·68	376·19	415·11	518·09
542·33	607·89	353·95	376·87	418·57	518·36
545·10	608·4	354·58	377·48	419·07	518·47
545·5	608·82	355·63	377·52	421·84	520·65
549·25	609·1	355·70	377·86	428·65	537·16
550·84	612·1	355·80	379·11	429·89	539·40
554·44	616·84	355·87	379·15	431·0	540·09
554·72	619·50	356·34	379·29	433·13	540·37
556·26	625·90	356·36	379·82	434·83	541·21
563·95	634·90	356·51	379·86	438·21	542·1
565·20	638·68	356·92	381·03	438·64	542·28
565·73	639·22	356·99	381·54	440·93	542·96
569·26	642·19	357·07	382·68	441·87	543·8
569·4	643·50	357·83	383·00	442·45	543·91
570·36	643·62	357·94	384·99	442·67	544·47
570·5	657·93	358·04	385·59	449·63	544·6
571·85	664·34	358·57	386·308	451·86	545·13
572·9	683·54	358·66	387·47	452·27	545·3
573·31	695·81	358·83	389·26	453·11	545·43
573·49		359·07	389·62	460·66	545·9
574·0	**Er**	359·58	390·27	467·31	547·0
574·72		360·59	390·39	472·27	547·93
574·84	298·55	360·74	390·54	476·26	548·11
577·69	326·479	360·94	390·63	482·07	549·73
578·47	331·242	361·24	393·70	485·74	549·90
579·40	333·15	361·25	394·44	493·12	551·3
580·73	334·31	361·658	394·80	495·69	552
583·20	334·43	362·01	395·64	497·13	552·8
583·4	336·67	362·78	396·45	497·95	553·35
584·44	336·80	362·80	396·63	498·77	554·48
585·56	337·27	362·93	397·26	499·60	554·58
585·73	338·20	363·32	397·30	500	555·17
586·8	338·37	363·46	397·36	500·24	555·70
588·4	339·96	363·63	397·67	500·7238	556·29
592·1	342·28	363·86	397·70	500·90	556·73
594·0	342·44	363·90	398·23	503·42	556·85

558·0	345·70	596·71	220·04	276·18	358·70
558·17	346·78	597·27	229·25	276·20	360·89
559·6	358·85	598·0	229·78	276·75	361·88
560	358·92	599·28	229·92	277·21	363·15
561·27	368·84	601·8195	237·36	277·82	364·78
562·8	372·49	602·90	238·20	278·81	367·99
563·4	381·96	604·46	239·56	281·33	368·31
564·5	390·71	605·73	245·76	282·33	368·75
565	393·04	608·38	246·26	282·56	370·56
566	397·19	609·93	246·51	283·24	370·78
566·4	412·96	618·81	246·89	284·40	370·92
567·4	420·505	619·50	247·29	285·18	371·99
568·4	443·553	623·0	247·48	293·69	372·26
569·3	452·25	626·22	247·98	294·134	372·76
570·48	459·40	626·69	248·327	294·79	373·33
570·6	462·72	628·0	248·81	295·39	373·49
570·81	466·1872	629·13	248·98	295·74	373·71
571·57	466·188	629·97	249·0644	296·53	374·34
572·33	474·0	633·57	249·1155	296·69	374·56
572·70	490·71	635·00	249·55	297·01	374·59
573·0	491·14	635·58	250·11	297·32	374·7
573·7	511·43	636·92	251·08	298·14	374·83
575·1	512·90	638·27	251·81	298·36	374·95
576·27	513·34	638·38	252·28	299·44	375·82
576·79	515	640·09	252·43	300·10	376·38
577·92	516·00	640·61	252·74	300·7	376·72
582·67	516·67	641·01	252·91	300·81	378·79
585·00	520·0	641·13	253·56	302·049	379·50
585·52	521·50	645·79	254·10	302·06	379·85
601	522·34	647·0	254·60	302·11	379·95
622·101	526·64	648	254·96	302·58	381·30
630·87	527·19	655·0	258·45	303·16	381·58
632·61	528·28	662·0	259·94	303·74	382·04
658·34	528·56	669·39	259·96	304·27	382·44
660·49	535·76	680·27	260·68	304·76	382·5884
	540·27	681·61	262·35	305·74	382·78
Eu	545·15	684·0	263·58	305·91	383·42
	545·29	686·45	266·79	344·06	384·04
305·89	548·86	686·46	267·91	344·10	384·11
310·61	551·05	702·0	269·01	344·39	385·00
311·14	553·0	710·65	270·66	346·59	385·08
321·05	554·74	733·0	271·90	347·55	385·64
321·28	557·03	749·0	272·09	347·65	385·99
321·37	557·71	758·0	272·36	347·67	386·55
323·51	558·00	776·5	272·80	349·06	387·25
324·14	564·58	814·0	273·36	349·78	387·80
324·60	565·5	863·0	273·73	351·38	387·8628
324·75	571·0		274·24	352·13	388·6284
331·44	576·52	**Fe**	274·41	352·60	389·57
332·22	578·37		275·01	352·62	389·9709
333·43	580·02	216·68	275·37	355·85	390·29
335·04	583·10	217·81	275·40	356·54	390·65
335·37	594	218·65	275·44	357·01	391·72
343·25	596·37	219·18	275·63	358·12	392·03

392·29	581·9	335·862	434·6626	525·4	583·2
392·79	586·8	336·22	437·20	525·9	584·4
393·0299	590·30	342·247	437·38	527·5	584·7
396·93	591·1	342·39	438·82	527·7	585·81
400·52	591·9	351·36	440·18	529·2	586·2
404·58	593·48	360·48	440·31	530·3	586·6
406·36	597·5	360·52	441·16	531·1	590·5
407·17	608·47	365·462	441·47	532·68	591·1
413·21	609	367·40	441·79	533·61	591·14
414·39	609·5	367·921	442·02	534·1	591·81
420·20	609·7	368·412	442·24	534·59	592·7
421·62	610·99	371·35	443·01	535·2	594·2
425·08	618·1	371·74	443·06	535·58	594·27
426·05	621·9	373·23	443·40	536·61	595·0
427·18	627·89	373·26	446·3	537·69	595·6
429·21	629·59	373·97	447·61	538·31	595·73
429·41	643·03	374·48	448·1	539·9	596·4
430·79	644·52	375·79	448·69	540·5	597·2
432·58	652·41	376·22	449·10	541·0	597·60
437·59	656·67	376·839	449·73	541·6	598·7
438·35	665·15	378·30	449·9	542·1	598·8
440·48	670·0	379·637	450·63	542·7	600·1
441·51	683·0	384·32	451·96	543·5	600·18
442·73	702·2	386·69	453·78	543·9	600·7
444·23	726·5	393·47	454·20	544·6	601
446·17	742·8	394·55	458·13	545·0	601·9
447·8	752·7	396·90	461·6	545·4	603·5
448·22	763·0	402·33	463·4	546·15	604·28
454·4	769·0	404·50	465·3	546·5	604·4
460·4	777·5	405·36	467·0	546·9	605·4
465·9	779·0	405·47	467·1	547·51	606·29
473·0	792·0	405·82	471·7	547·6	606·4
492·9	811	407·87	472·6	548·06	606·58
511·04	813·7	409·04	473·5	548·75	608·1
516·75	823·0	409·27	473·6	561·79	608·19
526·95	830·2	410·02	475·6	563·2	609
529·0	834·0	413·41	479·8	564·32	609·76
532·805	858·0	415·06	481·7	565·13	609·9
537·15		415·77	482·17	565·3	610·48
538·2	**Ga**	416·72	483·50	565·58	610·83
540·58		417·55	485·57	566·4	611·40
542·79	245·007	419·07	489·3	568·1	611·5
542·97	250·070	419·16	490·95	569·62	612
543	265·99	422·58	491·0	569·7994	613·2
553·14	271·97	426·01	492·8	569·9	613·6
554·32	287·424	426·60	494·9	571·59	614
558·28	294·36	426·70	501·50	573·4	615·2
561·40	294·42	427·41	509·9	574·6	615·6
562·13	403·30	430·63	510·0	575·13	615·73
562·41	417·21	431·38	510·34	575·6	617·2
564·7		432·11	511·7	577·1	617·6
567·9	**Gd**	432·5568	513·4	578·84	618·3
579·0		432·71	513·6	580·7	620·1
580·8	326·673	434·6459	524·7	581·9	620·30

361

621·2
622·1
623·2
624·2
624·53
625·21
626·3
626·7
627·3
628·4
629·0
629·42
630·6
631·2
631·60
632·9
633·59
633·8
635·2
673·07
682·82

Ge

204·169
204·376
206·520
206·865
209·423
219·87
249·796
259·25
265·12
265·16
269·13
270·96
275·46
303·9064
326·95

HCO

350

Hf

264·141
282·022
286·637
290·441
290·475
294·077

295·068
296·488
302·053
307·288
368·22
377·76

Hg

184·9
235·7
253·65
435·84
546·07

Ho

251·87
253·38
255·68
258·65
259·29
276·44
278·57
294·09
315·38
315·42
315·93
318·63
320·10
320·617
322·00
323·33
323·387
339·89
341·49
341·64
341·84
342·41
343·70
344·93
345·12
345·31
345·38
345·60
347·42
348·48
350·79
351·07
351·55
353·27
354·14
357·04

357·91
358·18
358·97
359·30
359·94
360·00
360·09
360·57
362·35
363·09
364·12
366·22
366·29
366·66
366·79
366·90
366·95
367·91
367·97
368·26
368·51
369·06
369·19
369·46
370·00
370·12
370·97
371·28
371·86
372·07
373·14
373·20
373·49
373·63
374·24
374·81
376·90
377·41
377·47
377·48
377·61
379·15
379·29
379·67
379·72
380·41
381·07
381·18
382·17
382·56
382·92
384·98
385·24
385·76

385·93
386·26
386·30
387·48
389·04
389·10
389·26
390·27
390·39
390·44
390·54
390·95
391·18
391·94
393·88
394·25
395·05
395·57
395·96
397·58
397·69
399·82
399·95
400·33
402·72
403·17
403·76
404·08
404·54
405·39
406·80
407·18
408·73
408·75
410·109
410·38
410·65
410·73
410·86
411·19
412·02
412·56
412·716
413·45
413·62
416·303
416·40
417·34
419·43
422·22
422·34
422·70
425·44
426·40

426·60
435·07
435·5
440·32
440·5
442·0
444·46
445·0
447·0
449·5
449·77
453·45
456·25
486·03
493·48
493·90
497·99
506·76
506·98
509·2
509·7
510·5
511·83
511·88
512·6
513·2
514·21
514·27
514·55
514·75
515·70
516·23
516·63
516·83
518·57
518·98
519·56
521·13
525·14
525·8
526·34
527·0
527·42
528·69
529·9
530·14
530·4
531·47
531·96
532·6
533·01
533·32
534·59
535·88

535·99
536·01
538·0
538·14
538·49
556·37
556·57
557·98
558·23
558·44
559·2
560·72
560·99
562·63
563·5
564·3
564·43
565·6
565·9
566·0
567·2
567·7
568·8
569·6
570·5
571·1
571·33
572·9
573·02
573·7
573·76
575·2
580·4
582·0
583·1
583·5
584·9
585·4
586·02
586·47
588·0
588·29
591·0
592·17
594·80
597·27
597·35
598·29
608·17
630·53
660·49
660·74
662·89
674·50

H₂O	422·41	352·20	578·26	406·03	460·73
	423·32	355·90	580·20	406·47	461·20
691·9	424·38	357·37	581·25	407·91	463·56
716·5	425·76	359·44	583·21	408·672	466·25
729·9	427·03	360·98	691·13	408·96	476·68
809·7	428·26	362·87	693·90	410·48	485·08
891·6	428·3	373·85	696·418	413·70	494·97
897·4	430·15	374·72	696·469	416·02	514·54
912·9	451·13	377·07	766·49	418·73	515·86
918·3		380·01	769·897	428·02	517·73
927·7	**IO**	390·25	850·35	435·6	517·9
933·3		391·54	850·52	437·2	520·27
938	435·6	393·48	890·22	437·6	520·4
944·0	448·8		890·40	437·97	521·18
948·5	458·7	**K**	959·56	438·0	522·8
955·9	469·4		959·78	438·2	523·42
961·0	484·5	299·22	1102·21	438·4	525·18
966·9	485	303·48	1169·02	438·76	525·3
973	496·4	310·20	1176·9	439·2	527·12
	513·1	310·22	1177·30	439·6	527·8
In	520·9	321·70	1243·2	440·0	530
	530·8	321·75	1252·2	440·5	530·3
256·02	531	344·67		441·0	532·6
271·03	549·5	344·770	**La**	441·46	532·8
271·39	553·3	404·414		441·8	535·4
275·39	573·0	404·721		442·3	538·1
277·54		463·85	333·74	442·31	538·25
293·26	**Ir**	464·158	354·5	442·8	540·56
303·9		464·217	355·3	443·3	543·1
303·93	208·882	474·09	355·64	443·8	543·31
325·61	209·263	474·43	355·9	444·3	545·51
325·86	237·277	475·39	356·09	444·8	545·8
341·8	247·51	475·45	356·10	445·3	548·43
345·7	250·26	475·74	356·61	445·8	550·13
349·9	250·298	478·689	357·44	446·4	550·90
353·5	254·397	479·108	360·44	447·13	553·3
355·6	261·13	480·016	360·81	453·22	553·5
359·9	263·942	480·519	361·1	453·62	553·63
362·4	263·97	484·988	361·14	454·02	556·1
363·8	266·198	485·61	361·308	454·47	558·8
365·1	266·48	486·361	361·47	454·91	560·0
367·9	269·42	486·98	361·5	454·95	560·01
369·8	279·73	494·20	361·9	455·36	560·24
372·7	279·77	495·08	362·01	455·83	562·7
375·8	282·44	495·60	362·26	456·30	562·86
379·0	284·97	496·50	362·3	456·79	565·23
399·3	288·116	508·42	364·953	457·00	565·4
409·9	292·48	509·71	370·97	458·1	568·0
410·18	294·32	509·92	389·86	458·5	568·11
410·476	322·078	511·22	392·75	458·9	570·7
412·64	336·85	532·32	394·91	459·4	573·2
414·25	343·70	533·97	400·76	459·8	574·06
415·49	344·90	534·30	401·53	460·27	576·18
416·53	351·36	535·95	403·72	460·3	576·93

		Li			
578·92	711·6		450·9	523·5	377·6
579·13	713·16		451·2	525·6	378·4
586·6	714·7	274·13	451·85	540·25	379·1
586·8	716·3	323·26	452·1	540·3	380·1
589	717·9	379·47	452·6	540·5	380·7
589·5	719·4	391·50	453·3	544·9	381·0
592·1	721·1	413·22	454·0	545·7	381·4
592·2	722·52	427·33	454·62	546·4	381·5
593·06	725·72	460·29	456·1	548·7	382·2
595·0	728·92	497·2	457·0	589·67	382·3
597·7	738·0	610·36	457·5	597·1	382·94
600·5	740·4	670·78	459·1	599·3	383·23
600·67	743	812·65	464·4	600·0	383·4
603·3	743·43		465·4	600·45	383·83
603·45	746·5	Lu	466·0	601·4	384·5
606·1	749·7		466·17	605·50	384·6
618·1	750·7	261·54	466·2	626·0	384·8
621·6	752·82	291·13	467·2	650·0	385·5
624·5	753·9	298·92	468·4	668·8	385·9
624·99	756·0	308·14	469·6	674·9	387·7
627·4	759·23	311·84	470·8	681·0	388·2
630·2	762·5	317·13	472·1	706·0	391·2
632·59	765·78	327·89	472·6		391·4
633·1	766·5	328·17	473·5	Mg	394·2
636·0	769·11	331·21	474·9		395·3
639·42	770·2	335·95	476·4	202·5	396·9
641·09	772·50	336·0	478·0	277·67	457·11
645·45	773·8	337·65	486·01	277·83	470·3
645·59	775·88	338·55	487·08	277·98	496·3
654·31	787·7	339·68	488·24	278·14	497·5
656·25	791·1	350·739	489·50	278·30	498·6
657·32	792	350·84	490·72	279·55	499·7
657·85	794·5	356·78	492·17	280·27	500
658·38	798·0	363·62	493·55	285·21	500·7
659·44	801·5	364·77	494·9 '	332·99	516·73
660·51	805·0	384·11	499·2	333·22	517·27
661·65	808·6	396·84	500·11	333·67	518·36
663·98	812·2	405·44	500·4	350·3	520·6
664·44	815·9	409·4	501·9	359·4	552·846
665·08	819·6	409·5	503·5	362·4	
666·14	823·31	409·6	504·1	362·7	Mn
677·13	827·07	410·7	505·8	364·60	
678·39	840·6	411·0	510·0	367·2	257·61
679·64	845·4	412·47	512·0	368·4	259·37
680·89	849·0	412·7	516·1	369·6	260·57
682·15	852·7	414·5	517·03	370·2	279·48
699·5	856·4	415·40	518·5	370·3	279·83
701·12	860·1	423·9	518·6	370·7	280·11
702·5	863·8	424·2	520·0	371·1	336·0
704·1	867·7	425·0	520·2	371·9	343·9
705·5	871·5	425·3	520·6	372·9	351·8
707·08	875·4	445·5	521·5	373·1	353·18
708·5	879·28	448·9	521·77	375·1	353·20
710·1	883·23	450·0	523·3	376·7	353·21

364

354·78	505·1	858·0	615·42	416·81	463·91
354·80	515·80	871·0	616·07	437·48	464·11
354·82	515·9		818·33	443·29	464·5
356·95	519·25	**Mo**	819·48	450·0	464·64
356·98	522·84			451·04	464·96
357·00	526·7	281·615	**Nb**	454·02	465·10
357·79	530·81	311·21		456·95	465·23
359·5	535·94	313·259		550	465·47
364·0	538·95	313·26	269·70		467·10
367·1	539·47	315·82	292·78	**Nd**	467·39
369·5	542·4	317·03	295·08		467·55
374·0	543·25	319·40	309·41	334·65	468·34
374·6	544·9	320·88	313·07	335·87	468·40
380·0	546·13	338·46	316·34	372·13	468·85
380·67	558·65	340·43	334·19	378·75	469·03
382·35	561·0	344·71	334·37	379·36	469·64
382·39	567·28	345·64	334·39	383·09	469·83
383·44	571·09	346·68	334·906	388·78	470·69
384·0	575·07	379·83	334·95	393·59	471·90
389·0	579·52	386·41	335·474	401·225	472·65
393·0	586·0	390·30	335·842	425·64	473·17
395·5	588·1	550	353·53	434·35	473·49
397·5	591·0	550·64	357·585	436·41	474·97
401·81	594·4	553·30	358·02	441·4	475·58
402·0	615·47	557·04	369·739	444·49	475·85
403·08	617·6	588·83	369·78	445·61	475·93
403·179	620·4	603·066	371·30	447·78	476·04
403·31	623·72		371·382	448·0	476·9
403·45	627·51	**Na**	372·62	448·09	477·019
404·14	631·47		372·723	448·18	477·22
404·5	649·81	268·04	373·98	449·73	477·84
404·88	652·38	285·28	374·178	451·1	477·94
405·55	655·77	285·30	374·23	452·72	478·74
405·89	659·47	330·23	375·955	452·98	480·66
406·35	663·55	330·30	376·064	454·20	483·0
407·0	688·0	439·01	378·70	454·82	483·56
407·92	691·3	439·34	378·748	454·91	483·66
407·942	695·3	449·42	379·01	455·96	485·33
408·294	699·6	449·77	379·12	456·04	485·53
408·362	704·4	454·17	379·81	456·18	485·95
423·51	724·1	454·52	380·29	456·73	486
423·53	726·8	466·485	382·488	458·66	486·67
445·16	730·1	466·859	405·89	459·46	486·92
446·20	734·0	474·79	407·973	460·38	487·979
460·54	738·4	475·18	410·04	460·98	488
475·40	765·6	497·85	410·09	462·0	488·17
476·15	767·3	498·28	412·38	462·19	488·38
476·24	769·7	514·909	413·71	462·42	488·50
476·585	772·9	515·36	413·75	462·65	489·10
476·643	776·7	554·52	413·944	462·79	489·32
478·34	801·0	568·26	413·97	462·9	489·69
482·35	810·0	568·82	415·25	463·12	490·15
497·6	836·5	588·995	416·36	463·42	490·18
501·3	855·0	589·59	416·46	463·72	490·72

490·77	572·92	639·76	680·09	741·1	301·20
491·00	574·91	640·4	681·9	742·0	301·91
491·34	574·96	641·0	683·14	744·0	303·79
491·46	577·21	642·5	683·9	746·7	304·50
492·45	577·61	643·0	684·23	749·8	305·08
494·48	578·49	644·21	684·63	752·9	305·43
494·67	578·82	644·54	685·7	755·6	305·76
495·02	580·00	644·8	686·47	759·2	306·46
495·07	582·67	645·3	688·0	762·2	308·08
495·25	583·91	645·42	688·7	763	310·16
495·47	588·97	645·57	689·37	779·2	310·19
495·9	592·12	646·4	689·71	785	311·41
496·33	596·14	648·56	690·42	799·6	313·41
496·97	597	649·3	690·8	800·4	323·30
497·54	597·4	649·89	691·8	821·6	323·46
497·6	599·0	650·0	692·42	830	324·31
498·86	599·8	650·15	693·1	838·0	324·85
501·45	600·00	651·58	694·0	866·1	327·11
502·71	600·2	651·8	694·26		328·27
502·94	600·76	654·16	694·5	**NH**	328·69
504·01	601·0	656·4	695·72		331·57
505·68	601·4	658·0	697·02	336·0	332·03
507·18	602·7	658·2	699·2	337·0	332·23
510·31	605·3	658·6	699·7		336·16
510·53	613	659·84	700·44	**Ni**	336·58
513·8	613·3	660	701·1		336·62
514·95	618·0	660·01	701·91	228·70	336·96
519·80	621·9	660·8	702·1	228·99	337·20
519·97	624·7	661·2	702·31	230·08	337·42
520·43	626·9	662·1	702·7	231·096	338·06
521·32	628	662·4	703·31	231·10	338·09
525·0	628·1	663·0	705·6	231·23	339·10
528·34	628·7	663·4	706·4	231·40	339·30
528·81	629·7	663·80	709·2	231·72	340·96
529·16	630·41	664·8	710·0	231·976	341·35
529·88	630·52	665·2	711·11	232·003	341·39
531·3	631·04	665·5	711·5	232·14	341·48
532·7	631·3	666·43	712	232·58	342·37
533·432	631·51	667·5	712·60	233·00	343·36
534·95	632·25	668·0	713·3	233·75	343·73
537·77	633·2	668·3	715·8	234·55	344·63
544	633·63	669·4	717·2	234·75	345·29
547·8	634·52	669·90	718·6	236·06	345·85
550	634·9	670·19	721·3	236·21	346·0
552·39	635·2	670·56	722·7	240·18	346·17
552·57	635·4	670·75	723	241·93	346·75
552·90	635·84	671·20	723·6	247·687	346·95
556·11	636·1	671·64	724·5	294·39	347·25
561·11	636·52	672·53	725·5	298·16	348·38
562·05	637·0	673·38	726·8	298·41	348·59
566·97	638·2	673·67	728·5	299·26	349·30
567·59	638·519	674·8	731·8	299·45	350·09
568·1	638·6	676·92	734·3	300·249	350·26
572·6	638·9	677·4	740·5	300·36	350·77

351·03
351·39
351·51
351·98
352·34
352·45
352·80
354·82
355·15
356·18
356·64
357·19
358·79
359·77
360·23
360·93
361·05
361·27
361·94
362·47
366·41
366·92
367·04
367·41
368·84
369·39
372·25
373·68
373·92
374·90
377·56
377·81
378·35
379·36
380·71
383·17
385·83
391·30
397·22
397·36
475·1
487·7
489
493·9
500·7
502·4
509·8
517·4
520
532·3
540·8
600
790·0
798·3

809·4
818·7
830·0
839·2
850·2
860·0
873·9
884·3
895·0
907·0
919·5

NO

214·91
215·49
223·94
224·54
226·28
226·94
236·33
237·02
244·00
244·70
247·11
247·87
255·00
255·90
258·75
259·57
267·14
268·0
271·32
272·22

OH

261·0
262·21
281·13
281·60
282·90
283·17
283·87
287·53
289·27
290
291·9
294·52
302·12
306·37
306·78
308·17
308·99

309·6
309·95
310·21
310·60
311·23
311·78
312
312·26
312·83
313
314·75
316·96
342·81
343·21
347·21
348·4

Os

263·7
264·4
271·46
280·69
290·91
301·80
303·07
305·87
323·206
326·229
326·79
330·16
352·86
355·98
356·09
375·25
378·22
413·58
426·085
442·04

P

177·495
228·82
229·49
230·17
230·69
231·37
232·06
236·73
237·52
237·99

238·3
238·35
238·79
239·63
245·46
246·42
247·79
251·87
252·94
253·40
253·56
254·04
255·33
255·493
259·57
520·0
524

Pb

216·99
217·00
217·56
220·35
223·74
224·69
233·24
238·88
239·38
239·96
240·19
241·17
244·38
244·62
247·64
257·73
261·37
261·42
266·32
280·20
282·32
283·31
287·33
357·27
363·96
368·347
373·99
387·8
391·0
395·5
398·80
403·8
405·78

414·6
420·64
422·9
431·7
441·0
445·4
450·9
455·4
465·8
470·6
474·9
481·7
485·1
498·4
500
513·8
516·2
525·8
533·1
535·4
540·5
545·9
561·8
567·8
591·1
616·1
625·1
634·2
642·8

Pd

244·79
247·44
247·64
276·31
302·79
324·27
325·16
325·88
328·72
330·21
337·30
338·07
340·46
342·12
343·34
344·14
346·08
348·12
348·98
351·69
355·31

357·12
360·95
363·47
369·03
371·89
379·92
383·23
389·42
395·86
408·73
416·98
421·30
447·36

Pr

390·84
410·075
416·0
416·30
438·0
440·88
451·962
452·0
453·23
454·12
455·22
460·0
463·56
463·98
464·69
466·09
467·0
467·48
468·78
469·57
471·31
473·0
473·06
473·66
474·41
480·81
482·0
489·61
490·69
491·40
492·0
492·45
493·60
493·97
494·03
495·13
496·02

497·49	602·0	690·64	842·7	304·26	335·09
497·57	602·6	692·4	843·2	306·47	358·71
497·64	603·6	693·0	847·7	320·40	359·16
501·85	603·8	695·0	848·9	330·19	420·19
501·97	604·20	697·62	849·4	331·50	421·56
502·0	605·5	698·72	851	332·38	496·7
502·69	606·09	699·1	855·5	340·81	498·3
503·33	607·17	700·55	862·2	342·79	501·7
504·38	609·06	701·81	930	348·34	502·3
504·55	609·8	703·9		348·53	503·7
505·34	610·2	709·5	**Pt**	362·81	507·58
508·71	613·0	712·03		363·88	508·90
513	613·5	712·9	204·937	369·99	513·2
513·3	614·64	713·8	206·75	381·87	515·07
513·34	616·3	724·55	214·423	396·64	516·49
513·7	618	727·9	214·77	411·87	519·53
513·98	618·20	729·68	216·517	416·46	523·40
515·7	619·56	732·02	217·467	444·26	526·00
519·44	620·80	732·7	235·71	444·55	532·24
519·53	625·9	734·3	243·67		536·26
522·80	626·31	735·04	244·01	**Pu**	539·06
530	626·7	737·63	245·097		543·15
535·2	627·1	740·1	246·74	387·85	557·88
536·7	627·89	741·32	248·717	420·647	564·81
537·8	627·9	743·1	249·01	420·8	565·37
538·0	629·80	743·65	249·58	473·5	572·44
538·2	631·8	745	249·85		607·08
539·8	631·83	748·5	250·85		615·96
541·4	633·6	753·9	251·56	**Ra**	620·63
546·02	634·14	759·41	253·92		629·83
559·7	635·2	766·3	262·80	381·44	630·63
561·0	635·7	766·82	263·93	440·5	761·89
561·2	636·3	770·8	264·69	459·2	775·77
562·8	637·7	771·4	265·09	468·23	780·023
564·5	639·5	782·7	265·95	471·8	794·76
566·2	645·8	786·6	267·71	475·0	
567·5	647·4	787·3	269·84	482·59	**Re**
569·1	647·9	788·5	270·24	520·0	
570·5	649·4	791·5	270·59	602·0	345·18
572·0	651·4	792·2	271·90	621·0	346·047
573·1	655	797·93	272·99	624·7	346·47
574·9	657·34	798·64	273·40	626·9	488·92
576·3	658·9	799·02	275·39	628·5	527·55
577·7	660·9	800	275·49	632·9	
579·1	663·1	804·8	277·17	634·9	
592·5	664·0	805·5	279·33	665	**Rh**
594·2	672·2	807·6	280·32	665·3	
595·0	681·8	817·3	283·03		265·90
596·3	682·2	818·42	285·3		271·75
597·2	685·1	823·4	289·39	**Rb**	271·85
598·3	688·22	823·53	289·79		326·31
599·1	688·7	828·04	292·98	322·80	327·16
600·5	689·3	838·6	299·80	322·91	328·06
601·9	690·15	839·21	300·23	334·87	328·36

368

				Sb	405·45
332·31	377·00	288·65	376·00		408·24
333·85	377·81	290·89	376·74		413·30
334·42	378·85	296·52	377·76	206·833	414·03
335·99	379·22	297·69	378·61	212·739	415·23
336·22	379·32	298·89	379·05	212·75	416·51
336·84	379·93	299·50	379·81	213·98	434·80
337·23	380·59	300·66	379·89	214·50	438·13
339·68	380·68	301·72	379·93	217·93	441·53
339·97	381·65	302·09	381·27	220·85	445·01
340·65	381·82	303·35	381·90	222·08	448·56
341·23	382·23	304·03	382·21	222·49	450·2
342·02	382·85	304·25	382·49	231·147	450·3
343·50	383·39	304·28	383·18	244·55	452·19
344·05	385·65	304·85	383·97	251·05	453·6
345·52	387·73	304·88	385·76	252·85	453·7
345·71	391·28	305·49	386·78	256·5	455·91
345·79	391·35	306·48	392·59	257·4	457·1
346·20	392·22	330·16	393·18	259·805	460·7
347·07	393·42	331·52	393·36	259·8062	464·44
347·48	394·27	333·96	395·02	261·23	467·3
347·89	395·89	336·85	395·04	265·26	470·7
349·87	397·53	338·95	397·94	267·06	472·8769
350·25	398·44	339·25	398·49	268·28	472·9226
350·73	399·56	340·17	402·22	271·89	473·40
352·80	399·61	340·93	406·41	272·72	473·76
353·81	408·28	341·74	406·45	276·99	474·10
354·19	408·43	342·83	407·67	287·79	474·2
354·39	412·17	342·86	408·06	302·98	474·38
354·95	412·89	342·95	409·78	317·5	475·31
356·41	413·53	343·08	410·17	323·25	477·85
357·02	419·65	343·27	411·27	326·75	477·9347
358·31	421·11	343·67	414·42		479·15
359·6194	428·87	343·84	416·75		485·81
359·7147	437·48	349·894	419·76	Sc	489·4
360·59	452·87	350·14	419·89		507·02
361·25	467·50	350·97	419·99		508·15
362·66	474·51	351·45	420·60	296·58	508·37
363·95	542·5	351·96	421·21	297·40	508·55
365·49		352·01	423·03	298·07	508·69
365·80		353·94	429·77	301·53	509·7
366·62		357·06	436·12	301·93	509·92
368·10	Ru	359·6179	437·22	303·07	513·37
369·07		359·98	438·54	325·57	517·10
369·236		363·49	438·56	326·99	520·9
369·55		366·34	439·04	327·36	530·19
370·09	261·21	366·95	439·10	361·38	534·29
371·30	264·30	367·83	441·00	363·07	534·97
373·53	265·13	371·93	443·98	390·75	535·6100
373·73	265·18	372·69	444·93	391·18	535·7
374·42	270·92	372·80	446·00	393·33	535·8
374·82	273·57	373·04	455·45	399·66	539·5
375·56	276·34	374·23	458·44	402·04	543·5
376·04	286·66	374·56	470·95	402·36	548·19
376·51	287·50	375·98	569·90	404·78	

548·46	627·63	**Se**	365·73	420·57	445·67
551·42	630·56		366·13	421·86	445·92
552·05	631·61	196·026	366·62	421·93	446·38
567·18	631·8	203·985	367·08	422·61	446·73
568·68	634·48	206·279	368·78	423·07	447·74
570·02	636·2	207·479	369·00	424·04	448·03
572·4	637·88	216·4	370·69	424·42	449·00
573·7	640·84		372·10	426·63	449·91
576·1	641·33	**Si**	372·20	427·18	450·33
576·5	642·0		372·81	428·22	451·13
577·27	642·4	220·797	373·07	428·28	452·25
577·3	643·7	221·088	373·12	428·35	452·31
577·6	644·62	221·54	373·71	429·67	452·74
579·8	644·81	221·667	373·92	429·8	453·24
580·2	645·8	223·63	374·54	429·91	453·38
581·1	646·2	227·72	374·85	430·12	453·75
583·7	647	229·89	375·64	431·28	454·6
584·0	648·54	234·24	376·00	431·38	455·0
584·91	649·6	234·43	377·33	431·95	455·66
587·8	650·0	236·45	378·26	432·44	456·67
588·8	650·3	236·57	380·39	432·51	456·95
591·8	651·8	238·79	380·98	432·61	458·1584
592·81	652·6	241·38	381·38	433·0	458·1737
595·9	653·53	243·63	381·83	433·14	459·67
596·9	655·8	245·90	382·29	433·41	459·83
600·2	657·6	248·19	383·28	433·61	461·12
601·7	659·1	248·68	383·44	433·93	461·5
603·62	660·1	249	384·62	433·99	462·94
605·5	661·0	250·69	384·67	435·08	464·54
606·4	661·8	250·99	384·87	435·58	464·80
607·3	664·5	251·43	385·32	435·78	464·94
607·9	665·44	251·61	385·45	436·29	466·35
609·2	666·10	251·92	385·85	436·59	467·0768
610·2	669·1	252·41	386·01	438·04	467·0833
610·99	670·1	252·85	387·74	438·62	467·21
611·0	670·6	256·38	388·17	439·33	468·15
611·6	673·8	258·71	392·52	439·73	468·87
614·0	674·82	264·48	395·18	440·11	471·0
614·9	675·2	266·90	396·21	440·31	471·61
615·4	678·7	269·37	397·46	441·15	471·6693
618·1	680·0	275·50	397·82	441·93	471·7088
618·81	682·4	278·05	399·00	442·05	471·8667
619·3	686·4	288·16	399·10	442·11	472·0129
621·06	690·6		399·83	442·33	472·84
622·0	692·2	**Sm**	405·45	442·43	475·07
622·94	693·7		406·23	442·96	476·0262
623·0	696·3	358·63	407·98	443·30	476·4
623·31	699·1	359·26	409·99	443·38	477·0191
623·94	700·5	360·42	412·52	443·43	477·0290
623·97	702·6	360·76	413·55	444·18	478·31
625·99	703·6	360·94	414·52	444·2271	478·5869
626·2	708·2	362·92	415·12	444·32	478·59
627·2	709·4	363·429	418·33	444·51	478·99
627·5	727·6	365·15	418·5	445·29	484·17

484·83	562·18	598·1	639·1	675·1	277·98
488·3779	562·60	598·8	639·6	675·61	278·50
488·3983	565·98	598·89	640·6	676·11	281·36
490·49	566·59	599·3	642·1	677·53	284·0
491·04	568·1	599·51	642·5	677·91	285·06
491·89	568·3	600·42	642·5909	679·6	286·33
492·40	569·8	601·9	643·9	680·29	291·35
494·63	570·62	602·72	645·0	680·8	300·91
497·39	570·67	603·0	645·6	681·54	303·28
497·59	571·60	603·5	645·68	682·0	303·41
503	572·55	604·49	646·21	682·78	317·5
504·42	572·93	604·7	646·9	683	320·58
504·95	573·29	606·5	647·30	685·39	322·98
506·09	574	607·0074	647·43	686·10	323·0
507·1187	574·64	607·1	647	687·2	326·2
508·83	575·54	607·9	647·82	690·12	326·23
511·71	577·13	608·14	648·5	691·82	326·24
512·21	577·37	608·41	648·53	715·5	329·18
515·72	577·83	608·9	649·88	718	332·3
517·27	577·92	609	651·0	722·9	332·35
517·54	578·83	609·99	651·24	731·0	333·06
518·70	579·09	609·9918	652·0	751·5	334·47
520·0591	580·0	610·84	652·5	759·0	335·14
521·07	580·02	611·3	652·80	819·0	337·47
525·1886	580·8	611·77	652·99		338·17
526·56	581·94	613·2	653·2	**Sn**	338·83
527·14	582·0	613·8	653·30		340·69
528·29	583·92	614·9	654·0	211·39	341
532·06	584·0	615·62	655·18	214·87	341·58
534·12	584·09	615·95	655·36	215·14	344·46
534·80	584·7	616·5	655·7	215·25	348·45
534·91	585·72	616·85	656·352	219·45	351·0
535·06	586·77	617	656·53	219·93	351·29
536·83	586·86	617·49	657	220·97	353·0
540	587·10	619·4407	657·5	223·17	354·0
540·36	587·41	619·5	658·0	224·61	354·24
540·52	587·59	620·0	658·053	226·89	357·45
541·63	589·51	622·9	658·5	228·67	357·5
542·15	589·89	623·8	658·89	231·72	358·5
545·30	590·26	624·3	660·0	233·48	358·54
546·67	590·60	625·6	661·90	235·48	361·47
548·54	590·90	627·33	663·4	238·07	362
549·37	591·63	632·8	664·59	240·82	367·77
549·82	592·4	633·5	664·76	242·17	369·14
551·11	592·70	633·9	665·1	242·95	372·12
551·21	594·09	634·45	665·6	248·34	375·23
551·61	594·3	634·8	666·92	249·57	377·93
552·0	594·7	635	667·14	254·64	380·10
554·89	595·0	635·3	668·0	254·66	380·27
555·04	595·52	636·1	670·02	257·16	380·3
556·13	595·85	636·74	671·65	259·44	381·7
556·34	596·21	637·4	672·59	266·12	383·32
558·81	597·21	637·9	672·8	270·65	386·3
561·7	597·93	638·8	674·43	276·18	386·49

387·76	362·70	582	**Ta**	383·40	419·67
388·42	364·60	585·2		388·51	420·10
389·93	367·15	594·3	243·27	388·82	420·37
391·95	369·02	595	255·94	390·13	420·64
392·0	370·85	597·0	260·78	390·79	421·35
394·76	375·24	604	260·82	391·54	421·51
395·10	377·53	606·0	260·863	393·23	421·75
397·87	382·24	607·7	264·747	393·76	421·91
398·39	384·56	608·5	266·189	394·29	422·42
401·9	389·40	609·0	266·21	394·36	423·18
402·6	389·71	609·6	268·51	395·01	423·28
407·9	391·83	610·1	271·467	395·04	423·53
410·89	391·96	611·4	275·83	395·18	425·52
411·22	393·71	620·0	277·588	395·73	425·82
412·16	396·16	624·0	331·88	396·26	426·36
412·2	398·34	626·0	340·69	401·00	426·63
414·4	400·94	628·3	360·74	401·08	429·83
421·77	407·77	630·6	457·43	402·37	430·72
424·0	413·50	635·9	466·11	402·47	431·04
426·2	416·7	636·2	481·27	403·23	431·8
430·3	421·55	639·4		403·88	431·88
434·3	428·1	641·9	**Tb**	403·94	432·02
436·5	430·3	645		404·71	432·648
437·9	430·54	646·0	349·49	405·41	433·21
441·1	434·6	646·5	350·91	406·03	433·65
442·9	440·0	651·2	351·66	406·15	433·84
445·2	442·1	652·8	351·89	407·01	434·06
449·9	452·3	660	359·89	407·12	434·25
452·2	454·4	661·4	363·89	408·12	435·68
452·47	456·5	663·3	364·53	408·66	442·31
454·1	460·733	665·6	366·96	409·22	447·1
457·0	469·3	669	368·71	409·40	447·3
459·1	472·23	672·0	369·35	410·34	448·6
461·2	474·19	674·5	370·01	410·53	449·30
463·4	481·19	680·0	370·11	411·25	457·1
466·1	483·21	682	370·13	411·28	458·7
485·0	487·63	682·0	370·30	411·99	459·0
496	496·23	689·259	370·50	413·01	460·7
516·0	523·86	704·0	370·84	413·24	461
516·5	525·69	707·0	371·17	413·53	461·5
563·0	548·09	722·0	373·15	413·90	461·9
609·5	550·42	734·7	374·36	413·97	472·7
	552·18	750·06	374·50	414·15	473·5
Sr	553·48	752·28	375·93	414·33	475·2
242·81	563·3	754·16	376·01	414·35	475·5
256·947	563·7	755·8	376·11	414·69	478·0
293·18	564·1	785·28	376·51	415·82	478·4
346·45	567·0	787·3	376·94	415·85	479·2
350·38	577·2	788·23	378·35	416·91	479·5
352·54	577·5	790·20	378·72	416·93	489·1
354·66	577·8	825·78	378·99	417·26	514·1
356·71	577·9	827·22	379·62	417·28	526·32
358·69	578·2	870·00	383·02	418·71	526·99
360·71	578·5	872·25	383·34	418·80	527·27

527·82	576·8	645·48	407·5	335·46	519·29
527·99	580·2	646·0	412·5	337·04	521·03
532·72	581·5	647·14	413·1	337·14	544·9
534·0	582·1	648·5	413·6	337·74	559·8
534·96	583·9	649·60	420·5	337·758	562
535·05	585·73	650	426·8	338·59	575·9
535·24	587·6	673·5	431·8	363·52	576·1
535·69	590·2	679	434·3	363·55	621·5
536	590·9		438·3	364·268	622·2
536·26	592·1	**Te**	440·8	365·35	668·1
536·86	592·5		446·0	365·81	671·4
537·26	593·3		448·7	366·89	673
537·85	594·0	214·275	455·6	367·16	674·7
538·68	594·04	225·9	461·6	372·98	678·1
539·62	596·0	238·32	464·0	374·10	705·7
539·99	596·15	238·576	469·9	374·106	709·0
540·5	596·6	253·07	471·1	374·164	712·7
541·5	597·9	276·9	477·4	375·28	715
542	598	334·50	480·2	375·36	715·90
542·8	598·15	334·6	486·3	377·16	819
543·27	599·9	338·27	494·1	392·45	930
543·6	601·4	338·3	497·5	392·98	
544·0	602·2	342·2	500·6	394·77	
544·33	602·6	342·24	503·5	394·86	**Tl**
544·52	603·45	346·38	511·8	395·35	
545·7	603·6	350·78	513·9	395·63	231·598
546·2	603·99	351·67	528·0	395·82	237·97
546·7	604·5	356·05	550·4	396·28	258·014
547·1	605·0	356·1	566·8	396·42	260·899
547·17	605·6	360·68		398·17	266·557
548·14	606·8	360·7	**Th**	398·24	270·9
550·12	607·1	363·6		398·97	271·067
553·38	607·66	366·15	329·05	399·86	276·79
553·76	607·8	366·20	329·203	400·89	282·616
555·0	607·90	371·08	353·875	402·45	291·83
555·83	608·0	371·4	411·27	430·05	292·15
556·63	609·1	371·8	576·055	430·10	·322·975
557·5	609·66	376·7		430·59	351·92
559·34	609·8	377·3	**Ti**	453·32	352·94
560·06	610·3	377·7		480·5	377·57
560·3	611·2	381·97	260·515	480·6	535·05
563·5	611·76	382·0	261·148	484·8	
563·9	612·2	383·5	264·11	484·9	
564·5	613·4	387·9	264·426	495·6	**Tm**
566·26	614·0	388·4	264·664	498·17	297·32
566·4	624·5	389·07	294·1995	499·11	297·339
568	629·2	393·4	294·826	499·95	304·68
568·1	634·1	394·06	295·61	500·0	308·11
569·9	634·9	395·2	318·64	500·3	313·126
570·2	635·0	395·4	319·19	503·99	317·26
572·73	635·2	400·3	319·99	506·46	317·28
573·0	639·54	400·75	323·452	516·7	318·05
573·86	643·61	401·0	334·18	516·8	323·37
575·51	645·0	406·0	334·904	517·37	324·15

329·10	388·34	531·24	733·5	309·31	386·48
329·90	388·73	532·47	814·0	310·2	386·76
334·99	388·745	532·79	846·0	310·23	387·50
336·26	389·66	532·92	861·0	311·07	389·01
338·50	391·64	533·9		311·84	389·28
339·31	394·92	534·7	**U**	312·11	390·22
339·75	395·81	535·2		312·53	390·98
339·78	399·66	535·42	288·96	313·03	409·05
341·00	404·44	536·1	346·63	313·3	409·26
341·25	409·481	536·5	348·937	318·34	409·54
341·26	410·58	536·9	350·007	318·40	409·98
341·66	413·83	536·95	350·734	318·54	410·21
342·93	415·86	537·02	351·462	319·07	410·51
344·15	418·76	537·44	355·08	319·39	410·98
344·30	420·37	538	356·66	319·80	411·18
344·73	422·26	539·2	358·48	320·24	411·51
345·36	424·22	539·6	363·82	320·74	411·64
346·22	427·17	540·05	365·916	321·24	412·35
346·75	435·99	540·23	367·007	326·32	412·80
347·66	438·64	541·38	389·036	326·77	413·20
348·09	439·44	541·5	394·38	327·11	413·44
348·73	459·90	542·5	404·27	327·16	417·94
349·99	468·19	543·4	405·005	327·61	418·25
350·00	472·42	547·3	406·255	328·33	418·98
351·40	473·33	548·7	411·61	329·81	420·98
351·76	474·5	548·8	415·39	350·56	430·62
353·55	475·0	550·04	434·169	367·57	430·98
353·79	477·5	552·03	502·739	370·36	433·0
356·38	481·5	552·36	550	370·47	433·28
356·73	483·0	553·0	591·54	370·50	434·10
356·98	484·02	553·1		377·86	435·29
358·60	485·0	555·77	**V**	379·03	435·59
359·86	489·2	556·6		379·36	436·80
360·87	489·3	556·73	250·778	379·49	437·92
362·42	489·8	557·16	251·165	379·99	438·47
363·84	490·3	557·78	251·195	380·34	439·00
364·67	490·94	558·1	251·714	380·75	439·52
370·026	493·75	558·86	251·96	380·85	440·06
370·13	494·6	563·14	252·62	380·96	440·66
371·79	496·23	564·0	253·018	381·34	440·76
373·41	506·08	567·58	257·4	381·78	440·82
374·40	511·39	568·4	292·36	381·82	440·85
375·18	522·6	574·9	303·382	381·99	441·64
376·13	522·83	576·30	304·31	382·14	441·99
376·19	523·0	576·42	304·36	382·20	442·15
378·11	525·61	578·25	304·49	382·28	442·60
379·52	525·7	580·00	305·08	382·32	442·85
379·576	526·74	589·15	305·22	382·85	442·98
379·85	528·2	593·18	305·36	384·04	443·61
380·77	529·4	596·43	305·4	384·07	443·78
382·63	530·02	597·12	305·63	384·44	444·17
384·08	530·27	644·0	306·05	384·73	444·42
384·80	530·59	646·02	306·64	385·53	445·74
388·31	530·71	720·0	307·38	385·58	445·97

446·03
457·71
458·04
458·63
459·41
481·3
482·745
483·16
483·24
485·148
485·8
486·47
487·547
488·16
490·4
495·1
501·1
505·7
505·8
510·5
522·82
532·45
537·33
542·51
546·93
551·73
556·77
561·80
562·76
566·99
569·85
570·36
570·70
572·38
572·70
573·7
578·64
583·8
588·9
594·25
599·76
603·97
608·14
608·64
609·02
611·16
611·95
613·91
619·16
619·91
621·38
621·63
622·45
623·07

623·32
624·28
624·31
624·72
625·18
627·46
628·51
629·28
629·64
630·33
636·04
641·85
647·8
653·1
653·28
658·89
664·64
670·60
676·62
689·40
695·2
697·9
701·1
703·8
707·0
707·02
710·0
710·8
713·17
737·4
739·4
740·5
741·80
743·4
743·53
744·40
745·4
746·60
747·0
747·2
749·2
751·7
753·4
785·1
786·5
789·7
791·9
792·85
793·97
794·77
796·1
797·34
798·21
800·0

852·09
853·8
857·4
859·7
860·5
862·5
864·2
865·79
866·8
870·2
873
892
1046

W

239·70
245·2
255·135
260·639
260·69
265·654
265·804
268·14
271·89
272·435
276·23
283·138
287·939
289·645
291·1
294·440
294·698
294·738
299·645
319·157
361·75
376·845
386·79
400·875
404·56
407·43
426·939
429·46
430·21
468·05

Y

294·84
296·49
297·45
298·42
302·17

302·228
304·53
324·228
355·26
358·45
359·29
360·07
361·10
362·09
371·02
377·01
377·43
378·87
396·84
403·98
404·76
407·737
408·37
410·23
412·83
414·28
416·75
417·41
423·59
449·6
450·05
452·5
452·6
452·72
452·77
456·2
460·37
460·4
464·37
465·02
467·48
467·63
470·7
474·5
476·09
481·82
484·20
486·83
487·0
502·5
505·0
507·8
511·2
515·3
515·4
527·5
530·4
546·64
552·75

558·18
563·04
569·8
571·38
573·1
574·7
576·42
578·3
580·0
581·9
583·8
584·2
585·9
586·0
587·6
589·4
591·22
593·1
593·9
595·6
597·2
598·8
599
600·4
602·0
603·7
605·4
607·3
608·9
609·7
610·8
611·5
612·7
613·2
614·8
615
616·51
618·2
620·0
621·8
622·26
625·1
625·58
627·50
629·55
631·6
631·84
633·55
633·8
635·2
637·0
638·7
640·20
640·6

642·4
643·50
646·82
648·5
650·12
651·83
653·6
655·4
655·73
657·3
661·1
663·1
668·760
671·9
679·37

Yb

246·44
246·5
267·19
267·265
297·05
328·93
328·94
328·98
346·44
368·79
369·41
373·47
388·0
398·80
399·4
415·0
419·0
444·2
447·6
451·5
451·8
457·4
459·0
468·2
470·7
471·6
472·5
473·3
473·8
474·2
474·9
475·6
476·0
476·5
476·8
477·4

477·8	524·7	587·0	334·56	339·19	571·81
482·4	525·4	602·0	334·59	350·9	572·40
484·0	529·6	622·0	468·01	351·96	574
485·0	532·5	648·91	472·22	354·76	574·81
485·8	534·0	654·0	481·05	360·11	575·4
495·6	541·1	679·96	520	362·386	577·85
496·1	542·2		636·23	383·59	580·92
496·6	543·8	**Zn**		386·38	597·77
497·1	544·0		**Zr**	388·54	602·13
497·5	546·0			389·03	622·94
498·1	546·9	213·856		389·13	626·09
498·5	554·2	271·25	257·14	468·78	629·28
501·2	555·0	281·05	260·86	555·17	634·49
514·4	555·65	307·2	271·15	562·9	637·83
517	572·5	307·59	298·539	563·4	641·23
517·4	573·3	328·23	301·175	564	647·37
518·0	575·8	330·28	302·95	565·8	650·81
518·3	586·7	334·50	327·305	566·3	654·30

References

Atlases and Tables

1. Allen C W 1963 *Astrophysical quantities* 2nd edn (London: Athlone Press)
2. Corliss C H 1967 *Revision of the NBS tables of spectral-line intensities below 2450 Å* (Washington, DC: NBS) Monograph no. 32 (suppl). Also in 1967 *Spectrochim. Acta* **23B** 117
3. Corliss C H and Bozman W R 1962 *Experimental transition probabilities for spectral lines of 70 elements* (Washington, DC: NBS) Monograph no. 53
4. Corliss C H and Tech J L 1968 *Oscillator strengths and transition probabilities for 3288 lines of FeI* (Washington, DC: NBS) Monograph no. 108
5. Darwent B 1970 *Bond dissociation energies in simple molecules* (Washington, DC: NSRDS, NBS) no. 31
6. Gatterer A, Junkes J and Frodl V 1945 *Spektren der seltenen Erden* (Specola Vaticana)
7. Gatterer A, Junkes J, Salpeter E W and Rosen B 1957 *Molecular spectra of metallic oxides* (Città del Vaticano: Specola Vaticano)
8. Gilbert P T Jr 1970/71 *Flame spectra of the elements* in *Handbook of chemistry and physics* 51st edn (Cleveland, Ohio: Chemical Rubber Co) p E-205. See also [387]
9. IUPAC 1972 *Nomenclature, symbols, units and their usage in spectrochemical analysis I. General atomic emission spectroscopy, Pure Appl. Chem.* **30** 653; 1978 *Spectrochim. Acta* **33B** 219
10. IUPAC 1976 *Nomenclature, symbols, units and their usage in spectrochemical analysis II. Data interpretation, Pure Appl. Chem.* **45** 99; 1978 *Spectrochim. Acta* **33B** 241
11. IUPAC 1976 *Nomenclature, symbols, units and their usage in spectrochemical analysis III. Analytical flame spectroscopy and associated non-flame procedures, Pure Appl. Chem.* **45** 105; 1978 *Spectrochim. Acta* **33B** 247
12. Jaffe H 1965 *Atlas of analysis lines* (London: Hilger and Watts)
13. *Landolt–Börnstein Physikalisch-chemische Tabellen* I. Band: *Atom- und Molekular-physik*, 1. Teil: *Atome und Ionen* 1950; 2. Teil: *Molekeln I* 1951; 3. Teil: *Molekeln II* 1951, 6th edn (Berlin: Springer-Verlag)
14. Meggers W F, Corliss C H and Scribner B F 1975 *Tables of spectral-line intensities. Part 1: arranged by elements; part 2: arranged by wavelengths* (Washington, DC: NBS) Monograph no. 145
15. Parsons M L and McElfresh P M 1971 *Flame spectroscopy. Atlas of spectral lines* (New York: Plenum)
16. Schäfer K and Synowietz C (ed) 1970 *D'Ans-Lax Taschenbuch für Chemiker und Physiker Band III: Eigenschaften von Atomen und Molekeln* 3rd edn (Berlin: Springer-Verlag)
17. Striganow A R and Sventitskii N S 1968 *Tables of spectral lines of neutral and ionized atoms* (New York: Plenum) (transl from Russian)
18. Stull D R and Prophet H (ed) 1971 *JANAF thermochemical tables* 2nd edn (Washington, DC: NSRDS, NBS) no. 37
19. Vedeneyev V I, Gurvich L V, Kondratiev V N, Medvedev V A and Frankevich Ye L 1966 *Bond energies, ionization potentials and electron affinities* (London: Edward Arnold) (Russian edn 1962 Moscow: Academy of Sciences of the USSR)

20. Weast R C (ed) 1977/78 *Handbook of chemistry and physics* 58th edn (Cleveland, Ohio: Chemical Rubber Co)
21. Wiese W L, Smith M W and Glennon B M 1966 *Atomic transition probabilities* vol I (Washington, DC: NSRDS, NBS) no. 4
22. Wiese W L, Smith M W and Miles B M 1969 *Atomic transition probabilities* vol II (Washington, DC: NSRDS, NBS) no. 22
23. Zaidel A N, Prokof'ev V K and Raiskii S M 1961 *Tables of spectrum lines* 2nd edn (London: Pergamon); 1961 *Spektraltabellen* 2nd edn (Berlin: VEB Technik)
24. *Zeiss' Flammenspektren* 1961 (Oberkochen: Carl Zeiss GmbH) no. A50-812

Bibliographies

25. Capacho-Delgado L, Lermond C A, Delumyea R, Malecki D and Rieke J 1969 *Atomic absorption bibliography* 2nd edn (Rochester, NY: Bausch and Lomb (Anal. Systems Div.))
26. Gilbert P T Jr 1964 *Advances in emission flame photometry* (a review with 510 references covering 1959–1964 in [139] p 193)
27. Masek P R and Sutherland I (ed) 1969–*Atomic absorption and flame emission spectroscopy abstracts* (bimonthly publication) (London: Science and Technology Agency)
28. Mavrodineanu R 1967 *Bibliography on flame spectroscopy—analytical applications 1800–1966* (Washington, DC: NBS) Miscellaneous publication no. 281 (with references to other, more specialised bibliographies)
29. Mavrodineanu R 1970 *Bibliography on flame spectroscopy—analytical applications* ch 13 in [77] p 651 (suppl to [28]; extending to 1968)
30. Scribner B F (and Meggers W F) 1941–*Index to the literature on spectrochemical analysis* (series of different parts covering successive periods from 1920 on) (Philadelphia, Pa: American Society for Testing and Materials)
31. Slavin S *An atomic absorption bibliography for 1968, 1969, 1970 and 1971* in: 1969 *Atom. Absorption Newsl.* **8** 8; 1970 **9** 13; 1971 **10** 17; and 1972 **11** 7. *Idem* for January–June 1972 in 1972 **11** 74; for July–December 1972 in 1973 **12** 9; for January–June 1973 in 1973 **12** 77; for July–December 1973 in 1974 **13** 11; for January–June 1974 in 1974 **13** 84; for July–December 1974 in 1975 **14** 1; for January–June 1975 in 1975 **14** 81; for July–December 1975 in 1976 **15** 7; for January–June 1976 in 1976 **15** 77; for July–December 1976 in 1977 **16** 7; (with D M Lawrence) for January–June 1977 in 1977 **16** 89; (with D M Lawrence) for July–December 1977 in 1978 **17** 7; for January–June 1978 in 1978 **17** 73
32. Slavin W *An atomic absorption bibliography for 1964, 1965, 1966 and 1967* in: 1965 *Atom. Absorption Newsl.* **4** 194; 1966 **5** 50; 1967 **6** 41; and 1968 **7** 11
33. Society for Analytical Chemistry (London) *Annual reports on analytical atomic spectroscopy*: vol I *1971* 1972 ed D P Hubbard; vol II *1972* 1973 ed D P Hubbard; vol III *1973* 1974 ed C Woodward; vol IV *1974* 1975 ed C Woodward; vol V *1975* 1976 ed C W Fuller; vol VI *1976* 1977 ed C W Fuller
34. van Someren E H S, (Lachman F and Birks F T) 1938–1973 *Spectrochemical abstracts* (series of different volumes I–XVIII covering successive periods from 1933–1972) (London: Adam Hilger)

Books

35. American Society for Testing and Materials (Philadelphia) 1968 *Methods for emission spectrochemical analysis* 5th edn
36. Angino E E and Billings G K 1972 *Atomic absorption spectrometry in geology* 2nd edn (Amsterdam: Elsevier)

37. Bingel W A 1967 *Theorie der Molekülspektren* (Weinheim: Verlag Chemie). 1970 *Theory of molecular spectra* (New York: Wiley Interscience)
38. Boumans P W J M 1966 *Theory of spectrochemical excitation* (London: Adam Hilger)
39. Breene R G 1961 *The shift and shape of spectral lines* 2nd edn (Oxford: Pergamon)
40. Burriel-Martí F and Ramírez-Muñoz J 1964 *Flame photometry, a manual of methods and applications* 4th edn (Amsterdam: Elsevier)
41. Christian G D and Feldman F J 1970 *Atomic absorption spectroscopy: applications in agriculture, biology and medicine* (New York: Wiley Interscience)
42. Clark G L (ed) 1960 *Encyclopedia of spectroscopy* (New York: Reinhold)
43. Dahl A I (ed) 1961 *Temperature, its measurement and control in science and industry* vol III part 2 (New York: Reinhold)
44. Dean J A 1960 *Flame photometry* (New York: McGraw-Hill)
45. Dean J A and Rains T C (ed) 1969 *Flame emission and atomic absorption spectrometry, I: theory* (New York: Dekker)
46. Dean J A and Rains T C (ed) 1971 *Flame emission and atomic absorption spectrometry, II: components and techniques* (New York: Dekker)
47. Dean J A and Rains T C (ed) 1975 *Flame emission and atomic absorption spectrometry, III: elements and matrices* (New York: Dekker)
48. Dvořák J, Rubeška I and Řezáč Z 1971 *Flame photometry, laboratory practice* (London: Butterworth) (transl from Czech)
49. Elwell W T and Gidley J A F 1966 *Atomic absorption spectrophotometry* 2nd edn (Oxford: Pergamon)
50. Fenimore C P 1964 *Chemistry in premixed flames* (Oxford: Pergamon and New York: Macmillan)
51. Finkelnburg W 1967 *Einführung in die Atomphysik* 11–12th edn (Berlin: Springer-Verlag)
52. Flügge S (ed) *Handbook of physics*: 1964 *Spectroscopy I* vol 27; 1957 *Atoms I* vol 35; 1956 *Atoms II* vol 36; 1959 *Atoms III and molecules I* vol 37/1; 1961 *Molecules II* vol 37/2 (Berlin: Springer-Verlag)
53. Fowler R H and Guggenheim E A 1949 *Statistical thermodynamics* 2nd edn (London: Cambridge UP) (reprinted 1965)
54. Fristrom R M and Westenberg A A 1965 *Flame structure* (New York: McGraw-Hill)
55. Fuchs N A 1964 *The mechanics of aerosols* (Oxford: Pergamon)
56. Fuchs N A and Sutugin A G 1970 *Highly dispersed aerosols* (transl from Russian by Israel Program for Scientific Translations, Jerusalem)
57. de Galan L 1971 *Analytical spectrometry* (London: Adam Hilger) (transl from Dutch)
58. Gaydon A G 1961 *Spectra of flames* in *Advances in spectroscopy* vol 2 ed H W Thompson (New York: Wiley Interscience) p 23
59. Gaydon A G 1968 *Dissociation energies and spectra of diatomic molecules* 3rd edn (London: Chapman and Hall)
60. Gaydon A G 1974 *The spectroscopy of flames* 2nd edn (London: Chapman and Hall)
61. Gaydon A G and Wolfhard H G 1970 *Flames, their structure, radiation and temperature* 3rd edn (London: Chapman and Hall)
62. Green H L and Lane W R 1957 *Particulate clouds: dusts, smokes and mists* (London: Spon)
63. Grove E L (ed) 1971 *Analytical emission spectroscopy* vol 1 parts 1 and 2 (New York: Dekker)
64. Herzberg G 1950 *Molecular spectra and molecular structure, I: spectra of diatomic molecules* 2nd edn (Princeton, NJ and London: Van Nostrand) (reprinted 1964)
65. Herzberg G 1966 *Molecular spectra and molecular structure, III: electronic spectra and electronic structure of polyatomic molecules* (Princeton, NJ and London: Van Nostrand)
66. Herzberg G 1971 *The spectra and structure of simple free radicals* (Ithaca, NY: Cornell UP)

67. Hoda K and Hasegawa T 1972 *Atomic absorption spectroscopic analysis* (Tokyo: Genshi Kyuko Bunseki, Kondanska)
68. Kaiser H and Menzies A C 1968 *The limit of detection of a complete analytical procedure* (London: Adam Hilger)
69. Kirkbright G F and Sargent M 1974 *Atomic absorption and fluorescence spectroscopy* (London: Academic Press)
70. Laidler K J 1955 *The chemical kinetics of excited states* (Oxford: Clarendon)
71. Lawton J and Weinberg F J 1969 *Electrical aspects of combustion* (Oxford: Clarendon)
72. Lewis B and von Elbe G 1961 *Combustion, flames and explosions of gases* 2nd edn (New York: Academic Press)
73. Lewis B, Pease R N and Taylor H S 1956 *Combustion processes* (Princeton, NJ: Princeton UP)
74. Lundegårdh H *Die quantitative Spektralanalyse der Elemente*: 1929 vol I; 1934 vol II (Jena: Fischer)
75. L'vov B V 1970 *Atomic absorption spectrochemical analysis* (London: Adam Hilger) (transl from Russian)
76. Malmstadt H V, Enke C G, Crouch S R and Horlick G 1974 *Electronic measurements for scientists* (Menlo Park, Calif.: Benjamin)
77. Mavrodineanu R (ed) 1970 *Analytical flame spectroscopy; selected topics* Philips Technical Library (London: Macmillan)
78. Mavrodineanu R and Boiteux H 1965 *Flame spectroscopy* (New York: Wiley)
79. Mitchell A C G and Zemansky M W 1961 *Resonance radiation and excited atoms* (London: Cambridge UP) (first printed in 1934)
80. Palmer H B and Beér J M (ed) 1974 *Combustion technology, some modern developments* (New York: Academic Press)
81. Parsons M L, Smith B W and Bentley G E 1975 *Handbook of flame spectroscopy* (New York: Plenum)
82. Pearse R W B and Gaydon A G 1976 *The identification of molecular spectra* 4th edn (London: Chapman and Hall)
83. Penner S S 1959 *Quantitative molecular spectroscopy and gas emissivities* (Reading, Mass: Addison–Wesley)
84. Pietzka G 1961 *Flammenspektrometrie* Ullmann's Enzyklopädie der technischen Chemie Band 2/1 (München–Berlin: Urban und Schwarzenberg)
85. Pinta M (ed) 1975 *Atomic absorption spectrometry* (London: Adam Hilger) (transl from French)
86. Pokhil P F, Maltsew W M and Saizew W M 1969 *Methods of investigation of combustion processes and explosions* (in Russian) (Moscow: Nauka)
87. Poluektov N S 1961 *Techniques in flame photometric analysis* (New York: Consultants Bureau) (transl from Russian)
88. Price W J 1972 *Analytical atomic absorption spectrometry* 2nd printing (London: Heyden)
89. Pringsheim P 1949 *Fluorescence and phosphorescence* (New York: Wiley Interscience)
90. Pruvot P 1972 *Spectrophotométrie de flammes* (Paris: Gauthier–Villars)
91. Pungor E 1967 *Flame photometry theory* (London: Van Nostrand) (transl from Hungarian)
92. Ramírez-Muñoz J 1968 *Atomic absorption spectroscopy and analysis by atomic absorption flame photometry* (Amsterdam: Elsevier)
93. Reynolds R J and Aldous K 1970 *Atomic absorption spectroscopy. A practical guide* (London: Griffin)
94. Rollwagen W 1970 *Chemische Spektralanalyse* 6th edn (Berlin: Springer-Verlag)
95. Rose J 1961 *Dynamic physical chemistry* (London: Pitman)
96. Rosen B 1970 *Spectroscopic data relative to diatomic molecules* in *International Tables of Selected Constants* vol 17 (Oxford: Pergamon)

97. Rubeška I and Moldan B 1969 *Atomic absorption spectrophotometry* (London: Butterworth) (transl from Czech by P T Woods)
98. Schuhknecht W 1961 *Die Flammenspektralanalyse* (Stuttgart: Enke Verlag)
99. Shuler K E and Fenn J B (ed) 1963 *Ionization in high-temperature gases* (New York: Academic Press)
100. Slavin W 1968 *Atomic absorption spectroscopy* (New York: Wiley Interscience)
101. Sommerfeld A *Atombau und Spektrallinien*: 1966 vol II 4th edn; 1969 vol I 8th edn (Braunschweig: Vieweg)
102. Sychra V, Svoboda V and Rubeška I 1975 *Atomic fluorescence spectroscopy* (London: Van Nostrand Reinhold)
103. Szabadvary F 1966 *Geschichte der analytischen Chemie* (Braunschweig: Vieweg)
104. Thorne A P 1974 *Spectrophysics* (London: Chapman and Hall and Science Paperbacks)
105. Thring M W 1962 *The science of flames and furnaces* 2nd edn (London: Chapman and Hall)
106. Tourin R H 1966 *Spectroscopic gas temperature measurement* (Amsterdam: Elsevier)
107. Unsöld A 1955 *Physik der Sternatmosphären* 2nd edn (Berlin: Springer-Verlag) (reprinted in 1968)
108. Voinovitch I A, Debras-Guédon J and Louvrier J 1962 *L'analyse des silicates* (Paris: Hermann)
109. Walsh A 1961 *Application of atomic absorption spectra to chemical analysis* in *Advances in spectroscopy* ed H W Thompson vol II (New York: Wiley Interscience) p 1
110. Weinberg J F 1963 *Optics of flames* (London: Butterworth)
111. Wells M J (ed) 1962 *Spectroscopy* (Oxford: Pergamon)
112. Welz B 1972 *Atom-Absorptions-Spektroskopie* (Weinheim: Verlag Chemie)
113. Willard H H, Merritt L L Jr and Dean J A 1958 *Instrumental methods of analysis* 3rd edn (Princeton, NJ: Van Nostrand)
114. Williams F A 1965 *Combustion theory* (Reading, Mass: Addison–Wesley)
115. Winefordner J D (ed) 1971 *Spectrochemical methods of analysis: quantitative analysis of atoms and molecules* (New York: Wiley Interscience)
116. Winefordner J D, Schulman S G and O'Haver T C 1972 *Luminescence spectrometry in analytical chemistry* (New York: Wiley Interscience)

Conference Proceedings

117. Abstracts of papers at *Int. Conf. on Atomic Absorption Spectroscopy, Sheffield, UK, 1969* 1969 (London: Adam Hilger)
118. Abstracts of papers at *4th Int. Conf. on Atomic Spectroscopy, Toronto, Canada, 1973* 1973 (The Spectroscopic Society of Canada)
119. Abstracts of papers at *5th Int. Conf. on Atomic Spectroscopy, Melbourne, Australia, 1975* 1975 (Melbourne: The Australian Academy of Science)
120. *Colloq. La Photométrie d'Absorption Atomique dans la Flamme, Gembloux, Belgium, 1968* 1968 (Gembloux: Centre de Recherches Agronomiques de l'État)
121. *Conf. on Limitations of Detection in Spectrochemical Analysis, Exeter, UK, 1964* 1964 (London: Adam Hilger)
122. *Conf. plénières du 3e Congr. Int. de Spectrométrie d'Absorption et de Fluorescence Atomique, Paris, 1971, Méthodes Physiques d'Analyse* 1971 (Paris: GAMS) numéro spécial
122a. Invited lectures of *Colloq. Spectroscopicum Int. XX and 7th Int. Conf. on Atomic Spectroscopy, Prague, 1977* 1977 (Prague: Sbornik VSCHT)
123. Papers of *3rd Int. Congr. of Atomic Absorption and Atomic Fluorescence Spectrometry, Paris, 1971* 1973 vol I and II (London: Adam Hilger)
124. Plenary lectures of *Int. Conf. on Atomic Absorption Spectroscopy, Sheffield, UK, 1969* 1970 ed R M Dagnall and G F Kirkbright (London: Butterworth). Also published in 1970 *Pure Appl. Chem.* **23** 1–143

125. Preprints of papers at *Colloq. Spectroscopicum Int. XVII, Firenze, Italy 1973* 1973 vol I–III (Associazione Italiana di Metallurgia)
126. *Proc. 14th Ann. Mid-America Spectroscopy Symp., Chicago, Illinois, 1963* in 1964 *Developments in applied spectroscopy* ed J E Forrette and E Lanterman vol 3 (New York: Plenum)
127. *Proc. Colloq. Spectroscopicum Int. VI, Amsterdam, 1956* 1957 ed W van Tongeren *et al* (Oxford: Pergamon)
128. *Colloq. Spectroscopicum Int. VII, Liège, Belgium 1958* 1959 *Revue Univlle Mines* **102** 9e série tome XV
129. *Proc. Colloq. Spectroscopicum Int. VIII, Luzern, Switzerland, 1959* 1960 ed H Guyer (Aarau: Sauerländer)
130. *Proc. Colloq. Spectroscopicum Int. IX, Lyon, France, 1961* 1961 vol I–III (Paris GAMS)
131. *Proc. Colloq. Spectroscopicum Int. X, College Park, Maryland, 1962* 1963 ed E R Lippincott and M Margoshes (Washington, DC: Spartan) (selected papers)
132. *Colloq. Spectroscopicum Int. XI, Beograd, Yugoslavia, 1963* (abstracts)
133. *Proc. Colloq. Spectroscopicum Int. XII, Exeter, UK, 1965* 1965 (London: Adam Hilger)
134. *Proc. Colloq. Spectroscopicum Int. XIII, Ottawa, Canada, 1967* 1967 (London: Adam Hilger) (invited papers in full)
135. *Proc. Colloq. Spectroscopicum Int. XIV, Debrecen, Hungary, 1967* 1967 ed E Gegus and K Zimmer vol I–III (London: Adam Hilger)
136. *Proc. Colloq. Spectroscopicum Int. XV, Madrid, Spain, 1969* 1970 vol I–II (London: Adam Hilger) (invited papers in full and summaries of contributed papers)
137. *Proc. Colloq. Spectroscopicum Int. XVI, Heidelberg, Germany, 1971* 1972 (London: Adam Hilger) (plenary lectures; preprints of contributed papers published in two volumes)
138. *Proc. Eur. Symp. of Combustion Institute, Sheffield, UK, 1973,* 1973 ed F J Weinberg (New York: Academic Press)
139. *Proc. 10th Natn. Analysis Instrumentation Symp., San Francisco, California, 1964* 1964 ed L Fowler *et al* (New York: Plenum)
140. *Proc. VIIth Symp. (Int.) on Combustion, London and Oxford, 1958* 1959 (London: Butterworth)
141. *Proc. VIIIth Symp. (Int.) on Combustion, Pasadena, California, 1960* 1962 (Baltimore, Md: Williams and Wilkins)
142. *Proc. IXth Symp. (Int.) on Combustion, Ithaca, NY, 1962* 1963 (New York: Academic Press)
143. *Proc. Xth Symp. (Int.) on Combustion, Cambridge, UK, 1964* 1965 (Pittsburgh, Pa: The Combustion Institute)
144. *Proc. XIth Symp. (Int.) on Combustion, Berkeley, California, 1966* 1967 (Pittsburgh, Pa: The Combustion Institute)
145. *Proc. XIIth Symp. (Int.) on Combustion, Poitiers, France, 1968* 1969 (Pittsburgh, Pa: The Combustion Institute)
146. *Proc. XIIIth Symp. (Int.) on Combustion, Salt Lake City, Utah, 1970* 1971 (Pittsburgh, Pa: The Combustion Institute)
147. *Proc. XIVth Symp. (Int.) on Combustion, Philadelphia, Pennsylvania, 1972* 1973 (Pittsburgh, Pa: The Combustion Institute)
148. *Proc. XVth Symp. (Int.) on Combustion, Tokyo, 1974* 1975 (Pittsburgh, Pa: The Combustion Institute)
148a. *Proc. XVIth Symp. (Int.) on Combustion, Cambridge, Massachusetts, 1976* 1977 (Pittsburgh, Pa: The Combustion Institute)
149. *Proc. Symp. on Flame Photometry, Atlantic City, NJ, 1951* 1951 *Spec. Tech. Publication* no. 116 (Pittsburgh, Pa: American Society for Testing and Materials)
150. Report of *Symp. on Trace Characterization, Chemical and Physical, Gaithersburg, Maryland, 1966* 1967 ed W E Meinke and B F Scribner (Washington, DC: NBS) Monograph no. 100

151. Selected papers from *9th Ann. Mid-America Spectroscopy Symp., Chicago, Illinois, 1968* in 1969 *Developments in applied spectroscopy* ed E L Grove and A J Perkins vol 7A (New York: Plenum)
152. Summaries of papers at *Atomic Absorption Symp., Prague, Czechoslovakia, 1967* (some papers were published in full in 1968 *Appl. Opt.* **7** no. 7)
153. *Symp. on Atomic Absorption Spectroscopy* at *71st Ann. Mtg of the American Society for Testing and Materials, San Francisco, California, 1968* 1969 (Philadelphia, Pa: American Society for Testing and Materials) *Spec. Tech. Publication* no. 443

Papers, Articles, and Dissertations

154. Agterdenbos J and Vlogtman J 1974 *Talanta* **21** 231
155. Alcock C B 1969 *Chem. Br.* **5** 216
156. Alder J F, Thompson K C and West T S 1970 *Anal. Chim. Acta* **50** 383
157. Alder J F, Thompson K C and West T S 1972 *Chemia Analit.* **17** 1091
158. Aldous K M, Bailey B W and Rankin J M 1972 *Anal. Chem.* **44** 191
159. Alger D, Kirkbright G F and Troccoli O E 1973 *Appl. Spectrosc.* **27** 177
160. Alkemade C Th J 1954 A contribution to the development and understanding of flame photometry, *Diss. Utrecht*
161. Alkemade C Th J 1960 in [129] p 162
162. Alkemade C Th J 1963 in [131] p 143
163. Alkemade C Th J 1966 *Anal. Chem.* **38** 1252
164. Alkemade C Th J 1968 *Appl. Opt.* **7** 1261
165. Alkemade C Th J 1969 in [45] ch 4 p 101
166. Alkemade C Th J 1970 in [77] ch 1 p 1
167. Alkemade C Th J 1970 in [124] p 73
168. Alkemade C Th J 1973 *Proc. Soc. Anal. Chem.* **10** 130
168a. Alkemade C Th J 1977 in [122a] p 93
169. Alkemade C Th J, Hollander Tj and Kalff P J 1965 *Combust. Flame* **9** 101
170. Alkemade C Th J and Hooymayers H P 1966 *Combust. Flame* **10** 306
171. Alkemade C Th J, Hooymayers H P, Lijnse P L and Vierbergen T J M J 1972 *Spectrochim. Acta* **27B** 149
172. Alkemade C Th J and Jeuken M E J 1957 *Z. Anal. Chem.* **158** 401
173. Alkemade C Th J and Milatz J M W 1955 *J. Opt. Soc. Am.* **45** 583; 1955 *Appl. Sci. Res.* **B4** 289
174. Alkemade C Th J, Smit J and Verschure J C M 1952 *Biochim. Biophys. Acta* **8** 562
175. Alkemade C Th J and Voorhuis M H 1958 *Z. Anal. Chem.* **163** 91
176. Alkemade C Th J and Zeegers P J Th 1971 in [115] ch 1 p 3
177. Allan J E 1961 *Spectrochim. Acta* **17** 467
178. Allan J E 1969 *Spectrochim. Acta* **24B** 13
179. Amos M D and Willis J B 1966 *Spectrochim. Acta* **22** 1325
180. Andrews G E and Bradley D 1972 *Combust. Flame* **18** 133
181. Armentrout D N 1966 *Anal. Chem.* **38** 1235
182. Armstrong B H 1967 *J. Quant. Spectrosc. Radiat. Transfer* **7** 61
183. Ashton A F and Hayhurst A N 1973 *Combust. Flame* **21** 69
184. Atsuya I and Glemser O 1969 in [117]
185. Avni R and Alkemade C Th J 1960 *Mikrochim. Acta* p 460
186. Bacis R 1971 *Appl. Opt.* **10** 535
187. Bahn G S 1958 in *Literature of the Combustion of Petroleum, Adv. Chem. Ser.* no. 20, *Am. Chem. Soc.* Washington DC, p 104
188. Bailey B W and Rankin J M 1969 *Spectrosc. Lett.* **2** 159
189. Bailey B W and Rankin J M 1969 *Spectrosc. Lett.* **2** 233
190. Baker M R, Fuwa K, Thiers R E and Vallee B L 1958 *J. Opt. Soc. Am.* **48** 576

191. Baker C A and Garton F W J 1961 *UKAEA Rep.* R 3490
192. Baker M R and Vallee B L 1959 *Anal. Chem.* **31** 2036
193. Barnes R H, Moeller C E, Kircher J F and Verber C M 1973 *Appl. Opt.* **12** 2531
194. Barnett W B, Fassel V A and Kniseley R N 1968 *Spectrochim. Acta* **23B** 643
195. Bartlet J C, Farmilo C G and Pugsley L J 1955 *Anal. Chem.* **27** 320
196. Bastiaans G J and Hieftje G M 1974 *Anal. Chem.* **46** 901
197. Bauer E, Fisher E R and Gilmore F R 1969 *J. Chem. Phys.* **51** 4173
198. Behmenburg W 1964 *J. Quant. Spectrosc. Radiat. Transfer* **4** 177
199. Behmenburg W 1968 *Z. Astrophys.* **69** 368
200. Behmenburg W and Kohn H 1964 *J. Quant. Spectrosc. Radiat. Transfer* **4** 163
201. Behmenburg W, Kohn H and Mailänder M 1964 *J. Quant. Spectrosc. Radiat. Transfer* **4** 149
202. Belcher R, Bogdanski S, Ghonaim S A and Townshend A 1974 *Anal. Lett.* **7** 133
203. Belcher R, Dagnall R M and West T S 1964 *Talanta* **11** 1257
204. Belcher R, Ranjitkar K P and Townshend A 1975 *Analyst, London* **100** 415
205. Bell J C and Bradley D 1970 *Combust. Flame* **14** 255
206. Bell W E, Bloom A L and Lynch J 1961 *Rev. Sci. Instrum.* **32** 688
207. Belyaev Yu I, Iwantsow L M, Karsyakin A V, Pham Hung Phi and Shemet V V 1968 *Zh. Anal. Khim.* **23** 980 (1968 *J. Anal. Chem. USSR* **23** 855)
208. Benetti P, Omenetto N and Rossi G 1971 *Appl. Spectrosc.* **25** 57
209. Bernstein R E 1955 *S. Afr. J. Med. Sci.* **20** 57
210. Berry J W, Chappell D G and Barnes R B 1946 *Ind. Engng Chem., Anal. Edn* **18** 19
211. Billings G K 1965 *Atom. Absorption Newsl.* **4** 357
212. Blades A T 1971 *Can. J. Chem.* **49** 2476
213. Bleekrode R 1968 *Appl. Spectrosc.* **22** 536
214. Boers A L, Alkemade C Th J and Smit J A 1956 *Physica* **22** 358
215. Bol'shov M A *et al* 1973 *Zh. Prikl. Spektrosk.* **19** 821
215a. Bol'shov M A, Zybin A V, Koloshnikov V G and Koshelev K N 1977 *Spectrochim. Acta* **32B** 279
216. Borgers A J 1965 in [143] p 627
216a. Boss C B and Hieftje G M 1977 in [122a] p 113
217. Bouckaert R 1973 Het gedrag van metaalatomen en metaalverbindingen in vlammen (with summary in English), *Diss. Leuven, Belgium* (partially published by R Bouckaert *et al* in [123] p 17 and in 1975 *Analusis* **3** 142)
218. Bouckaert R, D'Olieslager J and de Jaegere S 1972 *Anal. Chim. Acta* **58** 347
219. Boumans P W J M 1972 in [63] part 2 ch 6 p 1
219a. Boutilier G D, Bradshaw J D, Weeks S J and Winefordner J D 1977 *Appl. Spectrosc.* **31** 307
220. Bowman J A, Sullivan J V and Walsh A 1966 *Spectrochim. Acta* **22** 205
221. Bowman J A and Willis J B 1967 *Anal. Chem.* **39** 120
222. Bradley D and Ibrahim S M A 1975 *Combust. Flame* **24** 169
223. Bradley D, Jesch L F and Sheppard C G W 1972 *Combust. Flame* **19** 237
224. Bradley D and Sheppard C G W 1970 *Combust. Flame* **15** 323
225. Bratzel M P, Dagnall R M and Winefordner J D 1969 *Anal. Chem.* **41** 713
226. Braun W and Carrington T 1969 *J. Quant. Spectrosc. Radiat. Transfer* **9** 1133
227. Brewer L and Hauge R 1968 *J. Molec. Spectrosc.* **25** 330
228. Brewer L and Porter R F 1954 *J. Chem. Phys.* **22** 1867
229. Brewer L and Rosenblatt G M 1969 in *Advances in high-temperature chemistry* ed L Eyring vol 2 (New York: Academic Press) p 1
230. Broida H P 1951 *J. Chem. Phys.* **19** 1383
231. Broida H P and Shuler K E 1957 *J. Chem. Phys.* **27** 933
232. Brown R L, Everest D A, Lewis J D and Williams A 1968 *J. Inst. Fuel* **41** 433
233. Browner R F and Winefordner J D 1972 *Anal. Chem.* **44** 247

234. Browning J A 1958 in *Literature of the combustion of petroleum, Adv. Chem. Ser.* no. 20, *Am. Chem. Soc.* Washington DC, p 143
235. Bruce C F and Hannaford P 1971 *Spectrochim. Acta* **26B** 207
236. Buell B E 1963 *Anal. Chem.* **35** 372
237. Buell B E 1969 in [45] ch 9 p 267
238. Bulewicz E M 1956 *Nature* **177** 670
239. Bulewicz E M, James C G and Sugden T M 1956 *Proc. R. Soc.* **A235** 89
240. Bulewicz E M and Padley P J 1961 *Combust. Flame* **5** 331
241. Bulewicz E M and Padley P J 1963 in [142] pp 638, 654
242. Bulewicz E M and Padley P J 1970 *Combust. Flame* **15** 203
243. Bulewicz E M and Padley P J 1971 *Proc. R. Soc.* **A323** 377
244. Bulewicz E M and Padley P J 1971 *Trans. Faraday Soc.* **67** 2337
245. Bulewicz E M and Padley P J 1973 *Spectrochim. Acta* **28B** 125
246. Bulewicz E M and Sugden T M 1959 *Trans. Faraday Soc.* **55** 720
247. Burdett N A and Hayhurst A N 1975 in [148] p 979
248. Busch K W, Howell N G and Morrison G H 1974 *Anal. Chem.* **46** 2074
249. Butler L R P 1966 *Atom. Absorption Newsl.* **5** 99
250. Butler L R P and Fulton A 1968 *Appl. Opt.* **7** 2131
251. Calcote H F 1962 in [141] p 164
252. Calcote H F 1963 in [99] p 107
253. Calcote H F 1963 in [142] p 622
254. Calcote H F, Kurzius S C and Miller W J 1965 in [143] p 605
255. van Calker J 1960 in [129] p 150
256. Capacho-Delgado L and Manning D C 1966 *Spectrochim. Acta* **22** 1505
257. Capacho-Delgado L and Sprague S 1965 *Atom. Absorption Newsl.* **4** 363
258. Carabetta R and Kaskan W E 1968 *J. Phys. Chem.* **72** 2483
259. Carrion de Rosa-Brussin N 1971 Utilisation d'un générateur de plasma en spectrométrie atomique, *Diss. Lyon, France*
260. Cartwright J S, Sebens C and Slavin W 1966 *Atom. Absorption Newsl.* **5** 22
261. Castleman R A 1961 *J. Res. NBS* **6** 369
262. Caton R D and Bremner R W 1954 *Anal. Chem.* **26** 805
263. Chabbal R 1953 *J. Rech., CNRS* **24** 138
264. Chakrabarti C L, Katyal M and Willis D E 1970 *Spectrochim. Acta* **25B** 629
265. Chakrabarti C L, Lyles G R and Dowling F B 1963 *Anal. Chim. Acta* **29** 489
266. Champion P, Pellet H and Grenier M 1873 *C.R. Acad. Sci., Paris* **76** 707
267. Charles C W 1963 *J. Opt. Soc. Am.* **53** 1319
268. Charton M and Gaydon A G 1956 *Proc. Phys. Soc.* **A69** 520
269. Charton M and Gaydon A G 1958 *Proc. R. Soc.* **A245** 84
270. Chen C T and Winefordner J D 1975 *Can. J. Spectrosc.* **20** 87
271. Chenevier M and Lombardi M 1972 *Chem. Phys. Lett.* **16** 154
272. Chester J E, Dagnall R M and Taylor M R G 1970 *Anal. Chim. Acta* **51** 95
273. Clampitt N C and Hieftje G M 1972 *Anal. Chem.* **44** 1211
274. Coker D T and Ottaway J M 1971 *Nature* **230** 156
275. Coker D T, Ottaway J M and Pradhan N K 1971 *Nature* **233** 69
276. Collins G C and Polkinhorne H 1952 *Analyst, London* **77** 430
277. Cotton D H and Jenkins D R 1968 *Trans. Faraday Soc.* **64** 2988
278. Cotton D H and Jenkins D R 1969 *Trans. Faraday Soc.* **65** 1537
279. Cotton D H and Jenkins D R 1970 *Spectrochim. Acta* **25B** 283
280. Cotton D H and Jenkins D R 1971 *Trans. Faraday Soc.* **67** 730
281. Cowan R D and Dieke G H 1948 *Rev. Mod. Phys.* **20** 418
282. Cowley T G, Fassel V A and Kniseley R N 1968 *Spectrochim. Acta* **23B** 771
283. Cresser M S 1971 *Spectrosc. Lett.* **4** 275
284. Cunningham P T and Link J K 1967 *J. Opt. Soc. Am.* **57** 1000
285. Dagnall R M and Taylor M R G 1971 *Spectrosc. Lett.* **4** 147

286. Dagnall R M, Thompson K C and West T S 1966 *Anal. Chim. Acta* **36** 269
287. Dagnall R M, Thompson K C and West T S 1967 *Atom. Absorption Newsl.* **6** 117
288. Dagnall R M, Thompson K C and West T S 1967 *Talanta* **14** 557
289. Dagnall R M, Thompson K C and West T S 1968 *Analyst, London* **93** 153
290. Dagnall R M, Thompson K C and West T S 1968 *Analyst, London* **93** 518
291. David D J 1964 *Spectrochim. Acta* **20** 1185
292. Day R A 1970 *Appl. Opt.* **9** 1213
293. Dean J A 1958 *ASTM Spec. Tech. Publication* no. 238
294. Dean J A 1961 *Rec. Chem. Prog.* **22** 179
295. Dean J A 1964 in [126] p 207
296. Dean J A and Adkins J E 1964 in *15th Midwest Spectroscopy Symp., Chicago, Illinois*
297. Dean J A and Adkins J E 1966 *Analyst, London* **91** 709
298. Dean J A, Burger J C, Rains T C and Zittel H E 1961 *Anal. Chem.* **33** 1722
299. Dean J A and Carnes W J 1962 *Analyst, London* **87** 743
300. Dean J A and Carnes W J 1962 *Anal. Chem.* **34** 192
301. Dean J A and Simms J C 1963 *Anal. Chem.* **35** 699
302. Dean J A and Stubblefield Ch B 1961 *Anal. Chem.* **33** 382
303. Demayo A, Hunter L W and Kruus P 1968 *Can. J. Chem.* **46** 3151
304. Demers D R and Ellis D W 1968 *Anal. Chem.* **40** 860
305. Denton M B and Malmstadt H V 1971 *Appl. Phys. Lett.* **18** 485
306. Dippel W A 1954 A fundamental study of analytical flame photometry, *Diss. Princeton, NJ*
307. Dixon R N 1963 *Proc. R. Soc.* **A275** 431
308. Dixon-Lewis G 1970 *Combust. Flame* **15** 197
309. Doiwa A 1956 Das Problem der dritten Partner in der Flammenspektrometrie, *Diss. Frankfurt am Main*
310. Dolidze L D and Lebedev V I 1973 in [123] p 63
311. Dorman F H 1969 *J. Chem. Phys.* **50** 1042
312. Doty M E and Schrenk W G 1964 in [126] p 196
313. Dvořák J 1963 in [132]
314. Eckelmans V, Graauwmans E and de Jaegere S 1974 *Talanta* **21** p 715
315. Eckhard S and Püschel A 1960 *Z. Anal. Chem.* **172** 334
316. Edmondson H and Heap M P 1971 *Combust. Flame* **16** 161
317. Edse R and Lawrence L R 1969 *Combust. Flame* **13** 479
318. Eggertsen F T, Wyld G and Lykken L 1951 in [149] p 52
319. Ellis D W and Demers D R 1966 *Anal. Chem.* **38** 1943
320. von Engel A 1967 *Br. J. Appl. Phys.* **18** 1661
321. English P E and Hitchcock S S 1968 *J. Sci. Instrum.* Series 2 **1** 984
322. Eppendorf Gerätebau 1962 *Handbuch Flammenphotometer Eppendorf* (Hamburg: Wellingsbüttel)
323. Epstein M S, Rains T C and O'Haver T C 1976 *Appl. Spectrosc.* **30** 324
324. Eshelman H C and Armentor J 1964 in [126] p 190
325. Ewing J J, Milstein R and Berry R S 1971 *J. Chem. Phys.* **54** 1752
326. Falk H, Becker-Ross H and Schiller H 1971 *Ann. Phys.* **26** 166
327. Farber M and Srivastava R D 1973 *Combust. Flame* **20** 43
328. Farber M and Srivastava R D 1975 *High Temp. Sci.* **7** 74
329. Farley K R and Petersen H E 1966 *Bureau of Mines Report of Investigations, Washington, DC* no. 6820
330. Fassel V A, Curry R H and Kniseley R N 1962 *Spectrochim. Acta* **18** 1127
331. Fassel V A, Curry R H, Myers R B and Kniseley R N 1962 in [131]
332. Fassel V A and Golightly D W 1967 *Anal. Chem.* **39** 466
333. Fassel V A, Mossotti V G, Grossman W E L and Kniseley R N 1966 *Spectrochim. Acta* **22** 347
334. Fassel V A, Myers R B and Kniseley R N 1963 *Spectrochim. Acta* **19** 1187

335. Fassel V A, Rasmuson J O and Cowley T G 1968 *Spectrochim. Acta* **23B** 579
336. Fassel V A, Rasmuson J O and Cowley T G 1970 *Spectrochim. Acta* **25B** 559
337. Fenimore C P and Jones G W 1965 in [143] p 489
338. Feugier A and van Tiggelen A 1965 in [143] p 621
339. Fink A 1955 *Mikrochim. Acta* p 314
340. Fiorino J A 1969 The design and performance of a long-path oxy-acetylene burner for atomic absorption spectroscopy, *Diss. Iowa State University, Ames*
341. Fiorino J A, Kniseley R N and Fassel V A 1968 *Spectrochim. Acta* **23B** 413
342. Firman R J 1965 *Spectrochim. Acta* **21** 341
343. Fischer J and Kropp R 1960 *Glastech. Ber.* **33** 380
344. Fleming H D 1967 *Spectrochim. Acta* **23B** 207
345. Fontijn A 1974 *Pure Appl. Chem.* **39** 287
346. Fontijn A, Felder W and Houghton J J 1975 in [148] p 775
346a. Fontijn A, Felder W and Houghton J J 1977 in [148a] p 871
347. Fontijn A, Kurzius S C and Houghton J J 1973 in [147] p 167
348. Fontijn A, Miller W J and Hogan J M 1965 in [143] p 545
349. Foster W H and Hume D N 1959 *Anal. Chem.* **31** 2028
350. Foster W H and Hume D N 1959 *Anal. Chem.* **31** 2033
351. Fowler G N and Preist T W 1972 *J. Chem. Phys.* **56** 1601
351a. Fowler W K and Winefordner J D 1977 *Anal. Chem.* **49** 944
352. Frank C W, Schrenk W G and Melvan C E 1967 *Anal. Chem.* **39** 534
352a. Franklin M, Baber C and Koirtyohann S R 1976 *Spectrochim. Acta* **31B** 589
353. Fraser L M and Winefordner J D 1971 *Anal. Chem.* **43** 1693
354. Fraser L M and Winefordner J D 1972 *Anal. Chem.* **44** 1444
355. Friswell N J and Jenkins D R 1972 *Combust. Flame* **19** 197
356. Fukushima S 1959 *Mikrochim. Acta* p 596
357. Fukushima S 1960 *Mikrochim. Acta* p 332
358. Fulton A and Butler L R P 1968 *Spectrosc. Lett.* **1** 317
359. Fuwa K, Haraguchi H and Toda S 1972 in Abstracts of *IUPAC Int. Congr. on Analytical Chemistry, Kyoto, Japan*
360. Fuwa K, Haraguchi H, Okamoto K and Nagata T 1972 *Japan Analyst* **21** 945
361. de Galan L 1969 *Spectrochim. Acta* **24B** 629
362. de Galan L 1970 *Spectrosc. Lett.* **3** 123
363. de Galan L, McGee W W and Winefordner J D 1967 *Anal. Chim. Acta* **37** 436
364. de Galan L, Novotny I, Pickford C J and Wagenaar H C 1973 in [125] vol II p 747
365. de Galan L and Samaey G F 1969 *Spectrochim. Acta* **24B** 679
366. de Galan L and Samaey G F 1970 *Anal. Chim. Acta* **50** 39
367. de Galan L and Samaey G F 1970 *Spectrochim. Acta* **25B** 245
368. de Galan L, Smith R and Winefordner J D 1968 *Spectrochim. Acta* **23B** 521
369. de Galan L and Wagenaar H C 1971 in [122] p 10
370. de Galan L and Winefordner J D 1966 *Anal. Chem.* **38** 1412
371. de Galan L and Winefordner J D 1967 *J. Quant. Spectrosc. Radiat. Transfer* **7** 251
372. de Galan L and Winefordner J D 1967 *J. Quant. Spectrosc. Radiat. Transfer* **7** 703
373. de Galan L and Winefordner J D 1968 *Spectrochim. Acta* **23B** 277
374. Gatehouse B M and Willis J B 1961 *Spectrochim. Acta* **17** 710
375. Gaydon A G 1954 *NBS Circ.* **523** 1
376. Gaydon A G 1955 *Proc. R. Soc.* **A231** 437
377. Gaydon A G 1956 *Mém. Soc. R. Sci. Liège IV* **18** 507
378. van Gelder Z 1968 *Appl. Spectrosc.* **22** 581
379. van Gelder Z 1970 *Spectrochim. Acta* **25B** 669
380. Georgii H W 1954 *Z. Aerosol-Forsch.* **3** 496
381. Gibson J H, Grossman W E L and Cooke W D 1963 *Anal. Chem.* **35** 266
382. Gidley J A F 1964 in [121]
383. Gilbert P T Jr 1951 in [149] p 77

384. Gilbert P T Jr 1958 in *Pittsburgh Conf. on Analytical Chemistry and Applied Spectroscopy, 1958* (Fullerton, Calif.: Beckman Instruments Inc) report printed by Beckman Instruments, Inc. (1958)
385. Gilbert P T Jr 1960 *ASTM Spec. Tech. Publication* no. 269 p 73
386. Gilbert P T Jr 1961 *Analyzer (Beckman Instruments Inc.)* **2** 3
387. Gilbert P T Jr 1961 *Beckman Instruments Inc. Bull.* no. 753-A
388. Gilbert P T Jr 1962 *Anal. Chem.* **34** 1025
389. Gilbert P T Jr 1963 in [131] p 171
390. Gilbert P T Jr 1964 in [139] p 193
391. Gilbert P T Jr 1970 in [77] ch 5 p 181
392. Glass G P, Kistiakowsky G B, Michael J V and Niki H 1965 in [143] p 573
393. Gole J L and Zare R N 1972 *J. Chem. Phys.* **57** 5331
394. Goodfellow G I 1967 *AWRE Rep.* 0-87/66
395. Gouy A 1877 *C. R. Acad. Sci., Paris* **85** 70
396. Grant C L 1969 in [153] p 37
397. Green R B, Travis J C and Keller R A 1976 *Anal. Chem.* **48** 1954
398. Grove E L, Scott C W and Jones F 1965 *Talanta* **12** 327
399. Grunder F I and Boettner E A 1969 in [151] p 201
400. Gruzder P F 1967 *Opt. Spectrosc.* **22** 89
401. Güçer S and Massmann H 1973 in [125] vol I p 51
402. Gurvich L V and Ryabova V G 1964 *High Temp.* **2** 486
403. Gurvich L V and Ryabova V G 1965 *Opt. Spektrosk.* **18** 143
404. Gurvich L V, Ryabova V G, Khitrov A N and Starovoitov E M 1971 *High Temp.* **9** 261
405. Gurvich L V and Veits I V 1958 *Izv. Akad. Nauk* **22** 673
406. Gutsche B and Herrmann R 1972 *Z. Anal. Chem.* **258** 277
407. Gutsche B and Herrmann R 1974 *Z. Anal. Chem.* **269** 260
408. Gutsche B, Herrmann R and Rüdiger K 1972 *Z. Anal. Chem.* **258** 273
408a. Halls D J 1977 *Spectrochim. Acta* **32B** 221
409. Halls D J and Pungor E 1969 *Combust. Flame* **13** 108
410. Halls D J and Townshend A 1966 *Anal. Chim. Acta* **36** 278
411. Hambly A N and Rann C S 1969 in [45] ch 8 p 241
412. Hannaford P 1972 *J. Opt. Soc. Am.* **62** 265
413. Hannaford P 1975 in [119] paper B-32
414. Hannaford P 1975 in [119] paper P-7
415. Haraguchi H, Fowler W K, Johnson D J and Winefordner J D 1976 *Spectrochim. Acta* **32A** 1539
416. Haraguchi H and Fuwa K 1975 *Spectrochim. Acta* **30B** 535
417. den Harder A and de Galan L 1974 *Anal. Chem.* **46** 1464
418. Harrison A G 1969 *J. Chem. Phys.* **50** 1043
419. Hayhurst A N 1974 *IEEE Trans. Plasma Sci.* **PS-2** 115
420. Hayhurst A N and Kittelson D B 1972 *Chem. Commun.* p 422
421. Hayhurst A N and Kittelson D B 1972 *Combust. Flame* **19** 306
422. Hayhurst A N and Sugden T M 1967 *Trans. Faraday Soc.* **63** 1375
423. Hayhurst A N and Telford N R 1970 *Trans. Faraday Soc.* **66** 2784
424. Hayhurst A N and Telford N R 1972 *J. Chem. Soc. Faraday Trans. 1* 237
425. Hayhurst A N and Telford N R 1972 *Nature* **235** 114
426. Hayhurst A N and Telford N R 1974 *J. Chem. Soc. Faraday Trans. I* **20** 1999
427. Hegemann F and Osterried O 1963 *Glastech. Ber.* **36** 217
428. Held A and Stephens R 1975 *Can. J. Spectrosc.* **20** 10
429. van der Held E F M 1932 Meting der overgangswaarschijnlikheid 2p-1s voor Natrium door absolute intensiteitsmetingen aan vlammen, *Diss. Utrecht*
430. Herrmann R 1954 *Optik* **11** 505
431. Herrmann R 1957 *Z. Ges. Exp. Med.* **129** 55
432. Herrmann R 1961 *Optik* **18** 422

433. Herrmann R 1964 *Chemie Labor Betrieb* **15** 451
434. Herrmann R 1965 *Z. Anal. Chem.* **212** 1
435. Herrmann R 1971 in [46] ch 3 p 57
436. Herrmann R 1974 *Z. Klin. Chem. Klin. Biochem.* **12** 393
437. Herrmann R and Lang W 1961 *Z. Ges. Exp. Med.* **134** 268
438. Herrmann R and Lang W 1962 *Arch. Eisenhüttenw.* **33** 643
439. Herrmann R and Lang W 1962 *Optik* **19** 208
440. Herrmann R and Lang W 1962 *Z. Ges. Exp. Med.* **135** 569
441. Herrmann R and Lang W 1964 *Z. Anal. Chem.* **203** 1
442. Herrmann R, Lang W and Rüdiger K 1964 *Z. Anal. Chem.* **206** 241
443. Herrmann R and Schellhorn H 1952 *Z. Angew. Phys.* **4** 208
444. Hieftje G M 1975 in [119] paper P-5
445. Hieftje G M and Bystroff R I 1975 *Spectrochim. Acta* **30B** 187
446. Hieftje G M and Malmstadt H V 1968 *Anal. Chem.* **40** 1860
447. Hingle D N, Kirkbright G F and West T S 1968 *Analyst, London* **93** 522
448. Hinnov E 1957 *J. Opt. Soc. Am.* **47** 151
449. Hinnov E and Kohn H 1957 *J. Opt. Soc. Am.* **47** 156
450. Hinze J O 1949 *Appl. Sci. Res.* **A1** 273
451. Hobbs R S, Kirkbright G F, Sargent M and West T S 1968 *Talanta* **15** 997
452. Hofer A 1971 *Z. Anal. Chem.* **253** 206
453. Hofmann F W and Kohn H 1961 *J. Opt. Soc. Am.* **51** 512
454. Hofmann F W, Kohn H and Schneider J 1961 *J. Opt. Soc. Am.* **51** 508
455. Hollander Tj 1964 Self-absorption, ionization and dissociation of metal vapour in flames, *Diss. Utrecht*
456. Hollander Tj 1968 *Am. Inst. Aeronaut. and Astronaut.* p 385
457. Hollander Tj, Borgers A J and Alkemade C Th J 1956 *Appl. Sci. Res.* **5B** 409
458. Hollander Tj and Broida H P 1967 *J. Quant. Spectrosc. Radiat. Transfer* **7** 965
459. Hollander Tj and Broida H P 1969 *Combust. Flame* **13** 63
460. Hollander Tj, Jansen B J, Plaat J J and Alkemade C Th J 1970 *J. Quant. Spectrosc. Radiat. Transfer* **10** 1301
461. Hollander Tj, Kalff P J and Alkemade C Th J 1963 *J. Chem. Phys.* **39** 2558
462. Hollander Tj, Lijnse P L, Franken L P L, Jansen B J and Zeegers P J Th 1972 *J. Quant. Spectrosc. Radiat. Transfer* **12** 1067
463. Hollander Tj, Lijnse P L, Jansen B J and Franken L P L 1973 *J. Quant. Spectrosc. Radiat. Transfer* **13** 669
464. Hooymayers H P 1966 Quenching of excited alkali atoms and hydroxyl radicals and related effects in flames, *Diss. Utrecht*
465. Hooymayers H P 1968 *Spectrochim. Acta* **23B** 567
466. Hooymayers H P and Alkemade C Th J 1966 *J. Quant. Spectrosc. Radiat. Transfer* **6** 501
467. Hooymayers H P and Alkemade C Th J 1966 *J. Quant. Spectrosc. Radiat. Transfer* **6** 847
468. Hooymayers H P and Alkemade C Th J 1969 *Chem. Phys. Lett.* **4** 277
469. Hooymayers H P and Lijnse P L 1969 *J. Quant. Spectrosc. Radiat. Transfer* **9** 995
470. Hooymayers H P and Nienhuis G 1968 *J. Quant. Spectrosc. Radiat. Transfer* **8** 955
471. Huldt L 1948 Eine spektroskopische Untersuchung des elektrischen Lichtbogens und der Acetylen-Luft-Flamme mit besonderer Hinsicht auf ihre Anwendung als Lichtquelle für die quantitative Spektralanalyse, *Diss. Uppsala*
472. Huldt L and Knall E 1954 *Naturwissenschaften* **41** 421
473. Huldt L and Knall E 1954 *Z. Naturf.* **9a** 663
474. Huldt L and Knall E 1956 *Ark. Fys.* **11** 229
475. Huldt L and Lagerqvist A 1956 *Ark. Fys.* **11** 347
476. Hulpke E, Paul E and Paul W 1964 *Z. Phys.* **177** 257

477. Human H G C 1971 The influence of primary source line shape on the sensitivity and shape of working curves in AAS, *Diss. Pretoria*

478. Human H G C and Zeegers P J Th 1975 *Spectrochim. Acta* **30B** 203

479. Human H G C, Zeegers P J Th and van Elst J A 1974 *Spectrochim. Acta* **29B** 111

480. van der Hurk J 1974 Origin and excitation energy of visible alkaline earth bands in flames, *Diss. Utrecht*

481. van der Hurk J, Hollander Tj and Alkemade C Th J 1973 *J. Quant. Spectrosc. Radiat. Transfer* **13** 273

482. van der Hurk J, Hollander Tj and Alkemade C Th J 1974 *J. Quant. Spectrosc. Radiat. Transfer* **14** 1167

483. van der Hurk J, Hollander Tj and Alkemade C Th J 1975 *J. Quant. Spectrosc. Radiat. Transfer* **15** 113

484. Hurle I R 1964 *J. Chem. Phys.* **41** 3911

485. Ingle J D and Crouch S R 1972 *Anal. Chem.* **44** 777

486. Ingle J D and Crouch S R 1972 *Anal. Chem.* **44** 785

487. Jacobs P W and Russell-Jones A 1968 *J. Phys. Chem.* **72** 202

488. James C G and Sugden T M 1953 *Nature* **171** 428

489. James C G and Sugden T M 1955 *Nature* **175** 252

490. James C G and Sugden T M 1955 *Nature* **175** 333

491. James C G and Sugden T M 1958 *Proc. R. Soc.* **248** 238

492. Janin J and Bouvier A 1964 *Spectrochim. Acta* **20** 1787

493. Janin J, Bouvier A and Mathais H 1961 in [130] vol II p 282

494. Janin J, Roux F and d'Incan J 1967 *Spectrochim. Acta* **23A** 2939

495. Jansen B J 1976 Atomic spectral line profiles in flames, an experimental study, *Diss. Utrecht*

495a. Jansen B J and Hollander Tj 1977 *Spectrochim. Acta* **32B** 165

496. Jansen B J, Hollander Tj and Alkemade C Th J 1977 *J. Quant. Spectrosc. Radiat. Transfer* **17** 187

497. Jansen B J, Hollander Tj and Franken L P L 1974 *Spectrochim. Acta* **29B** 37

498. Jaworowski R J and Weberling R P 1966 *Atom. Absorption Newsl.* **5** 125

499. Jenkins D R 1967 *Spectrochim. Acta* **23B** 167

500. Jenkins D R 1968 *Trans. Faraday Soc.* **64** 36

501. Jenkins D R 1969 in [117]

502. Jenkins D R 1970 *Spectrochim. Acta* **25B** 47

503. Jenkins D R and Sugden T M 1969 in [45] ch 5 p 151

504. Jensen D E 1966 *J. Chem. Phys.* **51** 4674

505. Jensen D E 1968 *Combust. Flame* **12** 261

506. Jensen D E 1969 *J. Chem. Phys.* **51** 4674

507. Jensen D E and Jones G A 1972 *J. Chem. Soc. Faraday Trans. I* **68** 259

508. Jensen D E and Jones G A 1973 *J. Chem. Soc. Faraday Trans. I* **69** 1448

509. Jensen D E and Miller W J 1969 *Aerochem. Res. Lab., Princeton, NJ, Rep.* TP-223

510. Jensen D E and Miller W J 1970 *J. Chem. Phys.* **53** 3287

511. Jensen D E and Padley P J 1966 *Trans. Faraday Soc.* **62** 2132

512. Jensen D E and Padley P J 1966 *Trans. Faraday Soc.* **62** 2140

513. Jensen D E and Padley P J 1967 in [144] p 351

514. Johnson D J and Winefordner J D 1976 *Anal. Chem.* **48** 341

515. Johnson G M and Smith M Y 1972 *Spectrochim. Acta* **27B** 269

516. Johnson R W and Schrenk W G 1964 *Appl. Spectrosc.* **18** 144

517. Kahn H L 1968 *Atom. Absorption Newsl.* **7** 40

518. Kaiser H 1970 *Anal. Chem.* **42** no. 2 24A

519. Kaiser H 1970 *Anal. Chem.* **42** no. 4 26A

520. Kalff P J 1970 Alkaline earth compounds in flames and N_2–alkali energy transfer in molecular beams, *Diss. Utrecht*

521. Kalff P J and Alkemade C Th J 1970 *J. Chem. Phys.* **52** 1006

522. Kalff P J and Alkemade C Th J 1972 *Combust. Flame* **19** 257
523. Kalff P J and Alkemade C Th J 1973 *J. Chem. Phys.* **59** 2572
524. Kalff P J, Hollander Tj and Alkemade C Th J 1965 *J. Chem. Phys.* **42** 2299
525. Kallend A S 1967 *Combust. Flame* **11** 81
526. Kaskan W E 1959 *J. Chem. Phys.* **31** 944
527. Kaskan W E 1965 in [143] p 41
528. Kaufman F and Parkes D A 1970 *Trans. Faraday Soc.* **66** 1579
529. Kelly R and Padley P J 1967 *Nature* **216** 258
530. Kelly R and Padley P J 1969 *Trans. Faraday Soc.* **65** 355
531. Kelly R and Padley P J 1969 *Trans. Faraday Soc.* **65** 367
532. Kelly R and Padley P J 1970 *Trans. Faraday Soc.* **66** 1127
533. Kelly R and Padley P J 1971 *Trans. Faraday Soc.* **67** 740
534. Kelly R and Padley P J 1971 *Trans. Faraday Soc.* **67** 1384
535. Kelly R and Padley P J 1972 *Proc. R. Soc.* **A327** 345
536. Kerhyson J D and Ratzkowski C 1968 *Can. J. Spectrosc.* **13** 102
537. King I R 1959 *J. Chem. Phys.* **31** 855
538. King I R 1962 *J. Chem. Phys.* **36** 553
539. King I R 1963 in [99] p 197
540. Kirkbright G F and Ranson L 1971 *Anal. Chem.* **43** 1238
541. Kirkbright G F and Sargent M 1970 *Spectrochim. Acta* **25B** 577
542. Kirkbright G F, Semb A and West T S 1967 *Spectrosc. Lett.* **1** 7
543. Kirkbright G F and Troccoli O E 1973 *Spectrochim. Acta* **28B** 33
544. Kirkbright G F, Troccoli O E and Vetter S 1973 *Spectrochim. Acta* **28B** 1
545. Kirkbright G F and Vetter S 1971 *Spectrochim. Acta* **26B** 505
546. Kirkbright G F and West T S 1968 *Appl. Opt.* **7** 1305
547. Klaus R 1966 *Zeiss-Mitteil.* **4** (1. Heft) 26
548. Knewstubb P F and Sugden T M 1958 *Nature* **181** 474
549. Knewstubb P F and Sugden T M 1960 *Proc. R. Soc.* **A255** 520
550. Kniseley R N 1969 in [45] ch 6 p 189
551. Kniseley R N, Butler C C and Fassel V A 1969 *Anal. Chem.* **41** 1494
552. Knowles D J, Abachi M Q, Belcher R, Bogdanski S L and Townshend A 1975 in [119] paper C-12
553. Koirtyohann S R 1969 in [45] p 295
554. Koirtyohann S R and Pickett E E 1965 Paper at *Natn. Mtg Soc. Appl. Spectrosc., Denver, Colorado*
555. Koirtyohann S R and Pickett E E 1966 *Anal. Chem.* **38** 585
556. Koirtyohann S R and Pickett E E 1966 *Anal. Chem.* **38** 1087
557. Koirtyohann S R and Pickett E E 1967 in [134]
558. Koirtyohann S R and Pickett E E 1968 *Anal. Chem.* **40** 2068
559. Koirtyohann S R and Pickett E E 1971 *Spectrochim. Acta* **26B** 349
560. Koizumi H and Yasuda K 1976 *Spectrochim. Acta* **31B** 237
560a. Koizumi H and Yasuda K 1976 *Spectrochim. Acta* **31B** 523
561. Kolihova D and Sychra V 1973 *Anal. Chim. Acta* **63** 479
562. Konopicky K and Schmidt W 1960 *Z. Anal. Chem.* **173** 358
563. Konopicky K and Schmidt W 1960 *Z. Anal. Chem.* **174** 262
564. Kornblum G R and de Galan L 1973 *Spectrochim. Acta* **28B** 139
565. Krause H F, Fricke J and Fite W L 1972 *J. Chem. Phys.* **56** 4593
566. Krause L 1975 in *The excited state in chemical physics* ed J Wm McGowan (New York: Wiley) ch 4 p 268
567. Kropp R 1960 Störeinflüsse von Partnern in der Flammenspektrometrie sowie Methoden zu ihrer Eliminierung, *Diss. Frankfurt am Main*
568. Kuang-pang Li 1976 *Anal. Chem.* **48** 2050
569. Kuhl J and Marowsky G 1971 *Opt. Commun.* **4** 125
570. Kuhl J, Neumann S and Kriese M 1973 *Z. Naturf.* **28a** 273

571. Kumar A and Pandya T P 1970 *Indian J. Pure Appl. Phys.* **8** 42
572. Lagerqvist A and Huldt L 1955 *Naturwissenschaften* **42** 365
573. Lane W R 1951 *Ind. Engng Chem.* **43** 1312
574. Lang W and Herrmann R 1963 *Optik* **20** 391
575. Langmuir I 1918 *Phys. Rev.* **12** 368
576. Lapple C E 1960 *Chem. Engng* May p 1
577. Larkins P L and Willis J B 1974 *Spectrochim. Acta* **29B** 319
578. Lebedev V I and Dolidze L D 1969 in [117]
579. Letfus V 1966 *Opt. Spectrosc.* **21** 371
580. Levin L A and Budick B 1966 *Bull. Am. Phys. Soc. Ser. II* **11** 455
581. Lewis L L 1969 in [153] p 47
582. Lijnse P L 1972 Review of literature on quenching, excitation and mixing collision cross sections for the first resonance doublets of the alkalis, *Int. Rep. Physical Laboratory, University of Utrecht* i 398
583. Lijnse P L 1973 Electronic-excitation transfer collisions in flames, *Diss. Utrecht*
584. Lijnse P L 1974 *J. Quant. Spectrosc. Radiat. Transfer* **14** 1143
585. Lijnse P L and Elsenaar R J 1972 *J. Quant. Spectrosc. Radiat. Transfer* **12** 1115
586. Lijnse P L and Hornman J C 1974 *J. Quant. Spectrosc. Radiat. Transfer* **14** 1079
587. Lijnse P L and van der Maas C J 1973 *J. Quant. Spectrosc. Radiat. Transfer* **13** 741
588. Lijnse P L, Zeegers P J Th and Alkemade C Th J 1973 *J. Quant. Spectrosc. Radiat. Transfer* **13** 1033
589. Lijnse P L, Zeegers P J Th and Alkemade C Th J 1973 *J. Quant. Spectrosc. Radiat. Transfer* **13** 1301
590. Link J K 1966 *J. Opt. Soc. Am.* **56** 1195
591. van Loon J C 1972 *Atom. Absorption Newsl.* **11** 60
592. Lovett R J, Welch D L and Parsons M L 1975 *Appl. Spectrosc.* **29** 470
593. Lowe R M 1969 *Spectrochim. Acta* **24B** 191
594. Lücke W 1971 *Neues Jahrb. Mineral. Monatsh.* **10** 469
595. Lurio A 1964 *Phys. Rev.* **136A** 376
596. L'vov B V 1961 *Spectrochim. Acta* **17** 761
597. L'vov B V 1965 *Opt. Spectrosc.* **19** 282
598. L'vov B V 1969 *Spectrochim. Acta* **24B** 53
599. L'vov B V 1970 *Opt. Spectrosc.* **28** 18
600. L'vov B V 1970 in [124] p 11
601. L'vov B V 1972 *Revue du GAMS* **8** 3
602. L'vov B V, Katskov D A, Kruglikova L P and Polzik L K 1976 *Spectrochim. Acta* **31B** 49
603. L'vov B V, Kruglikova L P, Polzik L K and Katskov D A 1975 *J. Anal. Chem. USSR* **30** 545
604. L'vov B V, Kruglikova L P, Polzik L K and Katskov D A 1975 *J. Anal. Chem. USSR* **30** 551
605. L'vov B V and Orlov N A 1975 *Zh. Anal. Khim.* **30** 1653
606. L'vov B V, Polzik L K, Katskov D A and Kruglikova L P 1975 *Zh. Prikl. Spektrosk.* **22** 787 (English transl in *J. Appl. Spectrosc.*)
607. McCarthy W J, Parsons M L and Winefordner J D 1967 *Spectrochim. Acta* **23B** 25
608. McEwan M J and Phillips L F 1966 *Trans. Faraday Soc.* **62** 1717
609. McGee W W and Winefordner J D 1967 *J. Quant. Spectrosc. Radiat. Transfer* **7** 201
610. McGee W W and Winefordner J D 1967 *J. Quant. Spectrosc. Radiat. Transfer* **7** 261
611. McPherson G L 1965 *Atom. Absorption Newsl.* **4** 186
612. Malakoff J L, Ramírez-Muñoz J and Scott W Z 1969 *Appl. Spectrosc.* **23** 365
613. Manning D C 1966 *Atom. Absorption Newsl.* **5** 127
614. Manning D C 1967 *Atom. Absorption Newsl.* **6** 35
615. Manning D C and Heneage P 1967 *Atom. Absorption Newsl.* **6** 124
616. Manning D C and Slavin S 1969 *Atom. Absorption Newsl.* **8** 132

617. Manning D C, Trent D J, Sprague S and Slavin W 1965 *Atom. Absorption Newsl.* **4** 255
618. Manoliu C, Tomi B and Panovici A 1973 in [125] vol I p 412
619. Mansell R E 1968 *Appl. Spectrosc.* **22** 790
620. Mansell R E 1970 *Spectrochim. Acta* **25B** 219
621. Mansfield J M, Bratzel M P, Norgordon H O, Knapp D O, Zacha K E and Wineford-ner J D 1968 *Spectrochim. Acta* **23B** 389
622. Margoshes M 1959 *ASTM Spec. Tech. Publication* no. 259
623. Margoshes M 1962 in *Physical techniques in biological research* **4** 215 ed W L Nastuk (New York: Academic Press)
624. Margoshes M 1967 *Anal. Chem.* **39** 1093
625. Margoshes M and Vallee B L 1956 *Anal. Chem.* **28** 180
626. Marks J Y and Welcher G G 1970 *Anal. Chem.* **42** 1033
627. Marshall D and Schrenk W G 1968 *Spectrosc. Lett.* **1** 87
628. Marshall W R 1954 *Chem. Engng Prog. Monogr. Ser.* **50** no. 2
629. Marucic J and Voinovitch I A 1969 *Chim. Anal., Paris* **51** 537
630. Massmann H 1971 in [46] p 95
631. Massmann H 1974 *Angew. Chem. Int. Edn* **13** 504
632. Massmann H and Güçer S 1974 *Spectrochim. Acta* **29B** 283
633. Matousek J P 1973 in [125] vol I p 57
634. Mavrodineanu R 1960 in [129] p 15
635. Mavrodineanu R 1961 *Spectrochim. Acta* **17** 1016
636. Mavrodineanu R 1971 in [122] p 39
637. May K R 1945 *J. Sci. Instrum.* **22** 187
638. Mazing M A and Penkin N P 1966 *Opt. Spectrosc.* **21** 408
639. Menzies A C 1960 *Anal. Chem.* **32** 898
640. Menzies A C 1960 *Z. Instrumkde* **68** 242
641. Meshkova S B and Poluektov N S 1965 *Zh. Prikl. Spektrosk. Akad. Nauk Belorussk.* **2** 21
642. Miles B M and Wiese W L 1970 *Atom. Data* **1** 1
643. Millikan R C and White D R 1963 *J. Chem. Phys.* **39** 98
644. Mitchell D G and Johansson A 1970 *Spectrochim. Acta* **25B** 175
645. Mitchell R L 1960 in [45] ch 1 p 1
646. Moise N L 1966 *Astrophys. J.* **144** 774
647. Moldan B 1970 in [124] p 127
648. Mossholder N V, Fassel V A and Kniseley R N 1973 *Anal. Chem.* **45** 1614
649. Mossotti V G and Abercrombie F N 1972 in [137]
650. Mossotti V G and Duggan M 1968 *Appl. Opt.* **7** 1325
651. Mossotti V G and Fassel V A 1964 *Spectrochim. Acta* **20** 1117
652. Moutet A, Véret C and Nadaud L 1959 *Recherche Aéronaut., Paris* no. 68 9
653. Mukherjee N R, Fueno T, Eyring H and Ree T 1962 in [141] p 1
654. Mularz E J and Yuen M C 1972 *J. Quant. Spectrosc. Radiat. Transfer* **12** 1553
655. Nakahara T, Munemori M and Musha S 1972 *Anal. Chim. Acta* **62** 267
656. Nakahara T, Munemori M and Musha S 1973 *Bull. Chem. Soc. Japan* **46** 639
657. Nakahara T, Munemori M and Musha S 1973 *Bull. Chem. Soc. Japan* **46** 1166
658. Nakahara T and Musha S 1975 *Appl. Spectrosc.* **29** 352
659. Neumann S and Kriese M 1974 *Spectrochim. Acta* **29B** 127
660. Newman R N and Page F M 1970 *Combust. Flame* **15** 317
661. Newman R N and Page F M 1971 *Combust. Flame* **17** 149
662. Ney J 1966 *Z. Phys.* **196** 53
663. Nienhuis G 1973 *Physica* **66** 245
664. Nukiyama S and Tamasawa Y 1938 *Trans. Soc. Mech. Engrs, Japan* **5** 63
665. Okuyama M and Zung J T 1967 *J. Chem. Phys.* **46** 1580
665a. de Olivares D R 1976 Studies into tunable-laser-excited atomic fluorescence spectrometry, *Diss. Bloomington, University of Indiana*

666. Omenetto N, Benetti P, Hart L P, Winefordner J D and Alkemade C Th J 1973 *Spectrochim. Acta* **28B** 289

667. Omenetto N, Benetti P and Rossi G 1972 *Spectrochim. Acta* **27B** 453

667a. Omenetto N, Boutilier G D, Weeks S J, Smith B W and Winefordner J D 1977 *Anal. Chem.* **49** 1076

668. Omenetto N, Browner R, Winefordner J D, Rossi G and Benetti P 1972 *Anal. Chem.* **44** 1683

669. Omenetto N, Fraser L M and Winefordner J D 1973 *Appl. Spectrosc. Rev.* **7** 147

670. Omenetto N, Hart L P, Benetti P and Winefordner J D 1973 *Spectrochim. Acta* **28B** 301

671. Omenetto N, Hart L P and Winefordner J D 1972 *Appl. Spectrosc.* **26** 612

672. Omenetto N, Hatch N N, Fraser L M and Winefordner J D 1973 *Anal. Chem.* **45** 195

673. Omenetto N, Hatch N N, Fraser L M and Winefordner J D 1973 *Spectrochim. Acta* **28B** 65

674. Omenetto N and Rossi G 1969 *Spectrochim. Acta* **24B** 95

675. Omenetto N and Rossi G 1970 *Spectrochim. Acta* **25B** 297

676. Omenetto N and Winefordner J D 1972 *Appl. Spectrosc.* **26** 555

677. Omenetto N, Winefordner J D and Alkemade C Th J 1975 *Spectrochim. Acta* **30B** 335

678. Ottaway J M and Harrison A 1973 in [118] p 51

679. Padley P J, Page F M and Sugden T M 1961 *Trans. Faraday Soc.* **57** 1552

680. Padley P J and Sugden T M 1958 *Proc. R. Soc.* **A248** 248

681. Padley P J and Sugden T M 1959 in [140] p 235

682. Padley P J and Sugden T M 1962 in [141] p 164

683. Page F M and Woolley D E 1974 *Combust. Flame* **23** 121

684. Palermo E F and Crouch S R 1973 *Anal. Chem.* **45** 1594

685. Parsons M L, McCarthy W J and Winefordner J D 1966 *Appl. Spectrosc.* **20** 223

686. Parsons M L, Smith B W and McElfresh P M 1973 *Appl. Spectrosc.* **27** 471

687. Parsons M L and Winefordner J D 1966 *Anal. Chem.* **38** 1593

688. Patel B M, Reeves R D, Browner R F, Molnar C J and Winefordner J D 1973 *Appl. Spectrosc.* **27** 171

689. Pearce S J, de Galan L and Winefordner J D 1968 *Spectrochim. Acta* **23B** 793

690. Penkin N P 1964 *J. Quant. Spectrosc. Radiat. Transfer* **4** 41

691. Penkin N P and Shabanova L N 1963 *Opt. Spectrosc.* **14** 5

692. Penkin N P and Shabanova L N 1963 *Opt. Spectrosc.* **14** 87

693. Phillips L F and Sugden T M 1961 *Trans. Faraday Soc.* **57** 914

694. Pickett E E and Koirtyohann S R 1968 *Spectrochim. Acta* **23B** 235

695. Pickett E E and Koirtyohann S R 1969 *Anal. Chem.* **41** 28A

696. Piepmeier E H 1972 *Spectrochim. Acta* **27B** 431

697. Piepmeier E H 1972 *Spectrochim. Acta* **27B** 445

698. Piepmeier E H and de Galan L 1975 *Spectrochim. Acta* **30B** 211

699. Piepmeier E H and de Galan L 1975 *Spectrochim. Acta* **30B** 263

700. Pietzka G and Chun H 1959 *Angew. Chem.* **71** 276

701. Pigor E 1954 Zur flammenphotometrischen Bestimmung von Calcium, Natrium im Blutserum. Methodik und Beeinflussung durch Viscosität und Oberflächenspannung, *Diss. Giessen*

702. Pleskach L I and Beremzhanov B A 1969 *Izv. Akad. Nauk, Kaz. SSR, Ser. Khim.* 5

703. Poluektov N S and Nikonova M P 1959 *Zavod. Lab.* **25** 263 (transl in 1959 *Ind. Lab.* **25** 279)

704. Poluektov N S and Vitkun R A 1958 *Zh. Anal. Khim.* **13** 48

705. Poluektov N S and Vitkun R A 1961 *Zh. Anal. Khim.* **16** 260 (transl in 1961 *J. Anal. Chem. USSR* **16** 279)

706. Porter P and Wyld G 1955 *Anal. Chem.* **27** 733

707. Porter R P 1970 *Combust. Flame* **14** 275

708. Preist T W 1972 *J. Chem. Soc. Faraday Trans. I* **68** 661

709. Pritchard H and Harrison A G 1968 *J. Chem. Phys.* **48** 2827
710. Prothero A 1969 *Combust. Flame* **13** 399
711. Prugger H 1964 *Optik* **21** 320
712. Prugger H 1969 *Spectrochim. Acta* **24B** 197
713. Pueschel R F 1969 *J. Colloid Interface Sci.* **30** 120
714. Pungor E 1967 in [135] p 1125
715. Pungor E and Cornides I 1969 in [45] ch 3 p 49
716. Pungor E and Hegedüs A 1960 *Mikrochim. Acta* p 87
717. Pungor E, Hegedüs A, Konkoly-Thege I and Zapp E 1956 *Mikrochim. Acta* p 1247
718. Pungor E and Konkoly-Thege I 1956 *Magy. Kém. Foly.* **62** 225
719. Pungor E and Konkoly-Thege I 1958 *Acta Chim., Budapest* **13** 235
720. Pungor E and Konkoly-Thege I 1959 *Mikrochim. Acta* p 712
721. Pungor E and Konkoly-Thege I 1961 in [130] vol II p 296
722. Pungor E and Mahr M 1963 *Talanta* **10** 537
723. Pungor E, Tóth K and Konkoly-Thege I 1964 *Z. Anal. Chem.* **200** 231
724. Pungor E, Wesprémy B and Palyi M 1961 *Mikrochim. Acta* p 436
725. Püschel A and Eckhard S 1959 *Arch. Eisenhüttenwes.* **30** 731
726. Püschel R 1962 Über die erreichbaren Na-Konzentrationen in turbulenten H_2–O_2 Flammen, *Diss. Giessen*
727. Püschel R, Simon L and Herrmann R 1964 *Optik* **21** 441
728. Rains T C 1969 in [45] ch 12 p 349
729. Ramírez-Muñoz J 1970 *Anal. Chem.* **42** 517
730. Rann C S 1967 *J. Sci. Instrum.* **44** 227
731. Rann C S 1968 *Spectrochim. Acta* **23B** 245
732. Rann C S 1968 *Spectrochim. Acta* **23B** 827
733. Rann C S 1969 *Spectrochim. Acta* **24B** 685
734. Rann C S and Hambly A N 1965 *Anal. Chem.* **37** 879
735. Rasmuson J O, Fassel V A and Kniseley R N 1973 *Spectrochim. Acta* **28B** 365 (erratum: 1976 *Spectrochim. Acta* **31B** 229)
736. Rauterberg E and Knippenberg E 1940 *Z. Angew. Chem.* **53** 477
737. Rauterberg E and Knippenberg E 1940 *Z. Bodenk. Pflanzenernähr.* **20** 364
738. Reich H F and Grabbe F 1955 *Tonind.-Ztg. Keram. Rdsch.* **79** 127
739. Reid R W and Sugden T M 1962 *Discuss. Faraday Soc.* **33** 213
740. Reif I, Fassel V A and Kniseley R N 1973 *Spectrochim. Acta* **28B** 105
741. Reif I, Fassel V A and Kniseley R N 1974 *Spectrochim. Acta* **29B** 79
742. Reif I, Fassel V A and Kniseley R N 1975 *Spectrochim. Acta* **30B** 163
743. Reif I, Fassel V A and Kniseley R N 1976 *J. Quant. Spectrosc. Radiat. Transfer* **16** 471
744. Reif I, Fassel V A and Kniseley R N 1976 *Spectrochim. Acta* **31B** 377
745. Reigle L L, McCarthy W J and Campbell Ling A 1973 *J. Chem. Engng Data* **18** 79
746. Rice P A and Ragone D V 1965 *J. Chem. Phys.* **42** 701
747. Robin J 1969 *Revue du GAMS* p 303
748. Robinson J W 1961 *Anal. Chim. Acta* **24** 254
749. Robinson J W and Kevan L J 1963 *Anal. Chim. Acta* **28** 170
750. Rocchiccioli C and Townshend A 1968 *Anal. Chim. Acta* **41** 93
751. Roos J T H 1970 *Spectrochim. Acta* **25B** 539
752. Rothermel D L 1951 in [149]
753. Rubeška I 1969 in [45] ch 11 p 317
754. Rubeška I 1971 in [122] p 61
755. Rubeška I 1973 *Atom. Absorption Newsl.* **12** 33
756. Rubeška I 1974 *Spectrochim. Acta* **29B** 263
757. Rubeška I 1975 *Can. J. Spectrosc.* **20** 156
758. Rubeška I and Miksovsky M 1972 *Atom. Absorption Newsl.* **11** 57
759. Rubeška I and Miksovsky M 1972 *Colln. Czech. Chem. Commun.* **37** 440
760. Rubeška I and Moldan B 1967 *Anal. Chim. Acta* **37** 421

761. Rubeška I and Svoboda V 1965 *Anal. Chim. Acta* **32** 253
762. Rüdiger K, Gutsche B, Kirchhof H and Herrmann R 1969 *Analyst, London* **94** 204
763. Russell B J, Shelton J P and Walsh A 1957 *Spectrochim. Acta* **8** 317
764. Russell B J and Walsh A 1959 *Spectrochim. Acta* **15** 883
765. Rutgers G A W 1972 *J. Res. NBS* **76A** 427
766. Ryabova V G and Gurvich L V 1965 *High Temp.* **3** 284
767. Ryabova V G, Gurvich L V and Khitrov A N 1971 *High Temp.* **9** 686
768. Ryabova V G, Khitrov A N and Gurvich L V 1972 *High Temp.* **10** 669
769. Sachdev S L, Robinson J W and West P W 1967 *Anal. Chim. Acta* **37** 12
770. Sastri V S, Chakrabarti C L and Willis D E 1969 *Can. J. Chem.* **47** 587
771. Schäfer K and Staab K 1952 *Naturwiss.* **39** 375
772. Schallis J E and Kahn H L 1968 *Atom. Absorption Newsl.* **7** 75
773. Scheub W H and Stomsky C J 1967 *Atom. Absorption Newsl.* **6** 95
774. Schmidt W 1963 in [132]
775. Schofield K 1967 *Chem. Rev.* **67** 707
776. Schofield K and Sugden T M 1965 in [143] p 589
777. van Schouwenburg J Ch and van der Wey A D 1966 *Anal. Chim. Acta* **36** 243
778. Schuhknecht W 1953 *Optik* **10** 245, 269
779. Schuhknecht W and Schinkel H 1958 *Z. Anal. Chem.* **162** 266
780. Schuhknecht W and Schinkel H 1963 *Z. Anal. Chem.* **194** 161
781. Scott R H and Human H G C 1975 in [119] paper B-27
782. Shifrin N, Hell A and Ramírez-Muñoz J 1969 *Appl. Spectrosc.* **23** 365
783. Shimazu M and Hashimoto A 1962 *Science of Light* **11** 131
784. Simon L 1960 Emissions-, Absorptions- und Temperaturmessungen an der H_2–O_2 Flamme des Beckman–Brenners, *Diss. Giessen*
785. Simon L 1962 *Optik* **19** 621
786. Skene J F, Stuart D C, Fritze K and Kennett T J 1974 *Spectrochim. Acta* **29B** 339
787. Slavin W 1964 in [139]
788. Slavin W 1964 *Atom. Absorption Newsl.* no. 24 15
789. Slavin W and Manning D C 1963 *Anal. Chem.* **35** 253
790. Slavin W and Slavin S 1969 *Appl. Spectrosc.* **23** 421
791. Slavin W, Sprague S and Manning D C 1963 *Atom. Absorption Newsl.* no. 15 1
792. Slavin W, Venghiattis A A and Manning D C 1966 *Atom. Absorption Newsl.* **5** 84
793. Smit C and Alkemade C Th J 1963 *Appl. Sci. Res.* **10B** 309
794. Smit J, Alkemade C Th J and Verschure J C M 1951 *Biochim. Biophys. Acta* **6** 508
795. Smit J A and Vendrik A J H 1948 *Physica* **14** 505
795a. Smith B, Winefordner J D and Omenetto N 1977 *J. Appl. Phys.* **48** 2676
796. Smith R 1971 in [115] p 235
797. Smith R, Elser R C and Winefordner J D 1969 *Anal. Chim. Acta* **48** 35
798. Smith R, Stafford C M and Winefordner J D 1968 *Anal. Chim. Acta* **42** 523
799. Smith R, Stafford C M and Winefordner J D 1969 *Anal. Chem.* **41** 946
800. Smith R G, Craig P, Bird E J, Boyle A J, Iseri L T, Jacobson S D and Meyers G B 1950 *Am. J. Clin. Path.* **20** 263
801. Smith V J and Robinson J W 1970 *Anal. Chim. Acta* **49** 417
802. Smith W W and Gallagher A 1966 *Phys. Rev.* **145** 26
803. Smyly D S, Townshend W P, Zeegers P J Th and Winefordner J D 1971 *Spectrochim Acta* **26B** 531
804. Snelleman W 1965 A flame as a standard of temperature, *Diss. Utrecht*
805. Snelleman W 1967 *Combust. Flame* **11** 453
806. Snelleman W 1968 *Metrologia* **4** 117
807. Snelleman W 1968 *Spectrochim. Acta* **23B** 403
808. Snelleman W 1969 in [45] ch 7 p 213
809. Snelleman W, Rains T C, Yee K W, Cook H D and Menis O 1970 *Anal. Chem.* **42** 394
810. Snelleman W and Smit J A 1955 *Physica* **21** 946

811. Snelleman W and Smit J A 1968 *Metrologia* **4** 123
812. Sobolev N N 1957 in [127] p 310
813. Sobolev N N, Mezhericher E M and Rodin G M 1951 *Zh. Eksp. Teor. Fiz.* **21** 350
814. Spitz J and Uny G 1969 in [117]
815. Spitz J, Uny G, Roux M and Besson J 1969 *Spectrochim. Acta* **24B** 399
816. Sprague S, Manning D C and Slavin W 1964 *Atom. Absorption Newsl.* no. 20 1
817. Staab K 1953 Die Flammenphotometrie als analytisches Verfahren, *Diss. Karlsruhe*
818. Stafford F E and Berkowitz J 1964 *J. Chem. Phys.* **40** 2963
819. Stephens R and Stevenson R G 1975 *Spectrochim. Acta* **30B** 61
820. Stephens R and West T S 1972 *Spectrochim. Acta* **27B** 515
821. Stephenson G 1951 *Proc. Phys. Soc.* **A64** 458
822. Strasheim A, Strehlow F W E and Butler L R P 1960 *J. S. Afr. Chem. Inst.* **13** 73
823. Straubel H 1959 *Dechema Monogr.* **32** 153
824. Stupar J and Dawson J B 1968 *Appl. Opt.* **7** 1351
825. Sugden T M 1956 *Trans. Faraday Soc.* **52** 1465
826. Sugden T M 1962 in *Ann. Rev. Phys. Chem.* ed H Eyring **13** 369
827. Sugden T M 1962 in [111] p 137
828. Sugden T M 1963 in [99] p 145
829. Sugden T M 1965 *AGARD Conf. Proc.* no. 8 **1** 43
830. Sugden T M 1965 in [143] p 539
831. Sugden T M 1971 in *Rep. 10th Int. Conf. on Phenomena in Ionized Gases, Oxford* p 437
832. Sugden T M 1972 in [137] p 211
833. Sugden T M and Schofield K 1966 *Trans. Faraday Soc.* **62** 566
834. Sullivan J V and Walsh A 1965 *Spectrochim. Acta* **21** 721
835. Sullivan J V and Walsh A 1965 *Spectrochim. Acta* **21** 727
836. Sullivan J V and Walsh A 1968 *Appl. Opt.* **7** 1271
837. Svoboda V, Browner R F and Winefordner J D 1972 *Appl. Spectrosc.* **26** 505
838. Sychra V and Matousek J P 1970 *Talanta* **17** 363
838a. Sydor R J and Hieftje G M 1976 *Anal. Chem.* **48** 535
839. Syty A and Dean J A 1968 *Appl. Opt.* **7** 1331
840. Tako T 1961 *J. Phys. Soc. Japan* **16** 2016
841. Talmi Y, Crosmun R and Larson N M 1976 *Anal. Chem.* **48** 326
842. Thomas D L 1968 *Combust. Flame* **12** 541
843. Thomas D L 1968 *Combust. Flame* **12** 569
844. Thomas P E and Pickering W F 1971 *Talanta* **18** 123
845. Thompson K C 1970 *Spectrosc. Lett.* **3** 59
846. van Tiggelen A 1963 in [99] p 165
847. Townsend W P, Smyly D S, Zeegers P J Th, Svoboda V and Winefordner J D 1971 *Spectrochim. Acta* **26B** 595
848. Traving G 1968 in *Plasma diagnostics* ed W Lochte-Holtgreven (Amsterdam: North Holland) ch 2
849. Trent D and Slavin W 1964 *Atom. Absorption Newsl.* no 22 1
850. van Trigt C, Hollander Tj and Alkemade C Th J 1965 *J. Quant. Spectrosc. Radiat. Transfer* **5** 813
851. Tsuchiya S 1964 *Bull. Chem. Soc. Japan* **37** 828
852. Uny G, Guea Lottin J N, Tardif J P and Spitz J 1971 *Spectrochim. Acta* **26B** 151
853. Uny G and Spitz J 1970 *Spectrochim. Acta* **25B** 391
854. Vasilieva I A, Deputatova L V and Nefedov A P 1974 *Combust. Flame* **23** 305
855. Veenendaal W A 1969 Het onderzoek naar storingen bij de bepaling van Thallium in de vlam door middel van AAS en AES, *Diss. Amsterdam* (with summary in English)
856. Veillon C, Mansfield J M, Parsons M L and Winefordner J D 1966 *Anal. Chem.* **38** 204
857. Veillon C and Margoshes M 1968 *Spectrochim. Acta* **23B** 521
858. Venghiattis A A 1967 *Spectrochim. Acta* **23B** 67

859. Vickers T J, Cottrell C R and Breaky D W 1970 *Spectrochim. Acta* **25B** 437
860. Vickers T J and Winefordner J D 1972 in [63] part 2 ch 7
861. Vidale G L 1960 *US Dept Comm., Office Tech. Serv., PB Rep.* **148** 206, 878
862. Visser K, Hamm F M and Zeeman P B 1976 *Appl. Spectrosc.* **30** 620
863. Voinovitch I A, Legrand G, Hameau G and Louvrier J 1965 *C.R. Acad. Sci., Paris* **260** 5487
864. Vukanović D D 1958 *Bull. Inst. Nucl. Sci. 'Boris Kidrich'* **8** 43
865. de Waele M 1968 in [120] p 113
866. de Waele M 1969 Etudes des interférences sur la détermination par A.A.S. des éléments Ca, Mg, Co, Cu, Fe, Mn, Zn dans les extraits de sols et les végétaux, *Diss. Gembloux*
867. Wagenaar H C 1976 The influence of spectral line profiles upon analytical curves in atomic absorption spectrometry, *Diss. Delft*
868. Wagenaar H C and de Galan L 1973 *Spectrochim. Acta* **28B** 157
869. Wagenaar H C and de Galan L 1974 *Spectrochim. Acta* **29B** 211
870. Wagenaar H C and de Galan L 1975 *Spectrochim. Acta* **30B** 361
871. Wagenaar H C, Novotny I and de Galan L 1974 *Spectrochim. Acta* **29B** 301
872. Walsh A 1952/53 in *5th Ann. Rep. of the Commonwealth Scientific and Industrial Research Organization* p 121
873. Walsh A 1955 *Spectrochim. Acta* **7** 108
874. Walsh A 1963 in [131] p 127
875. Walsh A 1965 in [133]
876. Walsh A 1966 *J. N.Z. Inst. Chem.* **30** 7
877. Walsh A 1969 in [153]
878. Ward J, Cooper J and Smith E W 1974 *J. Quant. Spectrosc. Radiat. Transfer* **14** 555
879. Weichselbaum T E and Varney P L 1949 *Proc. Soc. Exp. Biol. Med.* **71** 570
880. West A C 1961 Variables in flame spectrophotometry, their control and elimination, *Diss. Cornell University, Ithaca, NY*
881. West A C 1964 *Anal. Chem.* **36** 310
882. West A C, Fassel V A and Kniseley R N 1973 *Anal. Chem.* **45** 1586
883. West A C, Fassel V A and Kniseley R N 1973 *Anal. Chem.* **45** 2420
884. West A C, Kniseley R N and Fassel V A 1973 *Anal. Chem.* **45** 815
885. West T S 1967 in [152]
886. West T S and Cresser M S 1973 *Appl. Spectrosc. Rev.* **7** 79
887. Whiting E E 1968 *J. Quant. Spectrosc. Radiat. Transfer* **8** 1379
888. Wiese W L 1963 in [131] p 37
889. Williams C H 1960 *Anal. Chim. Acta* **22** 163
890. Williams F A 1962 in [141] p 50
891. Willis J B 1960 *Spectrochim. Acta* **16** 259, 273, 551
892. Willis J B 1962 *Anal. Chem.* **34** 614
893. Willis J B 1967 *Pure Appl. Chem.* **17** 111
894. Willis J B 1967 *Spectrochim. Acta* **23A** 811
895. Willis J B 1968 *Appl. Opt.* **7** 1295
896. Willis J B 1970 *Spectrochim. Acta* **25B** 487
897. Willis J B 1970 in [77] ch 10 p 525
898. Willis J B 1971 *Spectrochim. Acta* **26B** 177
899. Willis J B 1971 in [122] p 83
900. Willis J B, Fassel V A and Fiorino J A 1969 *Spectrochim. Acta* **24B** 157
901. Willis J B, Rasmuson J O, Kniseley R N and Fassel V A 1968 *Spectrochim. Acta* **23B** 725
902. Winefordner J D 1963 *Appl. Spectrosc.* **17** 109
903. Winefordner J D 1967 in [150] p 565
904. Winefordner J D 1970 in [124] p 35
905. Winefordner J D 1971 *Acc. Chem. Res.* **4** 259

906. Winefordner J D and Elser R C 1971 *Anal. Chem.* **43** April 24A
907. Winefordner J D and Latz H W 1961 *Anal. Chem.* **33** 1727
908. Winefordner J D and Mansfield J M 1967 *Appl. Spectrosc. Rev.* **1** 1
909. Winefordner J D, Mansfield C T and Vickers T J 1963 *Anal. Chem.* **35** 1607
910. Winefordner J D, Mansfield C T and Vickers T J 1963 *Anal. Chem.* **35** 1611
911. Winefordner J D, Parsons M L, Mansfield J M and McCarthy W J 1967 *Anal. Chem.* **39** 436
912. Winefordner J D and Vickers T J 1964 *Anal. Chem.* **36** 161, 789
913. Winefordner J D and Vickers T J 1964 *Anal. Chem.* **36** 1939
914. Winefordner J D and Vickers T J 1964 *Anal. Chem.* **36** 1947
915. Winefordner J D and Vickers T J 1972 *Anal. Chem.* **44** 150R
916. Winefordner J D, Vickers T J and Remington L 1965 *Anal. Chem.* **37** 1216
917. Woldring M G 1953 *Anal. Chim. Acta* **8** 150
918. Woodward C 1969 *Atom. Absorption Newsl.* **8** 121
919. Woodward C 1971 *Spectrosc. Lett.* **4** 191
920. Yasuda K 1966 *Anal. Chem.* **38** 592
921. Yasuda K 1975 in [119] paper P-8
922. Zacha K E and Winefordner J D 1966 *Anal. Chem.* **38** 1537
923. Zaidel' A N and Korennoi E P 1961 *Opt. Spectrosc.* **10** 299
924. Zeegers P J Th 1966 Recombination of radicals and related effects in flames, *Diss. Utrecht*
925. Zeegers P J Th and Alkemade C Th J 1965 *Combust. Flame* **9** 247
926. Zeegers P J Th and Alkemade C Th J 1965 in [143] p 33
927. Zeegers P J Th and Alkemade C Th J 1966 in [133] p 290
928. Zeegers P J Th and Alkemade C Th J 1970 *Combust. Flame* **15** 193
929. Zeegers P J Th, Smith R and Winefordner J D 1968 *Anal. Chem.* **40** 26A
930. Zeegers P J Th, Townsend W P and Winefordner J D 1969 *Spectrochim. Acta* **24B** 243
931. Zeegers P J Th and Winefordner J D 1971 *Spectrochim. Acta* **26B** 161
932. Zettler H 1953 Die Methoden der Flammenspektrophotometrie und ihre Anwendungen, *Diss. Frankfurt am Main*
933. Zhitkevich V F, Lyutyi A I, Nesterko N A, Rossikhin V S and Tsikora I L 1963 *Izv. Vyssh. Ucheb. Zaved., Fiz.* no. 2 78
934. Zhitkevich V F, Lyutyi A I, Nesterko N A, Rossikhin V S and Tsikora I L 1963 *Opt. Spectrosc.* **14** 17
935. Zhitkevich V F, Lyutyi A I, Nesterko N A, Rossikhin V S and Tsikora I L 1963 *Opt. Spectrosc.* **14** 180
936. Zhitkevich V F, Lyutyi A I, Rossikhin V S and Tsikora I L 1963 *Opt. Spectrosc.* **15** 217
937. Zung J T 1967 *J. Chem. Phys.* **46** 2064

Index†

---

† The Index contains a list of authors, subjects, symbols and German terms (Ger.). R = Reference number.

Alkaline-earth metals—*contd*
  ionisation constants of, 82
  ionisation rate of, 86
  ionisation reactions of, 87
  line shifts of, 103
  in nitrous oxide–acetylene flame, 76
  partition functions of, 82
Alkaline-earth oxides, *see* individual oxides
  dissociation energies of, 75
  equilibrium of, 80
  in flames, 74, 100
  formation of, 80
Alkali nitrates, interference of, on molybdenum, 190
Alkali oxides (MO_2), rate of formation of, 78f
Alkali sulphates, interference of, on molybdenum, 190
Alkemade, C Th J, 10, 14, 15, 225, 310, R160–R176, R185, R214, R457, R460, R461, R466–R468, R481–R483, R496, R521–R524, R588, R589, R666, R677, R793, R794, R850, R925–R928
Allan, J E, R177, R178
Allen, C W, R1
Allowed transition, 92, 93, 97, 99
Aluminium, analytical curve of, 68, 169
  a-parameter of, 104f
  background emission of, 101
  determination of, reference, 15
  effect of hydrofluoric acid on, 184f
  energy levels of, 94, 110
  interference of, on calcium, 160, 184
    on lead, 186
    on magnesium, 189
    on molybdenum, 186f
    on vanadium, 187
  interference of fluoride on, 184f, 189
  ionisation of, 83
  ionisation energy of, 94
  in low-pressure nitrous oxide–acetylene flame, 26
  non-resonance atomic absorption of, 110
  oscillator strength of, 94
  oxidation rate of, 79
  partition function of, 110
  spectrogram of, 280f
  statistical weight of, 110
  volatilisation of, 58, 68, 184f
  wavelengths of, 354
Aluminium chloride (AlCl), molecular absorption by, 179
Aluminium hydroxide ($Al(OH)_2$), in flames, 74
Aluminium oxide (AlO), chemiluminescence of, 135

  dissociation energy of, 68, 75, 76
  in flames, 74, 76
  rate of formation of, 79
  spectrogram of, 281
Aluminium oxide (Al_2O_3), thermal radiation of, 101
  volatilisation of, 57, 58
A-Methode (Ger.), 195, 197
Amidogen (NH_2), background emission of, 150
Ammonium fluoride, interference of, on aluminium, 189
Amos, M D, 15, R179
Analoganzeige (Ger.), 215
Analogue, 196
Analogue reading, 214f
Analysenelement (Ger.), 196
Analysenergebnis (Ger.), 197
Analysenfunktion (Ger.), 196
Analysenkurve (Ger.), 196
Analysenlösung (Ger.), 217
Analysensubstanz (Ger.), 197
Analysis element, 196; *see* Analyte
Analyte, aspiration rate of, and emission, 45, 63, 64
  atomisation efficiency of, 61, 70, 77
  chemiluminescence of, 132–135; *see also* Chemiluminescence
  concentration of, in flame, 45, 61, 62f, 64, 65, 78, 108ff, 112f, 115, 121, 159–163
  definition of, 1, 196
  dilution factor of, in flame, 188
  distribution of, in flame, 44, 162f, 177, 185ff
  excitation of, *see* Excitation
  fraction atomised, 60f, 62, 70, 76, 77f
  ionisation of, 69, 81–89, 160ff; *see also* Ionisation (of analyte)
  losses of, during transport, 43f, 45, 50, 51, 60, 61
  molecular dissociation of, 69, 70–81; *see also* Dissociation
  quantity of, 214
  reaction of, with flame gases, 73–78, 79
  residence time of, in flame, 40, 53f, 56, 63
  volatilisation fraction of, 60f, 65–68
Analyte addition technique, 196
Analyte interference, 176, 177
  identification of, 192f
Analyte signal, 196
Analytical calibration curve, 196f, 218
Analytical calibration function, 157, 196f, 217f
Analytical curve, 157–174; *see also* Curve of growth
  in atomic absorption spectroscopy, 111ff, 158, 165–169

Atom—*contd*
  structure of, 92
  transitions in, 92ff, 107f, 112, 120f, 141, 143, 144
Atomabsorptionsspektroskopie (Ger.), 197
Atomarer Dampf (Ger.), 198
Atomdampf (Ger.), 198
Atomdampferzeuger (Ger.), 198
Atomdampferzeugung (Ger.), 198
Atomemissionsspektroskopie (Ger.), 197
Atomfluoreszenz (Ger.), 206
Atomfluoreszenzspektroskopie (Ger.), 198
Atomic absorption spectrometry, history of, 14
Atomic absorption spectroscopy, absolute absorption signal in, 123f
  analytical curve in, 158, 165–169
    theory of, 111ff, 166f
  apparatus, 3
  basis of, 1
  blank interference in, 179f
  cation interference in, 189
  comparison of, with atomic fluorescence spectroscopy, 142, 144
    with chemical methods, 5f
    with flame emission spectroscopy, 6–8
  comparison of laminar and turbulent flames in, 25
  with continuum source, 115–118, 123f, 165f, 180
  definition of, 197
  for determining atomic concentration in flame, 78, 112f, 115, 139
  with dye laser, 166
  effect on, of fine structure, 168
    of flame emission, 168
    of hyperfine structure, 168, 169
    of isotope shift, 168
    of lamp current, 167, 169
    of line shift, 115, 166
    of temperature, 114, 118
    of unabsorbed lines in lamp, 167f, 179
  elements determinable by, 4
  excitation interference in, 182
  in far ultraviolet, 154f
  flame background in, 147, 149, 154f, 191
  history of, 13f
  in hot plasmas, 110
  lateral-diffusion interference in, 185ff
  line intensity in, 108–115
  line profiles in, 110ff
  long-path, 30f, 100, 179
  modulated nebulisation in, 149, 168
  non-flame, 9, 10, 15, 110; *see also* Graphite rod atomiser, Graphite tube furnace,

Inductively coupled plasma
  at non-resonance lines, 110
  non-specific interference in, 180f
  with resonance monochromator, 140, 167, 174
  salt interference in, 190
  sensitivity of, definition of, 158
  specific interference in, 181ff
  spectral interference in, 179f
  volatilisation interference in, 184f
Atomic concentration in flame, 45, 61, 62f, 159–163; *see also* Analytical curve
  and absorption of light, 108ff
  by integration across flame, 64, 65
  measurement of, by AAS, 78, 112f, 115, 139
    by emission, 108
Atomic emission spectroscopy, basis of, 1f
  definition of, 197
Atomic fluorescence, 139–145, 206
  anti-Stokes, 142
  direct line, 141, 145, 173
  efficiency of, 126f, 137, 143, 144f, 203, 206, 214
  intensity of, 140, 142f
  non-resonance, 141f, 173
  polarisation of, 145, 180
  quenching of, 2, 143, 206
  radiant flux of, 144, 170
  radiation transport in, 145
  resonance, 140, 141, 142–145, 167, 170, 206
  saturation in, 140, 174
  stepwise line, 141f
  for temperature determination, 142
  tertiary radiation in, 145, 170
Atomic fluorescence spectroscopy, 139–145, 169–174
  analytical curve in, 142, 144f, 157, 169–174
  apparatus for, 3f, 142, 170
  argon-diluted flames for, 23, 36, 137, 140, 143
  background scattering in, 142, 145, 180
  basis of, 1f, 4, 130, 139ff
  bibliography, 140
  blank interference in, 180
  compared with AAS and FES, 144, 173
  with continuum source, 104, 140, 142, 144, 145, 170f, 173f, 180
  definition of, 198
  excitation interference in, 182
  history of, 13, 14
  with laser, 140f, 147, 174
  with non-resonance fluorescence, 142, 173
  non-specific interferences in, 180f
  pre-absorption in, 173, 174

Atomic fluorescence spectroscopy—*contd*
  primary source for, 4, 139ff, 142, 144, 170ff, 214
  quenching interference in, 184
  with resonance monochromator, 174
  saturation of excited level in, 140, 174
  self-absorption in, 142, 144f, 170ff
  self-reversal in, 126, 172f, 174
  specific interference in, 181ff
Atomic line, *see* Line, Line broadening, etc
  absorption intensity of, 108–115
  chemiluminescence of, 132ff; *see also* Chemiluminescence
  definition, 198
  spectra, theory of, 92–97
  thermal radiation intensity of, 105–108
Atomic transition, *see* Transition
Atomic vapour, 198
Atomisation, definition of, 2, 198
  efficiency, 61, 70, 203
  by flame radicals, 58
  fraction, 60f, 62, 70, 76, 77f, 206
Atomiser, *see* Flame
  definition of, 198
  graphite-rod, 172
  graphite-tube, 10, 16, 100, 101, 105, 179
  inductively coupled plasma, 10, 15, 161
  non-flame, 9, 10, 15, 110, 180, 199f
Atomiser-burner, 202
Atom line, *see* Atomic line, Line
Atomlinie (Ger.), 198
Atsuya, I, R184
Auflösung (Ger.), 216
Auflösungsvermögen (Ger.), 216
Aufstockungsverfahren (Ger.), 196
Ausgangsmaterial (Ger.), 197, 209
Ausreisser (Ger.), 210
Aussagewahrscheinlichkeit (Ger.), 201
Aussenkonus (Ger.), 217
Äussere Reaktionszone (Ger.), 217
Äussere Verbrennungszone (Ger.), 217
Auswertung (Ger.), 204
Avni, R, R185
Azimuthal quantum number, 95f

b, band peak or head, 307, 308
Baber, C, R352a
Bacis, R, R186
Background, 147–155, 198; *see also* Blank signal, Flame background
Background correction, in unpremixed flames, 41
Background interference, 177, 178, 179, 190f

Background (light) source, 3, 109, 198; *see also* Continuum source, Dye laser, Electrodeless-discharge lamp, Gas-discharge lamp, Hollow-cathode lamp, Resonance monochromator
  effect of continuum in, 168
  effect of current in, 167, 169
  effect of unabsorbed lines in, 167f, 179
  history of, 13f
  linewidth in, and analytical curve, 166f
    and atomic absorption spectroscopy, 114f
  radiance of, 123f
Background scattering in AFS, 142, 145, 180
Badger, R M, 14, 15
Bahn, G S, R187
Bailey, B W, R158, R188, R189
Baker, C A, R191
Baker, M R, R190, R192
Balanced reactions, 32f
Balancing, detailed, 136f, 145
Ballard, A E, 14, 15, 17
Band, 99, 219
Bandbreite (Ger.), 219
Bande (Ger.), 219
Bandenkopf (Ger.), 198
Bandenlinie (Ger.), 198
Band head, 99, 198
Band line, 99, 198
Band spectra, 97–100
  in absorption, 1, 69, 99f, 179
  analytical curve for, 163
  chemiluminescent, 135, 152f
  fluorescent, 145, 147
  intensity of, 108
Bandwidth, 219
Barium, energy levels of, 94
  fluorescence intensity of, 140
  interference of calcium on, 178, 179
  ion, atomic properties of, 94
  ionisation of, 84
  ionisation energy of, 94
  oscillator strength of, 94
  spectrograms, 246–249, 295
  wavelengths, 94, 355
Barium chloride (BaCl), molecular absorption by, 179
Barium fluoride (BaF), spectrogram of, 295
Barium hydroxide (BaOH), dissociation energy of, 75
  spectrograms of, 246, 249
  spectrum of, 100
Barium hydroxychloride (BaOHCl), 72f
Barium oxide, dissociation energy of, 71, 75
  in flame, 74
  spectrograms of, 246f, 249

Cu, copper solution, 309
Cunningham, P T, R284
Curry, R H, 310, R330, R331
Curve of growth, 104, 117, 120, 166; *see also* Analytical curve
CW, continuous-wave, 140
Cyanide ion (CN⁻), in flame, 32, 85
Cyanogen (CN), 31, 79
    absorption spectrum of, 100, 154
    background emission of, 151
    fluorescence of, 147
    in nitrous oxide–acetylene flame, 34, 76, 79
    spectrogram of, 301–304
    wavelengths of, 356
Cyanogen flame, *see* Oxycyanogen flame
Cyclohexane, desolvation rate of, 53

d, double line, 308
D, dimethylformamide, 309
Dagnall, R M, R124, R203, R225, R272, R285–R290
Dahl, A I, R43
Dark current, 201
Darwent, B, R5
David, D J, R291
Dawson, J B, R824
Day, R A, R292
Deactivation, 10; *see also* Quenching
    collisional, 131, 145
    equilibrium, 136
    in fluorescence, 143, 145
    by noble gases, 137
    in non-thermal radiation, 137f
Deaeration of solution, 46
Dean, J A, R44–R47, R113, R293–R302, R839
de Boers, F J, 10
Debras-Guédon, J, R108
Decrepitation, 58
De-excitation, *see* Deactivation, Quenching
de Galan, L, 125, R57, R361–R373, R417, R564, R689, R698, R699, R868–R871
Degenerate state, 93
Degraded band, 99
Degree of dissociation, 70ff
Degree of ionisation, 82f
de Jaegere, S, R218, R314
Dekadische Extinktion (Ger.), 195
Dekadischer Extinktionsmodul (Ger.), 195
Dekadisches Absorptionsmass (Ger.), 195
Delumyea, R, R25
Demayo, A, R303
Demers, D R, R304, R319
den Harder, A, R417

Denton, M B, R305
de Olivares, D R, R665a
Depolarising collisions, 145
Depression, 197, 201, 223
Deputatova, L V, R854
Derivative spectroscopy, 149
Desolvation, 43, 50–55, 201
    and convection, 55
    cooling effect of, 52
    in direct-injection burner, 54f, 60, 63–65
    explosive, 55, 58, 190
    in flame, 52–55
    fraction, 60f, 63–65, 206
    height of complete, 53f
    in spray chamber, 51f, 55, 60, 61
Desolvation interference, 177, 181, 201
Detailed balancing, 136f, 145
Detection limit, 201f
    and flame background, 147f
de Waele, M, R865, R866
Diatomic molecules, spectra of (reference), 308
Dieke, G H, R281
Diffusion, *see* Lateral diffusion
    of analyte in flame, 44, 61, 63
    of radiation, 126f
Diffusion flames, 23, 24, 25; *see also* Hydrogen diffusion flame
    definition of, 25, 202
    on direct-injection burner, 38
Digital, definition of, 202
Digitalanzeige (Ger.), 215
Digital reading, 215
Diluent, 202
Dilution factor, of sample by flame, 188
Dimers, metal, 73, 160
d'Incan, J, R494
Dippel, W A, R306
Direct-injection burner, 25, 202
    analyte concentration in, 65
    aspiration by, 45
    background emission of, 149
    background interference in, 178
    Beckman, 40f, 295
    desolvation in, 54f, 60, 63–65
    desolvation fraction in, 63–65
    dilution factor of, for sample, 188
    dissociation interference in, 182
    drop-size distribution in, 48, 50
    flame shape interference in, 181
    flame structure of, 37ff
    history of, 12
    nebulisation by, 38, 45, 48, 60
    observation height in, 54f, 63–65, 149
    reversed, 216

412

417

418

426

Nitrogen dioxide, emission spectrum of, 154
   recombination continuum of, 150
Nitrogen hydrides, *see* Amidogen, Imidogen
Nitrous oxide–acetylene flame, absorption by, 154
   alkaline earths in, 76
   aluminium in, 26, 68, 185f
   analyte molecules in, 73f
   background spectrum of, 151, 304
   boron in, 76
   burning velocity of, 21
   calcium hydroxide in, 179
   carbon-rich, 189
   chromium in, 76
   composition of, 34
   dissociation of oxides in, 76f, 78, 79, 185
   equilibrium in, 78, 79
   flashback of, 29
   fuel-rich, 21, 23, 26, 30, 31, 59, 73, 76ff, 85,
     185, 190, 304
   history of, 15
   interzonal region of, 30, 31
   ionisation of analyte in, 81, 83
   ions in, 32, 85
   lateral diffusion in, 162f, 185ff
   low-pressure, 26
   mixing ratio of, 21, 23, 31
   molybdenum in, 59, 68, 186f, 190
   noise of, 37
   observation height in, 76
   primary combustion zone of, 26
   radicals in, 31, 34
   red feather of, 30, 76, 151
   shielded, 76, 77, 186
     for short-wavelength work, 154f
   silicon in, 76
   spectrogram of, 304
   split, 151
   temperature of, 21, 23, 34
   titanium in, 76
   tungsten in, 74
   vanadium in, 68, 187
   volatilisation interference in, 185
Nitrous oxide–butane flame, atomic oxygen in,
   77
Nitrous oxide–carbon monoxide flame, burning
   velocity of, 21
   interzonal emission of, 153
   temperature of, 21
Nitrous oxide flames, premixed, 25
Nitrous oxide–hydrogen flame, background
   spectrum of, 150
   burning velocity of, 21
   oxide dissociation in, 76
   primary combustion zone of, 37

temperature of, 21
Nitrous   oxide–methylacetylene–propadiene
   flame, 24
Nitrous oxide–propane flame, atomic oxygen in,
   77
   temperature of, 21
NO, *see* Nitric oxide
NO^+, 28
Noble gases, *see* Argon, Helium
   in atomic fluorescence spectroscopy, 137, 143
   effect of, on analytical curve, 165
   excitation and deactivation by, 131, 137f, 143
   as flame diluent, 23, 24, 25, 36, 39, 81, 127,
     137, 140, 143, 154, 165
   and non-thermal radiation, 137, 165
Noise, 211; *see also* Drift, Precision
   acoustic, 36, 37, 38, 47
   in blank scatter, 199, 211
   drop-effect, 47
   flame background, 148
   flame emission, 33, 36f, 38f
   flicker, 37, 148
   nebulisation, 47
   shot, 36f, 47, 148
   sources of, 211
   white, 37
Noise level, 212
Noise spectrum, 37, 47, 148
Non-flame methods, 9, 10, 15, 110, 180; *see also*
   Graphite-rod atomiser, Graphite-tube furnace, Inductively coupled plasma, Sputtering
   chamber
Non-resonance fluorescence, 141f, 173
Non-resonance lines, absorption at, 110
   in atomic fluorescence, 141f, 173
   in background source, 167
   temperature effect on, 114
Non-specific interference, 177, 180f
Non-thermal radiation, 137ff, 165; *see also*
   Chemiluminescence, Suprathermal radiation
   in atomic fluorescence, 140
Norgordon, H O, R621
Normal curve, 212
Normal distribution, 202, 212, 221
Normalisator (Ger.), 208
Normalisatorsubstanz (Ger.), 208
Normalisierung (Ger.), 208
Normalverteilung (Ger.), 212
Novotny, I, R364, R871
Nozzle, nebuliser, 213, 220
   gas velocity in, 48, 49
   pressure in, 49
Nukiyama, S, R664
Null-Linie (Ger.), 198

432

Winkeldispersion (Ger.), 202
Wirkungsgrad der Atomdampferzeugung (Ger.), 206
Wirkungsgrad der Verdampfung (Ger.), 206
Wirkungsgrad der Verflüchtigung (Ger.), 206
Wirkungsgrad der Zerstäubung (Ger.), 203
Woldring, M G, R917
Wolfhard, H G, R61
Wollaston, W H, 11, 13, 16
Wood, R W, 13, 17
Woodward, C, R33, R918, R919
Woolley, D E, R683
Working curve, 197
Working range, 223
    in atomic fluorescence spectroscopy, 173
Wyld, G, R318, R706

x, sharp peak in a headless band, 308
Xenon lamp, as continuum source, 124, 145

y, rotational line or group, 308
Y, quantum efficiency of resonance fluorescence, 126, 143
Yasuda, K, R560, R560a, R920, R921
Yee, K W, R809
Ytterbium, energy levels of, 94
    ionisation energy of, 94
    oscillator strength of, 94
    spectrogram of, 253f
    wavelengths of, 375
Yttrium, wavelengths of, 375
Yttrium oxide (YO), absorption spectrum of, 100
Yuen, M C, R654

Z, partition function, 106, 109f
Zacha, K E, R621, R922
Zaidel', A N, 310, R23, R923
Zapp, E, R717
Zare, R N, R393
Zeegers, P J Th, 10, 21, 34, 76, 94, R176, R462, R478, R479, R588, R589, R803, R847, R924–R931

Zeeman, P B, R862
Zeeman modulation, 149
Zeiss flame spectrometer, 12
Zeitkonstante (Ger.), 222
Zemansky, M W, R79
Zero line, 198
Zero suppression, 223, 226
    for linearising analytical curve, 167
Zerstäuber (Ger.), 211
Zerstäuberdüse (Ger.), 220
Zerstäuberkammer (Ger.), 220
Zerstäuberwirkungsgrad (Ger.), 203
Zerstäubung (Ger.), 211
Zettler, H, R932
Zhitkevich, V F, R933–R936
Zimmer, K, R135
Zinc, a-parameter of, 105
    background of, in AAS, 154
    chemiluminescence of, 133
    energy levels of, 94
    fraction atomised, 62, 77
    interference of copper on, 179
    ionisation energy of, 94
    oscillator strength of, 94
    spectrograms of, 273f
    two-photon absorption by, 95
    volatilisation of, 67, 77
    wavelengths of, 376
Zirconium, wavelengths of, 376
Zirconium oxide (ZrO), dissociation energy of, 75
    in flame, 74
Zittel, H E, R298
Zuehlke, C W, 14, 17
Zufälliger Fehler (Ger.), 214
Zugabe (Ger.), 196
Zugabeverfahren (Ger.), 196
Zumischverfahren (Ger.), 196
Zung, J T, R665, R937
Zusatz (Ger.), 196
Zuverlässigkeit (Ger.), 196
Zwei-Zerstäuber Versuch (Ger.), 222
Zwischengas (Ger.), 208
Zwischenzone (Ger.), 208
Zybin, A V, R215a